ABC-TRIZ

Introduction to Creative Design Thinking with Modern TRIZ Modeling

Michael A. Orloff

ABC-TRIZ

Introduction to Creative Design Thinking with Modern TRIZ Modeling

With 11 learning modules and approximately 300 examples

3-in-1:
Book I: How to Learn to Invent
Book II: How to Become a Genius
Book III: Primary Instruments (Summary)

Springer

Michael A. Orloff
Academy of Instrumental Modern TRIZ
(AiMTRIZ)
Berlin
Germany

ISBN 978-3-319-29435-3 ISBN 978-3-319-29436-0 (eBook)
DOI 10.1007/978-3-319-29436-0

Library of Congress Control Number: 2016933241

Printed on acid-free paper

This Springer imprint is published by Springer Nature
The registered company is Springer International Publishing AG Switzerland

For my son Nikolai –
with love and respect for his
creative talent and
his superior teaching skills,
and for responsiveness and
patience he is instilling
not only in his students,
but also in me

This book is being published in a remarkable year.

It is the **60th anniversary** of the first TRIZ article (1956) and the **55th anniversary** of the first TRIZ-book (1961) by Genrikh Altshuller, founder of TRIZ.

The origins of TRIZ and ARIZ arose from this first article and book!

This year is also the **90th anniversary** of the birth of Genrikh Altshuller and his classmate and brother-in-arms, Raphael Shapiro, co-author of the first article.

The great is seen from a distance! Really, the concept and ideas of TRIZ—*Theory of Invention* of Genrikh Altshuller—are the source of an absolutely peculiar way of creative design thinking, systematic creative knowledge, and skills for the further development of civilization. TRIZ finds more and more users and followers!

Modern TRIZ carefully maintains historical continuity with the fundamental ideas of TRIZ initiators, facilitates understanding, and develops the effective application of these ideas.

Dear readers will see this succession in dozens of methodological parallels and applied examples in this book, **ABC-TRIZ**, and will read about Altshuller and Shapiro as well as other great persons.

Zhao YANG (China), MSc:

I am deeply convinced that Modern TRIZ should be intensively disseminated in China.

Adehi GUEHIKA (Cote d`Ivoire, now in Canada):

I think that e-education in Modern TRIZ could become my second or maybe first occupation.

Oktay TABAK (Turkey):

It is very interesting to model with MTRIZ any artifact and open the creative driver of its design.

Khairul KAMARUDIN (Malaysia), PhD, MSc:

Great MTRIZ e-courses! Interesting applications for green technologies are possible.

Eduardo ROCHA (Mexico):

This is invaluable MTRIZ-knowledge that I would like to bring not only in my engineering specialty, but also in every university and even in every school.

Camila MARTINEZ, Business Intelligence Analist (Brazil):

I would like to disseminate MTRIZ in Brazil. This is important for our companies and universities.

Prerak CONTRACTOR (India):

Modern TRIZ and e-courses on MTRIZ are absolutely irreplaceable for both engineers and managers!

Sugeng WAHYUDI, MSc (Indonesia, now in Singapore):

It is possible with MTRIZ to model historical development of systems as I realized it my Master thesis!

Khaled SHOAIB (Egypt):

Modern TRIZ is absolutely careful study for both initial training and applied design thinking!

El Gouna Center of Technical University Berlin,
Egypt, January 20th, 2016

Michael A. Orloff is a founder of the Academy of Instrumental Modern TRIZ (AIMTRIZ, Berlin, 2000), author of MTRIZ educational modeling with techniques "Extracting", "Reinventing" and "Meta-Algorithm of Invention T-R-I-Z".

The Academy of Instrumental Modern TRIZ has the partners in 18 countries and continues to develop educational Modern TRIZ society around the world.

His training and consulting are well known at many companies including SIEMENS (for *postautomation* and *electrotransportation* directions, Berlin, Germany), SAMSUNG Advanced Institute of Technology (for *system development*, Suwon, S. Korea), universities in China (till 1000 trainees in auditorium at 2-day trainings), Kazakhstan, Korea, Russia, etc. Till now about 500 innovative and inventive solutions (many of them are patented by customers) were made with the author's consulting and problem solving (on the order of customer or together with Think Tank Teams) at more than 120 enterprises.

He lectures MTRIZ at TU Berlin in partnership with Institute of Aeronautics and Astronautics (ILR) for European programs TEMPUS and ERASMUS MUNDUS and others, for international Master of Science programs "Global Production Engineering" and "Entrepreneurship and Innovation Management", and "Energy Engineering" (with attendees from faculties "Urban Engineering" and "Water Engineering") at Campus El Gouna, Egypt, as well as a part of some executive education programs at Center for Leadership Development Research (CLDR) of ESMT European School of Management and Technology (Berlin).

From 2015 he is a staff professor for MTRIZ e-learning at Chair of pedagogy and methodology of natural science education of the National Research Nuclear University MEPhI (Moscow Engineering Physics Institute). This textbook also includes the examples of pilot projects for teachers and students of high schools and universities in co-operation with MEPhI and UNESCO Institute of Informational Technologies in Education (IITE).

This book is addressed, first and foremost, to teachers and students of technical universities, but certainly it can be successfully used by teachers and students of all professions without limitation, as well as by high school students and teachers.

Modern TRIZ

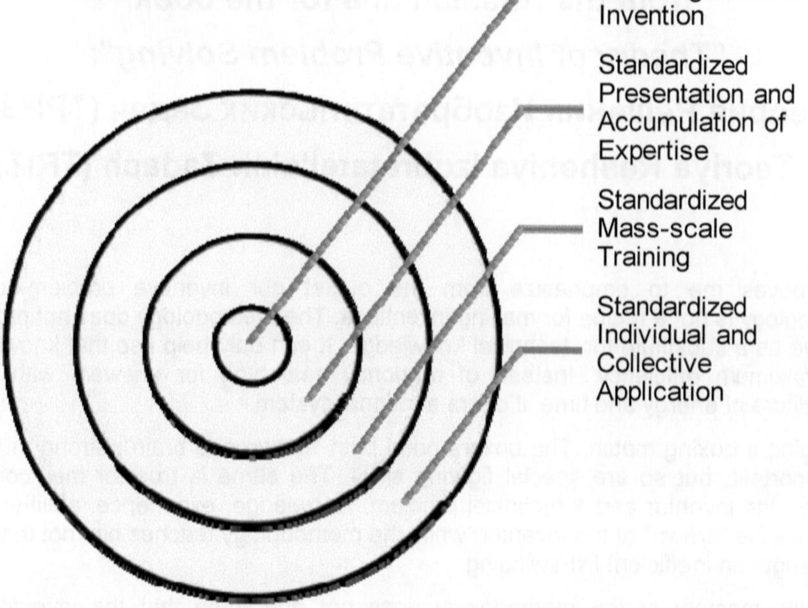

About the Title of the Book

The ABC series of books has always been respected.

They open doorways to the mysteries of the world—and they do that by teaching people to read and thereby partake of the wisdom and experience and the thoughts and feelings of other people. They help people cultivate understanding and empathy.

May this book become your creativity primer. May it help you discover modern TRIZ, which essentially is epitome of *modern invention theory*.

About the Title of the EASyTRIZ™ Training System

The title EASyTRIZ™ has two connotations:

1) it contains a reference to the adjective *easy*, and

2) it is based on a radiolocation abbreviation *EASy*, which stands for *Early Acquisition System* (target acquisition and locking-in as early as possible to develop an efficient reaction). With respect to training objectives, it is interpreted as *Easy Acquisition System for TRIZ: To obtain the right learning and skills as early as possible for efficient problem solving.*

"TRIZ" is an acronym
from the Russian title for the book,
"Theory of Inventive Problem Solving":
Теория Решения Изобретательских Задач (ТРИЗ)
– Teoriya Resheniya Izobretatel'skih Zadach (TRIZ)

It behooves me to emphasize from the outset our inventive problem-solving methodology is not a recipe for making inventions. The methodology does not replace, or serve as a substitute for, technical knowledge. It can only help use that knowledge with maximum efficiency. Instead of randomly searching for answers with huge expenditure of energy and time, it offers a rational system.

...Imagine a boxing match. The boxers need both "brawn and brain": strong muscles are important, but so are special fighting skills. The same is true for the "combat" between the inventor and a technical problem. Knowledge, experience, abilities – all these are the "brawn" of the inventor, while the methodology teaches him not to waste his strength on inefficient fist-swinging.

Naturally, mastery of the methodology does not guarantee that the inventor will exceed Popov or Edison. Well, by the same token mastery of a college physics course does not guarantee that the student will, with time, exceed Newton or Einstein.

To have great creative talent is not enough to create great inventions. Historical conditions have to be conducive, and many other things must occur at the same time to make that happen.

The methodology does not negate the role that can be played by abilities. It assumes that, to varying degrees, each human being has certain talents.

The methodology should be used primarily to resolve ordinary inventive problems – those that any inventor encounters on a daily basis.

The methodology helps develop [the] talents and put them to good use. [1]

Genrikh Altshuller

[1] G. S. Altshuller *How to Learn to Invent.* – Tambov: Tambov Book Publishers, 1961 (in Russian)

This course is an invitation to boost your power to discern:

- ***the beautiful in the practical,***
- ***the amazing in the mundane, and***
- ***the simple in the complex.***

The emblem of our Modern TRIZ Academy bears a biblical maxim[2]:

Be transformed by the renewing of your mind.

This course is invitation to change yourself,
to become more inventive and powerful,
so that you learn to overcome obstacles,
to transform the "impossible" into the possible,
and simply to rejoice
in creating beautiful ideas
and useful solutions.

---

2 Romans 12:2; the original Latin text is: Transformamini renovatione mentis. *Source: Nova Vulgata, Apostoli ad Romanos Epistula Sancti Pauli, 12 (2).* The fact that we quote the Bible does not mean that we are biased toward any particular religion, nor are we trying to proselytize the reader. It simply means that we are drawing wisdom from a specific historical literary source.

We cannot predict the future, but we can invent it.

Dennis Gabor
(1900-1979)

Inventor of holography,
Nobel Prize in Physics, 1971

(citation from the book "Inventing the Future", 1963)

The best way to predict the future is to invent it.

Alan Kay
(b. 1940)

Pioneer of object-oriented
programming and graphical interface

(citation from meeting of PARC, 1971)

ASSERTION OF NEW PARADIGM:

"Creative" vs "Analytic"

Three widely shared perspectives stand in relief:

1) **complexity is escalating,**
2) **enterprises are not equipped to cope with this complexity,**
3) **creativity is now the single most important leadership competency.**

IBM said that creativity is needed in all aspects of leadership, including strategic thinking and planning.

Capitalizing on Complexity: Insights from the Global Chief Executive Officer Study, July 2010

*This study is based on face-to-face conversations with more than **1,500** CEOs worldwide.*

IBM

Haydn Shaughnessy, contributor

What Makes Samsung Such An Innovative Company?

...it was TRIZ that became the bedrock of innovation at Samsung. And it was introduced at Samsung by Russian engineers whom Samsung had hired into its Seoul Labs in the early 2000s.

TRIZ is a methodology for systematic problem solving. Typical of its origins in Russia, it asks users to seek the contradictions in current technological conditions and customer needs and to imagine an ideal state that innovation should drive towards.

At Samsung even the subsidiary CEO has to take TRIZ training.

TRIZ is now an obligatory skill set if you want to advance within Samsung.

3rd of March, 2013

Forbes

Peter Fingar,
one of the industry's noted experts on BPM

Innovation as a Business Process

Clayton Christensen, Harvard Business School's professor: ...the creator of new value, innovation isn't hit-or-miss, trial-and-error lateral thinking, but a repeatable process. ***What is innovative about innovation today is the realization that it can be achieved systematically, and that the innovator is an obsessive problem solver.***

Howard Smith, CTO of CSC's "Innovation Center of Excellence in Europe": ...is now added something that may be a way of thinking, a set of tools, a methodology, a process, a theory or even possibly a deep science, but which may be gradually shaping up

as 'the next big thing.' It's called TRIZ.

...the antithesis of unreliable, hit-or-miss, trial-and-error, psychological means of lateral thinking. Its scientific, repeatable, procedural and algorithmic processes surprise all who first encounter them.

September 24, 2015

BPM Institute.org ™

A Peer to Peer Exchange for Business Process Management Professionals

Table of Contents

Book I How to learn to invent 1

ABC-TRIZ: Discover the Simple in the Complex 3

On the effectiveness of our skills 5

1 Miracle of Invention 7

1.1 Five "easy" tasks to warm up 7
1.2 Secret of the Miracle 12
1.3 Everyone Is an Inventor! 15
1.4 Workshop for Chapter 1 18

2 Toward the Modern TRIZ 19

2.1 TRIZ 19
2.2 A-Studio of Modern TRIZ 22
2.3 MTRIZ-*herbie*! 27
2.4 Workshop for Chapter 2 30

3 TRIZ Algorithms of Invention 31

3.1 Before TRIZ: Meta-Algorithm of Brainstorming 31
3.2 Meta-Algorithm of Invention ARIZ-1956 33
3.3 Meta-Algorithm of Invention T-R-I-Z (MAI T-R-I-Z 1995) 35
3.4 Workshop for Chapter 3 40

4 Modeling the Problem 41

4.1 Contradictions 41
4.2 Standard Contradiction (SC) 43
4.3 Radical Contradiction (RC) 47
4.4 Contradiction as an Attribute of Development 51
4.5 Workshop for Chapter 4 57

5 Extracting 59

5.1 Efficient Models in Each Artifact 59
5.2 Extracting as Key Teaching and Training Method in MTRIZ 61
5.3 Primary and Advanced Extracting 63
5.4 Workshop for Chapter 5 67

6 Inventing 71

6.1 Algorithm START T-R-I-Z™ 71
6.2 Solving the Standard Contradiction with Method BICO 74
6.3 Solving the Radical Contradiction with Method RICO 81
6.4 Workshop for Chapter 6 85

7 Reinventing 87

7.1 Reinventing as Fundamental Teaching and Training Methoc 87
7.2 Key Practical Procedures of Reinventing 90
7.3 Science and Art of Reinventing 93
7.4 Workshop for Chapter 7 116

8 Answers to Tests 117

Chapters 1 – 3 117
Chapter 4. Modeling the problem 118
Chapter 5. Extracting 130
Chapter 6. Inventing 132

Book II How to become a genius 133

To Seek and Not to Yield 135

TRIZonal Concept 138

9 Operative Zone 141

9.1 Reinventing of Operative Zone (OZ) 141
9.1.1 Definition of OZ 141
9.1.2 Reinventing with MAI T-R-I-Z 143

9.2 Transformation of OZ 147
9.2.1 Actors of OZ 147
9.2.2 Resources of OZ 152
9.2.3 Ideal Target Modeling 159
9.3 OZ makes this itself 165
9.3.1 Altshuller's Experiment-1: *Girl's Brilliant Solution!* 165
9.3.2 Altshuller's Experiment-2: *Boy's Brilliant Solution!* 169
9.3.3 Moscow Kremlin Stars: *Turn Harm into Good!* 174
9.3.4 Memories of the Future: *This is da Vinci!* 188
9.3.5 *Does Sir Norman Foster like da Vinci also?* 200
9.3.6 Water from Air: *The Magic Particles in OZ.* 205
9.4 Workshop for Chapter 9 210

10 Directed Development of Systems 217

10.1 System Development and Evolution 217
10.1.1 TRIZ Laws 217
10.1.2 String "Rail Sky Way" 225
10.2 Transfer of Experience 243
10.2.1 We learn by example: *Exemplis Discimus* 243
10.2.2 Professional Specialization 275
10.2.3 The Nature invents! 279
10.3 Advanced Reinventing Cases 286
10.3.1 Originated from SIEMENS 286
10.3.2 *Per Aspera Ad Astra!* 294
10.3.3 *Per Aspera Ad Orbis:* Into Space… on a Wheel! 316
10.3.4 *Let's do it!* 318

11 The Origins of Creative Personality 323

11.1 Josef von Fraunhofer 323
11.2 Werner von Siemens 328
11.3 Konstantin Tsiolkovsky 337
11.4 Jacob Perelman 340
11.5 Genrikh Altshuller 344
11.6 Steve Jobs 350
11.7 Richard Branson 353
11.8 Anatoly Yunitskiy 357
11.9 *It's Your Choice!* 362

Book III Primary Instruments (Summary) 365

S1 TRIZ 367

S2 Invention is an in-depth look 368

S3 Mentality levels of idea generating 369

S4 Directed problem-solving with TRIZ 370

S5 System tendency to ideality 371

S6 Limitation of main system characteristic 372

S7 Breakthrough of main system characteristic 373

S8 Levels of invention 374

S9 Invention complexity 375

S10 Meta-Algorithm of Invention T-R-I-Z (MAI T-R-I-Z 1995) 376

S11 Modern TRIZ: Standardization of Training, Practice, and Problem Solving on the Basis of MAI T-R-I-Z 377

S12 Reinventing on the base of MAI T-R-I-Z 378

S13 Inventing on the base of MAI T-R-I-Z 379

S14 Standard Contradiction 380

S15 Radical Contradiction 381

S16 START: Simplest TRIZ-Algorithm of Resourceful Thinking 382

S16.1 START: integrated scheme 382
S16.2 START-pass through Standard Contradiction 383
S16.3 START-pass through Radical Contradiction 384

S17 Operative zone 385

S18 Resources 386

S19 Resolving of Standard Contradiction on the method BICO (Binary In Cluster Out) 387

S20. Example "Swimmer" (reinventing) 388

S21 A-matrix 390

S21.1 List of 39 Plus and Minus factors 390
S21.2 Table of A-matrix 391

S22 As-catalogue 394
- S22.1 List of 40 navigators (specialized transformations) 394
- S22.2 Table of As-catalogue (text form) 395
- S22.3 As-catalogue (with pictures) 399

S23 Resolving of Radical Contradiction on the method RICO (Radical In Cluster Out) 409

S24 Example "Diver" (reinventing) 410
- S24.1 Solution-pass through Standard Contradiction 410
- S24.2 Solution-pass through Radical Contradiction 412

S25 Afs-catalogue 414

S26 Af-catalogue for four fundamental models with examples 415

S27 Simple form for Extracting-1 418

S28 Examples of form of Extracting-1 419

S29 "START-form" for reinventing 420

S30 Example "Ice for a Drink" at "START-form" 421

S31 Brief Junior-form for Extracting and Reinventing 422

S32 Example "Fischer' Dowel" at brief Junior-form 423

S33 "Junior-form" for reinventing (two pages) 424

S34 Example "Leonardo da Vinci's Bridge" at "Junior-form" 426

S35 As-catalogue (with reinventing) 428

S36 Af-catalogue for four fundamental models with reinventing 508

S37 Terms and abbreviations 516

S38 Main Web sites for AIMTRIZ 516

Book I
How to learn to invent

Ascension changes a man:
it changes his value system.

Anyone is capable of making each step along the path of that *Ascension*, and it would be a crime to waste one's entire life marking time. [3]

Genrikh Altshuller

[3] Compiled by the author from the following book: G. S. Altshuller, I. M. Vertkin *How to Become a Genius: Life Strategy for a Creative Personality.* – Minsk, Belarus, 1994 (in Russian)

The author gave the subtitle to first "book-part" according to the title of first outstanding book[4] (above) by Genrikh Altshuller in memory of the TRIZ founder, thus emphasizing the continuity of the ideas of TRIZ.

4 G.S. Altshuller *How to Learn to Invent*. – Tambov: Tambov Book Publishers, 1961 (in Russian)

ABC-TRIZ: Discover the Simple in the Complex

Presentation[5] of the book by Dr. Val Tsourikov, founder of the company Invention Machine Corp.

Despite the broad publicity that **TRIZ (Theory of Inventive Problem Solving)** has gained worldwide during the last 25 years, it remains inaccessible to many engineers and managers let alone college students. One of the reasons for that is the lack of simple practical textbooks presenting TRIZ in an easy-to-understand manner suitable not only for experienced specialists but also for college students and senior school children.

The author has already released a book[6] about "classical" TRIZ (targeting engineers as its primary audience), which was first published in 2002 to 2003 and repeatedly reprinted by Springer Publishing Company in Berlin and New York, by SOLON Press Publishers in Moscow, and by Science Press Publishers in Beijing, China, in 2010. His newest textbook,[7] "Modern TRIZ," is the most comprehensive work for basic TRIZ instruments.

The book you are holding in your hands is the fruit of a logical attempt by the author to approach another, much larger, and potentially much better motivated target audience: school children and college students. A detailed but simple presentation of TRIZ primary models with clear coherent examples—these are the things that readers and incipient users need by way of quick and proper introduction to TRIZ.

Michael's acquaintance with TRIZ dates back to 1963 when he studied at the Minsk Polytechnic secondary school and was as young as today's senior school children, the intended beneficiaries of this book. Later, as he became first a student and then an employee of our institute,[8] he applied TRIZ models to make his first inventions based on instructions promulgated by our patent bureau experts. Those inventions eventually evolved into the first algorithms and catalogs of the future TRIZ.

In the 1980s, the author supported the laboratory in which I started to create an *Invention Machine* intelligence system that was at least partially based on TRIZ, and in the mid 1990s he helped us to disseminate that software in a number of countries.

Now, having tried his hand at being a TRIZ user, trainer, consultant, and developer, the author wishes to address a critically important audience—school children and school teachers, college students, and professors—with an introductory book that reflects his understanding of TRIZ and sums up his TRIZ training experience.

---

5 The book is presented by Dr. Valery Tsourikov, the designer of trailblazing intelligent software *Invention Machine* (from mid 1980s) based on TRIZ-modeling, pioneering in the world, and the founder of the world-renowned company *Invention Machine Corp.*, Boston, USA.

6 To date there have been published 3 German editions (2002-2006) and 4 Russian editions (2006-2015); English edition: **M. Orloff** (2003, 2006) *Inventive Thinking through TRIZ: A Practical Guide*. Springer-Verlag Inc., New York, 350 pp., 2nd edition, 2006, ISBN-10 3-540-33222-7, ISBN-13 978-3-540-33222-0.

7 **M. Orloff** (2012) *Modern TRIZ: A Practical Course with EASyTRIZ Technology*. Springer-Verlag Inc., New York, 468 pp., 2nd edition, 2012, ISBN-10: 3642252176, ISBN-13: 978-3642252174

8 The author, like Valery Tsourikov, is a graduate of the Minsk Radio-Technical Institute (now Belarus State Information Science and Radio-Electronics University).

For many years, the author—using the phrase "discover the simple in the complex" as his motto—has been seeking to develop a mass-scale TRIZ-training methodology. His modern TRIZ techniques of extracting and reinventing have become effective tools for both beginners and professionals to understand TRIZ ideas and to master the practical TRIZ instruments.

The importance of this special part, which is devoted to Altshuller's **Theory of Developing the Creative Personality (TDCP)**, may be underlined by the paraphrased words[9] of Steve Jobs: *...When you first start off trying to solve a problem, the first solutions you come up with are very complex, and most people stop there. But if you keep going, and live with the problem and peel more layers of the onion off, you can oftentimes arrive at some very elegant and simple solutions. Most people just don't put in the time or energy to get there.* Genrikh Altshuller clearly understood the great importance of TRIZ tools for creativity as well as the irreplaceability of developing the constructive personality, i.e., preparing a person to uphold his ideas despite opposition and counteraction of the circumstances as mentioned by Steve Jobs.

Unfortunately, Genrikh Altshuller did not have time to complete TDCP. Therefore, it is particularly valuable that the author of "ABC-TRIZ" has developed the impressive examples of character formation of famous persons, known and less known, for the readers. There is immense personal, psychological, social, and ethical value in life stories and character sketches of people endowed with outstanding individual qualities that are indispensable for the attainment of socially significant creative goals. They rely primarily on autobiographic essays and statements of their protagonists both historical and contemporary.

The author gave subtitles to the first two books according to the titles of two outstanding books[10]—the first one and the last—of Genrikh Altshuller in the memory of the TRIZ founder, thus emphasizing the continuity of the TRIZ ideas.

The "ABC-TRIZ" is worth your effort!

I wish the readers success in mastering both the basics of TRIZ and the messages of TDCP.

Valery Tsourikov

January 2010 – June 2015

---

[9] "On the design of the iPod" as quoted in Newsweek, 14 October 2006 – https://en.wikiquote.org/wiki/Steve_Jobs

[10] G.S. Altshuller *How to Learn to Invent.* – Tambov: Tambov Book Publishers, 1961 (in Russian); G.S. Altshuller, I.M. Vertkin *How to Become a Genius: Life Strategy for a Creative Personality.* – Minsk, Belarus, 1994 (in Russian)

On the effectiveness of our skills

Here is a story that will help us to properly appreciate efficiency of TRIZ. In I bought a poster (fig. 1,a), one of those that people in Germany often put up on walls in offices and corridors for the sake of a joke. The text exhilarated me with its unexpected wisdom and paradoxical poignancy. You can judge for yourselves – just look at my translation in fig. 1,b, which, although somewhat awkward and unwieldy for an English-speaking reader, preserves the structure of the original to point out the "stress" word combinations. But, is the hint clear?

Indeed, schools and universities sometimes teach us using antediluvian methods designed by teachers who themselves had been students, shall we say, ages ago, i.e., very much "yesterday," as it says in the poster! And we are supposed to believe that these teachers will prepare their pupils for solving future problems?

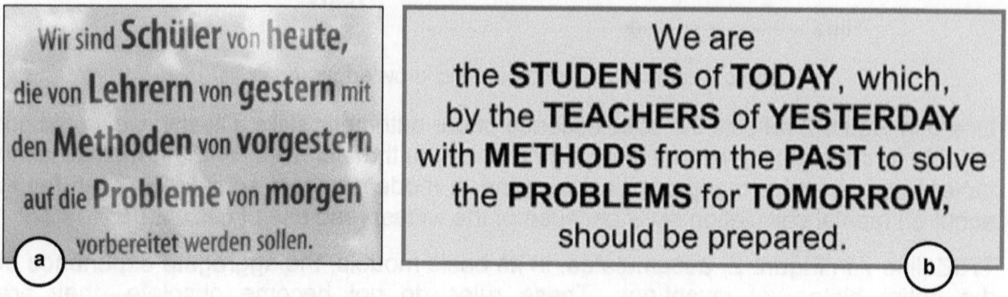

fig. 1. About the "Learning for the Future"!

So maybe TRIZ is also obsolete?! And they keep trying to teach us using nothing but "past inventions"?! Now that is simply outrageous! And of these are all legitimate questions!

TRIZ is not obsolete! On the contrary, it is still not yet fully mature!
TRIZ is up to date! But the methods that have been used to teach TRIZ, and that are still being used everywhere, have become obsolete!!

Now is the time to turn to an interesting illustration of the obsolescence of basic knowledge and skills (fig. 2) that I dug up in 1995 from a German magazine called *Machine Engineering Market* and then complemented with a TRIZ-knowledge obsolescence line,[11] and now with an "arithmetic line."

Experts have established that in the modern world some knowledge changes very fast, e.g., computer technologies and software, while other knowledge changes slowly, e.g., school knowledge.

The thing is that school knowledge contains sustainable fundamental information that has persisted for decades or even centuries and remains essential in virtually all walks of life. It includes ABC reading skills, basics of practical arithmetic and practical algebra, and basics of physics, chemistry, biology, history, astronomy, literature, music, etc. Special professional knowledge in many areas evolves very fast, but it goes without saying that even in those areas there is sustainable fundamental knowledge.

---

[11] M. Orloff *(2003, 2006) Inventive Thinking through TRIZ: A Practical Guide.* – Springer, NY, 2006

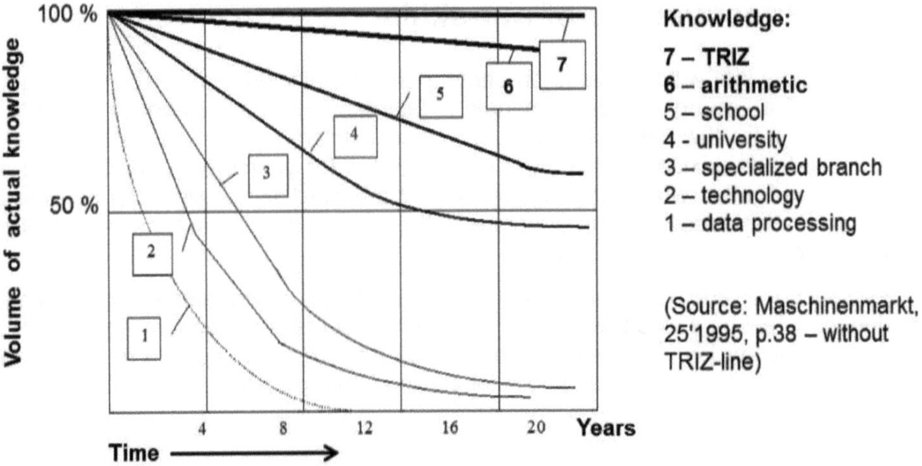

fig. 2. The obsolescence of basic knowledge and skills

Line 6 in Figure 2 reflects the obsolescence of our arithmetic skills. Clearly, we would not be able to do anything without knowing numbers, arithmetic rules, and the multiplication table. But even here many people lose (and in modern school sometimes even fail to acquire!) mental calculation skills because of the widespread use of calculators.

TRIZ (line 7 in Figure 2) accumulates, in its basic models, the aggregate experience of the entire history of inventions. These rules do not become obsolete—they are invariants! So, to remain capable of "mental calculation" by TRIZ rules, you must keep exercising your thinking ability by developing your TRIZ modeling and designing skills as they apply to ideas.

People often ask me how efficient TRIZ is. And sometimes I respond with a counter-question: How efficient is arithmetic? The efficiency of arithmetic depends on who uses it and how, i.e., on the quality of skills, on the extent of proficiency, etc. The same is true for the efficiency of TRIZ: It is a fusion of theory and practical tools, and the efficiency of its application depends entirely on you.

Now is a good time to recall the outstanding Russian composer and pianist, Sergey Rakhmaninov, who used to say: *If I don't play for one day, I notice that; if I don't play for two days, my family members notice that; if I don't play for three days, everybody notices that!*

Now, do not forget that he was a great pianist! This means that even masters need *ongoing training.* If a master neglects his training, he inevitably loses part of his mastership. Therefore, he must train throughout the active phase of his professional career. This philosophy also underlies training in all oriental martial arts.

Indeed, at the decisive moment, when you need to be mobilized and concentrated, it is too late to ask yourself whether you have done everything that you could have done in the period leading up to that moment.

It should also be noted that TRIZ, as a science, is still very young. TRIZ will necessarily improve. Just think of physics or mathematics: Before reaching their current state-of-the-art level, they were improved upon by many thousands of researchers toiling for at least the last 200 years.

1. Miracle of Invention

Ascension changes a man: it changes his value system.
Anyone is capable of making each step along the path of that Ascension, and it would be a crime to waste one's entire life marking time.[12]

Genrikh Altshuller

1.1. Five "Easy" Tasks to Warm Up

In most cases, the assignments in this course were once solved without TRIZ.

Thus, it can be possible for you too, to solve or invent the problems no matter what skills, experience, or creative talent you have.

Now, the questions are: Can we extract any creative experience from solving these or similar problems? Can these experiences then be generalized and furthermore represented as models or practical recommendations for solving similar tasks in the future?

TRIZ – *and only TRIZ!* – says "*yes.*"

If we extract creative models from formerly solved problems and thereby learn how to solve similar ones, does it not lead to the conclusion we are facing at the beginnings of a new theory of creativity, namely, the theory of inventing?

To this question TRIZ again says "yes." By definition, TRIZ is indeed ***the theory of inventing.*** TRIZ is constantly developing new models, definitions, and ideas that support the further development of the theory. Still, there is a *primary knowledge* with which TRIZ once started and without which TRIZ as a theory cannot exist. Everyone who wants to acquire skills in innovation and inventive thinking needs this knowledge.

The study of the primary (basic and advanced) models is absolutely necessary to aquireme the skills needed to apply these models. The models are universal and independent from the field of their application, although some of the models are still too "technocratic." They can be adapted to any kind of situation or problem.

The introductory practical complex of **three books** under the common title, **ABC-TRIZ™**, is devoted to studying the primary models of classical TRIZ. During the last several decades many primary models of TRIZ with improved definitions have accumulated. They are underpinned with new examples and, most importantly, new educational techniques that are supported by a new instrumental software package. This whole complex shall be referred to as the educational system in the basics of modern TRIZ or, for short, Modern TRIZ or, even shorter, MTRIZ.

Now let us shift our attention to the assignments. We would like to kick things off with a very beautiful creative assignment, one of those we call "benchmark assignments," i.e., assignments whose solutions can serve as models in terms of their effectiveness and

12 Compiled by the author from the following book: G. S. Altshuller, I. M. Vertkin *How to Become a Genius: Life Strategy for a Creative Personality.* – Minsk, Belarus, 1994 (in Russian)

M.A. Orloff, *ABC-TRIZ*, DOI 10.1007/978-3-319-29436-0_1

exquisiteness. We will also demonstrate problems at a "child's level"; one of these was told to us by the author of TRIZ, Genrikh Altshuller.

We should remember that none of these problems would have been solved without the effort and creative talent of the ones who solved them. Many of the names of such solution-finders have been lost throughout our civilization's history. However, we are still making use of the fruits of their creativity.

Meanwhile, we also offer tasks that are still far from being solved. You can apply your talent and knowledge—as well as your TRIZ skills—to deal with the problems you will find in this book.

Problem P1. Genrikh Altshuller's Experiment.

This problem[13] was solved by a bunch of kindergarten kids. Not all of them, of course, only the most talented and inventive. And now I present it to you with slight modifications as a little joke. As the saying goes, many a true word is spoken in jest... To find out what is true and what is not, we will use a short "comic strip" with appropriate explanations.

Two ropes are hanging from the ceiling in the playroom (fig. 1.1). You need to grab them with both your hands. But the ropes are too far apart: If you grab one of them, you cannot reach the other!

fig. 1.1. Original problem situation with two ropes

The original position is shown in fig. 1.2,a. This shows the state "is." We need to get a required position as shown in fig. 1.2,b. This shows the state "should be," which is necessary to achieve.

You come up with an idea: Swing one of the ropes (1.2,c)! But by the time you grab the second rope and get back to your original position, the first rope swings back out of reach (1.2,d). The problem is that the first rope is too light and its swing range is small.

---

[13] G. S. Altshuller *To Find an Idea: Introduction to the Theory of Inventive Problem Solving*. – Novosibirsk, Science Publishers, 1986 (in Russian, later editions exist).

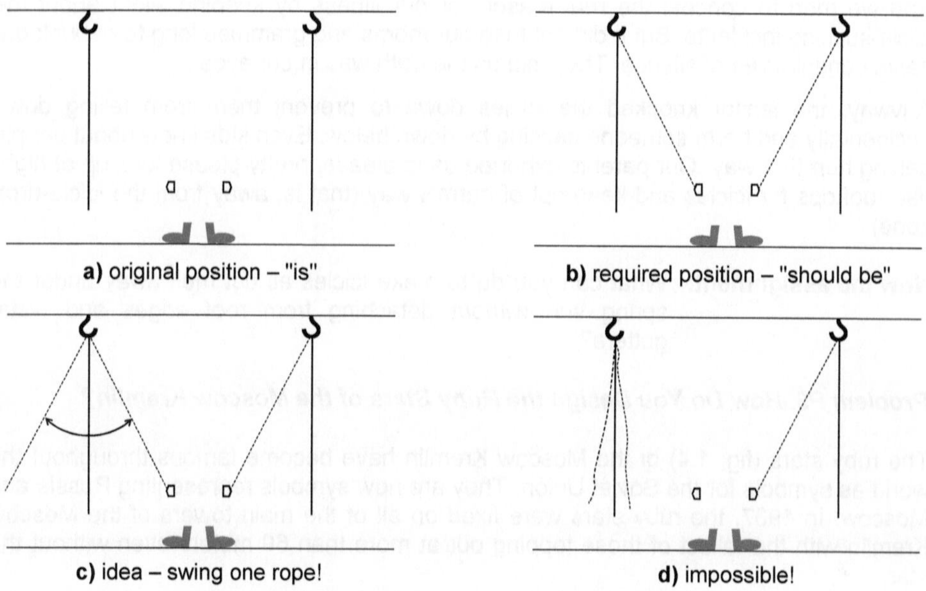

fig. 1.2. Rethinking the problem situation

According to Genrikh Altshuller, at this point many children broke out crying and had to be consoled with sweets. They stopped before an "unsolvable" problem," before an "insurmountable" barrier. Maybe they were not given enough time to think. Or they felt shy in the presence of the experimentalist and kindergarten teachers. Maybe... But some of them DID come through despite everything! How could it be done?

Problem P2. How to Make Icicles Fastened to Eaves Gutter?

From my childhood, I have retained a vivid image of a quaintly bizarre tableau: In spring or during a thaw, our janitor (in those times, janitors were caretakers minding the courtyards of one or several adjacent high-rises) climbed to the roof of the house and knocked down icicles hanging off roof edges and water gutters (fig. 1.3). Some of those icicles were as long as, or longer than, one meter.

fig. 1.3. Icicles hanging off roof

We children liked them because they had fancy shapes and sparkled in the sun, because they wept icy-cold tears, and because from time to time they crashed to the ground shattering into myriad glittering fragments.

We could also break off the smaller icicles adorning the barn roofs by climbing snow banks or low fences. Then we sucked on those icicles like they were candies, and they seemed so sweet... Only later that same day many of us were "rewarded" for our antics with soar throats and high fevers.

And we tried to conceal the real reason for our illness by keeping silent about the icicle-sucking incidents. But it did not take our moms and grammas long to debunk our naïve conspiracies of silence. They told us the truth was in our eyes...

Anyway, the janitor knocked the icicles down to prevent them from falling down accidentally and harm someone passing by down below. Even kids knew about people getting hurt that way. Our parents exhorted us to please, pretty-please look up at high-rise rooftops for icicles and keep out of harm's way (that is, away from the icicle-drop zone).

Now the assignment: What can you do to make icicles all but melt away under the spring sun *without* detaching from roof edges and water gutters?

Problem P3. How Do You Design the Ruby Stars of the Moscow Kremlin?

The ruby stars (fig. 1.4) of the Moscow Kremlin have become famous throughout the world as symbols for the Soviet Union. They are now symbols representing Russia and Moscow. In 1937, the ruby stars were fixed on all of the main towers of the Moscow Kremlin with the tallest of those topping out at more than 60 meters even without the star.

fig. 1.4. Ruby Star of the Moscow Kremlin

The stars are made of special ruby-colored glass containing selenium and other additives ("selenium ruby") and semitransparent milk-colored glass fastened with a special stainless steel mesh. The strips of steel mesh follow the contour of the stars and make them extremely robust.

The main stars have a diameter of approximately 5 meters and a surface of 6 square meters. In other words, in a strong wind there is a great probability that the star could fall off the tower.

With the star weighing approximately a ton, it was a serious task for the engineers who planned and constructed the stars to guarantee their reliability and safety under strong wind conditions.

Assignment: How can one guarantee reliable protection of the stars in case of heavy storms and at the same time keep the construction simple and the expenditure of energy efficient?

Problem P4. How Do You Make a "Magic Faucet"?

This faucet[14] is a well-loved tourist attraction in Aqua Park on Tenerife, one of the largest Spanish Canary Islands.

The stream of water issuing from the faucet (fig. 1.5), which apparently hangs in midair, astounds even those who know how this thing works.

There are many similar dynamic sculptures in various other cities and countries.

Assignment with a prompt: How would you go about creating such an "illusion"?

Also remember this: If a phenomenon, trick, hocus-pocus, or circus attraction exists, this means that all resources required to make this "miracle" happen are available!

fig. 1.5. Magic faucet

The challenge is to figure out—or reinvent—the construction used to demonstrate the trick! So your task is: "reinvent" the "magic faucet."

Problem P5. How to Convince People to Go in for Physical Culture?

We have all heard about how important it is to do morning exercises, to go in for sports, or to walk or jog outside to get fresh air. We have all read about the damage done by smoking and substance abuse. But here is a question: Have we heeded this good advice? Do we follow important health recommendations?

Alas, the answer is often far from optimistic.

---

[14] You can find the answer at http://sobrecadiz.com/2008/07/11/el-parque-acuatico-aqualand-bahia-de-cadiz and some other Web sites. But do not hurry to do that! Give it the old college try first!

Still, health care ministries and enthusiastic members of numerous health societies keep trying to attract people's attention to burning health issues to make them quit their pernicious habits such as smoking, drinking, endless mindless TV surfing...

I remember, from my distant childhood, that in many places in the streets, educational establishments, and industrial enterprises one could see huge posters proclaiming, in huge letters, various useful mottos like this one:

"If you want to be healthy—Harden yourself!"

fig. 1.6

Or in such propaganda poster from the 1950s in fig. 1.6.

Clearly, not everybody can become as strong as the father (or maybe older brother) of the "pioneer" (member of children's organization in the former USSR) shown in this poster.

And the task with a prompt:

How else can we change the design and information content of posters to encourage people to go in for physical culture? Maybe, this needs to be done with humor and inventiveness?

1.2. Secret of the Miracle

TRIZ is important because it provides shortcuts to solving extremely complex problems. But its true value lies in the fact that it gives you the key to the miracle! The key to comprehending things that lurk in the deepest recesses of the unknown and can be discovered only by those armed with proper tools and skills. Of course, there is always a small probability that the miracle will be discovered by accident.

After you complete the training and behold the miracle, it becomes easy for you to deal with unexpected problems, and you begin to understand and appreciate the power of the new problem-solving method and experience the joy of being able to generate vibrant nontrivial ideas.

TRIZ miraculously fuses method and art!

Method yields itself to description and learning. Method opens the treasure chest of efficient ideas. Miracle is born when your talent is enriched with *TRIZ thinking*, when you overcome the impossible, when you are overwhelmed with the beauty of the idea.

Do you doubt that in our enlightened times it is still possible to *join method and miracle?* Then look at the picture (fig. 1.7) in the following page. If you have no method, you may never see what is hidden deep inside that picture.

Example 1.1. Magic Eye Poster

Here is the method that must be used to view this picture: View it in a well-lit room and turned 90 degrees clockwise. Bring the picture close to your face until the image becomes blurred, then start to gradually move it away looking *beyond, or through, the sheet rather than on its surface.* And at some point you will see a miraculous transformation of the image into a 3D shape!

fig. 1.7. Before you proceed, learn the "secret" of this poster:
Turn the sheet 90° clockwise and try to look *through* it rather than *at* it!

TRIZ works the same way!!! The secret is to *look deep inside the problem, rather than on its surface!*

As Goethe[15] put it:

> *to place oneself at the level of objects is to learn;*
> *to take objects in their depth is to invent.*

When you perceive the three-dimensional depth of the image hidden in the picture, you are sure to experience the ***miracle of discovering*, the *miracle of understanding*!**

It may happen that your first attempts to penetrate the pseudo-3D world of the *Magic Eye* picture will be unsuccessful. However, the pleasure derived from watching such pictures will increase as you hone your rapid self-adjustment skills.

The same happens as you study TRIZ. First, you will not see any breathtaking miracles behind the methods you will be learning, and your assignments will appear either too simple and uninteresting or, conversely, incomprehensible—just like that picture if viewed hastily and superficially.

But after you have independently completed at least several dozen assignments, after you have learnt to bravely tackle any challenging problem, *you will become initiated to the mystery and craft of TRIZ.*

I had already been ascending TRIZ heights for several dozen years when in 1992 my German boss and partner brought and hung in the corridor of our joint-venture's office in Minsk a strange huge poster densely filled with blue-green wavy lines. The poster had been made with *Magic Eye* technology, of which we had never heard before. For approximately one week, both the poster and the boss became the objects of general ridicule. Then the boss told everybody to take a closer look at the poster because it harbored a secret. He also said that whoever learned the secret must not share it with the others but should instead tell him, and only him, in private because the first "discoverers" would get prizes, and he wanted the competition to be a fair one.

Some of approximately 50 staff engineers and programmers became almost obsessed with the poster, but others passed by it with affected indifference. Ultimately, not everybody got it, and nobody got it the first time.

It was then that I noted the similarity between really "seeing" the *Magic Eye* pictures and mastering TRIZ. And I am positive that to rise to the level of master in any activity, one must not only learn the method but also discover the mystery of the chosen profession however modest it may appear. *Because in the absence of the miracle, no profession can possibly be an inexhaustible source of joy…*

I made my acquaintance with TRIZ more than 50 years ago (in 1963) using as my guide the first book[16] written by Genrikh Altshuller, the founder of TRIZ. In that small book, TRIZ had neither a name nor any meaningful toolset or structure! It has taken me 3 years (2007 to 2009) to write this small book for you. This is my "payment" for my interest in TRIZ, for my ascension to TRIZ. This is also the time of my communion with TRIZ and my joyful investigation of its mysteries!

---

15 Johann Wolfgang von Goethe (1749 – 1832) – outstanding German thinker, poet, philosopher and naturalist.
16 G. S. Altshuller *How to Learn to Invent.* – Tambov, Tambov Book Publishers, 1961 (in Russian)

Admiration for the principles, simplicity, and depth of TRIZ is still alive in me as a feeling of continuous contemplation of the miracle—the miracle of the beauty inherent in inventive thinking and the miracle of human talent begetting new ideas.

It may take some of you 3 weeks to read this book, while others may need just 3 days. Divide the years I spent on writing this book by the time it takes you to read it, and you will appreciate the huge advantage you are getting in your study of TRIZ. Your time costs will be dozens, even hundreds of times lower than mine!

But there is an even greater prize: By the time you finish this book, you will have internalized, in a concentrated form, the cognitive experience accumulated by hundreds of thousands of inventors and talented people!

And, naturally, you can always hope that your discovery of the miracle of TRIZ will be accompanied by even more intense admiration than that produced by the *Magic Eye* pictures.

Of course, mastering TRIZ will require years of hard work, and your success in its application will depend on your talent and character. Still, you can complete your ascension to TRIZ much faster than did the author of this book. And this means that you may go further and create, over the course of your life, even more wonderful ideas or maybe just one idea, but an idea worthy of a genius. May it be so!

Incidentally, the poster I told you about featured the majestic Statue of Liberty. It weightlessly hovered above the bay, and behind it the City of New York emerged from the water clothed in thick morning mist, grazing the sky with its two unforgettable towers.[17]

It did not take everybody too long to learn to see that 3D image freely and effortlessly, to *behold the miracle* with only minimum "self-adjustment."

I wish you to attain similar freedom in beholding the miracle every passing day: The *freedom of creative thinking according to TRIZ.*

1.3. Everyone Is an Inventor!

I often start my lectures and workshops with these words. Professional inventors do not always concur. Initially. Deep in his heart, each professional believes that he is a particularly talented individual. Incidentally, this is true! But I will get back to that later.

Man is a very motivated and single-minded creature! He is also endowed with a quality that makes him an individual—the *freedom of choice!* He is free to choose his way of life, education, professional pursuits, musical works, books, purchases, actions, etc. Billions of people—men and women, experienced professionals and students, gray-haired sages and small children—make individual choices every day. Billions of choices, including *choices of thoughts, words, and actions*, coalesce into what we call "life."

Choices often appear to be made unconsciously, automatically. In reality, the human brain always goes through the full "invention cycle" (meta-algorithm of modern TRIZ):

---

[17] The twin towers of the *World Trade Center* in New York were destroyed on September 11, 2001, as a result of a terrorist attack; there were more than 3,000 victims, innocent people who had led ordinary human lives and had done no harm to their murderers.

1) Man experiences a need. This provides him with motivation for his actions.

2) He formulates (invents!) a purpose, the attainment of which is tantamount to meeting that need.

3) He devises (invents!) methods that he can apply to attain that purpose, and finds (invents!) ideas and solutions.

4) He compares and selects the best ideas. Sometimes he finds no solution, and then he repeats the cycle changing the original methods or even the purpose.

Even people who are very lazy and not very brave are inventive. Moreover, they frequently have to be extra inventive to get what they want with minimum effort.

Children (adults, too) can be very resourceful when they invent excuses to justify their mistakes or missteps! Among other things, to convince themselves to give up their attempts to attain "unattainable" objectives. To prove to themselves and to others that they are not to blame for... well, for whatever went wrong!

It is true that choices can be extremely difficult. Sometimes we lack information. Sometimes we lack resources (materials, time, and money). Sometimes it is just not clear what objectives need to be achieved to resolve the situation. And in all such cases – without exception – what we deal with is *contradictions!*

We get to our destinations by overcoming contradictions! This is one of the key TRIZ postulates.

Here are some simple (and, perhaps, somewhat facetious) examples of "typical" contradictions (fig. 1.8), which may give you a pleasantly tingling feeling of superiority because they have nothing to do with you.

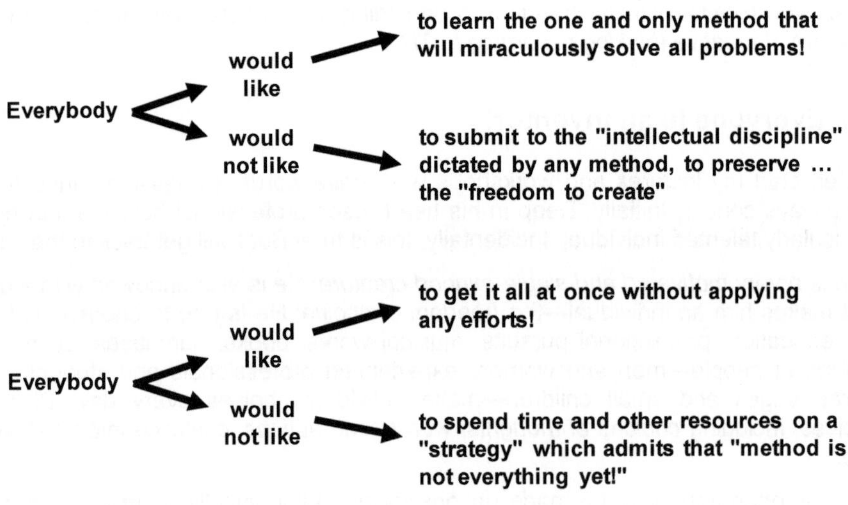

fig. 1.8. Schematic representation of some "fundamental" contradictions

Well, do not judge yourselves too harshly if you recognize yourselves in these examples! It is very *natural to seek to attain the "ideal result" with a minimum expenditure of resources!* Mind this caveat, though: It is not important where you are going, but it IS important how you get there.[18]

Moreover, it turns out that with the situation becoming more and more convoluted and complicated, it becomes increasingly acceptable, from a psychological point of view, to rely on a strategy where the task apparently *"solves itself"!* Where there is no need to resort to additional resources! Where all *required resources are extracted from the problem situation!*

This does not mean that there are no unsolvable problems. There are plenty. But very often we capitulate *before we solve the problem.* Because we have no requisite knowledge of *"here and now."* Because we fail to use more efficient methods to find ideas for possible solutions.

Man invents both the purpose and the way to achieve it!

Man does what is deemed undoable! He puts together what is deemed incompatible! And he performs that feat in unexpected, often inconceivable ways.

What makes this possible—knowledge or art, natural laws or intuition, assiduous work or inspiration, blind luck or routine experience and common sense? Can we find methods that may be used to invent ideas and substantiate choices? Can we recommend those methods if not to everybody (after all, each human being is unique) then to as many people as possible?

TRIZ gives positive answers to all these questions. TRIZ is a study of both *practical examples* and *theoretical methods.*

Knowledge of practical examples facilitates the internalization of methods. But *examples must not replace methods!* It is hardly possible to find examples covering every possible life situation. In real life you cannot copy answers as you may have copied them from your friend's exercise book when you were a school child. Besides, *examples become "obsolete"* and lose their appeal, whereas *methods continue to work!*

TRIZ methods promote creativity. Knowledge of problem-solving techniques enhances your abilities and increases your freedom.

Freedom to generate creative ideas and attain your objectives.

If you accept fundamental TRIZ ideas and skills and adapt them to your needs, *you will be able to:*

1. *Radically save time and resources* needed to generate an idea for a solution. Or several ideas, which is even better!
2. *Minimize the risk of making a serious mistake.* And not be afraid of difficulties no matter what happens but instead persistently *improve your odds and transform drawbacks into advantages.*
3. *Obtain the strategic freedom of choice in any situation* for increasing the probability of successfully implementing your idea.

[18] From *Tight-Rope Artist* (Канатоходец, 1967), a song by Yuri Kukin (Юрий Кукин, Russian bard, b. 1932).

Naturally, like any teacher or advisor, I will be *there for you* rather than do things *instead of you.*

So it all depends on you. You will need to exercise. If possible, on an ongoing basis. You will need to do TRIZ etudes—See an experimental A-studio etude in the next section. And the last two pages of this preface following that little joke will tell you *how* you need to exercise to do something meaningful in your lives.

1.4. WORKSHOP FOR CHAPTER 1

Answer these questions yourself with (or better without) a book:

1. Why may we consider mastering TRIZ as personal ascension?
2. Could we dig up creative knowledge from the existing inventions and innovations?
3. Is TRIZ a universal panacea to solving any problem?
4. Does the application of TRIZ require logical thinking only?
5. Does TRIZ teach us to take objects in their depth to invent?
6. Could we perceive the miracle metaphor of deep seeing with the Magic Eye pictures?
7. When was the first book by Genrikh Altshuller published?
8. What are the most common negative psychological contradictions in learning?
9. Do you think that TRIZ can save your much time and resources in problem solving, minimize the risk of making a serious mistake, and obtaining the strategic freedom of successful choice in any situation?
10. Come back to these questions after end of reading this book and answer them anew!

2. Toward the Modern TRIZ

The work is creative if its result is the creation of something new. The very same process of creating something new can be deeply conscious and systematic. ***The methodology of invention is not accidental godsend, but a natural stage in the development of technical creativity.***[19]

Genrikh Altshuller

2.1. TRIZ

TRIZ was founded in Russia, starting from the end of the 1940s, by the inventor, writer, and independent scientist Genrikh Saulowitsch Altshuller.

The first steps toward the development of what would later become TRIZ were made by him together with his ex-classmate Raphail Borisovich Shapiro.[20]

Genrikh Saulowitsch Altshuller (1926-1998) – founder[21] of TRIZ

The notion of the *"inventive problem"* in TRIZ was defined as a problem whose efficient (acceptable) solution still awaits because obvious solutions would be unsatisfying or because they do not exist and would be impossible to obtain using existing professional methods.

Definition of **"Inventive Problem"**	**"Inventive Problem"** (Inventive Assignment) – A problem containing *a contradiction* in the form of incompatible requirements and/or properties that appeared through an irregular development of (1) the parts of the system or of (2) the given system and its surroundings that cannot be solved by adequate methods and means.
Definition of **"Invention"**	**Invention (inventive idea, inventive solution)** – An idea that abolishes the contradiction and thereby the problem containing that contradiction.

[19] From: G. S. Altshuller *How to Learn to Invent.* – Tambov, Tambov Book Publishers, 1961 (in Russian)

[20] *The Psychology of Inventive Creativity.* – ***Journal "Questions of Psychology"***, No. 6, Moscow, 1956; coauthors: Genrikh Saulowitsch Altshuller (1926-1998) and Rafail Borisovich Shapiro (1926-1993)

[21] Picture: magazine ***Der Spiegel***, June 2005

M.A. Orloff, *ABC-TRIZ*, DOI 10.1007/978-3-319-29436-0_2

> The most fundamental concept introduced by TRIZ is that a new inventive problem can be solved on the basis of experience accumulated in inventions made earlier in various other areas!

To identify such experience, G. Altshuller and his followers analyzed dozens of thousands of inventions. As a result, they discovered "inventive techniques" (transformation models) most frequently encountered in real inventions. They also discovered two basic types of contradictions whose resolution engendered new inventions: *the technical ("standard" in MTRIZ) contradiction and the physical ("radical" in MTRIZ) contradiction.*

Definition of **TRIZ**	***Theory of Inventive Problem Solving (TRIZ) – TRIZ is a constructive methodology*** *for generating ideas and solving problems, primarily when projecting engineering systems*, on the basis of contradiction models, as well as the methods to solve them, which were extracted from previous known inventions.
Constructive methodology – addition to definition of TRIZ	TRIZ is a ***constructive methodology*** that includes practically reproducible models and methods that allow the development of new inventions as well as the teaching of the process, the models, and the methods of creating inventions.

TRIZ rests on the foundation of the following basic provision:

System Platform	*Technical systems evolve in accordance with certain laws that can be cognized and applied to designing a new creative technology… thereby turning the development of technical systems into a precise science.*[22]

The structure of TRIZ is shown in fig.2.1.

In their first article, titled *The Psychology of Inventive Creativity*, G. Altshuller and R. Shapiro formulated **three fundamental premises** found in each creative solution of each new problem:

> 1. Formulation of the problem and determination of the contradiction that prevents resolution of the problem by applying the usual methods already familiar to engineers.
>
> 2. Elimination of the cause of the contradiction with a view to attain a new, i.e., higher technical effect.
>
> 3. Transforming the other system components to bring them into conformity with the modified component.

[22] Altshuller, G. *To Find an Idea. Introduction to Theory of Inventive Problem Solving.* – Novosibirsk, Nauka (Science) Publishers, 1986 (there are also later editions).

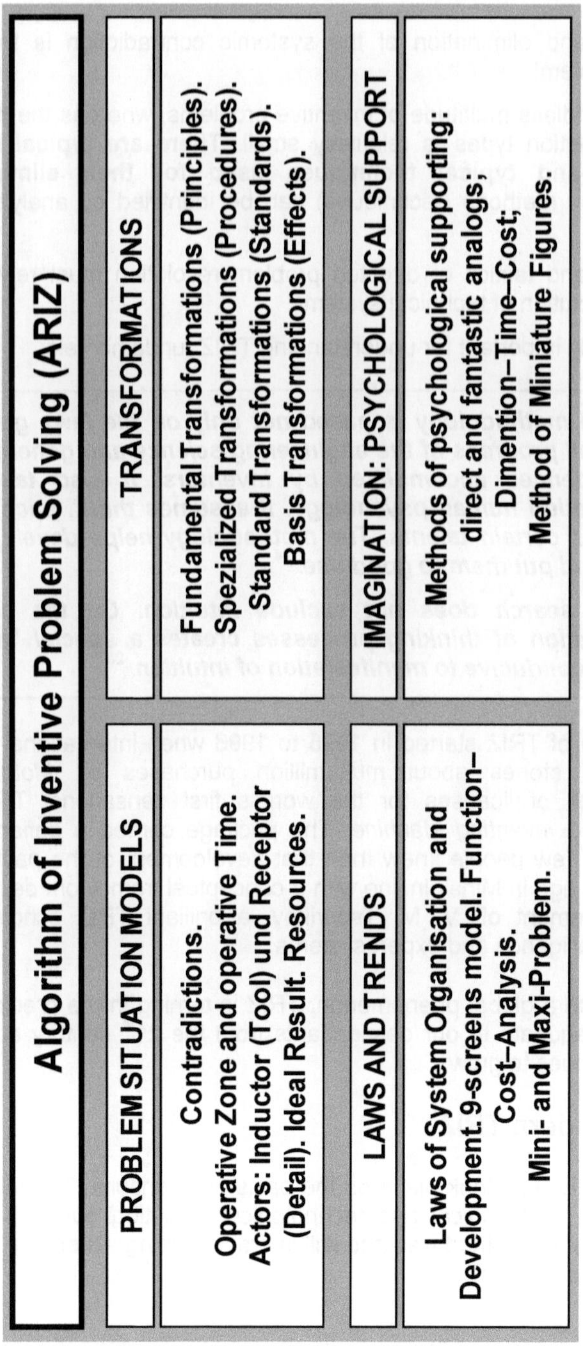

fig. 2.1. Structure of classical TRIZ

Subsequently G. Altshuller developed **three principles to resolve inventive problems:**

1. Identification and elimination of the systemic contradiction is the key to resolving the problem!

2. "There is an endless multitude of inventive problems, whereas the number of systemic contradiction types is relatively small. **There are typical systemic contradictions—and typical techniques used for their elimination.**[23]" Problem-resolution methods (techniques) can be identified by analyzing great inventions.

3. The strategy and tactics of directed problem resolution must rely on laws governing the evolution of technical systems.

The next point is critically important for understanding TRIZ fundamentals:

Psycho-logical Platform	***Invention methodology is based not only on the laws governing the overall progress of the engineering science and generalization of experiences accumulated by inventors. It also takes into consideration human psychology. It assumes that... each human being has certain talents. The methodology helps develop those talents and put them to good use.***[24] ***Directed search does not exclude intuition. On the contrary, normalization of thinking processes creates a special "attitude" which is conducive to manifestation of intuition.***[25]

Rapid global expansion of TRIZ started in 1995 to 1996 when international business newspapers published stories about multi-million purchases by *Motorola* and *Mitsubishi* of thousands of licenses for the world's first sensational TRIZ-based software package called *Inventing Machine.* The package carried a defiant slogan: *Software That Invents.* Few people knew then that development of the package had started almost 10 years ago in Minsk in line with a conceptual framework designed by, and under the management of, V. M. Tsourikov, a brilliant TRIZ scholar and a specialist in artificial intelligence and expert systems.

Today TRIZ has become a global phenomenon. TRIZ is turning into a precious asset that we are going to bequeath to our descendants from the 21st century and the 3rd millennium. And it continues to grow.

2.2. A-Studio of Modern TRIZ

The development of civilization took place on the basis of inventions. The invention of a new idea is not only to technical and technological evolution but also to social evolution. The main creative method was and still is brainstorming. Question: Does this thinking work in our current times?

23 G.S. Altshuller *How to Learn to Invent.* – Tambov, Tambov Book Publishers, 1961 (in Russian)

24 Defined by the author (M.O.) based on G. S. Altshuller *How to Learn to Invent.* – Tambov, Tambov Book Publishers, 1961 (in Russian)

25 G.S. Altshuller *Invention Algorithm.* – Moscow, Moscow Worker Publishers, 1973 (in Russian)

The development of civilization in our time becomes more technocratic. Technical artifacts themselves are becoming more diverse and complex. The creation of new ideas for the development of technics and technologies requires more effective thinking. The new thinking must be even more intense. **However, most importantly, creative thinking should be more constructive.** A new creative technology of effective thinking is needed in order to link motivation with objective knowledge and talent with systematic management of the inventive process.

Modern TRIZ creates opportunities to overcome negative stereotypes of poorly organized thinking. In this case, MTRIZ does not negate intuition and talent, associative thinking, and conjecture. These creative components will remain at the heart of personal creativity, albeit in a new way, in conjunction with MTRIZ.

MTRIZ provides a systematic way of applying TRIZ by learning techniques and patterns for well-organized thinking.

According to TRIZ, invention is the path (fig. 2.2) from the available state of the artifact ("is" or "exists") to a future state ("should be" or "needs to be") with the help of transformation models acting as thinking navigators.

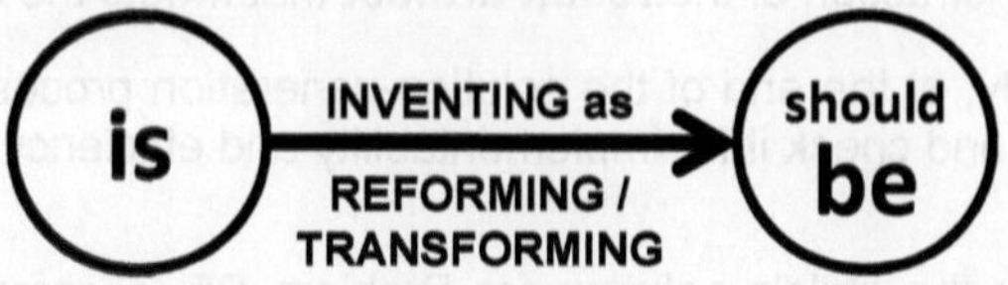

fig. 2.2. Invention as system transition from the state "is" to the state "should be"

The application of TRIZ can be illustrated as follows (fig. 2.3).

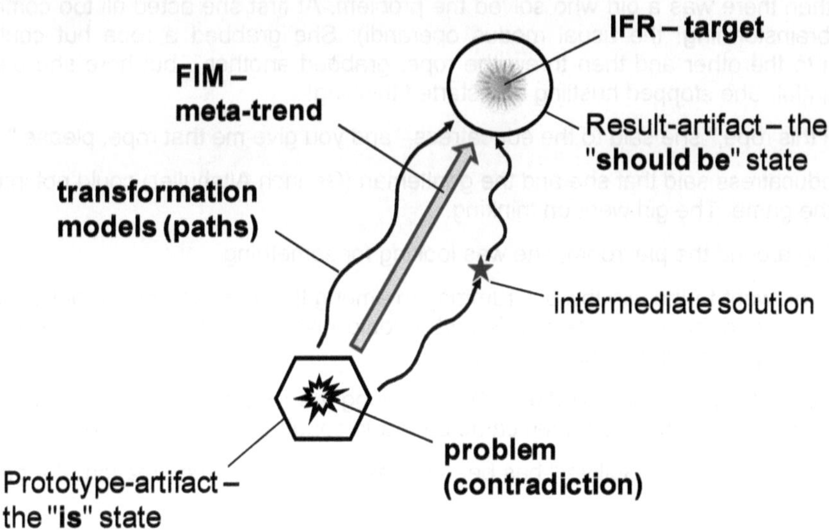

ATTENTION: You should know this scheme by heart!

fig. 2.3. Simplest scheme of resolving the problem according to TRIZ

A simplified method of constructing the ideas of solutions on TRIZ (fig. 2.3) is as follows:

1. The solution begins with the formulation of **contradiction(s)** as a **problem model.**
2. Then you set **a target** – an **ideal final result (IFR).**
3. Then you formulate one or several **functional ideal models (FIMs)** setting the direction towards the IFR (**the "meta-trend"**).
4. It is necessary to choose the **transformation models**, i.e., specific **paths,** to the target. The number of test and **intermediate solutions** (indicated by an asterisk) is reduced multiple times by using TRIZ tools (catalogues, tables [matrices], etc.)!
5. Finally (and this is the most critical in importance and most creative moment!), you must go through this path to invent specific solutions that change the existing construction of the **prototype-artifact** for forming a new construction of the **result-artifact** that meets the required IFR.
6. Naturally, at the end of the solution-generation process, we must verify the idea(s) and check it for implementability and efficiency.

Now we will study the child's solution for Problem P5 to compare it with the TRIZ approach! It will help introduce us to TRIZ modeling and problem solving step-by-step.

Example 2.1. First child's solution[26] to Problem P5.

And then there was a girl who solved the problem. At first she acted all too commonly (my brainstorming! the usual modus operandi): She grabbed a rope but could not reach to the other and then threw the rope, grabbed another... but here she become thoughtful. She stopped hustling and started thinking!

"I pull this rope," she said to the educatress, "and you give me that rope, please."

The educatress said that she and the gentleman (Genrich Altshuller) could not interfere with the game. The girl went on thinking.

Looking around the playroom, she was looking for something.

Then she went to the window sill, rummaged among the toys, and pulled out a tattered doll. A second "person" was needed who could give her the rope! And the girl has found this "person" in the form of a doll!

She tied the rope to the doll (fig. 2.4) and swung the resulting "pendulum," ran to the second rope and grabbed it, then came back and caught the swinging doll.

So, a very big candy "Gulliver" has been honestly earned for the creation of a genuine miracle!

---

[26] G. S. Altshuller *To Find an Idea: Introduction to the Theory of Inventive Problem Solving*. – Novosibirsk, Science Publishers, 1986 (in Russian, later editions exist).

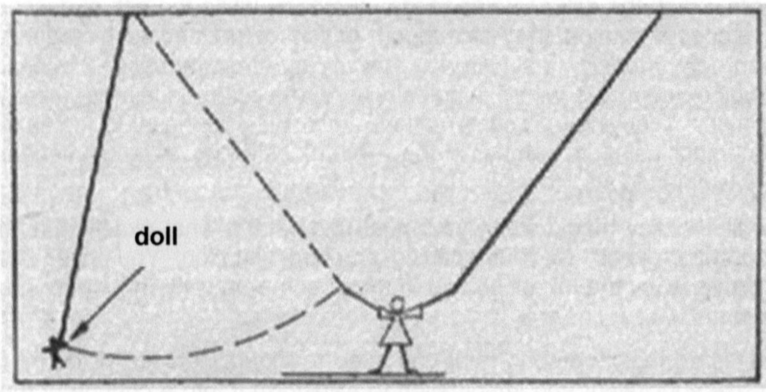

fig. 2.4. First child's solution to Problem P5

Example 2.2. First TRIZ's solution to Problem P5.

Now let us start the "magic" problem-solving process according to TRIZ We will solve (model) Problem P5 together with the kids from kindergarten!

And now imagine that you are already well aware of TRIZ, at least from the volume ABC-TRIZ!

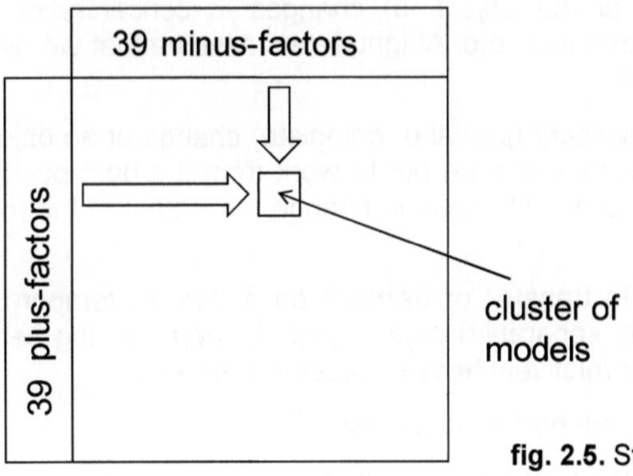

Purpose: increase rope swing range. To do that, we must increase swing time. *Main deficiency of the "system":* the rope quickly loses momentum because it is thin and too light. Air resistance quickly stops its swinging.

Now let us go to our "magic" adviser (fig. 2.5) – *Altshuller's matrix*, or *A-matrix*[27].

fig. 2.5. Structure of A-matrix

First we must give our "adviser" a brief presentation of the problem in the form of a *formal contradiction*: "We want this—our purpose or *plus-factor*, but we cannot do it because of that—the obstacle or *minus-factor*". The "classical" A-matrix (fig. 2.5) contains 39 *plus-factors (lines)* and 39 *minus-factors (columns)*.

If we select one plus-factor and one minus-factor, they will point (see arrows) to an A-matrix cell that contains "advice," i.e., problem-solving models, and we are talking here about *any problem* where contradictions can be described by the selected pair of factors.

[27] see S21.2 Table of A-matrix

To be more exact, the cell contains model numbers, while the actual descriptions are placed in a special catalog, the As-catalog,[28] which features 40 such models. These 40 problem-solving *models* (*navigators, specialized transformations*) constitute the "magic" toolbox complete with examples! There are such models as "inverse action," "dynamization," "copying,", "segmentation," etc. All you have to do is select the requisite number using its number in the A-matrix cell, look at recommendations, and examine examples to see how it works.

Naturally, like in any magic story, you must understand what exactly this model tells you to do with respect to your specific problem situation! You must interpret the prompts provided by the model so that you can come up with an idea for a solution to your problem!

The model is a magic metaphor that you need to "figure out" (just like in a fairy tale!) and then use your talent and imagination to transform it into a change of the situation or object resulting in removal of the problem. Now let us look at how we can tell our "adviser, i.e., the A-matrix, about the contradictions that exist in the situation at hand.

How about this: *The rope must swing for a long time, but it rapidly loses the energy we transfer to it.* This informal description of just one of the possible contradictions can be matched to the following A-matrix inputs: plus-factor *23 Functional time of the moveable object* and minus-factor *39 Loss of energy.* The A-matrix "advises" us to apply models *01 Change in the aggregate state of the object*, *02 Preliminary action*, and *18 Mediator.*

Let us try to figure out what those models "advise" us to do. The "composite sketch" of a possible solution looks like this:

01 Change in the aggregate state of the object: b) changes in concentration or consistency, degree of flexibility, temperature, etc. Alright, let us assume that we need to **increase the weight** of the rope;

02 Preliminary action: a) previous necessary (partial or complete) change of an object; b) prepare objects in advance so that they can be put to work from the best position and are available without loss of time. Aha! We need to change the weight of the rope **in advance**;

18 Mediator: a) use another object to transfer or transmit an action; b) temporarily connect an object with another (easily separable) object. Well, it keeps getting better and better! We need to **connect** some relatively **heavy object** to the rope!

...I hope you remember that the playroom had a lot of toys!

This is where inventing thinking kicks in: You can take an adequately heavy toy, say, a large doll, tie it to one of the ropes in advance, and swing that rope as a pendulum! Then you should run and grab the second rope, get back to the central position, and catch the first rope.

That's it! The "mediator" has worked!

So, in looking for a solution, we passed through the following four phases: *problem, problem model (contradiction), solution model (method), and solution!* This is a simplified "classical" TRIZ cycle, which is illustrated in many books by the following chart (fig. 2.6).

---

[28] see As-Catalogue (sections S22.2, or S22.3, or S35)

Of course, in reality it is much more complicated than that. Still, TRIZ is based on contradictions and methods (models) of their resolution. The purpose of this book is to help you study those models and acquire initial skills of their correct application.

That is why it is called the "ABC" of TRIZ.

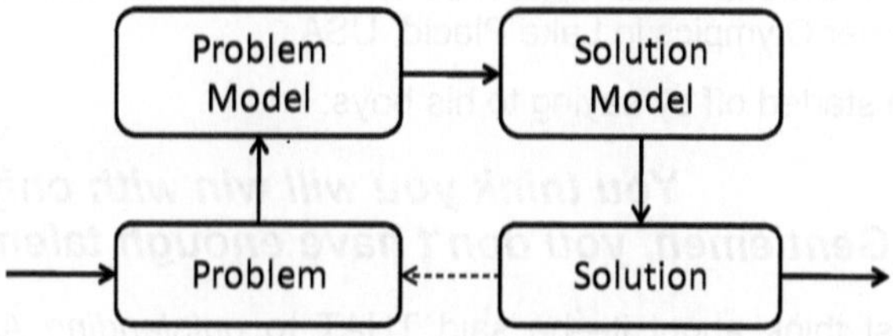

fig. 2.6. Simplified problem-solving scheme according to classical TRIZ

The primary TRIZ toolset includes contradiction models and contradiction-resolution models based on various A-catalogs as well as the A-matrix used to select models in one of the catalogs. The author called this primary toolset the "A-studio"[29] in honor of G. S. Altshuller.

The basic A-studio models constitute the "ABC" of TRIZ.

Naturally, this book is not as simple as a primer you would use to learn to read letters, syllables, and words. However, the problems you will learn to solve are much more complex than the kindergarten experimental study. And you are probably more than three times (!) older than the kids in the kindergarten where that experiment was staged. So you have completely different tasks in store for you.

Based on many years of TRIZ-teaching experience, the author hopes that this book will prove to be not only useful—but also interesting and even funny—and fascinating in a way that any exploration, including self-exploration, and quest for magic powers can be deemed fascinating.

2.3. MTRIZ-*herbie*!

I address this section to my young readers to tell them that to be successful in achieving what one desires requires perseverance and will; indeed, much of the training is very hard work, not amusement.

The book you hold in your hands gives you the standard initial minimum knowledge to qualify for MTRIZ Junior and Practitioner certification under the *EASyTRIZ™* program (www.modern-triz-academy.com).

However, that alone is not enough: You also need training and character to master and use that knowledge and keep going despite possible failures.

In conclusion of this preface, I would like to tell you a story[30] that not only confirms the importance of standard training and vigorous practice but also reminds us that in anything we do our success depends to a large extent on our motivation, perseverance, and will to pursue and attain our goals.

[29] M. Orloff *Inventive Thinking through TRIZ*, part "A-studio: Algorithmic Navigation of Thinking"

[30] The author started this book concurrently with his book *Modern TRIZ*. He deemed it possible to recount this story in both books as he believes it to be very significant.

Herb Brooks[31] was appointed to coach the US national ice hockey team for the 1980 Winter Olympics in Lake Placid, USA.

He started off by saying to his boys:

You think you will win with only your talent...
Gentlemen, you don't have enough talent to win on talent alone.

Just think about it—he said THAT to *outstanding American players* hand-picked to become one team, one fraternity, one family!

To boost the starting speed and increase the endurance of his players, he invented and implemented a method which was later dubbed *herbie* based on the diminutive version of his first name. The method basically boiled down to this[32]: During each training session, the players had to start from their goal line (fig. 2.7), sprint for the blue line on their half of the playing field, stop dead in their tracks as soon as they reached it, and skate back to the goal line; then stop, turn around, and immediately run for the middle red line and back to their goal line; then stop, turn around, and immediately run for the blue line on the other half of the playing field and back to their goal line; and finally, stop, turn around, and immediately run for the other team's goal line across the entire playing field—and then back to their goal line! And they had to do it hundreds of time!

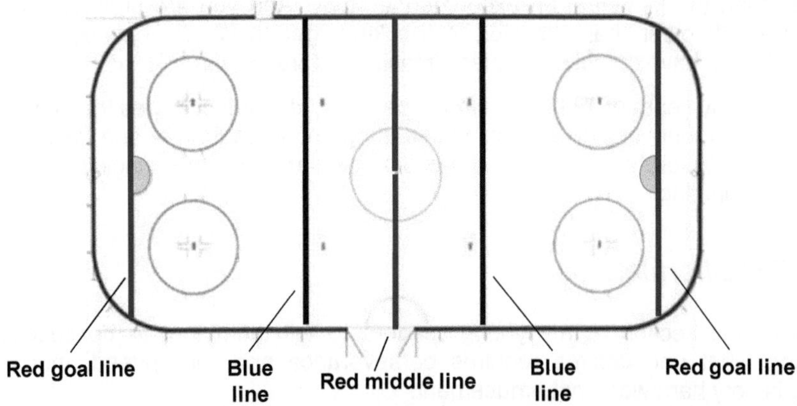

fig. 2.7. The ice-hockey rink is 30 meters wide and 60 meters long!

Just ask yourselves, would you be able to withstand such a drill?!

If the answer is *no*, then maybe this book is not for you, and you need to find something else to do with your time, maybe something simpler.

If the answer is *yes*, then rest assured that the *EASyTRIZ*™ program is not all that difficult for those who are able and willing to work hard.

---

[31] Herbert Brooks (1937-2003) – outstanding US ice hockey coach.

[32] Herb Brooks quote: *Red line, back. Blue line, back. Far blue line, back. Far red line, back. And you have 45 seconds to do it. Get used to this drill. You'll be doing it "a lot." Why? Because the legs feed the wolf, gentlemen. I can't promise you we'll be the best team at Lake Placid next February. But we will be the best conditioned. That I can promise you.*

Prior to his appointment, Herb Brooks said to the members of the US Olympic Committee: *The Americans have to change their style of playing and adopt the hybrid style of the Canadians and the Russians!*

In the semifinal match with the invincible (!) USSR team, his boys scored a 4:3 victory. Many remember that match as if it were the final because it was more memorable, even though the US–Finland final (4:2) was equally dramatic. The US national ice hockey team became the 1980 Olympic champion.

Herb Brooks wrote in his memoirs[33]: ***We were a fast, creative team that played extremely disciplined without the puck.***

Thus the victory was made *before the game*!

And only then was it made *during the game*!

The match (popularly known as the *"Miracle on Ice"*) became an outstanding sporting event because it showed what could be achieved by individual mastery and teamwork based on a *well-designed intensive practice regimen* multiplied by single-mindedness and purposefulness.

I remember that match…

Individuals and teams that understand and skillfully apply Modern TRIZ are capable of similarly great achievements.

The *EASyTRIZ™* training program can become a catalyst of success for anyone who is willing to succeed and is prepared to go through *MTRIZ-herbie* to train to be successful.

I sincerely wish you luck!

Michael Orloff

Berlin, January 2011 – October 2015

P.S. I have just remembered another Herbert Brooks maxim—one that is particularly important for me—*Give the Game Back to the Players.*

This brings to mind another dream of mine. I want basic TRIZ training to start as early as possible but in any event not later than during senior school years, better still before children go to school.

I want children to perceive training as an exciting game, a joyful and fascinating examination of the surrounding world. I want that joy and fascination to persist throughout their adult lives.

TRIZ must do more for children: It must come to schools, kindergartens, and families.

At the beginning of this post scriptum, I quoted the words spoken by Herbert Brooks after he had led his team to its victory in the Olympics. He wanted to give the game back to the kids, to revive the joy of the game, to prevent the game from being destroyed by thoughtless commercialism and equally thoughtless "fanism."

33 www.herbbrooksfoundation.com

I would like to finish by another Brooks quote; just pretend that this is said not about ice hockey but about TRIZ:

> ***the courage to be unusual, without compromise, innovative, highly prepared, maximally successful through hard work and persistence, to have the courage to dream, to be unafraid of failures, to respect others, both students and teachers... unrelenting, insubmissive, contradictious and hard working and all of this with a great love for all humans, both the players and volunteers.***

2.4. WORKSHOP FOR CHAPTER 2

Answer these questions yourself with (or better without) a book:

1. When was first the TRIZ article published?
2. What is an inventive problem?
3. What is an inventive idea?
4. What is the most fundamental concept of TRIZ?
5. What is the definition of TRIZ?
6. Why is TRIZ a constructive methodology?
7. What is the system platform in TRIZ to obtain new inventions?
8. What are three fundamental premises to solve any problem?
9. What are three principles to resolve inventive problems?
10. What is the psychological platform in TRIZ to obtain new inventions?
11. Is an invention, what should be the system transition from the state "is" to the state "should be"?
12. Can you draw and explain the simplest scheme of resolving a problem according to TRIZ?
13. Can you tell your friends the story about an inventive little kindergarten girl who solved the "problem with two ropes"?
14. Can you demonstrate to your friends the solution of the "problem with two ropes" using the A-matrix and the transformation models from the As-catalogue?
15. What is a simplified problem-solving scheme according to classical TRIZ?
16. What is *MTRIZ-herbie*, and what do you think about it?

3. TRIZ Algorithms of Invention

> In the broad sense of the word,[34] an algorithm is any program of a planned course of actions. The program of inventive problem solving is called an algorithm in this sense.
> To effectively solve inventive problems... [we] need a heuristic program that allows [us] to replace the brute force of options with directing motion targeted in the area of solution(s).
>
> ***The algorithm discards the wrong steps.***
>
> ***The algorithm does not eliminate the need of the inventor to think.***
>
> *Genrikh Altshuller*

3.1. Before TRIZ: Meta-Algorithm of Brainstorming

All known creative techniques can be practically generalized as "*brainstorming.*" This statement is not entirely precise, but at least it is brief. It is also fairer than the method of *trial-and-error* because it could be renamed *trial-and-success*, which would make one's relation to this method an easier one but would at the same time increase the psychological inertness when applying it.

Still, the conception is the same as in brainstorming, in which every kind of attempt is understood as *guessing* the idea and "jumping" on it while the so called "creative hunting" happens. It is like an agonizing wandering through the foggy atmosphere of "creative hunting" because such a search lacks any kind of navigators and points of orientations.

Let us now take a look at a "classical" diagram[35] (fig. 3.1), which describes "creative thinking" in the style of brainstorming as it is defined above.

The first question arising from looking at the diagram: Can this diagram help with problem solving?

Answer: No, it cannot.

Understanding this diagram increases the efficiency of problem solving as much as understanding the operating principles of one's digestive system increases the joy of eating—just like in the famous joke.

Then a second question comes up: Does this diagram have any practical use?

You will find the answer to this question below.

In reality, it is not the diagram that is inefficient: The entire ideological platform, for which the diagram is only a generalized ambassador, is *a meta-algorithm*.

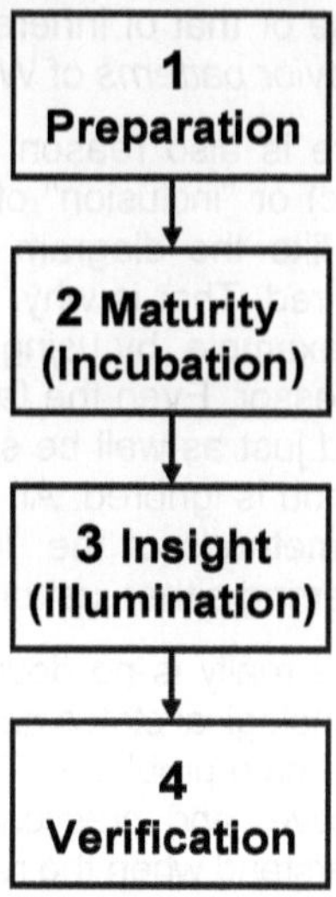

fig. 3.1. Diagram of creative process according to Wallace (1926).

---

[34] Compiled and arranged here by M.O. from: G.S. Altshuller, *Invention Algorithm.* – Moscow Worker, 1973. – (Г.С.Альтшуллер, *Алгоритм изобретения.* – Московский рабочий, 1973)

[35] G. Wallas (1926) *The Art of Thought, Harcourt Brace*, New York

M.A. Orloff, *ABC-TRIZ*, DOI 10.1007/978-3-319-29436-0_3

The diagram simply describes the four *basic psychological meta-states* of a human being (very generalized states) when solving *any problem.*

This allows many authors, i.e., successors of Wallace, to define his description as the *general model of creative thinking.*

But from what point of view? From the point of view of *a psychologist who is an external spectator.*

The content of such *an external observation* is clearly characterized by Genrikh Altshuller in an example,[36] in which a person (psychologist) watches the movements of a boat without knowing anything about its destination, the river's conditions, or the steering method of the boat, but is trying to relate the changes in the boat's movement to anything at all. For example, he tries to relate the position of the sun to the boat's course even though it is apparent that the captain is simply blinded and therefore changes course.

Thousands of researches have described and commented about the "movement of the boat" out of inertness or beloved tradition. They were not considering the content of the problem, the area of expertise to which the problem belongs, or the laws and models from this field.

Still, Wallace's diagram is of great practical meaning thanks to a prominent fact, which has been proven with scientific reliability:

> *With psychological precision, the diagram describes man's behavior during the search for solutions to complicated problems.*

At least, in all the cases in which the solution cannot be found on the basis of common sense or that of inherent instinctive automatisms, the search for a solution follows the *behavior patterns* of Wallace's diagram.

There is also reason to believe that finding the solution on the basis of knowledge (logic) or "inclusion" of a necessary inherent or acquired automatism (instinct) occurs just like the diagram describes it except that a relatively short amount of time is required. That is why, and it is the only why, it is usually said that a solution is found. For example, by using logic, as if the human mind is completely equivalent to a digital processor. Even the fact that the digital processor requires time to calculate and that it could just as well be searching for a solution by the preliminary selection of a solution method is ignored. All of this leads to the severe methodological mistake of separating the methods of the "logical" from the "instinctive" search for a solution by radically preferring either just one or the other.

There really is no doubt (and we will see that more precisely in the following text) that psychological states can and should be examined to create models supporting the solving of intricate problems. The examinations would serve only the purpose to receive reliable, effective, and practical recommendations for problem solving. It is also important to understand when the recommendations can, or cannot, be of effective use.

The results of years-long research and uncounted sources of "creative knowledge" prove unsatisfying because of the existence of just as many approaches, styles, and basics for an "explanation" and a "theory" of creativity.

---

[36] G. Altshuller (1988) *Creativity as an Exact Science*. Gordon and Breach, New York (Translated by Anthony Williams from Russian – Г. Альтшуллер, *Творчество как точная наука*. – Советское Радио, Москва, 1979)

The complete conformity of one *fundamental property* is more puzzling than the differences between these "directions": They're all nothing but one and the same *method of trial-and-error* in hundreds of varying "author-variants."

Genrikh Altshuller, in his unwillingness to compromise, wrote[37] about this issue:

> *There is a great inertness about the traditional conceptions of the "trial-and-error" method to be the one and only thinking mechanism of creativity.*
>
> *For a thousand years people have been solving creative problems with the "trial-and-error" method. For a thousand years the idea that there is not and can not be another method had time to take root and fix itself.*
>
> *The idea of "creativity" itself eventually merged with technological problem solving by intuitive searching examinations.*
>
> *Inspiration, intuition, inherent talents and lucky moments were considered to be the eternal attributes.*

3.2. Meta-Algorithm of Invention ARIZ-1956

ARIZ[38]—Algorithm of Inventive Problem Solving—is a method that contains practical instructions for the transformation from a state of "is" to the state of "should be" and is illustrated in a scheme (fig. 3.2), the form and character of which are algorithmic. The first ARIZ was published in 1956.

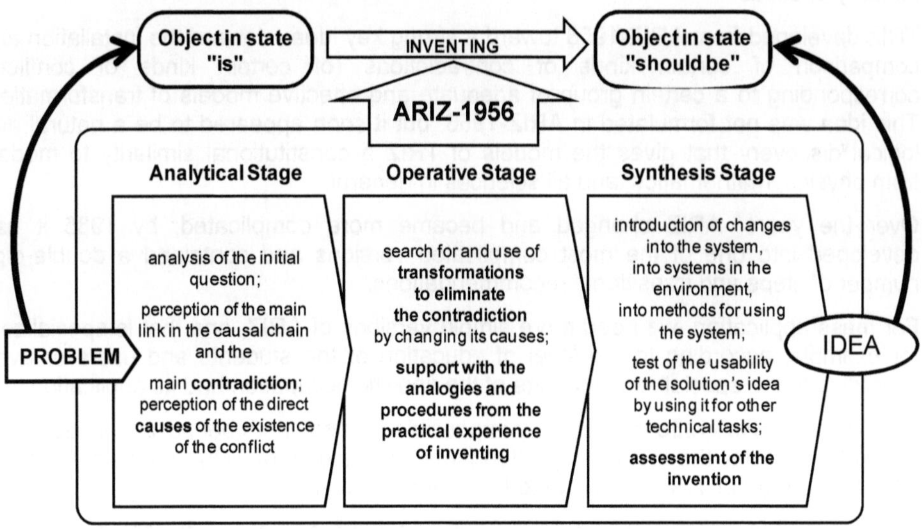

fig. 3.2. The first meta-algorithm of invention: ARIZ-1956

37 G. Altshuller (1986) *How to Find an Idea* (Russian, not translated – Г. Альтшуллер, *Найти идею.* – Наука, Новосибирск)

38 ARIZ is an acronym for the Algorithm of Inventive Problem Solving from Russian: Алгоритм Решения Изобретательских Задач (АРИЗ) – Algoritm Resheniya Izobretatelskih Zadach (ARIZ)

Back then Genrikh Altshuller and Rafail Shapiro published a three-phase scheme, the first "algorithm of invention," called ARIZ–1956, which became not only the foundation of contradiction-and-transformation models but also the core of growth for the future TRIZ.

The most prominent difference of this *meta-navigator of thinking, or meta-algorithm,* compared with other similar schemes and theories consisted of its *instrumentality* and *constructiveness.*

Instrumentality means that the three main phases of ARIZ-1956, i.e., the analytical, the operative, and the synthetic, were already then equipped with the first models and recommendations for practical changes (transformation) of any object while at the same time allowed the *constructiveness* (goal-oriented and effective) for ARIZ-1956.

By the year 1961, G. Altshuller had already done research on 10,000 inventions from 43 patent classes!

The idea of the possibility to extract inventive procedures (transformation models, navigators) was fully proven.

The author of future TRIZ wrote: "...of course every engineering problem is individual in its own way. In every problem there is something unique.

With the aid of an analysis, we get the opportunity to penetrate the essence – to the ***system contradiction*** and its causes. Immediately the situation is changed. Opportunities to do a creative search appear by using a defined ***rational scheme***.

There is no magic formula but there are ***procedures*** which are sufficient for the majority of cases."

TRIZ developed from ARIZ-1956 toward a strong key idea – toward the installation and comparison of certain kinds of contradictions (of certain kinds of conflicts) corresponding to a certain group of adequate and effective models of transformation. This idea was not formulated in ARIZ-1956, but it soon appeared to be a natural and logical discovery that gives the models of TRIZ a constitutional similarity to models from physics, mathematics, and all sciences in general.

Over the years, ARIZ changed and became more complicated; by 1985 it had developed into one of the most complicated versions and contained a double-digit number of steps and transitional recommendations.

For mass application, we need more simple versions of ARIZ, one that is specialized, for example, according to the level of education of the students and users or even according to the specialized language of the specific occupation and its application.

> To study ARIZ-1956 is practically impossible for beginners because it is necessary to first master the instrumental (fundamental) models of TRIZ. Genrikh Altshuller also stated that ARIZ is designed for a "master level."

That is why the experiences of TRIZ education have so far proven that learning the fundamental TRIZ models on the basis of simple schemes, such as the Meta-Algorithm of Invention and others that have been elaborated in Modern TRIZ, is a necessity.

When predicting and approaching the mass application of TRIZ, one can predict that it will be used not only in engineering but basically in all other fields, i.e., with no limits from today's point of view.

Yet regardless of an entire series of versions of ARIZ and Meta-Algorithms, the most perfect ARIZ awaits to be an unrivalled ideal that probably can be endlessly approximated.

In every ARIZ, there will always be a fragment left over that cannot be put into an algorithm, an irrational component that cannot be eliminated and that is just as important as all the other components merely manifesting themselves in terms such as beauty and harmony, intuition, and ... lucky circumstance.

3.3. Meta-Algorithm of Invention T-R-I-Z (MAI T-R-I-Z 1995)

We make a great number of decisions automatically, one could say intuitively, without thinking about them for too long. It appears to us that we do not use any logical "algorithm" for the formation of a solution idea. Furthermore, it seems that in a greater number of cases we do not apply any special knowledge from, for example, physics, chemistry, or other fields.

But is this so? What if we are applying the "algorithm" so quickly that we are not able to recognize the step-by-step character of our reasoning? And what if in many situations the most simple knowledge is enough for us to make a decision and our inertness makes us believe that it is also enough for more complicated problems.

The year-long praxis shows that too many problems are too complicated for those who are trying to solve them, and that they can not find a solution for months and years, even though such problems can be solved by using TRIZ within a countable amount of hours or days. This experience is equal for different countries and continents, for companies with different profiles, and for different objects.

To construct a solution for more intricate problems, might we need more effective schemata and models?

Maybe the problems themselves appear to be complicated because we don't posses approved models and schemata for how to solve them?

Let's take a look at a simple example from everyday life.

fig. 3.3. Attention: the glass of tea is hot!

Example 3.1. A Glass of Hot Tea

You are preparing a glass of tea; therefore, you need hot water. After a couple of minutes the tea is ready, and you could drink it if you did not have to hold the glass in your hands, thereby burning your fingers (fig. 3.3). Let us remark that a human being is able to drink tea of a temperature that the skin of his fingers cannot withstand.

What could you do in this situation? Very simple: You can either grab the glass on its upper rim, which provides a somewhat insecure grasp, or you could hold it safely after wrapping a paper towel around it.

Other solutions are also possible, but for a simple analysis these two should suffice.

Indeed it seems as if there has not been a step between realizing the problem and finding a solution for it.

Now we shall take a mental "camcorder" and change the scale by "zooming"—together we will diagnose and observe a few intermediate steps in the development of the solution-finding process.

***1.** First of all, we noticed that we could drink the tea but that it was impossible to hold the glass because it was too hot.*

What kind of moment is this? Answer: It is the first realization of the problem situation and confrontation with the conflict: We would be able to drink the tea from the glass but cannot hold the glass of tea in our hand.

At least two key motives appear here: (1) the ambition to reach the *main target*, namely, to drink the tea; and (2) the previous choice of the *trend* of how to reach the target. That is, we somehow have to grab the glass with our hand without burning ourselves on the glass.

We shall label this first step with the Latin letter **"T"** like in the words *target* and *trend.*

2. *From experience, we know that a glass is usually not filled all the way up to the rim, so we can try to carefully grab the glass where the temperature is lower because the rim of the glass is cooled by the air surrounding it. Additionally, we can use a paper towel.*

What happened here? The answer to this "trend" is this: We have separated the conflicting properties of the space! It is true that a part of the glass (the upper rim) is colder and can be touched with the fingers in order to drink from the other part of the glass which is hot, thus keeping the tea from cooling down.

Eventually though, *the separation of the conflicting properties in the space* is nothing but a generalized model of thousands of similar decisions we formerly took automatically and instinctively in similar situations. That means that at some point we have developed and internalized this behavioral "pattern."

In our given example, we have simply abstracted the problem to a known model and a known solution method (by using the procedure of *reducing* from the complicated to the simple).

The same is valid for a solution on the basis of the model "mediator"! It is really the paper towel that mediates between the fingers and the glass. Here there is also a reduction taking place because the solution process tries to simplify by leading us toward the known behavioral pattern.

Let us imagine that you cannot grab the glass at its upper rim because it is filled all the way, and you can not find a single paper towel.

The initial problem situation can be transformed into a thought, into a model, in the form of a standard contradiction type, which can basically be found everywhere: *We want to drink the tea but cannot hold the glass with our fingers because we will burn them.*

The problem can be formulated as a more radical contradiction: *The glass must be hot because there is hot tea in it, but it must not be hot so that one is able to drink from it.*

So we finally realized the radical contradiction, which is the essence of the problem. We label this second step with the Latin letter **"R"** after the key word "*reducing.*"

***3.** We produce changes in time, space, structure, material, etc.*

What happens during the first step? Answer: *The invention of the idea for solving the radical or the standard contradiction means the invention of the idea for the problem's solution!*

You can not wait until the glass has cooled down, which means that the resource of time is not available.

You can not use any other part of the glass, which means that the resource of space is not available.

However, you could change the structure of the objects by using a paper towel as a mediator . . . but a paper towel is not available.

It is possible to replace the paper towel by any other material, e.g., by folding a sheet of paper into a strip and wrapping it around[39] the glass! This would already be your very own personal little invention and problem solution for the given situation.

However, in our analysis we are not interested in the invention of something meaningful; we are interested in the thinking process itself and in the determination of its first steps.

With certainty this step can be called the *invention* or *inventing* of the solution idea. We label this step with the Latin letter **"I."**

***4.** Before making a final decision, you should evaluate the effectiveness of one or more ideas.*

What happens before the transition to the actions? Answer: you check the ideas. You have managed (if you did not rush!) to check the possible conditions and consequences of the application of the idea by using the following method more or less.

At first you have checked the quality of your first idea in your mind. You have focused your attention to the method of how to hold the glass. You can try to hold the glass on its upper rim, but this would be too dangerous. You immediately switched your attention to the part of the glass where it is being touched and you mentally "sensed" that one could burn oneself. However, you thought it would be worth the risk and tried to hold the glass.

Now your thinking has switched scales (zooming) within the "camcorder's" observation of the situation and in terror you "saw" that you could ruin your suit. If you wouldn't be able to hold the glass, the hot tea would spill on your beautiful silver grey suit!

Immediately your thinking has zoomed to the scale of your examination of the situation, and with even greater terror you suddenly remember that the reception in your boss's office is coming up in 1 hour where you are supposed to be awarded as ... the outstanding inventor of the company! But you have no suitable suit to wear! And you won't make it to your own award ceremony!!

Your thinking successively managed to zoom switching from your fingers on the hot glass to your ruined suit and from your ruined suit to the reception at your bosses and so on, switching back and forth constantly. You did think of that, didn't you?

---

39 Do you remember the carton cup for tea or coffee at a small coffee bar, e.g. *Starbucks*, that is girdled round by a paper strip that had been prepared in advance?

This means that nobody, except hasty people maybe, can get very far without undertaking the step of examining the idea from different "scales" or taking into account the different requirements.

fig. 3.4. Enjoy your tea!

That is why the crucial thinking "operation" of this step and the step itself can metaphorically be called "*zooming.*" We shall label this action according to its name with the Latin letter **"Z."**

Based on the two models of transformation known to you, namely "preliminary action" and "mediator," I would like to give you some friendly advice. In order to not experience the just described "horror," you might want to use a tea glass holder when you prepare and later enjoy a glass of hot tea (fig. 3.4).

Surely you have already noticed that the first letters of the designations of the different stages of **Meta-Algorithm of inventing (MAI),** which we have used in this example, match the name of the theory TRIZ! This hint is helpful for beginners to memorize the stages of MAI.

In Modern TRIZ, T-R-I-Z is another name for MAI.

The stages of **MAI T-R-I-Z** (fig. 3.5) precisely reflect the most important technological process of TRIZ and also express the essence of TRIZ as a theory of inventing.

Thereby every process of generating an idea to solve a problem contains the **four key steps:**

1) analysis of the initial problem situation and setting a goal (targeting and trend);

2) construction of a preferably simple model of the problem (reduction);

3) one's own creation of an efficient idea (invention);

4) examination of the idea from different scales and surroundings to evaluate the effectiveness (zooming).

MAI T-R-I-Z, developed in 1995, addresses beginners who can in this way better memorize the algorithm's stages for its similarity with the theory TRIZ.

We shall again remark that the "Meta" part designates the very abstract and generalizing level of MAI; it defines only the widest steps of problem solving and does not contain specific models or instructions for every step.

The steps themselves, though, are obligatory and characteristic when solving any problem or creating any invention.

When fulfilling each of MAI's steps with specific models and recommendations, MAI is functionally unraveling and transforming into a specific practical algorithm of inventing. The type (designation) of practical algorithm is being defined by what type of model has been employed. Still, in every practical algorithm the four important steps of MAI can be singled out precisely.

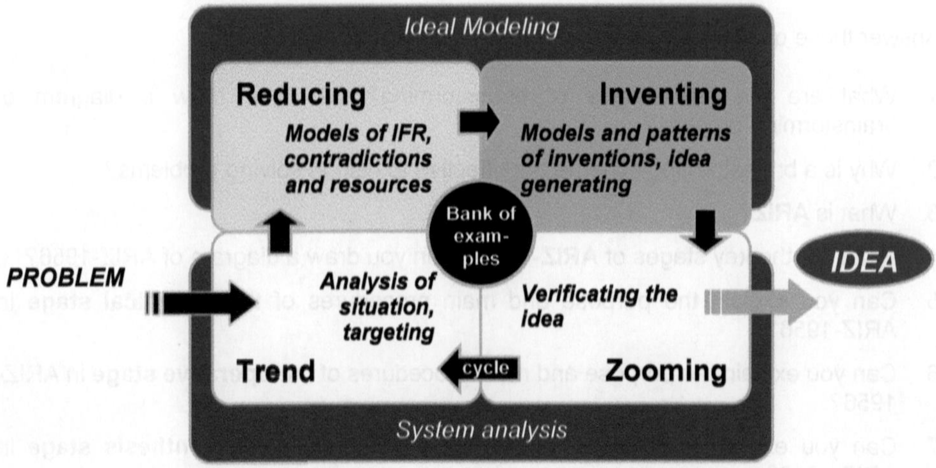

fig. 3.5. Diagram of the Meta-Algorithm of Inventing T-R-I-Z

(The arrow "cycle" shows the cyclic transition to the repeating run through of MAI in the case of a previous cycle not having led to a satisfying solution)

The fundamental value of MAI is that with its help, the important stages of TRIZ-based problem solving are easily made clear and understandable on first sight already in the first lesson. MAI being equipped with TRIZ's instrumental arsenal turns into an effective navigator of inventive thinking, even for beginners.

MAI is effective for training assignments as well as for solving actual practical problems. To close the paragraph, we shall remark that conceptually MAI is very close to the first algorithm of inventive problem solving, ARIZ-1956.

The interpretation possibilities and the spread of TRIZ in connection with any problem solution to any applied sphere in principle know no limits. This means that on the basis of TRIZ, a general theory of creativity can very well be developed as a theory of problem solving and generating effective ideas.

The same can be done with MAI, which can be adapted to the solution of a practically unlimited circle of problems and to the generation of effective ideas in any applied sphere. The applied trend of MAI is exclusively determined through the equipment of its stages with the concrete practical models of diagnostics, reduction, transformation, and verification.

In this book, which has been designed for beginners, the starter's version of MAI T-R-I-Z is used and is accordingly limited to the elementary universal models of TRIZ for solving contradictions.

3.4. WORKSHOP FOR CHAPTER 3

Answer these questions yourself with (or better without) a book:

1. What are the key stages of brainstorming? Can you draw a diagram of brainstorming process?
2. Why is a brainstorming scheme not effective to help in solving problems?
3. What is **ARIZ**?
4. What are the key stages of **ARIZ-1956**? Can you draw a diagram of ARIZ-1956?
5. Can you explain the purpose and main procedures of the **analytical stage** in ARIZ-1956?
6. Can you explain the purpose and main procedures of the **operative stage** in ARIZ-1956?
7. Can you explain the purpose and main procedures of the **synthesis stage** in ARIZ-1956?
8. Why do you think the Meta-Algorithm of Invention T-R-I-Z is a return to ARIZ-1956?
9. Can you explain the purpose and main procedures of the **trend (target)** stage in MAI T-R-I-Z?
10. Can you explain the purpose and main procedures of the **reducing stage** in MAI T-R-I-Z?
11. Can you explain the purpose and main procedures of the stage **Inventing** in MAI T-R-I-Z?
12. Can you explain the purpose and main procedures of the **Zooming stage** in MAI T-R-I-Z?
13. Can you tell your friends the story about inventive solutions for problem "Glass of Hot Tea" according MAI T-R-I-Z? You may use the book to illustrate your story.

4. Modeling the Problem

The algorithm of inventive problem-solving ... based on identification and elimination of ***the contradiction – the main factor that restrains further development of the system.*** [40]

Genrikh Altshuller

4.1. Contradictions

There are many ways to define and regard contradiction models. However, we shall concentrate on the definitions which are of great relevance to the basics of the classical TRIZ. However, during the expanded courses we will also examine other models.

Definition of **"Contradiction"**	**Contradiction** – model of a system conflict that reflects the incompatible demands on the functional properties of an object.
Definition of **"Binary Model of Contradiction"**	**Binary model of contradiction**—or more simply *binary model* or *binary contradiction*—reflects the incompatibility conflict between two properties (factors) only.

The graphic binary contradiction is presented in fig. 4.1.

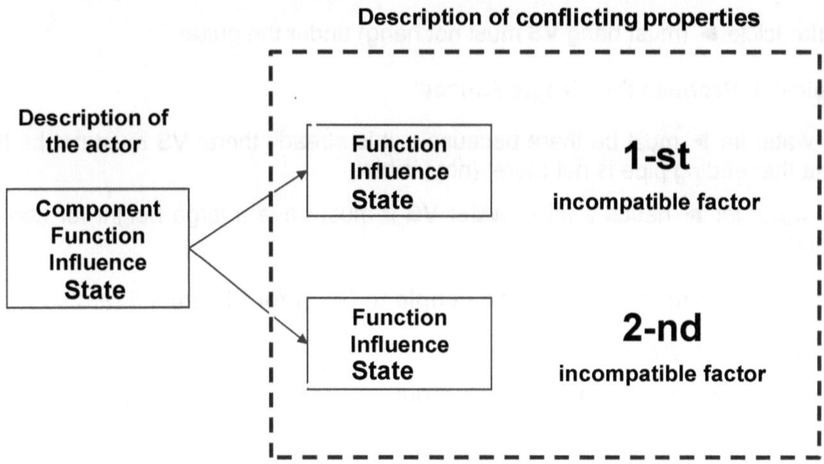

fig. 4.1. Generalized graphical presentation of binary contradictions

---

[40] From: G.S. Altshuller, I.M. Vertkin (1994) *How to Become a Genius. Life Strategy of a Creative Personality.* – Minsk, Belarus (in Russian)

M.A. Orloff, *ABC-TRIZ*, DOI 10.1007/978-3-319-29436-0_4

Any intricate, complex, multifactored conflict can be shown in the form of the composition of binary models. Hereafter, the main binary "key" contradiction must be found, the solution of which is a necessary condition for the resolving to the complex composition model.

Let us formulate the contradictions for first five problems, P1 through P5, in an informal style (but of course close to the TRIZ style presented later in the text).

Example 4.1. Problem P3. Ruby Stars of the Moscow Kremlin

A) The star must have a ***large surface area*** to be visible from afar, BUT (VS) this leads to ***large windage (sail area)*** and ***low reliability*** in strong wind.

Formula: Star ► large surface area **VS** low reliability

The **"VS"** means here **"versus."**

B) The star must ***be large*** (i.e., have a large surface area to be visible from afar), BUT (VS) it must ***be little*** (to have a slight windage (sail area) and to be resistant to strong wind).

Formula: Star ► large surface area **VS** small surface area.

Example 4.2. Problem P2. Icicles Frozen to Gutters

A) The icicle has to be fastened securely to gutter **VS** thaw-related damaging factors (because of increasing temperature of the gutter surface).

Formula: Icicle ► fastened securely **VS** temperature

B) The icicle *must hang under the gutter* until it melts **VS** the icicle *must not hang under the gutter* as it detaches itself from the water gutter when it warms up

Formula: Icicle ► (must hang **VS** must not hang) under the gutter

Example 4.3. Problem P4. "Magic Faucet"

1) The water jet ► must be there because... it is already there! **VS** must not be there because the feeding pipe is not there! (not visible!)

2) The water jet ► needs a lot of water **VS** it must have a large height for non-stop water flow!

Example 4.4. Problem P5. Convince People to Go in for Physical Exercise

3) The poster ► is useful **VS** boring

4) The poster ► must attract interest about its usefulness **VS** the poster does not attract interest because it is boring

Example 4.5. Problem P1. Genrikh Altshuller's Experiment

5) Two ropes ► (must be close **VS** do not have to be close over the long distance) – to the child

6) Two ropes ► one rope swinging for a long time **VS** the rope quickly loses its swinging energy

4.2. Standard Contradiction (SC)

Many philosophers and researchers who examined various creative methods noted that *contradiction is the heart of the problem*, but before Genrikh Altshuller nobody turned that notion into a "universal key" to be used to open and resolve the problem itself!

Only TRIZ turned the contradiction into a constructive model complete with instruments designed to transform that model to eliminate the contradiction.

To invent is to remove the contradiction!

Example 4.6. Training of Long-Distance Swimmers (for the sake of brevity, let us call this example **"Swimmer"**)

In fig. 4.2, you can see the general view of the Olympic basin in Peking for the water competitions, which was constructed for the Olympic Games in 2008. It is very beautiful and complete in function construction. The walls and the roof seem to consist of thousands of soap bubbles which, united in the large structure, looks like a gigantic foam cloud.

Problem situation

Assume that in the pool swimmers are training a long-distance swim of, let us say, 5, 10, or more kilometers. Usually such swimming units take place in the sea, in rivers, or in large lakes. Obviously it would be better to train in the pool. Without any dependency on the weather, the training can be realized all year long.

However, the problem in a normal pool (as seen in fig. 4.2 и 4.3,a) is that the swimmer constantly reaches the walls of the pool where he must turn and thus is not able perform an uninterrupted straightforward movement.

fig. 4.2. Olympic Pool in Beijing.
The Beijing National Aquatics Center "Water Cube"
(Illustration from www.china.org.cn/english/olympic/241055.htm)

The movement technique does not correspond to the real conditions of the upcoming competition. The athlete's rhythm and breathing are disrupted.

Compared with the track of a normal pool, the long-range swim might seem infinitely long. We could thus say that we need an infinitely long pool for the training.

Hearing about this problem, some listeners in the audience of a seminar will quickly remember the known solution. All there is left to do then is to applaud them for knowing the correct solution and to say that we will now be looking at how to come to such a solution with the help of TRIZ models.

Often different suggestions are presented, which is characteristic for brainstorming: The construction of a round pool (fig. 4.3,b). When this solution is first proposed, the participants of the seminar will most likely not come up with the control solution. They will insist that their proposed solution is easy to implement.

I would then object that the diameter of a round pool is much larger than the 50m-side-length of a rectangular pool. What I usually get for an answer is: "*Yes, a round pool would probably be larger, but it has a simple shape and it can easily be built.*"

I would eventually try to bring in my last "saving" objection: If the athlete keeps swimming into one direction, one of his arms would end up being longer than the other.

Thereupon, I hear proposals about a figure-eight shaped basin (fig. 4.3,c), a shape rather tricky to realize. With the solution quality obviously decreasing, the first idea is taken up again and passionately supported.

a)

b)

c)

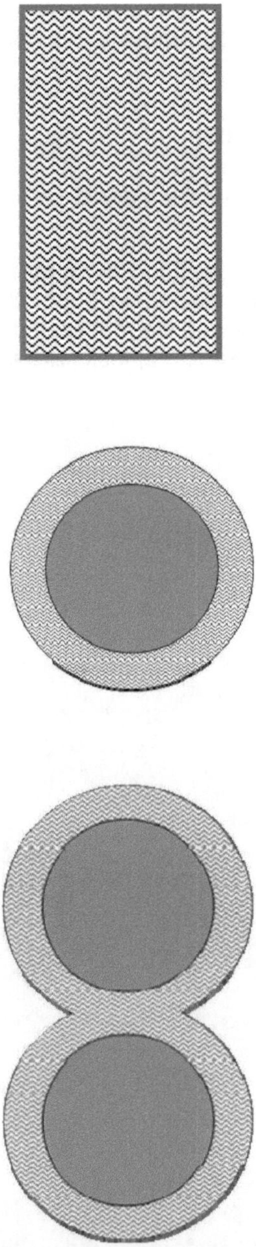

fig. 4.3. Traditional (a) and possible (b, c) pool shapes for swimming "infinite" distances

Standard Contradiction as a Problem Situation Model

So, we need a pool with track of an infinite length.

The problem consists in the fact that such a basin will probably have a complicated shape, e.g., a round or a figure-eight shaped basin. In any case, the rectangular shape is the less difficult one of the three. We find ourselves in a situation in which we cannot build a new pool without finding a solution to the problem.

The generic model of the problem can be illustrated as follows: Two different features of the hypothetic pool are facing two incompatible requirements with one of the requirements, i.e., the prolongation of the track complicates the shape of the pool.

The common pool has a simple rectangular shape, but it cannot be very long. Thus, if we improve one feature, we downgrade the other feature.

The model of the problem representing the incompatibility of two requirements to different properties can be reasonably described as typical or *standard*, because each such model emerges in a "standard" fashion in virtually each problem situation.

Standard Contradiction **(1)**	***Standard Contradiction*** (in classical TRIZ: *technical contradiction*) – binary (of two-factor) model that reflects incompatible requirements between *two different functional features* of an object (or several conflicting objects).
Standard Contradiction **(2)**	***Standard Contradiction*** – Two-factor model in which one of the factors corresponds with and supports the most important feature of the system (positive *trend-factor* or *plus-factor*), whereas the other factor does not correspond to this feature or counteracts it (negative *problem-factor* or *minus-factor*).

Let us formulate, at a very simplified and entry level, as a sample "benchmark" the standard contradiction for this example modeling the conflict between the property "swimming track length" and the property "pool shape" (fig. 4.4,a).

The text (formula) format is shown in fig. 4.4,b.

A generalized representation of standard contradiction is provided in fig. 4.4,c.

ATTENTION: try to solve the problem in **Example 4.6. "Swimmer"** independently.

If you hit a snag, keep reading the lessons.

The solution through the standard contradiction is provided in section **8. Answers.**

Do not look it before making your own attempts!

This kind of problem situations is often encountered in practice. You will easily be able to name a few examples. By the time you have read this book, you will have gone through dozens of such examples.

If you complete training and receive MTRIZ Junior or maybe Practitioner certification, you will have mastered a virtually perfect technique that you can use to model and resolve such contradictions in any situation.

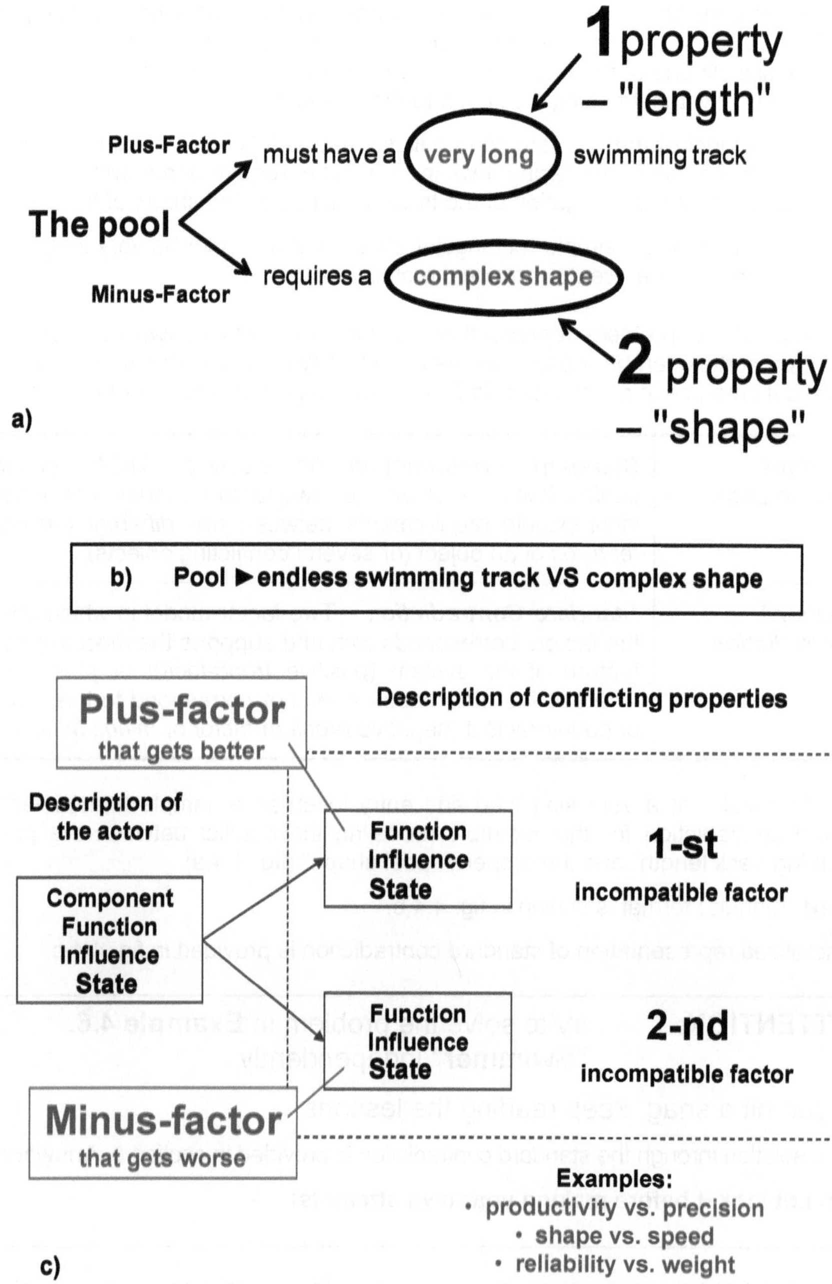

fig. 4.4. Definition of standard contradiction

4.3. Radical Contradiction (RC)

Example 4.7. Training on diving board and diving tower (for the sake of brevity, let us call this example **"Diver"**).

In fig. 4.5, a complete view of the diving tower and the diving boards of the Olympic swimming pool in Beijing is shown.

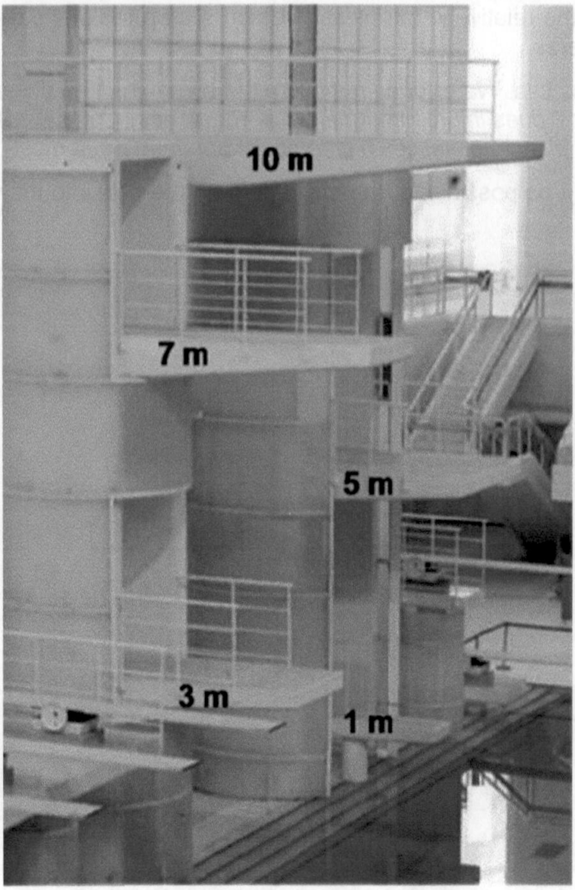

fig. 4.5. Diving tower and giving boards in the Olympic Swimming Pool in Beijing (from the website www.china.org.cn/english/olympic)

Problem situation

Competitive swimmers practice modern intricate dives with acrobatic pirouettes and somersaults for years. In fig. 4.6,a, you can see a dive by Dmitri Sautin, a 2-time champion of the Olympic Games, a 5-time World Champion, and 12-time European Champion. He has participated in the Olympic Games five (!) times, and he is the only athlete who has 8 Olympic awards in diving.

The critical moment of the dive is the plunge into the water. It's a good thing if the plunge happens as precisely as seen in fig. 4.6,b (Sautin) and 4.6,c. But even here the weight bearing on the body, especially on the wrists, is so high that it leads to occupational diseases. Suffice it to say that Dmitri Sautin was not spared from surgeries or long healing periods of his joints. In an interview he once said: "*So much has been cut open and out that I am unable to jump from the 10m-tower because my back hurts too much.*"

The consequences may be even more dire if the diver goes into water the wrong way—even if in the relatively "harmless" fashion shown in fig. 4.6,d—but, say, from a height of 10 meters!

The bottom line is this: Water *does not injure* the diver when he dives from a height of, say, 1 meter, but it *does injure* him if he dives from a height of, say, 10 meters. In other words, *the same object possesses diametrically opposed properties!*

How then would it be possible for the divers to train the dive into the water without any danger?

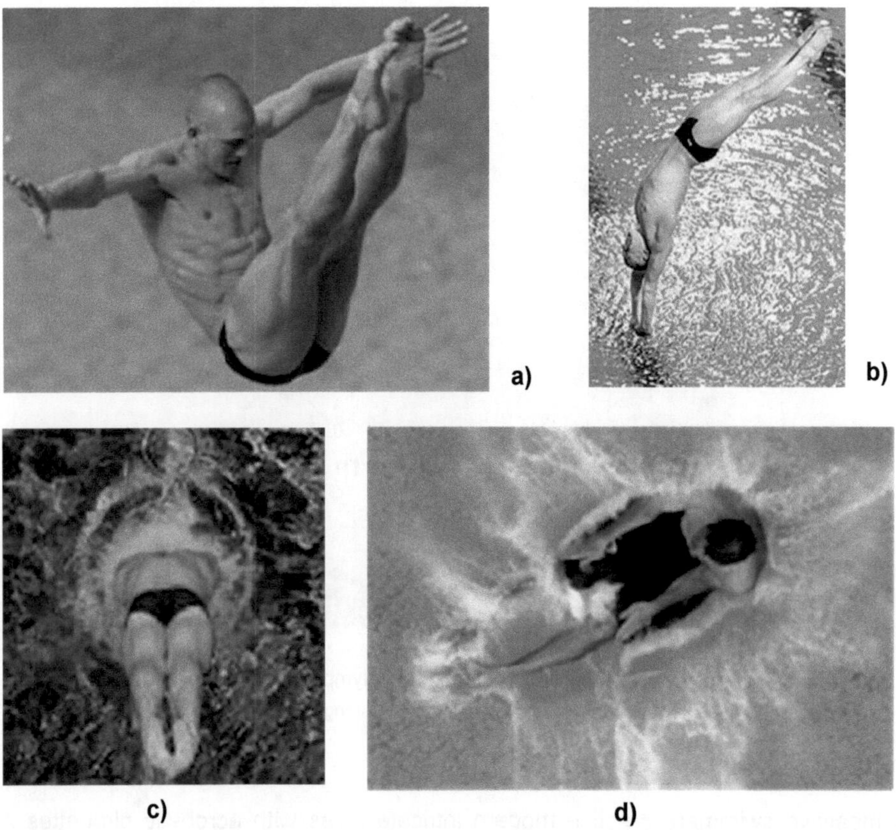

fig. 4.6. Plunge into water

(Illustrations from the Web sites www.moikompas.ru/compas/dmitriy_sautin and www.newizv.ru/news/2009-04-08/107653/)

Radical Contradiction as a Problem Situation Model

At first we will create a standard contradiction of the problem in the following variant: The height of the jump *must be increased* but that *raises water resistance* to the plunge, which has a negative effect on the athlete. The analysis shows that with the increase of the jump's height, the velocity with which the athlete plunges accelerates. We cannot change the height and the plunge velocity.

Then the solution has to be found in the... water! But how? With increasing height, the impact on the human body when hitting the water surface is only slightly different from the impact when hitting concrete or stone.

Can the water be made "soft"? At least for the short moment when the body hits the surface, especially in case the dive was not correct?

We can generalize the problem as follows: One and the same feature (water density) is facing two hypothetically different and additionally contrary features: The water must be "hard," which corresponds with the natural characteristics when a body hits a water surface from a large height, and the water must be "soft" so that injuries can be avoided.

Thus, two different demands, which exclude one another, are being requested of the space on the water surface, i.e., the water must be soft and hard at the same time.

Radical Contradiction (1)	***Radical Contradiction*** (in classical TRIZ: *physical contradiction*) – binary-contradiction model in which contrary, i.e., excluding one another, requirements are being requested from one and the same feature of one and the same construct (component, resource, function, effect, condition etc.).
Radical Contradiction (2)	**Radical Contradiction** – binary two-factor model where the first factor reflects one demand as a plus-factor and the second factor reflects the same demand as a minus-factor such that both factors represent the same property of the same construct (*component, resource, function, action, state, etc.*), but they are incompatible.
Addition	In a Radical Contradiction: a) both factors may be necessary for the main useful function of the artifact, b) "minus-factor" is the undesirable state of property that must be transformed into a plus-factor only, c) "minus-factor" opposes development of the main property.

This contradiction might metaphorically be called ***contradiction of Hamlet*** with a line in his famous speech, **"To be or not to be – that is the question!"**

ATTENTION: Try to solve the problem in **Example 4.7. "Diver"** independently.
If you hit a snag, keep reading the lessons.
Solution through standard contradiction is provided in section **8. Answers.**
Do not look it before making your own attempts!

The *radical contradiction*, which lies in the core of the problem is precisely defined. The conflict within the model is extremely exacerbated, and a solution seems impossible at first sight (see Section 8. Answers). Such a contradiction is also referred to as *radical*!

The illustrative example of a radical contradiction is shown in fig. 4.7,a; the text (formula) format is shown in fig. 4.7,b; and a generalized representation is provided in fig. 4.7,c.

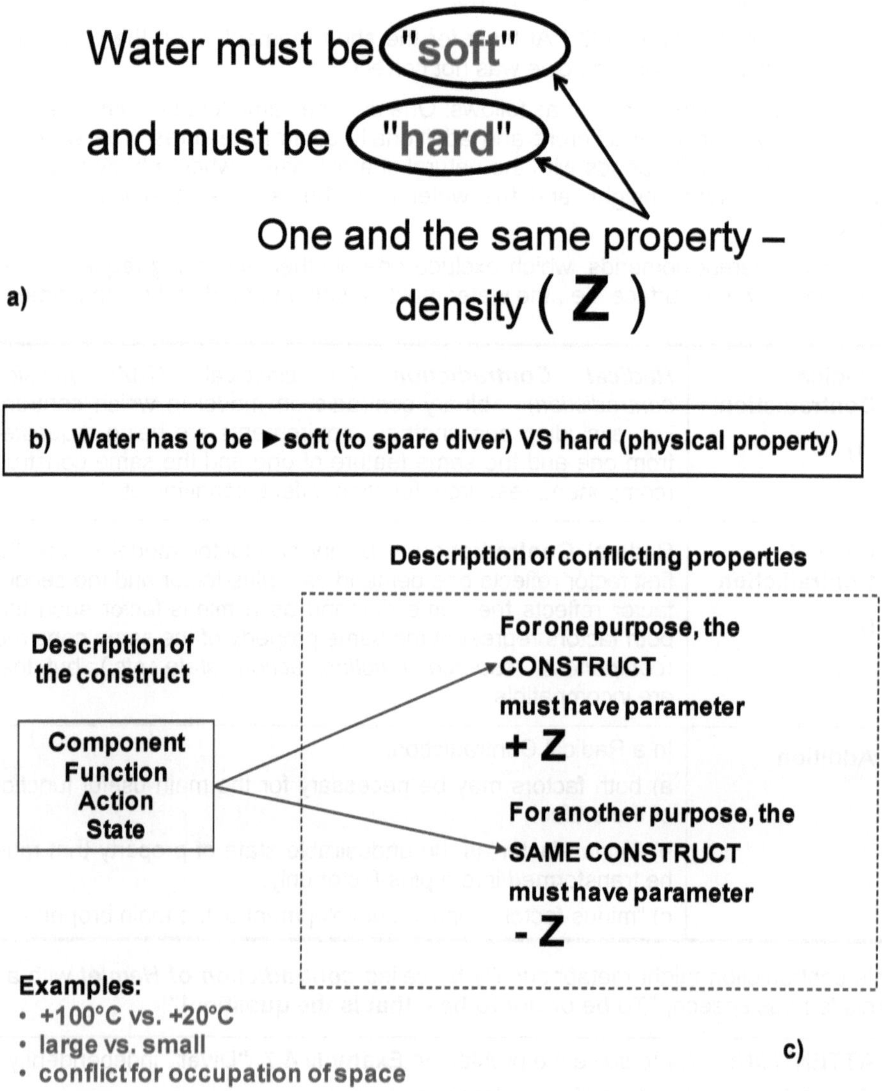

fig. 4.7. Definition of radical contradiction

4.4. Contradiction as an Attribute of Development

The aim of TRIZ is to create and use efficient models for inventing efficient ideas. Solving problems with embedded contradictions requires special supra-professional creative knowledge and skills, which have now been presented in an integrated form, i.e., as a certain unity of concepts, theoretical models, and practice-oriented technologies, only in TRIZ.

No system (product, technology, organization, conflict situation) can develop without resolving contradictions. That is why problems defined in terms of developmental contradictions always represent a challenge for creatively minded engineers and managers and require full mobilization of their knowledge and creative abilities and special psychological concentration.

The characteristic of the idea's effectiveness is in its core, according to System Addition no. 1, a complex of important demands on the quality of the inventive idea.

Property: Effectiveness – "system" addition (1) to the definition "invention"	The inventive idea excels itself by its high effectiveness due to the following preferences: 1. clears the problem that hinders the system's development; 2. creates an object with new properties that are necessary for higher systems; 3. opens up an opportunity for further development of the system and/or of the surrounding systems; 4. usually solves the problem with the least amount of means (resources) possible; 5. requires the least amount of small changes within the system itself or within bordering systems.
Property: Disappearance of the "Miracle" Phenomena – "psychological" addition (2) to the definition "Invention"	The inventive idea is: 1. not obvious until it has been discovered; 2. perceived as an exceptional and shining phenomena "miracle" when it is found; 3. perceived as an ordinary and simple phenomena after it has been explained.

To understand the particularities of the emotional reception to an idea, according to System Addition no. 2, is a warning and a protection towards at least two dangers:

> 1) a too high (very rarely too little) evaluation of the idea discovered by its author; one should not forget that objective evaluations take place only after the practical realization of an idea, i.e., after the real experiment;
>
> 2) a too little (very rarely – too high) evaluation of the idea through people (society); of course a better evaluation depends on a successful use of the idea in practice.

Let us take a look at a few examples that illustrate the above-mentioned terms.

Example 4.8. Invention of the Bicycle

The bicycle can be regarded as an inventive idea that found its beginning amongst the known means of transportation by the use of the wheel, e.g., the war chariot, cart (for goods), or carriage (for people).

The bicycle as an inventive idea radically differs in three points compared with its potential predecessors.

The first, i.e., "energetic," difference of the bicycle from its predecessors lies in the idea of using this construction only through human energy, namely, with the help of leg power of the one who moves forward using this very construction. We should annotate that a cart or a carriage used to be moved by an additional energy source, e.g., a horse.

The second, i.e., "functional," difference is the change of the principle of locomotion, namely, the dynamic stability of the free-rolling wheel.

The third very radical, i.e., "engineering and construction," difference is the placement of the wheels, which are connected with a conjoining construction frame with one behind the other on one level.

The original inventive or problem situation can be formulated as the necessity of human locomotion on Earth by using exclusively his legs (or hands or the whole body, which is just as possible). The main factor of aim can be recognized as an increase in the speed of locomotion and/or a decrease of energy consumption for the locomotion.

However, the direct use of the known means of transportation did not seem very effective. One could get a wagon to move by running and jumping on it in order to travel for a short distance. To really get the wagon going, a great effort has to be made, and the wagon loses speed quickly because of its high weight. This idea could be developed with the parameters being changed to the point of the idea for a roller.

The core of the problem can thus be formulated in the form of the following contradiction: The goal of using a wagon (which we shall consider as a prototype or analogue) for human locomotion by using exceptionally the human's own energy is to increase the speed of locomotion (*plus-factor, goal parameter of development*), which is impeded though by the wagon's weight (*minus-factor as obstacle when reaching the first factor or direct disadvantage of the prototype*). The initial situation and the solution idea are demonstrated more clearly in the following chart (fig. 4.8).

#	Problem Description	Contradiction		Problem Solution	Time-line
		(+)-factor	(-)-factor		
1	To accelerate human locomotion on Earth by using muscle power	**Increase of speed;** **decrease of energy consumption**	**High weight of known carriages with two, three, or four wheels**	Comte de Sivrac invented a two wheeled scooter which was made of a wooden frame and had no handlebar (France)	1790

fig. 4.8. Contradiction as a model of the initial problem situation shortly before the invention of the idea of the bicycle

The first bicycle was a scooter (fig. 4.9) where the "bike rider" was sitting in the saddle and pushing with his legs from the floor to get moving.

Inventive models that prompted the author to invent the solution idea can hypothetically be defined as follows: copying of a freely rolling wheel; joining of elements designed to perform the same function; copying of animal riding (to make the first "bicycles" more comfortable, they were fitted with soft cushions or even small "real" saddles).

fig. 4.9. Bicycle-toy for children (copy) – model of de Sivrac's bicycle

(The side part that was covering the wheels is eliminated due to a better illustration)

The given invention is a pioneer invention and allows the appearance of a new form of engineering systems of human locomotion on Earth, even though the basic schema of the wheel's placement has not changed in 200 years. That's 2 whole centuries!

NOTE: Recently proofs have been showing up that the information about Comte de Sivrac and his "pioneer" invention is a facetious falsification. For us, though, something else is important, namely, the spread of two-wheeled bikes for children in the late 17th and beginning 18th century, which is related to the description of Comte de Sivrac's invention. What counts for us is the principle of locomotion with the means of transport shown in fig. 4.9, no matter what its name is or who invented it. By the way, such a toy can still be bought nowadays and indeed with no handlebar on the front.

Example 4.9. Invention of the Handlebar and the Maneuverable Front Wheel.

Obviously, one of the severe disadvantages of the "first bike" was indeed the steering. Making the bicycle turn could only be done with the complicated method of leaning the body sideways into the direction of the turn.

The reason for this complication is found in the connection of all the bicycle's elements in one single fixed construction.

Today it seems hard to imagine that almost 30 years lie between the invention of the first bicycle and its equipment with the first primitive mechanisms to move the front wheel and moreover that the schema of a bicycle's steering has principally remained unchanged over the last 200 years.

The description of the problem and the contradiction modeling are shown in the chart in fig. 4.10.

In this invention, the front wheel is clasped by a "fork" whose top is connected to a rod (fig. 4.11), which goes through an opening in the bicycle's frame and is attached to a transverse bar (the "handle bar"). The core inventive model that prompted the author to invent the steering wheel and the handle bar can be hypothetically defined as simplified copying of the steering mechanism of a chariot or cart. That mechanism had been known for a long time.

#	Problem Description	Contradiction		Problem Solution	Time-line
		(+)-factor	(-)-factor		
2	Difficulty of turning to change the direction of motion	**Improvement of control when turning**	**Fixed connection of all the construction's elements, the absence of steering elements**	Baron Karl von Drais invents steering by using a mobile front wheel (Germany)	1817

fig. 4.10. Contradiction as a model of the problem and the initial problem situation shortly before the invention of steering the front wheel

a)

b)

fig. 4.11. Model (a) and real exemplar (b) of Karl Drais bicycle
(author's photograph from the Museum of Technology in Vienna, Austria – on the right)

Despite the improvement of steering, the principle of locomotion, i.e., pushing off using one's legs, did not change (fig. 4.12)!

fig. 4.12. Locomotion with Karl Drais' bicycle
(http://www.daociyou.com/BicycleEvolution3.htm)

Example 4.10. Invention of Pedals

Today it seems hard to imagine that the movement the bicycle was accomplished by pushing off with one's legs. History holds the legend that Karl Drais – walked! – all the way through Germany and France in this manner to advertise his great work.

Of course this method of locomotion is neither explicitly economic from the energy consumption point of view nor very consistent in its movement.

The descriptions of the corresponding problem and the contradiction modeling are shown in chart 4.13.

#	Problem Description	Contradiction		Problem Solution	Time-line
		(+)-factor	(-)-factor		
3	Locomotion by pushing off from the ground with one's legs requires a great use of energy and has a slow and inconsistent nature	**Increase of motion comfort; decrease of use of energy; increase of velocity**	**Absence of transfer of energy from man to means of transportation**	Philip Fischer invents pedals, which were linked to the front wheel. He also enlarged the front wheel by 2 meters to increase the drive per pedal circumvolution (Germany)	1830

fig. 4.13. Contradiction as a model of the initial problem situation shortly before the invention of pedals

The key creative idea here is apparently copying by analogy, for example, a water-well winch or any other similar hoisting mechanism with one or two handles.

fig. 4.14. Exemplar of one of the first pedal bicycles

(photograph taken by the author during an automobile exhibition of Renault in Berlin)

So, two basic variants of incompatibility can be discriminated in this chamber.

The contradictions are an inherent part of the development of systems. New demands (claims) are imposed in the application of the systems. The contradictions reflect the incompatibility of the new requirements and the existing or proposed properties of the object.

For example, standard contradiction model problem situations are sown in examples 4.8-4.10, namely (in a simplified form),

> 1) In example 4.8: The need to *"increase speed"* conflicts with the property *"weight"* of the prototype;
>
> 2) In example 4.9: The need to *"improve steering"* contradicts the state *"rigid connection of elements"* in the construction of the prototype;
>
> 3) In example 4.10: The need to *"improve comfort of movement"* contradicts the state *"lack of energy transfer"* in the construction of the prototype.

In particular, radical contradictions can be formulated for problem situations presented in examples 4.8-4.10 as follows:

> 1) In example 4.8: The requirement *"small weight"* (to facilitate acceleration) conflicts with the property *"large weight"* (as it applies to the prototypes, for example, a cart);
>
> 2) In example 4.9: The *"non-rigid connection"* (for steering) contradicts the state *"rigid connection of elements"* (to improve structural integrity);
>
> 3) In example 4.10: *"Use legs to push off from the ground"* (to start the "bicycle" wheels rolling) contradicts the requirement *"do not use legs to push off from the ground"* (because there is no known method to start the wheels rolling to ensure continuous motion of the "bicycle").

These incompatibilities exist temporarily and have been removed (2) by the invention of an effective change of an object and attainment of a new object with the same principle of action or sometimes (2) by changing the principle of operation of the object.

If a set of requirements cannot be met by the obvious way, there is a problem situation that is difficult to resolve. Solving the problem in such situations requires the invention of nontrivial, often striking, solutions with unexpected ideas possessing high efficiency.

So, the contradiction can be seen as a model problem. Contradictions indicate the roots of the problem.

The roots of the problem are the incompatibility of requirements and properties of the object, apparent or real, but always conditioned objectively (physically) and leading to a decrease of the efficiency of the system or even to the impossibility of implementing the main useful functions of the system.

Only two types of system contradictions—standard and radical—are enough for modeling any conflict, any problem. These contradictions reflect both the objective content of the uneven development of systems and the psychological, subjective perception and description (formulation) models of conflict in the form of contradictions.

The exact formulation of the contradiction is not an easy operation, and it requires considerable experience and, of course, the necessary skills. However, the further course of the solution depends on the exactness of contradiction formulation, from what it reflects.

4.5. WORKSHOP FOR CHAPTER 4

Exercise 1. You know now that point A) of Examples 4.1 and 4.2 at the beginning of this chamber contains a standard contradiction and point B) a radical contradiction.

The task is to define the type of contradiction in points 1 through 6 of Examples 4.3 through 4.5.

Exercise 2. Wall Paper and Carpet Knife. Previously the entire knife was tossed out after its blade became blunt. Under normal household conditions, it was very difficult to regenerate a sharp cutting edge. How can we increase the service life of a knife?

SC for the carpet knife:

Action, state, object	
Carpet knife	
(+) - factor	**(-) - factor**

RC for the carpet knife:

Action, state, object	
Carpet knife	
+ Z	**- Z**

Exercise 3. Protection from Car Theft. Almost all known car antitheft alarm systems have one common deficiency: After alarm activation, a car thief still has some time to disappear with a car from the scene of the crime.

SC for the car:

Action, state, object	
Car	
(+) - factor	**(-) - factor**

Exercise 4. Protection of Motorbike Rider in the Event of a Crash. To be protected in the event of a crash, the motorbike rider can wear a thick jacket and thick trousers. However, such clothes will constrain his movements and make it more difficult to steer the motorbike.

RC for the motorbike rider's clothing:

Action, state, object	
Motorbike rider's clothing	
+ Z	**- Z**

Exercise 5. Aircraft Carrier. During construction of the first aircraft carriers, a problematic situation appeared, the essence of which can be described as follows. Many planes should be stowed aboard the aircraft carrier. However, the wings of the aircraft are large (large wingspan). It is not possible to place the planes on the deck and in the hold compact. The length of the wings cannot be reduced because the wingspan is determined according to the tactical and technical requirements for the airplane.

SC for the aircraft carrier:

Action, state, object	
Aircraft carrier	
(+) - factor	**(-) - factor**

SC for the airplane:

Action, state, object	
Airplane	
(+) - factor	**(-) - factor**

RC for the aircraft carrier:

Action, state, object	
Aircraft carrier	
+ Z	**- Z**

RC for the wing:

Action, state, object	
Wing	
+ Z	**- Z**

5. Extracting

Methodology[41] *of invention* (even when initial acquaintance) increases [your] "creative arsenal", including ***dozens of methods (models)*** which together constitute ***a rational system of solving to problems***.

Genrikh Altshuller

5.1. Efficient Models in Each Artifact

Extracting is the first fundamental method of teaching based on the Modern TRIZ methodology. Each student will use this method throughout his or her creative life. To introduce it, let us use a simple example.

Example 5.1. The Corner of a Personal Planner's Page (completion: Chapter 6)

In fig. 5.1, we see the perforated corner of a personal planner. The perforation serves the purpose to easily and neatly tear off the corner of a page.

The everyday tearing off of the corners creates a clearly visible line between the pages that still have their corners and the ones with their corners already torn off. The perforation line is used instead of a bookmark, to be able to open the planner on the current date which is written on the still existing corner of the page (as you can see in fig. 5.1, the current date is "December 7th").

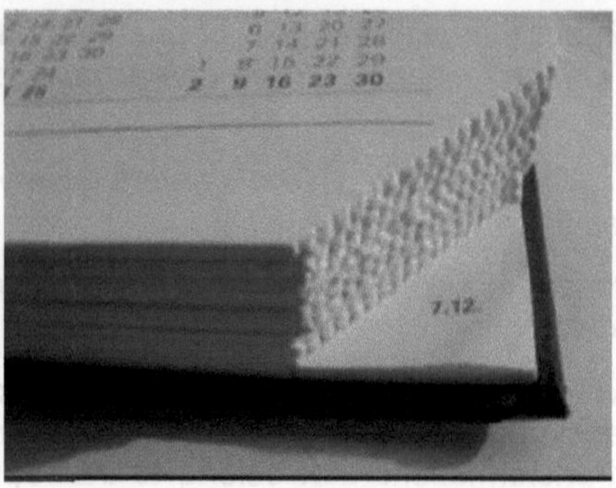

fig. 5.1. Page's corner of a diary or a personal planner

Obviously, before perforation was invented, an orderly tearing off of the corners was difficult. The tearing line was uneven and an orderly tearing off of a corner took a lot of time.

Question: *What creative process (model) has been used here for the creation of the idea to implement perforation?*

---

[41] From: G.S. Altshuller, I.M. Vertkin (1994) *How to Become a Genius. Life Strategy of a Creative Personality.* – Minsk, Belarus (in Russian)

M.A. Orloff, *ABC-TRIZ*, DOI 10.1007/978-3-319-29436-0_5

To give a correct answer must seem impossible to the ones who do not know TRIZ because the question itself is not understandable.

At first glance, everything seems simple because we usually only see the applied engineering process, in this case the perforation.

Let us ask a leading question: *What was the perforation made for?*

The answer can be formulated in this manner: To tear off the page corner in a consistent and orderly way.

Another question: *How does perforation help?*

Answer: The perforation gives the general tearing line and leaves only little paper parts between the holes, which makes the process of tearing fast and neat.

Already now one of the most effective creative models, which at some point had been discovered and accumulated in TRIZ, can be stated as *"Preliminary Action".*

The creative method that corresponds to this model is in TRIZ stated as follows:

> **Preliminary Action:**
>
> a) implement changes to the object in advance (completely or at least partly);
>
> b) place the objects in advance in such manner that they can manifest their effect from a suitable place and without loss of time.

The connection to the cited example is obvious. Indeed, with the paper perforated in advance, it is possible to tear off the corner of the page faster and much more accurately.

Additionally we shall remark that with the help of this method, the following standard contradiction has been cleared: *Tearing off the corner must be executed quickly, but this leads to an uneven tearing line.*

Or like in this variant: *When tearing off the corners slowly a neat tearing line is possible, but this requires too much time.*

In classical TRIZ, 40 primary models (specialized transformations[42]) have been stated and put in so called As-catalogue.[43]

Each model was discovered by the method of extraction, which we today are able to define. Namely, for every studied object (invention), not as much has been studied about the engineering transformations that belong to the "applied noosphere" of the artifact but rather the creative transformations belonging to the "creative noosphere" of the artifact.

Only "effective" models have been included in the catalogue, i.e., those that were most often found and that were found in many objects.

Here we ought to remark that many out of the 40 classical models are closely related to engineering transformations and that their interpretation onto different fields of application asks for adaptation, sometimes quite a complicated nature.

---

[42] **M. Orloff** (2003, 2006) *Inventive Thinking through TRIZ: A Practical Guide*. Springer-Verlag Inc., New York, 350 pp., 2nd edition, 2006, ISBN-10 3-540-33222-7, ISBN-13 978-3-540-33222-0

[43] see As-Catalogue (sections S22.2, or S22.3, or S35)

A student should learn all 40 models in order to recognize the sense from each single model and to be able to apply it just on hearing it.

The method of extracting is primarily designed for a faster acquirement of the basic models of transformation.

The creative content of this method is aimed at discovering the unusual and the "invisible" within the familiar and the usual. The studies, when using the method of extracting, are of a playful nature.

Extracting always leads to quite impressive as to the discretion of unexpected creative ideas in the apparently most simple objects and events.

We invite you to use the method of extraction and to define new models of transformation in any sphere and for any object. This will be especially valuable when completing certified work.

5.2. Extracting as Key Teaching and Training Method in MTRIZ

As we can see, extracting is a process to draw out creative models from objects from natural surroundings or artificial objects. To correctly accomplish this kind of work. we will use as a platform the following definition.

Definition **"Extracting"**	**Extracting** – to draw out transformation models from any (given) artifact using any (given) source of information describing innovative ideas and objects.
Addition 1	**TRIZ-extracting** – drawing out of TRIZ-models of transformation from objects.
Addition 2	Extracting has the aim to study the fund of transformation models and illustrative examples *quickly and correctly*.
Addition 3	Extracting's first super-goal is to develop the skills to *understand the functional and creative evolution of the artifact.*
Addition 4	Extracting's second super-goal is to *discover new transformation models* and to develop the systematic classification and organization of all models of transformation.

During the procedure of extracting, one must stick to a certain schema, which is presented in fig. 5.2.

The general description of the extracting process consists of the following:

1) First we shall take any artifact and choose a real or an imaginary (virtual) prototype for it, assuming the prototype to be a predecessor of the artifact in the history of its constructional development. Ideally the prototype possesses the same useful function (intent, purpose of use) as the artifact, but still we can only employ an object as a prototype that shares only a couple of the functions matching the ones of the artifact.

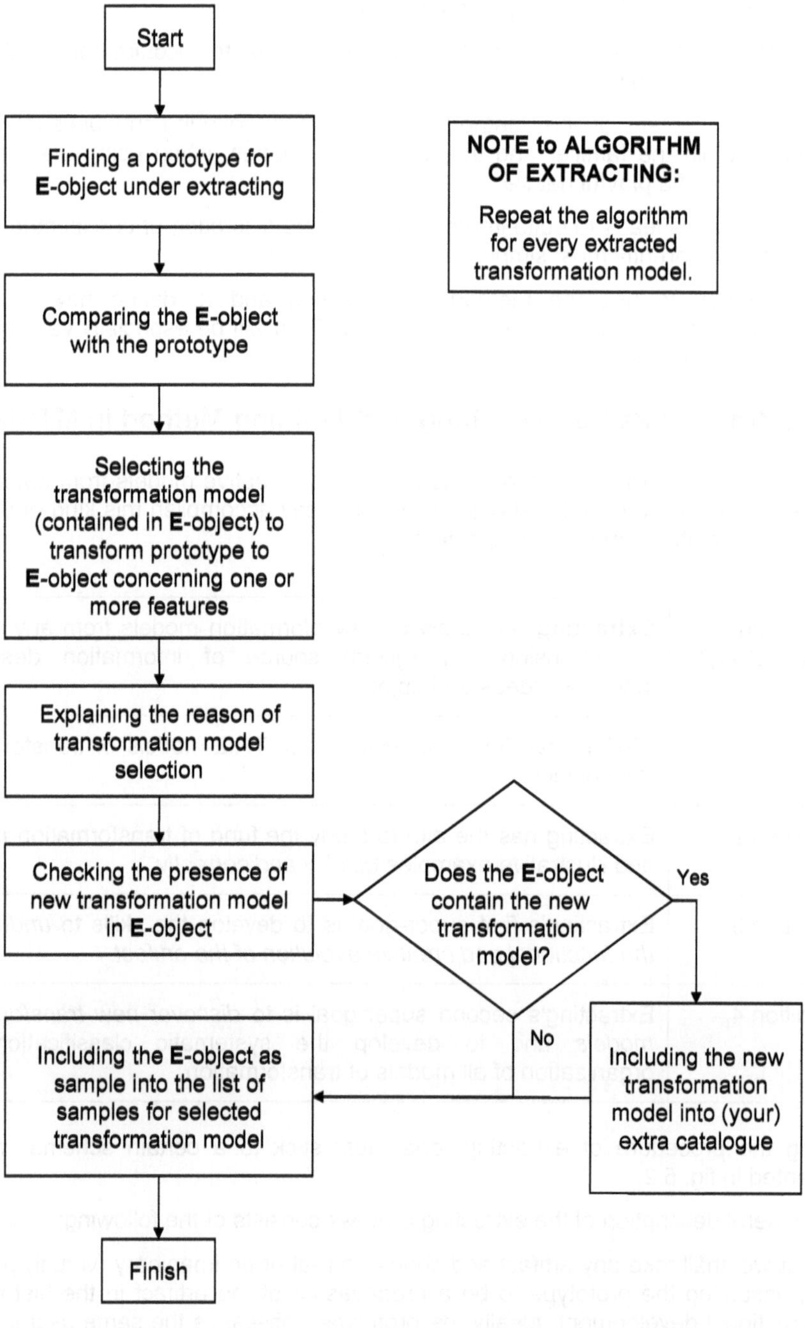

fig. 5.2. Algorithm of the extraction of transformation models

2) Afterward the positive changes inherent in the artifact must be examined and compared with the properties of the prototype.

3) The identified changes are, of course, of a constructive nature. Still, the essence of extracting is to find the creative transformation model, objectively taking part in the change, from every identified constructive change.

4) At first those extracting models are being carried out that can be found in the early edition of the TRIZ-catalogue of transformation models. If some changes cannot be convincingly described with the given transformation models, then a new model can be described, named, and included in your personal creative catalogue and further on be sent to the administrators of the course to have the definition evaluated.

Here is an explanation to the algorithm using the first example: Any personal planner that has no perforation can be used as a prototype.

Then it becomes clear that the applied method of "preliminary action," namely, to prepare the perforation beforehand, allows the fast and neat tearing off of the page's corners.

Two basic variants of extracting are being differentiated:

1) "Extracting-1" – extracting of all objective transformation models that are partly or completely taking part in the transition from the prototype to the artifact;

2) "Extracting-2" – extracting of the dominating transformations that are necessary for such a transition and which should be connected with certain eliminated contradictions.

Three characteristics can help to determine the dominating transformation:

> First, the transformation takes part entirely in the integral transformation.
>
> Second, every such transformation is sufficient to describe the basic inventive idea.
>
> Third, when tentative eliminating such a transition, the description of the transition from prototype to artifact is impossible or absolutely incomplete.

Like all highly qualitative models, transformation models require a certain amount of experience in interpretation and extracting.

5.3. Primary and Advanced Extracting

Example 5.2. Ice for a Drink: Extracting-1 (completion: in Chapter 6)

To cool drinks (juices, cocktails etc.) for a longer period of time, ice cubes can be put into them. When the ice cubes melt, the taste of the drink changes as the relative quantity of the water increases while the quantity of the drink shrinks.

The solution to this "conflict" is well known: the use of ice inside plastic or metal containers. The water or a different liquid inside the container is being frozen before its use. The containers can be manufactured having different amusing forms and figures (fish, flowers, fruits etc. fig. 5.3). After using them, the containers can be cleaned and reused.

Because the containers can be used several times, water used for the ice cube preparation is being saved.

fig. 5.3. Reusable ice figurines

The results of the extracting are illustrated in a special chart (fig. 5.4).

At the same time, the following models prove to be important for the characteristics of various properties of engineering solutions:

"01 Change in state of matter"

"05 Extracting"

"08 Periodic function"

"10 Copying"

"12 Local quality"

"18 Mediator"

"25 Use of flexible covers and thin films"

"34 Matryoshka"

Consider an alternative realization of this example in the EASyTRIZ software package.

LC*	No.	Navigator	Substantiation for the Extracting
++	01	**Change in the aggregate state of an object**	Repeated freezing
+	02	**Preliminary action**	Preliminary preparation
+	03	**Segmentation**	Many figurines
	04	**Replacement of mechanical matter**	
++	05	**Separation**	The property of cooling has been introduced (exists); the property of water mixing with the drink has been eliminated (is missing).
	06	**Use of mechanical oscillations**	
	07	**Dynamization**	
++	08	**Periodic action**	Multiple usage
+	09	**Change in color**	The figurines can be colored
++	10	**Copying**	Copies of many objects
	11	**Inverse action**	
++	12	**Local property**	Cooling of the drink's dosage
	13	**Inexpensive short-life object as a replacement for an expensive long-life one**	
+	14	**Use of pneumatic or hydraulic constructions**	The figurines float
+	15	**Discard and renewal of parts**	Recreation of ice inside the figurines
	16	**Partial or excess effect**	
	17	**Use of composite materials**	
++	18	**Mediator**	The container is the mediator between the ice and the drink.
	19	**Transition into another dimension**	
+	20	**Universality**	Arbitrary forms
+	21	**Transform damage into use**	Cost of materials of container – Production of various figurine shapes
	22	**Spherical-shape**	
	23	**Use of inert media**	
	24	**Asymmetry**	
++	25	**Use of flexible covers and thin films**	The containers can consist of thin foil
+	26	**Phase transitions**	The melting of the ice supports the low temperature of the drink
	27	**Full use of thermal expansion**	
		....	
	33	**Quick jump**	
++	34	**Matryoshka (nested doll)**	The ice inside the container and inside the drink
	35	**Unite**	
		....	
	40	**Uninterrupted useful function**	

fig. 5.4. Results of Extraction-1 for Example 5.2 Ice for a Drink

* LC – Level of Conformity: (++) – complete for effective navigators; (+) – for navigators with partial effectiveness

The reader can see for himself that the **key creative models** for the creation of an inventive idea are the methods "18 Mediator" and "25 Use of flexible covers and thin films."

Actually the container plays the role of the mediator between ice and drink. That is why it is difficult to determine which of the methods is more important for the overall solution. It is possible that the method "18 Mediator" should be slightly prioritized with the explication that the container can just as well be non-flexible and not made of thin foil but rather out of hard metal – such solutions are also known.

The identification of dominating transformations is the purpose of Extracting-2.

It should also be noted that the aims of Extracting-2 are not limited to those described previously. Another important task at this stage is **to define the contradictions** removed by the application of key transformations.

ATTENTION. The contradiction is being stated concerning the prototype and not the studied artifact!

The extracted navigators were not applied to the artifact but to the prototype.

The artifact is the result of the transformation. The object of the initial problem situation and the carrier of the conflicting properties is the prototype.

That is why the transformation models are taking part in the transition from the prototype to the artifact and why they were meant to clear the contradictions of the prototype.

It is easy to see that key transformations are selected from among dominating transformations based on the Extracting-1S results table. In the "software-less" "manual" variant, only those models from the table which have the rating "++" are transferred to the new table. In the new table, whose length by definition cannot exceed five lines, we select the key transformations that most affect the solution idea–formation process.

Example 5.3. Ice for a Drink: Extracting-2

Then we build, for each key transformation, a model of the conflict, which is presented as the contradiction removed by such transformation. If the solution is devised on the basis of several transformations working as one, they may have one common contradiction model.

The Extracting-2 results table for the example under analysis is shown in fig. 5.5. Consider also an alternative realization of this example in the EASyTRIZ software package.

Interestingly, in this case, we could have swapped the contradictions, or they both could have been placed in each cell, because both related transformations can be viewed as "key transformations", i.e., transformations that have "all it takes" to generate the main solution idea.

LC*	No.	Navigator	Substantiation for the Extracting
++	01	Change in the aggregate state of an object	
++	05	Separation	
++	08	Periodic action	
++	10	Copying	
++	12	Local property	
++	18	Mediator	**Radical Contradiction:** the ice must be in the drink to cool it, and it must not be in the drink not to change its taste.
++	25	Use of flexible covers and thin films	**Standard Contradiction:** when melting, the ice cools the drink but also changes its taste.
++	34	Matryoshka (nested doll)	

fig. 5.5. Extracting-2 results for Example 5.2 Ice for a Drink

* LC – level of compliance; key models contain descriptions of the relevant contradictions

NOTE. In the future, attentive readers may well identify other new models in these practical examples. If and when that happens, it will suffice to describe and name the model, establish the frequency of its occurrence in other objects and solutions, and send it to the Modern TRIZ Academy to be verified and possibly included into the As-catalog of models. Such proposals will considerably improve their authors' certification odds.

5.4. WORKSHOP FOR CHAPTER 5

fig. 5.6. How to use a scrunchie.

Task 1. The Scrunchie "Operation"

Let us consider a very creatively invented procedure – the fastening of a ponytail with an elastic hair tie, i.e., a "scrunchie"!

This is how it is done: First the scrunchie is placed on the wrist of one hand, which is then used to grab the hair tail; then the scrunchie is pulled off the wrist with the free hand (fig. 5.6).

Your assignment: Perform the Extracting-1 procedure.

Task 2. Structural Organization of Our Clothes

Do not be intimidated by the "scientific-sounding" title—it is just a small joke. The assignment is this: Identify the main TRIZ principles in accordance with which our clothes are made. Write them down as Extracting-1 results.

Use the following set of objects as a prompt: What do an onion, a hen egg, a head of cabbage, and a wrapped sweet have in common?

Task 3. Live Flower Carpet

When in Germany, I once visited a huge shop selling everything a gardener could ever want. My attention was drawn to a large, long box with a drawing of a flower carpet. The flowers looked very real, as if they were growing right out of the carpet. Besides, from the drawing it was clear that the carpet, an "indoors" item if I ever saw one, was spread out (fig. 5.7) in a real garden!

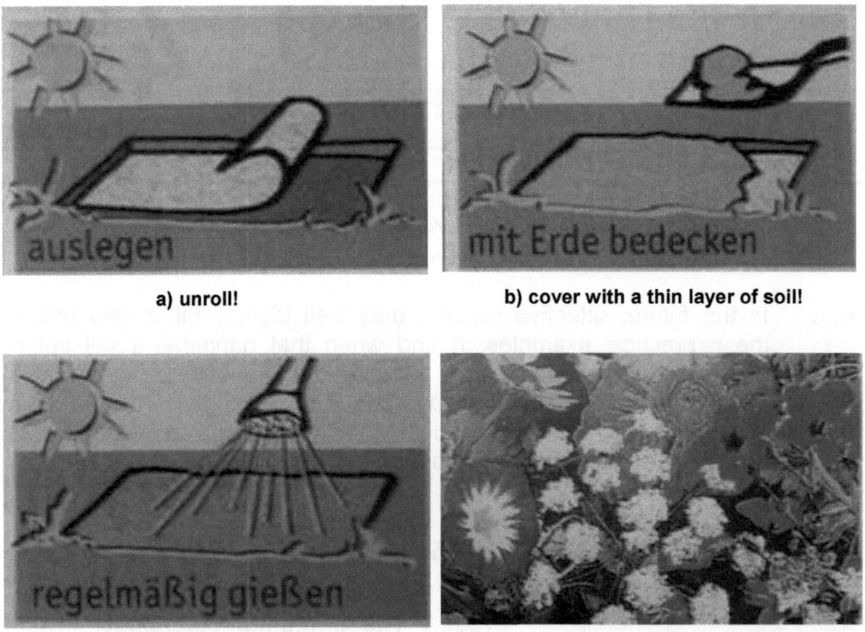

a) unroll! **b) cover with a thin layer of soil!**

c) water! **d) enjoy!**

fig. 5.7. Carpet boasting real live flowers

It turned out that the shop advertised "carpet strips" of varying lengths made of special soil with a porous base made of some spongy material. The pores in the soil were filled in advance with seeds of different flowers and with fertilizer, and the drawing showed what the "carpet" would look like after all seeds germinated and grew into flowers (fig. 5.7,d). To crown it all, clients could devise their own patterns, whereupon professional gardeners selected flower varieties required to recreate those patterns and make customized flowerbed carpets!

Do the Extracting-1 procedure for this artifact: A carpet for… future live flowers.

ADDITIONAL ASSIGNMENTS. Make a list of other artifacts, natural objects, fairytale heroes, and other literary objects, perform the extracting, and send the most interesting results to us to be incorporated into the MTRIZ Pool databank.

Task 4. "Mobile Traffic Light"

Figure 5.8 features a mobile stoplight that is temporarily installed at a busy crossing during rush hours. Please perform Extracting-1 for thaw construction and operation of this object.

fig. 5.8. Mobile traffic light at a road crossing in Harbin, China

Task 5. "Fan-Cap"

The figure 5.9 shows a pretty Chinese girl, an ice cream vendor. The peak of her sun cap has an inbuilt battery-operated fan. Perform Extracting-1 for the construction and solution for this cap.

fig. 5.9. Ice-cream vendor near the monument on the shore of the Sungari river, Harbin, China

Task 6. Compare your results for Task 4 and Task 5.

Tasks 7 – 9: Define one (or two) dominating transformation model (specialized navigator) in each solution.

Task 7. "Ultrasonic whistles"

An American firm has produced special ultrasonic whistles that are mounted on the front bumper. At speeds over 50 km/h under influence of the air stream, the whistles emit ultrasonic signals that are frightening for animals but inaudible to people. The higher the speed, the "louder" the signal.

fig. 5.10. Ultrasonic whistles

Task 8. "Avalanche rescue airbag"

German entrepreneur Peter Aschauer suggested a new rescue device: an avalanche airbag made of very thin and bright orange nylon. The airbag is carried in a small backpack and is inflated by compressed nitrogen from a small cylinder activated by the user when he finds himself in danger of being buried under snow.

fig. 5.11. Avalanche rescue airbag

Task 9. "Magic packing"

The Firm Sealed Air Corporation (USA) has developed highly elastic polyethylene bags that come in all sizes. Mechanical or thermal impact triggers the production of polymer foam, which is then evenly distributed throughout the interior of the bag.

fig. 5.12. Magic packing

6. Inventing

> *Algorithmic methodology*[44] regards the process of inventive problem solving as ***a sequence of operations to identify, clarify and overcome the contradictions.***
>
> ... [but] we should not think that after reading the text of the algorithm, we can immediately solve any problem. After reading the description of the techniques *sambo*[45], do not just go to the contest.
>
> ***Practical skills are developed on educational tasks.***
>
> *Genrikh Altshuller*

6.1. Algorithm START T-R-I-Z ™

In this course, which is basically designed for beginners, we use the start version of MAI T-R-I-Z, which is restricted to the use of primary universal TRIZ models to resolve standard contradictions.

The name **START** is an acronym for "**S**implest **T**RIZ-**A**lgorithm of **R**esourceful **T**hinking" (fig. 4.7) and was made as a mnemonic aid for better memorizing. **T-R-I-Z** indicates the stages of the algorithm.

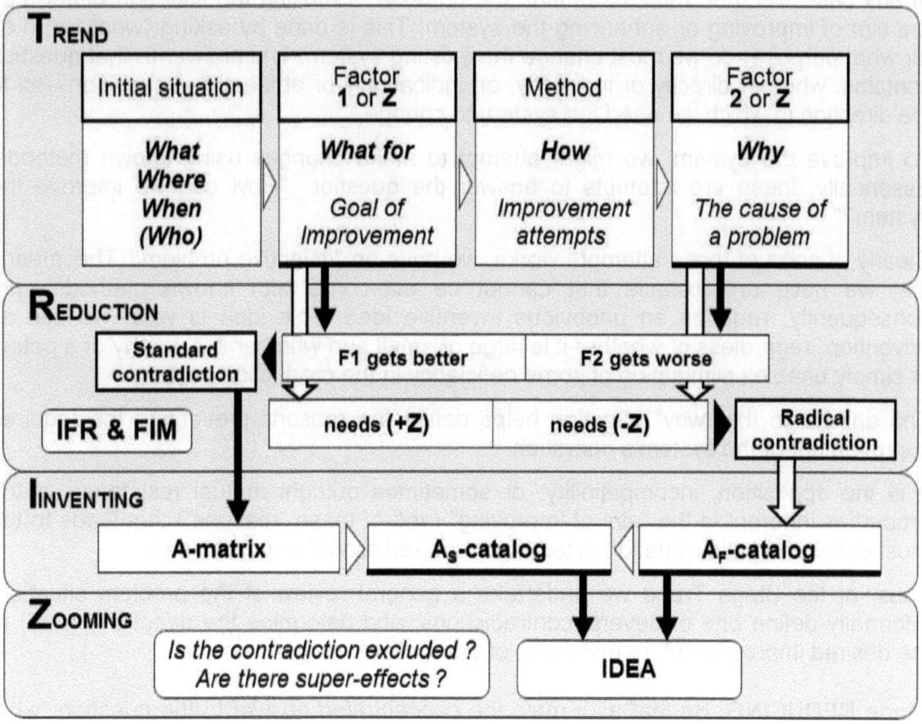

fig. 6.1. START T-R-I-Z™ – the simplest invention algorithm

---

44 Compiled and arranged here by M.O. from: G.S. Altshuller, *Invention Algorithm*. – Moscow Worker, 1973. – (Г.С.Альтшуллер, *Алгоритм изобретения*. – Московский рабочий, 1973)

45 Russian combat and sport art of "*self-defense without arms*"

M.A. Orloff, *ABC-TRIZ*, DOI 10.1007/978-3-319-29436-0_6

The main difference between START and MAI T-R-I-Z lies in the fact that each START stage is filled with specific procedures enabling development of solutions based on TRIZ models. This transforms the "meta-algorithm" (i.e., the maximally integrated, aggregated, and, accordingly, abstract "path") into an executable practical algorithm that directs, organizes and disciplines the work performed to create new solutions.

ATTENTION:This scheme must be committed to memory!

The general comment to the scheme is as follows.

Stage TREND. Resolution of any problem starts with an analysis of the initial situation or, if you will, with diagnosing the problem. Accordingly, in the extended Modern TRIZ version this stage is also called "diagnostics."

The analysis naturally involves determination of "what," "where," and "when" occurs in the system so that the quality of outcomes produced by such system and, accordingly, the efficiency of its operation and makeup, become suspect

Answers to these questions[46] must produce a sufficiently accurate and concise description of the initial problem situation.

In any case, we then must determine the purpose of making the situation better, i.e., the aim of improving or enhancing the system. This is done by asking "what for" (i.e., for what purpose) do we must change the existing system. The answer to that question contains, whether directly or indirectly, an indication—or at least a supposition—as to the direction in which we need the system to change.

To improve the system, we might attempt to make changes using known methods. Essentially, these are attempts to answer the question: "How can we improve the system?"

Clearly, if none of these attempts works, we have an "inventive problem." This means that we have an obstacle that cannot be overcome with known methods and, consequently, requires an unobvious inventive idea. This idea is what we call an "invention" regardless of whether it is large or small and whether it is worthy of a patent or simply enables elimination of some deficiency in the production process.

The answer to the “why” question helps define the reasons preventing the required improvement of the system’s operation.

It is the opposition, incompatibility, or sometimes outright mutual resistance of the properties inherent in the "aim of improving" (and of these "reasons") that leads to the must define the initial situation in terms of "contradictions" and "problems."

Thus, at the Stage Trend we undertake a general review of the problem situation, informally define one or several contradictions, and determine the direction (trend) of the desired improvement, or evolution, of the system.

Stage REDUCING. Be that as it may, the *concentrated answer* to the question "what for"—as posed at the Stage Trend—is formulated in the form of the “Ideal Final Result”

[46] The use of leading, or "controlling", questions was suggested by Quintilian, Roman rhetorician, in the first century AD – see "Inventive Thinking through TRIZ" by M. Orloff, Section 4.1 "Inventing the Theories of Inventing."

(IFR). This IFR substantially shapes the trend, i.e., the purpose and pattern, of the future system transformation.

The IFR may be recorded separately, or incorporated into the formulation of the "Functional Ideal Model" (FIM). Implementation of the FIM is aimed (trend!) at the attainment of the IFR.

Methodologically, the FIM and the IFR are fixed at the Stage Reducing as teleological benchmarks for all subsequent modeling and generation of ideas with the FIM being reduced to certain a standard form.

However, the key task at this stage is to reduce the informal contradiction to some standard format.

For starters, let us recall that all contradictions are divided into two types (have a closer look at fig. 6.1):

1) Standard contradiction: Each attempt to improve one property of the system results in deterioration of another property of the same system.

2) Radical contradiction: Diametrically opposed requirements are imposed on the same property. In fig. 6.1, this conflict is represented in the form of two opposed requirements as they apply to the same property **Z**: For one purpose, **Z** must be increased (trend **+Z**) while for the other purpose, **Z** must be decreased (trend **-Z**).

Both formal contradiction models have their "own" means and "paths" of resolution.

The "path" to resolution of a standard contradiction is discussed in part "**6.2. Solving the SC according to the Method BICO—Binary In Cluster Out,**" while the "path" to resolution of a radical contradiction is presented in part "**6.3. Solving the RC according to the Method RICO—Radical In Cluster Out**".

Stage INVENTING. At this stage, the generation of ideas is supported with transformation models borrowed from catalogs created in TRIZ. In Modern TRIZ, this stage is also called "transformation."

To resolve standard contradictions, we use *specialized* transformation models from the "A_S-catalog" (see S22.2), while the A-matrix helps us select suitable models (see section S21.2).

On the whole, we believe that the creation of an efficient idea is the fruit of fusing talent with knowledge and creative thinking skills.

Stage ZOOMING. In Modern TRIZ, this stage is also called "verification," which points to the forthcoming examination of the quality of newly discovered ideas. These ideas must be "verified," i.e., checked for implementability and efficiency.

The use of the metaphorical name "zooming" is determined by the nature of the main verification procedures, namely, by the must verify the ideas at several levels of examination. Thus, it is necessary to assess the quality of the solution at the level of the operating zone, i.e., the area which originally hosted the cause of the problem.

To do this, we must "zoom up" and assess the quality of the solution, say, at the level of the entire system. Then we might want to take an even broader look at the level encompassing the interests of the user for whom the artifact (system) was created in the first place.

Last, but not least, in many cases it is possible (and often necessary) to measure the impact of the idea at an even higher and more general level, for example, at the level of the society or even at the level of the whole area of nature.

These transitions are similar to changing the focus—*zooming!* —in a photo or video camera to change the "field of vision," i.e., the amount of space that we see through the lens. The metaphorical name of the fourth and final stage of MAI T-R-I-Z originates from this functional idea.

6.2. Solving the Standard Contradiction with Method BICO—Binary In Cluster Out

Of course everybody is now familiar with the popular solution for the day planner.

The previously examined Example 5.1 demonstrates the most simplified MTRIZ-analysis as the idea that the "invention" of the solution has been "created" based on the selection of several transformation models (navigators) after a complete check of all the navigators from the As-catalog or from the software EASyTRIZ, which you are working with right now.

We can now, with the help of **Method BICO (Binary In Cluster Out)**, present this solution in a variant based on the very first and very simple method of TRIZ, i.e., the selection of navigators from the A-catalogue with the help of the A-matrix.

The difference now is that we draw the chosen plus- and minus-factors from the A-matrix to the respective standard meanings. The chosen standard factors become inputs for the lines and columns of the A-matrix for the selection of matrix cells with navigator numbers that are statistically recommended for standard contradiction solving, which can be presented with exactly these standard plus- and minus-factors.

The chosen plus- and minus-standard factors define the *binary input* of the A-matrix, and the chosen cell contains the cluster (group, majority) of the recommended navigators' numbers. This means that to obtain the sought-after result, we receive this cluster of navigators' numbers as the outcome of the A-matrix (*cluster out*). An example of such a process is presented in fig. 6.2.

Example 6.1. The Corner of a Personal Planner's Page (BICO-Solution).

Let us cite the summarized description of the process and results of problem solving to **Example 5.1. The Corner of a Personal Planner's Page**.

Stage 1. Trend (Diagnostics)

To comfortably open the day planner on the page of the actual date (the goal), one can step by step and day by day tear off the page corner of the past date. Then the page correlating to the actual date is always the first page with the un-torn corner. The page is well visible because all of the torn off corners are missing and reveal the actual date.

The problem is that it is hard to tear off the paper corner neatly. The direction of the problem solution (the trend) consists of finding a method that helps the user of the day planner to tear off the corner accurately.

How can the user be helped to tear the corner off quickly and accurately?

Remark: Other trends are not being examined in this training assignment.

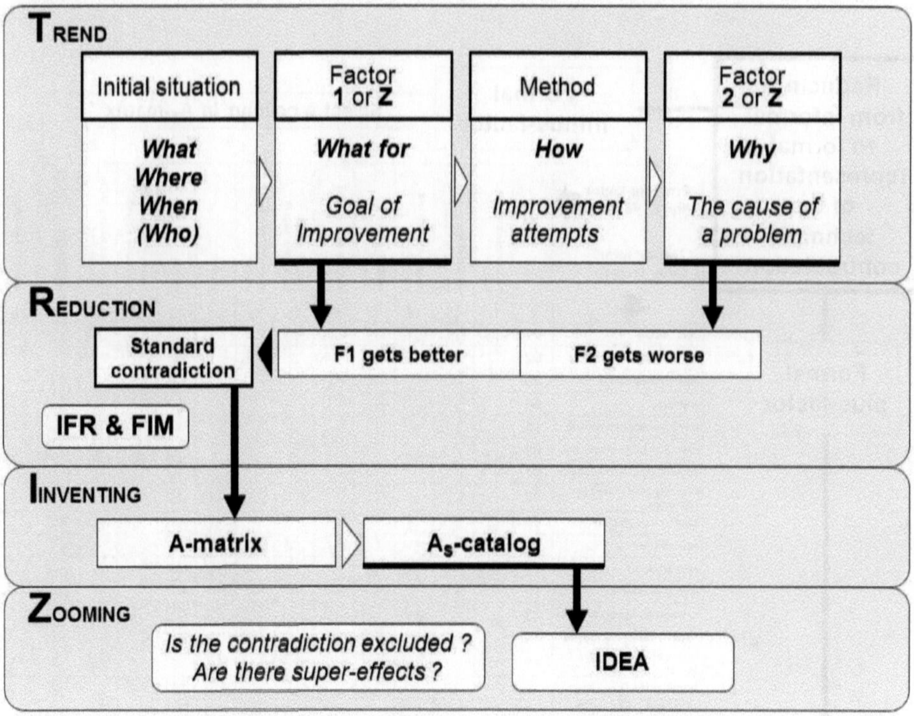

fig. 6.2. The path in START to solve to a standard contradiction

Stage 2. Reducing (Reformation)

Let us reformulate the problem (reformation) and imagine it in the form of generally known standard contradictions:

- the corner must be torn off quickly but the result is an uneven tearing line;
- if the corner is torn off slowly, the tearing line turns out to be more accurate (but this takes too much time).

Let us select **non-formal standard contradiction**:

the corner must be torn off quickly,
but this leads to an uneven tearing line.

It is very important to state an **ideal final result**, for example "ideal" would be a tearing method in which: [*the page corner can be torn off quickly and accurately*].

To define a method of inventing, we use the A-matrix. Therefore, we must choose formal plus- and minus-factors from the A-matrix.

Stage 3. Invention (Transformation)

The requirements for getting a "straight line" can be connected with the plus-factor **21 Shape**. The negative circumstance, too much time is needed that in order to obtain the "correct form" can be connected with the minus-factor **25 Loss of time** (fig. 6.3).

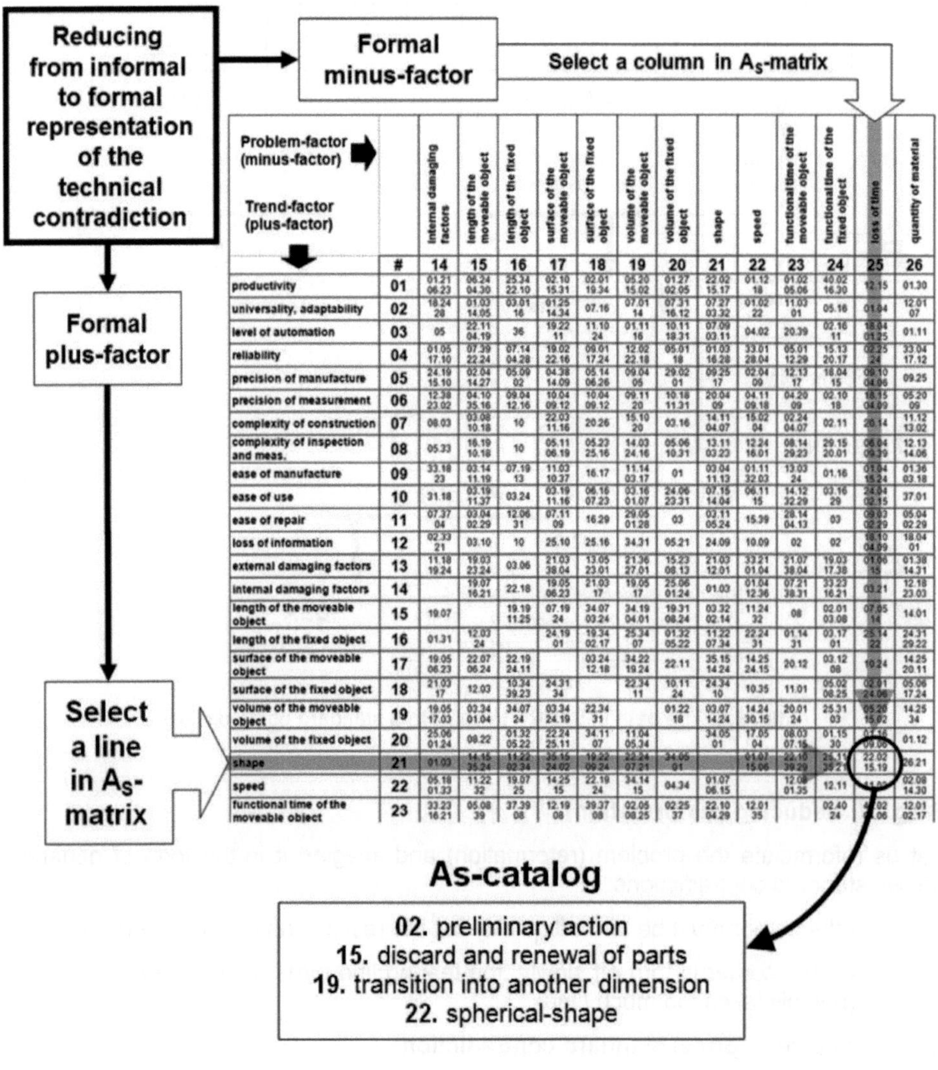

fig. 6.3. Example of selection process of navigators from the A-catalogue based on the creation of a formal standard contradiction in the A-matrix

On the intersection of **line 21** and **column 25** of the A-matrix, there is a cell from which we choose the following navigators:

02 Preliminary action – doing a part of the work in advance;
15 Discard and renewal of parts – replacing the used part with a new one;
19 Transition into another dimension – using a space of advanced dimension (e.g., 3D instead 2D);
22 Spherical-shape – using curved and round forms and trajectories.

NOTE. In the EASyTRIZ software, this selection is "hidden" in the program, and we can immediately see only the result of the selection—the content of the chosen cell that is the cluster 02, 15, 19, 22.

The model 02 Preliminary Action in section (a) contains a rather constructive hint:

02 Preliminary action	a) execute the necessary changes of the object in advance (completely or at least partly); b) place the objects in advance in a convenient spot so they can be reached easily and flash into action without losing any time.

The general solution idea based on the model (fig. 6.4): *beforehand little cuts or holes are being made where the tearing line should later appear.*

Stage 4. Zooming (Verification)

The main contradiction has been cleared.

The problem was solved without any negative effects.
You can find the realization variant of the example in the EASyTRIZ software.

Concerning the verification of the solution on the basis of zooming, we can add the following:

1) *On the product level* (day planner), the new solution assures an excellent form (of different arts as in fig. 6.4) of the tearing line and obtaining an accurate look for the day planner independent from the number of torn-off corners.

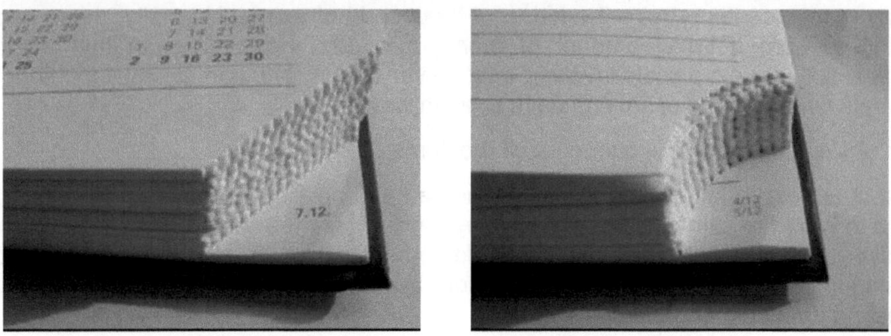

fig. 6.4. Page corners of a diary or personal planner

2) *On the user level*, it indeed became more convenient to use the day planner, and the contradictions were cleared almost completely (the reader can now think for himself why we say "almost" and not "completely").

3) *On the company level*, which produces the day planner, a complication of technology and facilities has taken place for the introduction of a new operation procedure.

4) *On the market level*, the new day planner can now be more expensive than the prototype and all expenses of the company are being passed off to the buyer of the planner; however, the company's strategy could also be more farsighted.

To research and discover the world of TRIZ models is an entertaining, exciting, and useful pastime! And it is also EASY!

To do this, you do not need a special laboratory. On the contrary, you must develop a habit of seeing TRIZ models in surrounding artifacts, extracting such models from such artifacts, and to model an invention! So your laboratory is wherever you are. In other words, it is the entire world around you.

Here are some simple examples.

Example 6.2. "Inflatable Crowd".

Knowing my interest in stories where people win in the face of adverse circumstances, two years ago my elder son gave me a disc with a movie called *Seabiscuit*. Indeed, both the story and the movie about a well-known jockey and his horse *Seabiscuit* were powerful and exciting.

Trying to find out more about the movie, I discovered a wonderful trick that its makers used to film a large crowd of people at the hippodrome. Naturally, I immediately decided that it should be offered to my students as a reinventing example.

And this is what they made of it.

Trend

To fill large spaces with people, movie-makers have to invite many extras, which is very challenging in administrative terms as well as very expensive.

Many mass scenes require the hiring of thousands of extras. Normally, those people willing to earn a small amount of money per day. However, from the director's perspective, the extra costs involved may be very significant, particularly if the shooting extends over a period of several days or even weeks.

And this is only the first in a long list of problems.

First, it is next to impossible to gather the same people day in day out. Second, they have to wear the same clothes. Third, they have to take the same positions (in the hall, stadium, field, etc.). The last three factors together create extremely complex management issues (supervision, invitation, assembly, placement, briefing, etc.). The shooting drags on, and the costs go up.

The overall direction of search for ideas must be formulated so as to mitigate or eliminate all of these problems. So, what can be done?

Reducing

To find a solution, we must model the problem as a standard contradiction.

The conflict exists between the following properties: increasing the space occupied with people (the crowd) and increasing size of the crowd *versus* complexity of management and high costs (increasing expenses).

These informal factors can be replaced with the following formal factors from the A-matrix:

Plus-factor *19 Volume of the moveable object*: The crowd moves, occupies large shooting space.

Minus-factor *08 Complexity of inspection and measurements*: It is difficult to organize and control the crowd.

The A-matrix cell located at the intersection of the line corresponding to the selected plus-factor and the column corresponding to the selected minus-factor contains the numbers of transformation models that in the past have been frequently used to resolve this type of contradiction: In this A-matrix variant, these are numbers 10, 14 and 24.

The problem object to be transformed is the "crowd."

Inventing

Model *10 Copying* recommendation: a) use a simplified and inexpensive copy instead of an inaccessible, complicated, expensive, inappropriate, or fragile object.

Idea: use dummies.

fig. 6.6. "Inflatable Crowd"

Model *14 Use of pneumatic and hydraulic constructions* recommendation: Use gaseous or fluid parts instead of fixed parts in an object: parts that can be blown up or filled with hydraulic fluid...

Idea: the dummies will be inflatable[47] (fig. 6.6).

Model *24 Asymmetry*: a) move from a symmetrical shape of an object to an asymmetrical one.

Idea: to make the crowd look more real, intersperse dummies with a small number of "real" actors and extras.

Zooming

Advantages:

1. Inflatable dummies are cheap.

2. Dummies are easy to "manage" and can fill any space.

3. Dummies can be dressed in any clothes.

4. It is possible to use wigs and masks.

Disadvantages:

1. Dummies are static, so some extras have to be used.

2. Dressing, placement, removal, undressing, etc. of dummies is a labor-intensive process.

3. In this example, zooming – in addition to its metaphorical meaning – can be quite literal: We must evaluate acceptability of the solution for shooting at different zoom levels. The cameraman must make sure that the dummies are not filmed up close and that the crowd is not filmed from the same point for a long period of time or viewers may notice the static and artificial nature of the scene.

The main thing, however, is this: THE contradiction has been eliminated together with the original problems associated with filming large crowds.

Clearly, such simple dummies can only be useful for recreating a static crowd (or some part of a static crowd).

Direct enhancement of the idea: transformation of the dummies into "robot-like" creatures that can make simple movements, for example, rock from side to side.

There is also another apparent enhancement: The use of computer graphics makes it possible to reproduce any movement of any number of figures "placed" inside any space. In this case, the copying method is used at its best, too.

Taking into account the computer-animation method, the "inflatable" figure metaphor will be quite apt. The mathematical model of such figure is a discrete frame "wearing" an outer "shell."

Last but not least, the "parting shot": Dummies for the *Seabiscuit* movie were provided by an American company with a very fitting name—*Inflatable Crowd.*

---

[47] Reproduced from http://www.flickr.com/photo_zoom.gne?id=296959242&size=l.

6.3. Solving the Radical Contradiction with Method RICO – Radical In Cluster Out

It is our goal to develop skills for the construction of descriptions of any invention and of any interesting creative solution in the format of MAI T-R-I-Z. Hereby the creation of such an interesting idea modulates and presents itself as the process of the invention's creation with the help of the methods and models of TRIZ.

To resolve the radical contradiction, it is necessary to use the path in START as shown in fig. 6.7.

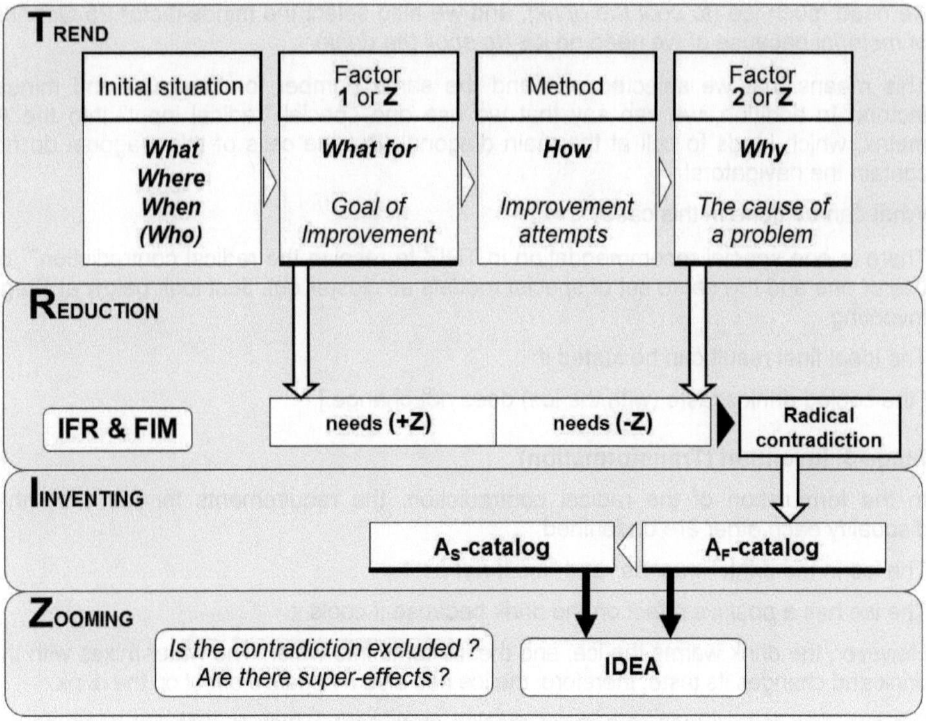

fig. 6.2. The path in START to solve the radical contradiction

Let us cite the summarized description of the process and results of problem solving for **Example 5.2. Ice for a Drink** from the previous chapter 5 on extracting.

Example 6.3. Ice for a Drink (RICO Solution)

Stage 1. Trend (Diagnostics)

To keep a drink – juice, cocktail, etc. – cool for a longer period of time, the drink is filled with ice cubes. However, as the ice melts, the drink's taste changes because the relative amount of the water enlarges within the shrinking amount of the drink.

How can a cool drink be prepared with ice added without changing the drink's taste while it is being drunk?

Stage 2. Reduction (Reformation)

Let's formulate a radical contradiction:

The ice inside the drink – must be there to cool the drink and
must not be there because it leads to a change of the drink's taste during its consumption.

We can use the A-matrix to select some plus- and minus-factors, but we will get in this case a confused situation: We select the plus-factor *25 Quantity of material* because we need much ice (*to cool the drink*), and we also select the minus-factor *25 Quantity of material* because of we need no ice (*to spoil the drink*).

This means that we selected one and the same number for the plus- and minus-factors! In addition, we can say that we use one special "radical input" into the A-matrix, which leads to cell at the main diagonal. But the cells of this diagonal do not contain the navigators!

What can be done in this case?

There is one special recommendation in TRIZ to resolve the radical contradiction[48] by use of one and the same set of special models as cluster out. Just look below at Stage Inventing.

The ideal final result can be stated if:

[the cooled drink's taste (with the ice) does not change]

Stage 3. Invention (Transformation)

In the formulation of the radical contradiction, the requirements for ice cube that disqualify each other are underlined:

The ice in the drink "*must be*" and "*must not be*."

The ice has a *positive effect* on the drink because it cools it.

However, the drink warms the ice, and the ice turns into water. The water mixes with the drink and changes its taste; therefore, the ice has also a *negative effect* on the drink.

We can also take a look at a more precise statement, which is different because the objects taking part in the conflict are named more precisely.

The ice is nothing but frozen water so we are dealing with water and not with ice. It changes its state from a solid one (called ice) into a liquid one (ordinary water). This means that in reality it is the water itself, the object producing the conflicting properties "cooling of the drink and change of composition," which changes from one state to the other.

So the radical contradiction can also be illustrated in a slightly different way:

The ice inside the drink - must be in the form of ice and
must not be in the form of water during the transition from ice to the liquid state when melting.

---

[48]See Af-catalogue in S26 or-and S36

Therefore, we are dealing with water only in two different states.

There are only four main possibilities (conditional Cluster Out) to separate the conflicting properties in the problem situation with a radical contradiction:

1) *in* ***space*** – one property is realized within one area of space while the opposite one is implemented in the other space;

2) *in* ***time*** – one property is being realized within one time interval while the opposite property is being realized within a different time interval;

3) *in* ***structure*** - one part of the system owns one property and the entire system owns an opposite property;

4) *in* ***material/energy*** – the material or the energetic field (or its elements) for one goal have one property and for the other goal an opposite property.

The system (the drink) must have in a total the following properties:

- *a specific taste (composition);*
- *be cool.*

One part of the drink that cools it can possess the following properties:

- *be of a different composition than the drink;*
- *be cool, just like the drink (be in direct contact with the element).*

Consequently there are first two properties in conflict.

Solution (see Afs-catalog in section S25 or advanced Af-catalogues in the author's book[49] or follow the option RICO in the software EASyTRIZ with the outcome to have the fundamental transformations supported with the help of the specialized ones):

1) According to the "fourth" fundamental navigator—"Separation in the material"—and the specialized navigator 38 *Homogeneity: to construct the co-acting objects from one material,* ice figurines from the same drink can be produced.

2) According to the "third" fundamental navigator—"Separation in structure"—and the specialized navigator 18a *Mediator. using an in-between object for the transfer and the transmission of effect,* the ice figurines (no matter what material they are made of) can be stored in small containers made of, for example, plastic or metal.

3a) According to the "second" fundamental navigator—"Separation in time"—and the specialized navigator 02 *Preliminary action: complete or partial beforehand realization of the necessary action* (in addition to the previous points), the ice figurines can be frozen *in advance.*

3b) According to the "second" fundamental navigator—"Separation in time"—and the specialized navigator 02 *Preliminary action* (as an addition to section 1), the ice cubes, from the same drink, can be frozen during the process of preparing the drink. With the aid of ice machines, this can be accomplished within a few minutes the drink is consumed.

---

49 M. Orloff, *Modern TRIZ* (sections 14.2.4, 14.2.5 and 14.2.6)

Stage 4. Zooming (Verification)

Zooming *at the level of the glass containing the drink and the drink itself*: The main contradiction is eliminated. The drink will be cooled, and its taste will remain unchanged regardless of which solution is selected.

Zooming *at the level of the consumer of the drink*: Some consumers may like "disappearing" same-drink figurines, whereas others may want their "non-disappearing" figurines to keep floating in the drink! In this case, we need to perform a "marketing experiment."

Zooming *at the level of the manufacturer of the drink*: The process of drink manufacturing becomes more complex; it is necessary to acquire new equipment (for example, a quick-freeze machine) and/or materials (multiple-use freeze–unfreeze figurine molds).

Zooming *at the level of the manufacturer of ice figurine molds*: There emerges a new business opportunity: the mass production of ice-figurine molds that can be used to freeze or cool drinks.

IN CONCLUSION:

The most important result of this chapter is the development of a meta-algorithm of inventing in his first instrumental version of START.

SMART is a fundamental instrumental and model method for learning and for practical application of Modern TRIZ.

The main value of START is that the process of creating any effective solution can be modeled and presented in a standard START format as if the decision had been made on the basis of primary instrumental models and methods of TRIZ.

This approach is an effective platform for the rapid and proper teaching the beginners in the original technique of applying TRIZ models.

All of this is explained with following key factors:

- The logic of inventive problem solving in the first approximation is well displayed with START and is absolutely understandable for beginners.

- All examples of training courses at MTRIZ Academy and in all information sources—including software—are structured and presented in the same START format.

- All students from the first days or even hours of training can easily learn the START format and accurately applying it to create similar descriptions in the exercise of reinventing educational or any other artifacts chosen by them.

- As will be seen later in this book, a simple standard START format is indispensable and highly effective for accumulating the innovative and inventive experience of any artifact as well as to solve real problems and to generate effective solutions.

6.4. WORKSHOP FOR CHAPTER 6

Exercise 1. Enter the correct answers (several correct answers are possible) about MAI T-R-I-Z and START T-R-I-Z.

Stage **Trend** is intended for the following purposes:

#	Purpose	Yes	No
01	To structure the initial problem situation		
02	Assign responsible person for solving the problem		
03	Identify informal contradictions		
04	Determine the direction (trend) of search of solutions		

Stage **Reducing** is intended for the following purposes:

#	Purpose	Yes	No
01	Identify suitable formal models of contradictions		
02	Select the model of transformation of the A-matrix		
03	Determine the technology of the future solutions		
04	Identify IFR and FIM		

Stage **Inventing** is intended for the following purposes:

#	Purpose	Yes	No
01	Calculate the economic efficacy of solution		
02	Suggest ideas of solutions based on models from As-catalogue		
03	Suggest ideas of solutions based on models from the Af-catalog		
04	Suggest ideas of solutions based on resources		

Stage **Zooming** is intended for the following purposes:

#	Purpose	Yes	No
01	Check whether the conflict is eliminated		
02	Develop technical documentation for the project		
03	Check how could it be possible to develop a solution		
04	Check whether there are positive and negative effects		

Exercise 2. Can you repeat independently the solution of the **Problem P1. Genrikh Altshuller's Experiment** as shown in **Example 2.2**?

Exercise 3. Would you like to solve the **Problem P2. How to Make Icicles Fastened to Eaves Gutter**? Try to do it!

Exercise 4. Would you like to solve the **Problem P3. Ruby Stars of the Moscow Kremlin**? Try to do it!

Exercise 5. Would you like to solve the **Problem P4. "Magic Faucet"**? Try to do it!

Exercise 6. Would you like to solve the **Problem P5. To Convince People to Go in for Physical Culture**? Try to do it!

Exercise 7. Can you repeat independently the solution shown in **Example 3.1. A Glass of Hot Tea**?

Exercise 8. Try to solve independently the problem in **Example 4.6. Training of Long-Distance Swimmers.**

Exercise 9. Try to solve independently the problem in **Example 4.7. Training on diving board and diving tower.**

Exercise 10. Try to solve independently the problem in **Example 4.8. The Invention of the Bicycle.**

Exercise 11. Try to solve independently the problem in **Example 4.9. Invention of the Handlebar and the Maneuverable Front Wheel.**

Exercise 12. Try to solve independently the problem in **Example 4.10. Invention of Pedals.**

Exercise 13. Try to solve independently the problem in **Exercise 2. Wall Paper and Carpet Knife** of subchapter 4.5. WORKSHOP FOR CHAPTER 4.

Exercise 14. Try to solve independently the problem in **Exercise 3. Protection from Car Theft** of subchapter 4.5. WORKSHOP FOR CHAPTER 4.

Exercise 15. Try to solve independently the problem in **Exercise 4. Protection of Motorbike Rider in the Event of a Crash** of subchapter 4.5. WORKSHOP FOR CHAPTER 4.

Exercise 16. Try to solve independently the problem in **Exercise 5. Aircraft Carrier** of subchapter 4.5. WORKSHOP FOR CHAPTER 4.

7. Reinventing

Extracting and Reinventing based on the Meta-Algorithm of Inventing is a modern "TRIZ-Tomograph" of inventive thinking. It works like a "Time Machine": reinventing allows us to explore creative thinking of any inventor from any time and era!

Ruwim Kisselman,

PhD in engineering, inventor, head of the Inventors Club[50] "Schöpfer" ("The Creators" – German) from Bonn (Germany), organizer of the yearly seminars of the Modern TRIZ Academy in the German Museum in Bonn[51]

7.1. Reinventing as Fundamental Teaching and Training Method

Reinventing is a fundamental educational method in Modern TRIZ.[52] The content of reinventing is the reconstruction and description of all stages of an invention's creation with the involvement of TRIZ models and recommendations at each stage.

When reinventing, beginners are studying problems with known solutions illustrated in the format MAI T-R-I-Z in order to understand how these problems would have been solved on the basis of TRIZ and how in the future similar problems could be solved by using both MAI T-R-I-Z as well as TRIZ.

Definition of **"Reinventing"**	**Reinventing** – *modeling (reconstruction, reproduction, renewal) of the invention process.*
Addition 1	**TRIZ-Reinventing** – *modeling of the invention process* on the basis of TRIZ models.
Addition 2	The initial reinventing has the first goal to allow students to *quickly and correctly acquire* the algorithm of inventive problem solving in the format of MAI T-R-I-Z.
Addition 3	The second and supreme mission of reinventing is to reliably prepare students to work autonomously on *any new practical problem.*

MAI T-R-I-Z is only a general framework and a general navigator for inventive problem solving. It becomes a practical instrument only in connection with the models and methods of TRIZ, which equip the stages of MAI T-R-I-Z and thereby turn it into a certain variant of ARIZ – Algorithm of Inventive Problem Solving.

Extracting and reinventing are fundamental tools of a researcher in Modern TRIZ as well as the fundamental set of tools for learning the basics of Modern TRIZ. In this book, we look at the earliest models of TRIZ.

---

50 Association of Inventor Clubs created in Germany earlier within the framework of the former INSTI Program (Innovationsstimulierung – state program for the support of innovative movement in Germany

51 www.deutsches-museum.de

52 **M. Orloff** *Inventive Thinking through TRIZ: A Practical Guide*; part *"9. From What exists to What's coming"*

M.A. Orloff, *ABC-TRIZ*, DOI 10.1007/978-3-319-29436-0_7

For this, we start from the "other end," i.e., from reverse analysis, from results for problems already solved, from already-existing inventions, from any artifact, which always contains a creative idea. We will learn from the experience of all civilization.

If we have some kind of artifact that is of interest to us, then it usually has a prototype—an artifact serving the same purpose but with some flaws that have disappeared in the result-artifact. Therefore, invention always leads from the prototype-artifact to the result-artifact (fig. 7.1).

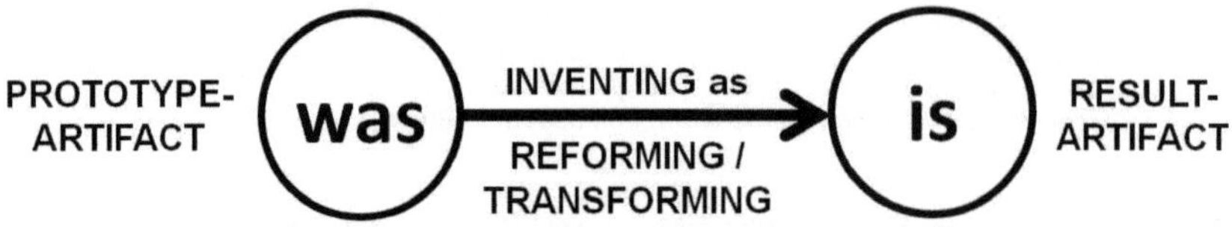

fig. 7.1. Reinventing as modeling the system transition from the state "was" toward the state "is"

The difference between figures 2.2 and 7.1 is that in fig. 7.1 we know the solution-result. We can compare the result with the prototype and extract patterns of change that have led to this result. In this case, study of the invention process could be realized as **modeling the invention process** (for one reason in particular that we do not know all the components of the invention process). This modeling as "reconstruction" has been called **"Reinventing."**

Reinventing is a basic and effective method of training in modern TRIZ. This simple technique of reinventing development and learning of any known inventions through reinventing is the way to "learn how to invent."

And first of all we are interested in the following aspects:

- types of problems in the prototype-artifacts;

- types of creative transformations underlying any invention;

- types of purposes for the invention of effective ideas;

- types of those components of prototype-artifacts (resources) that are subject to real change to create a new artifact;

- the composition and interaction of those components of the prototype that lead to problems, which you need to eliminate by inventing the effective idea.

It should be noted that we have access to pretty much the only technical knowledge of underlying result. However, for example, the aesthetic evaluation and motivation that led to this result are usually not known.

Similarly, when explaining the method for producing the invention, we can only approximately describe the required models according to TRIZ that led to the result in the concrete situation. Yet it makes great and fundamentally important progress compared with any other "theory of creativity."

Reinventing is a fascinating process. The games of interest and research are combined in reinventing. What could be more fun?!

The algorithm of reinventing is shown in fig. 7.2.

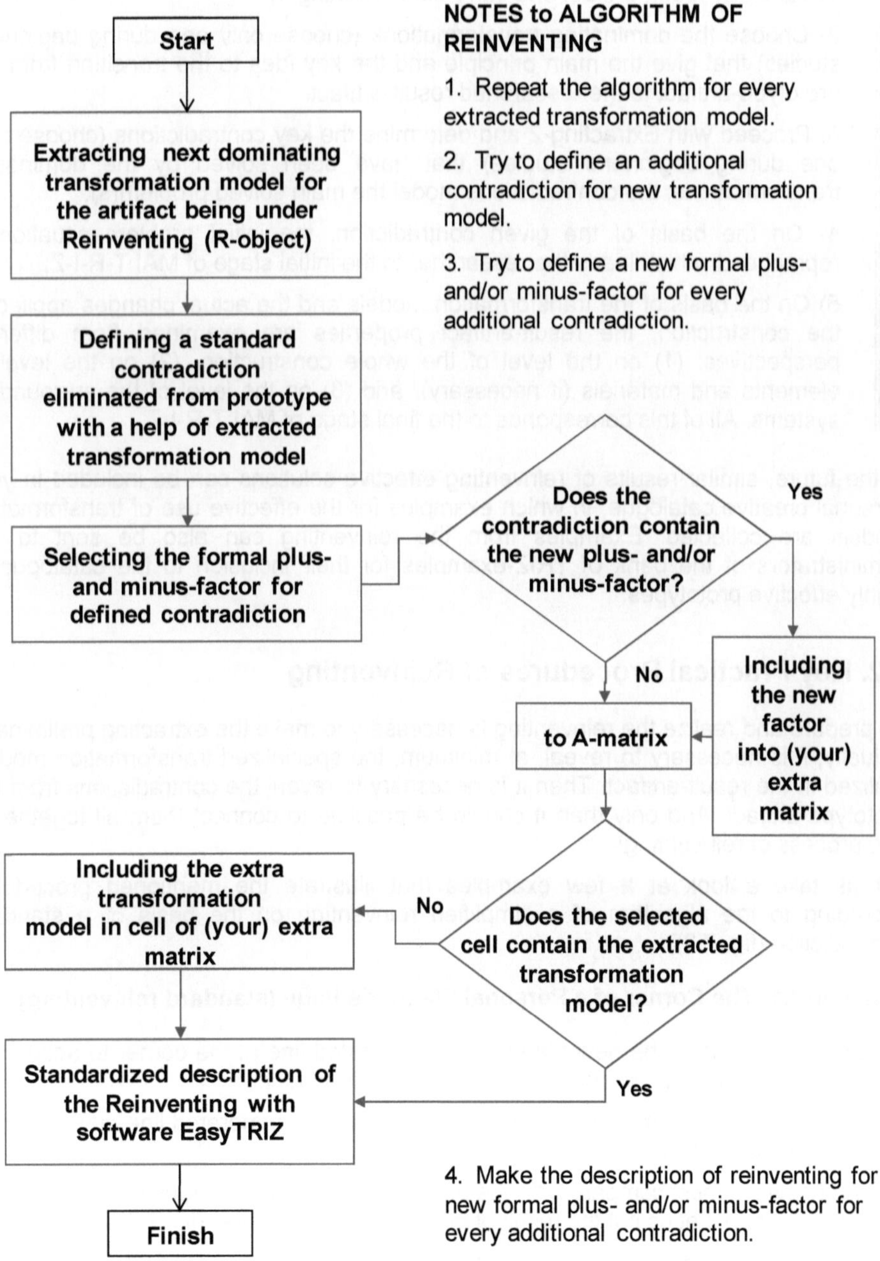

fig. 7.2. Algorithm of a simplified reinventing on the basis of a standard contradiction

The general description of the process of reinventing consists of the following:

1) Take any result-artifact and carry out Extracting-1.

2) Choose the dominating transformations (choose only one during beginner's studies) that give the main principle and the key idea to the transition from the prototype-artifact to the researched result-artifact.

3) Proceed with Extracting-2 and determine the key contradictions (choose only one during beginner's studies) that have been solved by the dominating transformations; the contradictions model the main solved problem(s).

4) On the basis of the given contradiction, the initial problem situation is reproduced, which leads the researcher to the initial stage of MAI T-R-I-Z;

5) On the basis of the transformation models and the actual changes applied to the construction, the result-artifact properties are examined from different perspectives: (1) on the level of the whole construction, (2) on the level of elements and materials (if necessary), and (3) on the level of the surrounding systems. All of this corresponds to the final stage of MAI T-R-I-Z.

In the future, similar results of reinventing effective solutions can be included in your personal creative catalogue, in which examples for the effective use of transformation models are collected. Examples from the reinventing can also be sent to the administrators of the bank of TRIZ-examples for their inclusion to the catalogue of highly effective prototypes.

7.2. Key Practical Procedures of Reinventing

To prepare and realize the reinventing is necessary to make the extracting preliminary! Actually, it is necessary to reveal, at minimum, the specialized transformation models realized in the result-artifact. Then it is necessary to reveal the contradictions from the prototype-artifact. And only then it should be possible to connect them all together in one process of reinventing!

Let us take a look at a few examples that illustrate the mentioned procedures according to the algorithm of a simplified reinventing on the basis of a standard contradiction (fig. 7.2).

Example 7.1. The Corner of a Personal Planner's Page (standard reinventing)

Imagine that we have the new planner with perforated line at the corner to accurately and quickly tear a small part of page. This is a result-artifact.

First, we must recall some similar planner without such a perforated line on the pages. It is a prototype-artifact.

OK! And now we must compare both the planners and extract the transformation models from the new one. Let us put the result of this operation into tables (fig. 7.3 and 7.4).

You can see that the navigator ***02 Preliminary action*** is the dominating model (marked with two pluses in fig. 7.3).

The standard contradiction in the table for Extracting-2 is resolved mainly with the fundamental *transformation in time*: ***preliminary making*** of a line of small holes that will facilitate tearing the corner of page.

The radical contradiction is resolved mainly with the fundamental *transformation in space*. We must understand that the preliminary ***perforated line*** changes the space of the pages making their shape different – not plain but with small holes aligned in a line. Thus, this space change in the form of line of small holes will facilitate tearing the corner of page.

Extracting-1			
LS	**No.**	**Navigator**	**Reason**
++	02	preliminary action	doing a part of the work in advance
+	12	local quality	it must be guaranteed that the tearing line runs exactly where it is supposed to and has the requested form
	16	partial or excess effect	it is possible that the page must be slightly mis-torn in some spots

fig. 7.3. Extracting-1 (in brief) for Planner

Extracting-2 for standard and radical contradictions			
LS	**No.**	**Navigator**	**Reason**
++	01	Separation in space	Radical Contradiction: to get a nice tearing line (with loss of time) **VS** do not get a nice line (with saving time)
++	02	Separation in time	Standard Contradiction: the corner must be torn off quickly **VS** the result is an uneven tearing line

fig. 7.4. Extracting-2 (in brief) for Planner

After extracting, we are able to incorporate all the results into special table for standard modeling the reinventing (fig. 7.5).

The results have been incorporated into the ***standard reinventing form*** for making reinventing **without software EASyTRIZ** at the computer.

It is possible to change a little bit this form in terms of the size of fields and in their composition on the page.

> The most important thing for "software-less" work (i.e., without software EASyTRIZ) is the possibility to put any necessary models, extracted from the result-artifact, into the fields for "navigators."

This result could be demonstrated at some meeting of a think-tank team, put into an archive of some project, or be used at certifying (e.g., distantly) it in the Academy of MTRIZ.

So, there are **TWO KEY PROCEDURES** for making a whole process of reinventing:

1) Extracting, and

2) Reinventing itself.

TREND

In order to replace a book mark the corners of the pages with the dates passed can be torn off. Then the page with the actual date is always visible and the organizer can easily be opened on the required page.

At the same time it is difficult to tear off the corner accurately and quickly. You have to take your time to get a good result and even if you do so, there is no guarantee for a neat outcome. What could be proposed for an improvement of the situation?

REDUCTION

FIM: X-resource, together with available or modified resources, and without producing any negative effects, guarantees attainment of the following **IFR:**

[*tear off the corner accurately and quickly* **].**

Standard Contradiction

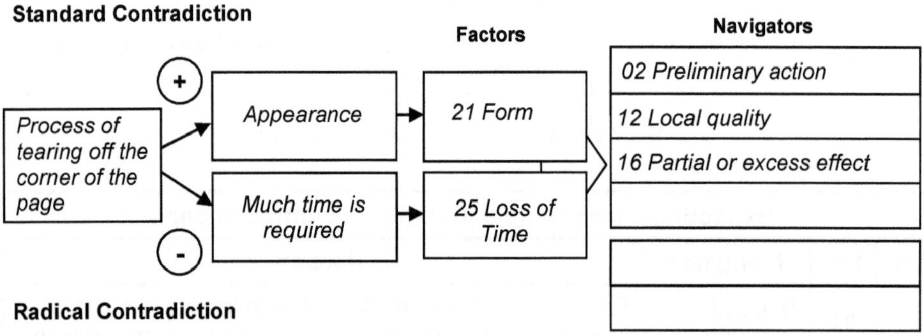

Radical Contradiction

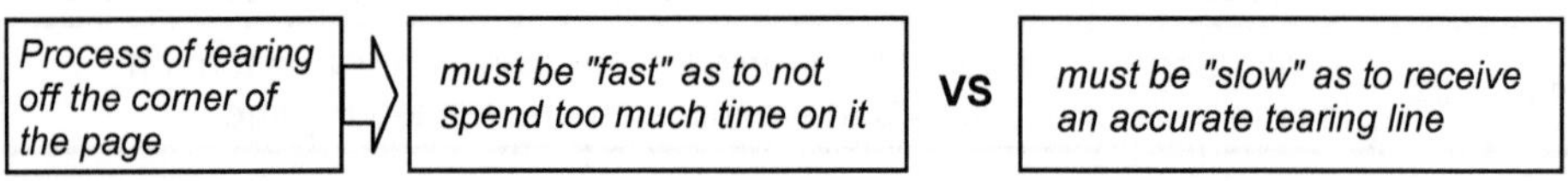

INVENTION

Dominating model: 02 Preliminary Action – *a) change the object in advance (completely or at least partly).* ***Key idea:*** *preliminary making a line with set of small holes that will facilitate tearing the corner of page.*
Auxiliary models: 12 Local quality and 16 Partial or excess effect – *making the holes.*

ZOOMING

Are the contradictions eliminated? - Yes. - ~~No.~~

Super Effects: -

Negative Effects: *Constraint on application of idea – the recommendation can only be realized by the manufacturers of the personal organizer.*

BRIEF DESCRIPTION

To guarantee an accurate appearance of the personal organizer after a quick tearing off of the corner of the page, a perforation of the corner is made in advance which provides easier tearing. The navigator ***02 Preliminary action*** *is dominant.*

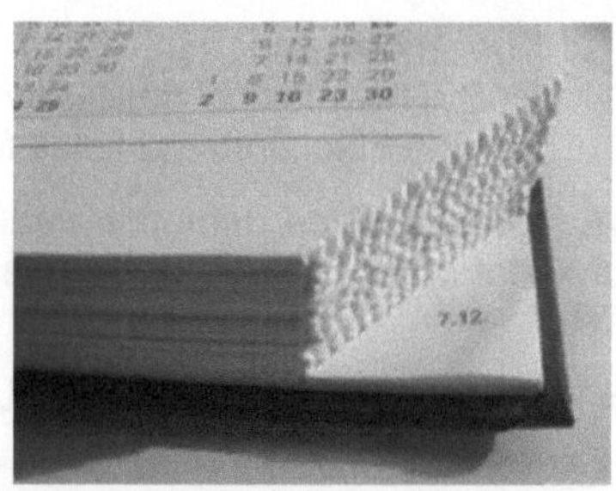

fig. 7.5. Reinventing for planner

7.3. Science and Art of Reinventing

There are many variants to make and record the reinventings. For example, it is possible to create a short version to demonstrate the operation with specialized transformation models without formulation of a radical contradiction and sometimes even a standard contradiction. Or, on the contrary, it is possible to introduce into the resulting description the **new models** suggested by you after your extracting in accordance with your experience **if these new models are absent in the A-catalogue**.

Example 7.2. "Inflatable Crowd" (a simplified and very short reinventing)

We omit the first phase extracting and immediately show the result of the second phase in the standard format of reinventing. We believe that all of the simulation details are completely clear here. We note only that there is no radical contradiction, and the standard contradiction is given in the form of formula. All this is possible in the "paper" format. We have used here the formal models from the A-matrix and the As-catalog.

TREND

To fill large spaces with people, movie-makers have to invite many extras, which is very challenging in administrative terms and very expensive. What could be done?

REDUCTION

FIM: X-resource, together with available or modified resources, and without producing any negative effects, guarantees attainment of the following **IFR:**

[*to get thousands of extras with easy control* **].**

Standard Contradiction

Initial informal version:

Thousands of extras ► big crowd with movable figures **VS** complexity of control

Formal version and "recommendation" from cell of A-matrix (solution of formula):

Thousands of extras ► **(+) 19** Volume of the moveable object **VS (-) 08** Complexity of inspection and measurements **= 10, 14, 24**

INVENTION

Dominating models: ***10 Copying*** and ***14 Use of pneumatic and hydraulic construction*** for ***key idea*** – *to use the inflatable dummies!* ***Auxiliary model: 24 Asymmetry*** – *combining dummies with actors and extras.*

ZOOMING

Are the contradictions eliminated? - Yes.

Negative Effects: *impossible taking close frame.*

BRIEF DESCRIPTION

To fill large spaces with people, movie-makers have invented use of inflatable dummies according to navigators ***10 Copying*** *and* ***14 Use of pneumatic and hydraulic construction****.*

fig. 7.6. Reinventing for "Inflatable Crowd"

If you did not find the examples above to be even remotely interesting, or did not think them particularly fascinating, will you please take a look at two more in figures 7.7 and 7.8—my last hope!

Try to extract contradictions and transformation models from both advertising posters, which I photographed on the streets of Berlin.

Example 7.3. I Wanna Go There!

There is a very simplified and very short reinventing!

I personally liked that this poster in fig. 7.7 (it was approximately 4 by 6 meters like the next one I saw in Minsk) because it immediately draws your attention to the "impossible contradiction" between the age of the young man and the way he is "comfortably" cuddled in the arms of a slim young woman, like a child in the arms of his mother! I guess he must be her husband. Also, you can see at once that he is throwing a tantrum like a child, pointing with his "grownup finger" at the address of the Vienna Motor Show and apparently demanding to be taken there.

You see how long it takes to describe in words what is caught in a blink of an eye in a good poster like this one? Still, you can easily continue studying its structure... There are many TRIZ models there.

Extract them!

And what a radical contradiction is there?

OK, what about this one?

Man ► should be adult (to drive a car) **VS** should be a child (to be on the hands of women)

fig. 7.7. "Where the Man Becomes a Child Again" (now THAT is a good slogan!)

Why is this picture paradoxical? Because it uses the metaphor of the stereotypic behavior of adult man acting like a child in reality by seeing the exhibition poster!

How could a designer depict this image and situation?

To use a fundamental transformation ***in space*** to place a smaller figure of an adult man in the arms of a women and ***in time*** to represent this man as a child mixing paradoxically two ages – that of a capricious child who could be in a woman's arm and that of an adult man being directed toward a poster about a son coming car exhibition!

Example 7.4. If You Wanna Be Healthy...

And in conclusion of this section, we would like to show you yet another poster – this one from a very serious "author", the Ministry of Health Care of Germany[53] (fig. 7.8). This poster extols healthy way of life and "agitates" for sport pursuits (in moderation, naturally).

This is like an answer to one of the "easy" problems formulated in the first pages of this book – "Problem P5. How do we convince ourselves to go in for physical culture?"

fig. 7.8. Sport: "Support well-being and good mood" with at least "30 minutes on the move—and better yet, 3 times a week"—is what they recommend instead of popping pills

So, you decide which of these posters (in fig. 1.6 and fig. 7.8) better motivates you to at least do your daily "morning jerks" and whether they have the same "humor charge." However, our "scientific" research is not over yet!

<u>Radical contradiction</u> Poster has ► to agitate (to bring the benefit)
VS not agitate (to bring a joy)

The solution is to combine both requests—**benefit AND joy**—in one picture with paradoxical components!

Here two fundamental models were applied:

1) transformation ***in space*** to place soccer balls (which of course are much larger in reality!) in the small cells of the pack of medicine pills, and

2) ***in structure*** to combine paradoxically just soccer balls, improbably reduced in size, and very small pill cells.

That's inventive advertising solution!

[53] Das Bundesministerium für Gesundheit der Bundesrepublik Deutschland

Example 7.5. Icicle Taming.

This solution for **Problem P2** was proposed by student Li Xiaotian from China. He was born and grew up in the Henan Province in central China, the seat of the famous Shaolin Monastery, the birthplace of that art of kung fu.

Li told us that in his childhood he saw this problem solved by his grandfather. In the village where he lived, many did this to prevent icicles from falling off the roofs. So he remembered his grandfather's solution and presented it in the form of an MAI reinventing.

TREND

In spring or during a thaw, rooftops often become adorned with ice growths – icicles. Large icicles are dangerous for passers-by as they may drop down. PROBLEM: what do you do to prevent icicles from falling off roofs.

REDUCING

FIM: X-resource, together with available or modified resources, and without making the object more complex or introducing any negative properties, guarantees attainment of the following **IFR**:

[*icicles are securely attached to the roof*].

Standard Contradiction

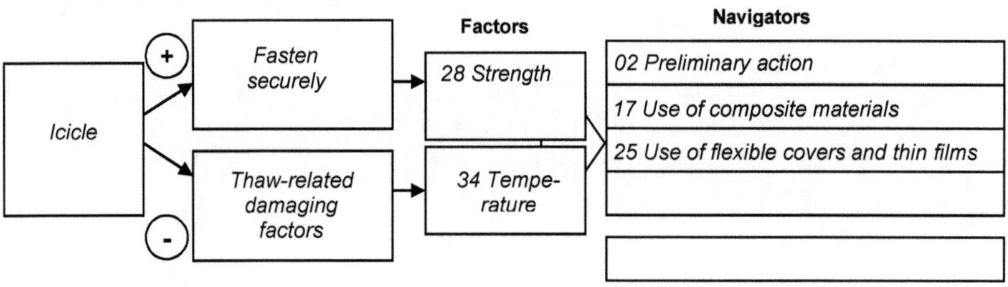

Radical Contradiction

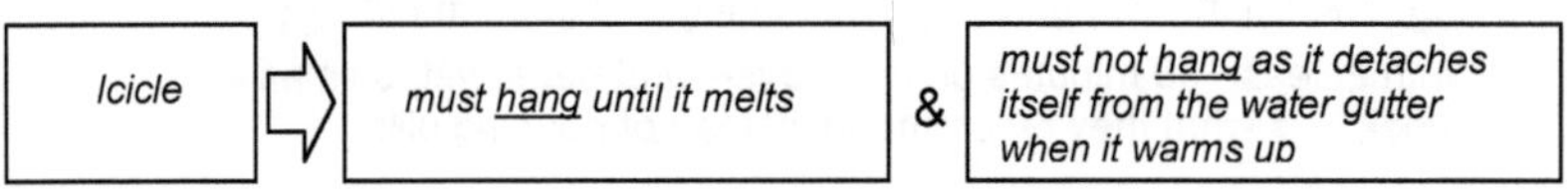

INVENTING

***Key idea – 17 Use of composite materials**: move from homogeneous materials to combinations.*

***Key "grandfather" idea:** a strong rope is tied to the water gutter from below. It freezes into the ice when an icicle is formed.*

Additional model 02 Preliminary action: fasten the rope in advance.

ZOOMING

Have the contradictions been removed? – Yes. – ~~No~~.

Super-effects: *none*

Negative effects: *none*

Figure

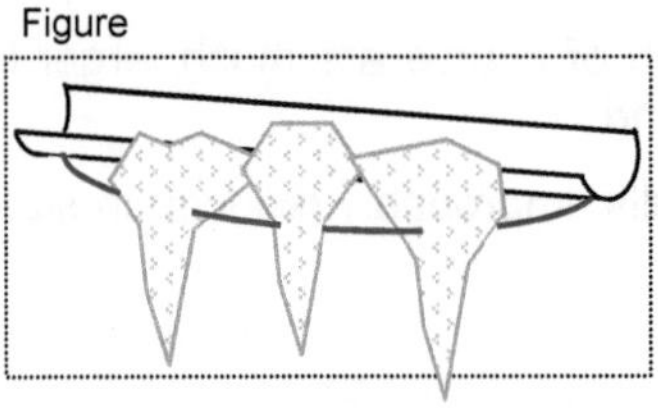

BRIEF DESCRIPTION

A strong rope is tied to the water gutter in advance; as a result, as icicles start to form there emerges a "composition" of ice and the rope frozen into the ice. Icicles are held fast virtually until they completely melt away. Dominating navigators: 17 Use of composite materials and 02 Preliminary action.

fig. 7.9

Example 7.6. Prove You Are Right.

This example was created by student Prerak Contractor from India. First he found this system at a gadget site called GIZMODO. He liked the system, and found its maker, an English company called TTB Industries.

Then it became clear that such systems can go a long way in enforcing universal compliance with traffic rules. At the workshop, participants formulated the following simple and efficient idea: All cars should be fitted with onboard "black boxes" to record major motion parameters and make panoramic environmental videos.

TREND

It often happens that the innocent party in a car accident cannot prove that he or she is right because there are no witnesses. Even post-accident photographs are not always helpful in establishing what exactly happened prior to the accident. What can you do?

REDUCING

FIM: X-resource, together with available or modified resources, and without making the object more complex or introducing any negative properties, guarantees attainment of the following **IFR**:

[*reliable proof of what really happened*].

Standard Contradiction

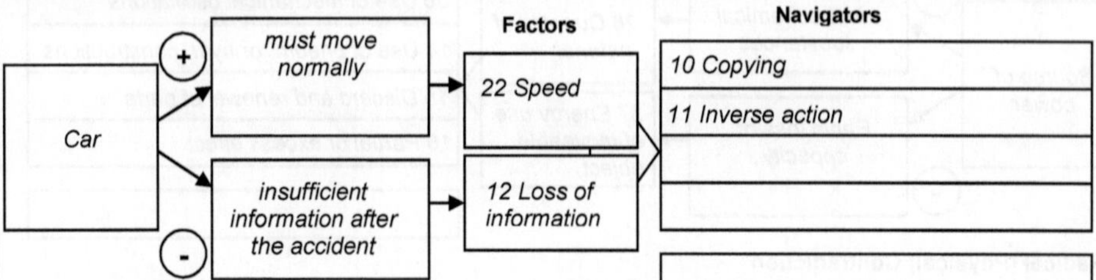

Radical Contradiction

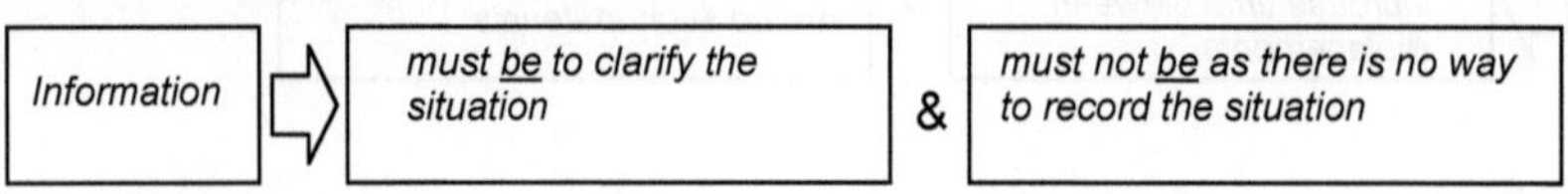

INVENTING

Key model: 10 Copying*: commence uninterrupted recording of the immediate environment in the course of movement.*

Additional model 11 Inverse action: when viewing the recording, it is possible to "go back in time" to see important events as they occurred.

ZOOMING

Have the contradictions been removed? – Yes. – ~~No~~.

Super-effects: *together with acceleration sensors, we receive a "black box" for the car – which is even better, as it may promote universal compliance with traffic rules among the drivers.*

Negative effects: *none*

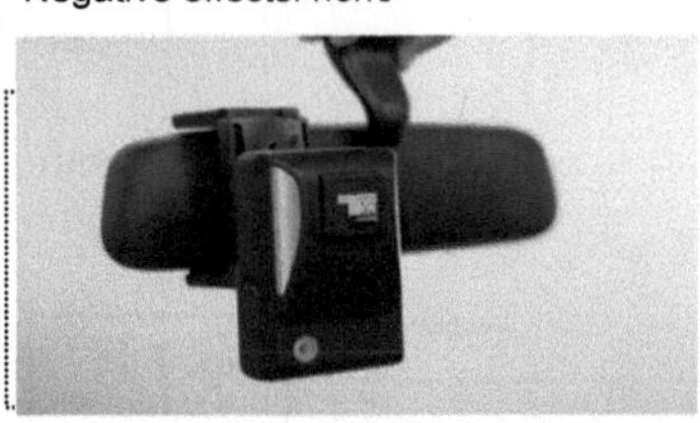

BRIEF DESCRIPTION

According to navigator 10 Copying, it is recommended to conduct uninterrupted recording of the immediate environment of the car PLUS diverse motion information, such as speed, braking, coordinates, etc.

Such recording is made with the help of the T-EYE system manufactured by the British company TTB Industries.

fig. 7.10

Example 7.7. Accumulators Instead of Batteries.

This example was designed by student Carolina Olave from Columbia.

She used reinventing to illustrate the transition from the first nonrenewable electrochemical sources of electric power (batteries) to those that are renewable and rechargeable. (accumulators). She also believes that in terms of TRIZ modeling this transition is closer to model "15 Discard and renewal of parts".

TREND

The usual batteries for portable devices (players, cameras, mobile phones, etc.) are relatively inexpensive, but after they have run down you have to throw them away. Then you need to buy new batteries – and with time the costs start to run high. What can you do?

REDUCING

FIM: X-resource, together with available or modified resources, and without making the object more complex or introducing any negative properties, guarantees attainment of the following **IFR**:

[*long useful life of a source of electric power*].

Standard (Technical) Contradiction

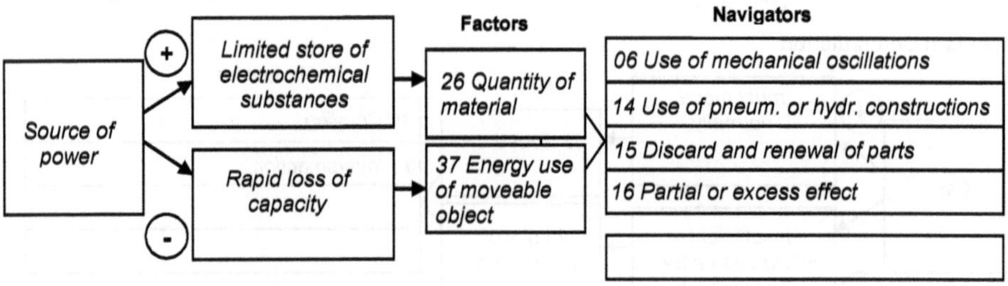

Radical (Physical) Contradiction

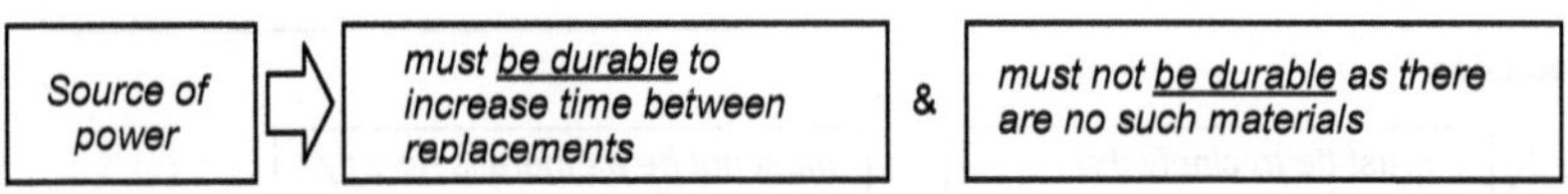

INVENTING

***Key model: 15 Discard and renewal of parts**: b) used parts of an object should be immediately replaced during work.*

Idea for a solution: replacement of batteries with accumulators by changing the electrochemical medium (material) in the source of power.

ZOOMING

Have the contradictions been removed? – Yes. – ~~No~~.

Super-effects: *reduced negative environmental impact because the number of sources of power being recycled decreases.*

Negative effects: *none*

BRIEF DESCRIPTION

According to navigator 15 Discard and renewal of parts, it was suggested that one-time electrochemical sources of electric power be replaced with those that can be recharged – accumulators. Lithium-ion accumulators have the highest efficiency ratio among modern accumulators for portable devices.

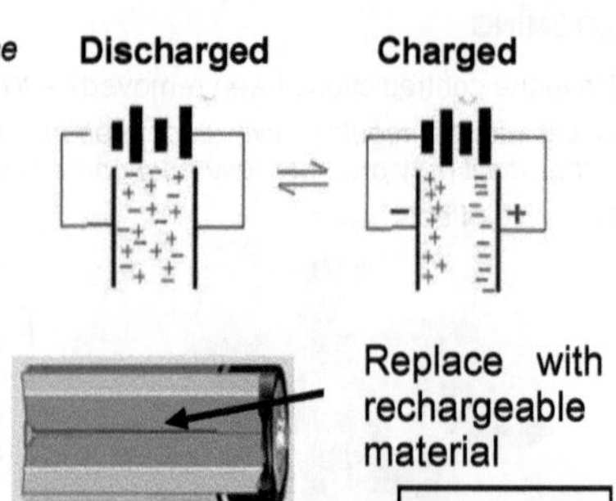

fig. 7.11

Example 7.8. Flexible Carving Board.

This example was developed by student Chien-Chih Tseng from Taiwan. She found it in a German magazine "Die moderne Hausfrau" ("The Modern Housewife").

All she had to do to turn that artifact into a teaching aid was to formulate a reinventing example showing how such board could have been invented if its makers had known TRIZ.

TREND

When a housewife chops vegetables, potatoes or condiments, it is sometimes difficult to take them to the frying pan or casserole without losing some on the way. PROBLEM: what do you do to prevent chopped food from falling off the carving board?

REDUCING

FIM: X-resource, together with available or modified resources, and without making the object more complex or introducing any negative properties, guarantees attainment of the following **IFR**:

[*chopped food cannot fall off the carving board*].

Standard Contradiction

Carving board
(+) *Food must not fall off in transit* → Factors: *27 Loss of material*
(−) *Small storage area* → Factors: *17 Surface of moveable object*

Navigators
- *01 Change in the aggregate state*
- *02 Preliminary action*
- *05 Separation*
- *31 Use of porous materials*

Radical Contradiction

Board ⇒ *must be big to prevent food from falling off* & *must be small to take less space in storage*

INVENTING

Key model: 01 Change in the aggregate state*:*
b) changes in concentration or consistency, in the degree of flexibility, in temperature, etc.

Additional model 05 Separation: add a new function to the board – "hold chopped food".

Key idea: *make the board from a flexible material.*

ZOOMING

Have the contradictions been removed? – Yes. – ~~No~~.

Super-effects: *none*

Negative effects: *none*

BRIEF DESCRIPTION

According to models 01 Change in the aggregate state and 05 Separation, the carving board must be made of a flexible material so that it can be rolled into a trough to carry chopped food products.

fig. 7.12

Example 7.9. iPhone in a Winter Glove.

This example was made by student Adehi Guehika from Canada. He was born in the African country Côte d'Ivoire (Ivory Coast).

He proposed to integrate a mobile telephone (in this case an iPhone manufactured by Apple) into winter gloves. He even drew a sketch of such a glove. Apparently, in Canada he found that topic to be of special personal significance!

TREND

In heavy frost it is difficult to use your mobile phone without gloves because your fingers freeze, while using the phone when wearing gloves is extremely inconvenient, if indeed possible. PROBLEM: how do you use your mobile phone in frosty weather?

REDUCING

FIM: X-resource, together with available or modified resources, and without making the object more complex or introducing any negative properties, guarantees attainment of the following **IFR**:

[*it is convenient to use mobile phone in frosty weather*].

Standard Contradiction

Mobile phone

(+) *Use in frosty weather* → Factors: *10 Ease of use*

(-) *Without gloves – it is cold; in gloves – it is inconvenient* → *34 Temperature*

Navigators

10 Copying

11 Inverse action

13 Inexpensive short-life object as...

Radical Contradiction

Mobile phone	⇨	*has to be used <u>without gloves</u> to dial numbers and hold the phone*	&	*has to be used <u>in gloves</u> not to freeze fingers off*

INVENTING

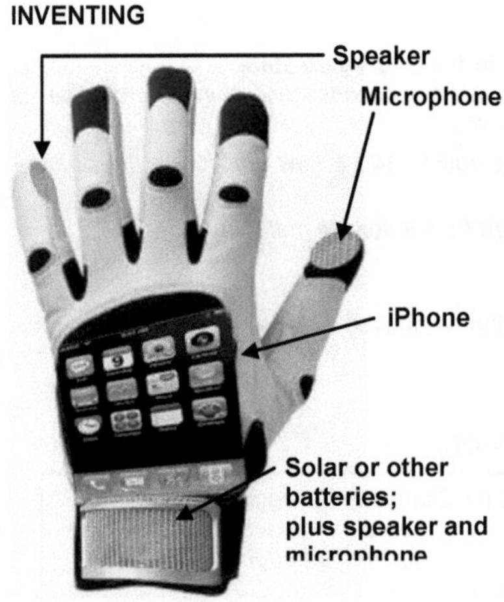

***Key idea 10 Copying**: a) use a ... copy instead of an inaccessible ... object.*

Additional model 11 Inverse action: instead of an action prescribed by the conditions of an assignment (operate without gloves), complete a reverse action (operate in gloves).

***Key idea:** integrate the phone into gloves.*

ZOOMING

Have the contradictions been removed? – Yes. – ~~No~~.

Super-effects: *multiple uses are possible.*

Negative effects: *none*

BRIEF DESCRIPTION

According to models 10 Copying and 11 Inverse action student Adehi Guehika proposed to integrate a mobile phone and winter gloves.

fig. 7.13

Example 7.10. "White board" instead "Black board".

This example was proposed by student Sara Hassan from Iraq. This work is very interesting from a methodological perspective because it illustrates the presence of radical creative ideas in familiar artifacts around us.

TREND

In ordinary classrooms the teacher uses chalk to write or draw on a usual (black) board. This takes a lot of time, and the quality of drawings is often inferior. PROBLEM: what do you do to improve the quality and productivity of the work performed by the teacher?

REDUCING

FIM: X-resource, together with available or modified resources, and without making the object more complex or introducing any negative properties, guarantees attainment of the following **IFR**:

[*superior quality of illustrations used during lessons and lectures*].

Standard Contradiction

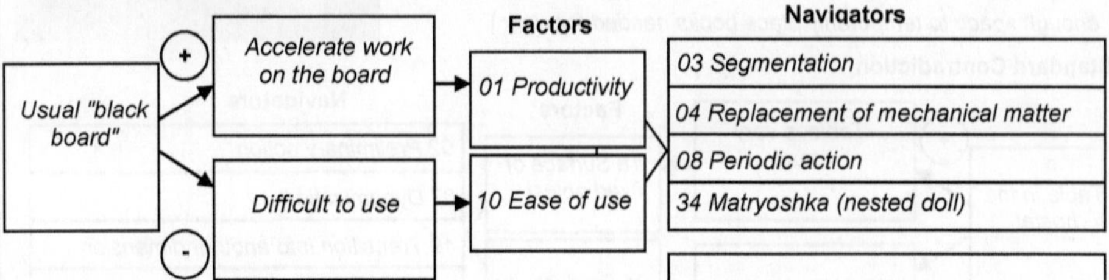

Radical Contradiction

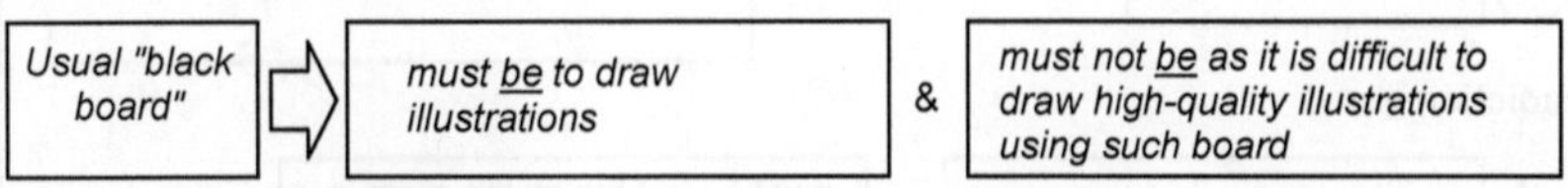

INVENTING

Key idea 04 Replacement of mechanical matter*: a) replace mechanical schemes with optical, acoustic, or olfactory schemes; b) use electrical, magnetic, or electromagnetic fields to interact with objects; c) replace static fields with dynamic fields, fixed fields with fields changing in time, unstructured fields with fields with a specific structure;*

Key idea: *1) screens and projectors; 2) electronic "white boards"; 3) computer classes.*

Model 34 Matryoshka (nested doll) is also used: it is possible to "insert" images into the "board" without limitations.

ZOOMING

Have the contradictions been removed? – Yes. – ~~No~~.

Super-effects: *multiple uses possible.*

Negative effects: *none*

BRIEF DESCRIPTION

According to model 04 Replacement of mechanical matter, classrooms are equipped with screens and projectors, electronic "white boards" and computers.

fig. 7.14

Example 7.11. "Table top" in "Drawer in wardrobe".

This example was designed by student Ayeh Hosseini from Iran. She came up with this efficient solution when she found herself in a cramped student hostel. Later she turned it into a reinventing example with her inimitable humor. To get the "table," she measured the dimensions of the drawer, had the panel cut out for her in a construction materials shop, and then easily and precisely inserted the "table top" into her wardrobe drawer.

TREND

Once a student needed more space to place her books, notebook and other things she needed for work. The room in the student hostel had no large table. PROBLEM: what can be done in such situation?

REDUCING

FIM: X-resource, together with available or modified resources, and without making the object more complex or introducing any negative properties, guarantees attainment of the following **IFR**:

[*enough space to temporarily place books needed for work*].

Standard Contradiction

			Factors	Navigators
Table in the hostel	(+)	*Table is very small*	*18 Surface of fixed object*	*02 Preliminary action*
				07 Dynamization
	(-)	*Work slows down*	*01 Productivity*	*19 Transition into another dimension*
				34 Matryoshka (nested doll)

Radical Contradiction

Table in the hostel ⇒	*must* <u>*be big*</u> *to make work comfortable*	&	*must* <u>*be not big*</u> *as the room is small, and there is no space for a large table*

INVENTING

Key idea 19 Transition into another dimension: *a) an object is shaped so that it can move or is placed not only in a linear fashion, but also in two dimensions, meaning on a surface. It is also possible to improve the transition from a surface to a three-dimensional space; b) do construction on several floors; tip or turn the object on its side; use the back of the space in question.*

Second key model 34 Matryoshka (nested doll): *a) an object is inside another object that is also inside another, etc.; b) an object runs through a hollow space in another object.*

Key idea: *make an additional temporary table top ... in a wardrobe drawer!*

ZOOMING

Have the contradictions been removed? – Yes. – ~~No~~.

Super-effects: *a lesson in "survival" and enterprise!*

Negative effects: *none*

BRIEF DESCRIPTION *According to models 19 Transition into another dimension and 34 Matryoshka (nested doll), the temporary table top is made in a wardrobe drawer.*

fig. 7.15

Example 7.12. Useful Soap Bubbles.

This example was made by student Leonid Nasledskov from Russia. He used numerous examples from a wonderful web site www.membrana.ru as prototypes and artifacts for his certification work and then went back to the site several times to verify his data.

TREND

Advertising agencies often install huge roadside advertising boards featuring corporate logotypes, but drivers cannot see them clearly because they move too fast. PROBLEM: how do you make logotypes visible for many people and from large distances?

REDUCING

FIM: X-resource, together with available or modified resources, and without making the object more complex or introducing any negative properties, guarantees attainment of the following **IFR**:

[*logotype is visible from a large distance*].

Standard Contradiction

Logotype

(+) *must be visible from a large distance* → *16 Length of fixed object*

(-) *difficult to make many large boards* → *01 Productivity*

Factors | **Navigators**

10 Copying

22 Spherical-shape

25 Use of flexible covers and thin films

34 Matryoshka (nested doll)

Radical Contradiction

Logotype ⇨ *must be visible from a large distance to maximize impact on the potential* & *must not be visible from a large distance as it is difficult to make large advertising boards*

INVENTING

If the authors of the invention knew TRIZ, the winning idea might have been suggested to them by the following "bouquet" of properties:

22 Spherical-shape and 25 Use of flexible covers and thin films *– how is this different from a soap bubble?*

Also: 10 Copying and 34 Matryoshka (nested doll)*: many soap bubbles placed inside a vessel can ... become a logotype soaring high in the sky!*

ZOOMING

Have the contradictions been removed? – Yes. – ~~No~~.

Super-effects: *a lesson in "survival" and enterprise!*

Negative effects: *none*

BRIEF DESCRIPTION

Inventors working for the company "Snowmasters", Alabama, US, designed a machine making soap bubble figures filled with helium. Dominating navigators: 10 Copying and 25 Use of flexible covers and thin films.

Illustrations: http://www.flogos.net/events.htm; http://www.membrana.ru/articles/business/2008/04/18/151600.html

fig. 7.16

Example 7.13. Everyone Is Happy – Kid, Mom, Pop!

This example was developed by student Khairul Kamarudin from Malaysia. She is an industrial designer.

This reinventing is one of her certification works.

TREND

A simple "hands-free baby holder" (see picture) is well known. However, the holder shown here cannot be resized, which makes it somewhat inconvenient.

In theory, it is possible to use buckled straps, but this introduces sharp elements – which may lead to trouble.

PROBLEM: how do we make the baby holder more convenient?

REDUCING

FIM: X-resource, together with available or modified resources, and without making the object more complex or introducing any negative properties, guarantees attainment of the following **IFR**:

[*resizable hands-free baby holder*].

Standard (Technical) Contradiction

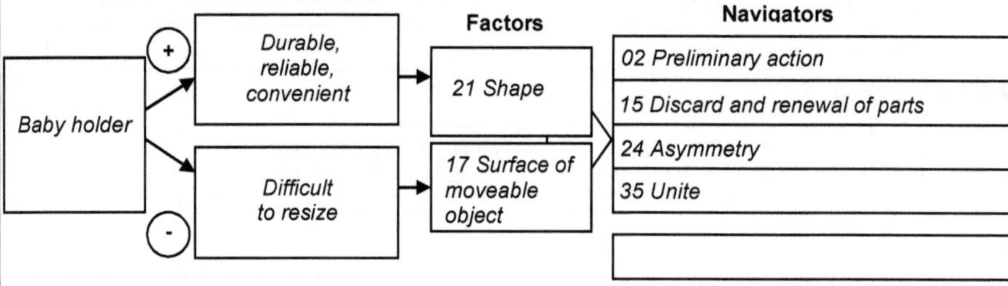

Radical (Physical) Contradiction

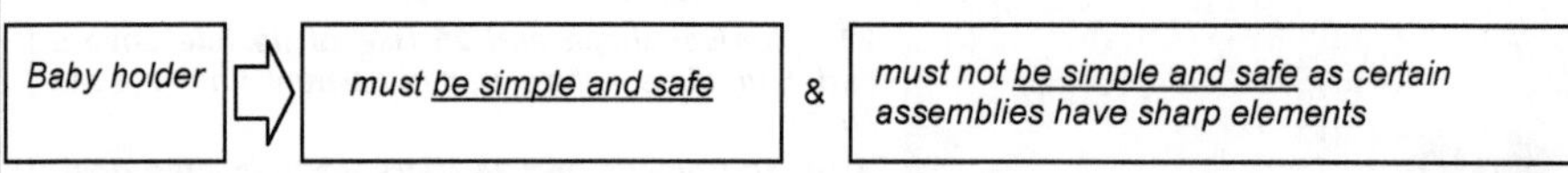

INVENTING

Extracting helped to identify the following transformation models offering suitable contradiction resolution options:

24 Asymmetry: *a) move from a symmetrical shape of an object to an asymmetrical one; b) increase the degree of asymmetry if the object is already asymmetrical;*

35 Unite*: a) unite similar objects or objects for neighboring operations.*

Solution: make a "holder" which has no sharp elements or straps!

This can be a special cloth cape held by two rings threaded through each other and kept together by the cloth which is stretched by the baby's weight!

fig. 7.17,a

Examples of "holders" which can be used in different situations and for different applications are shown below.

The key element here is a two-ring clasp – it takes some time to learn to handle it right…

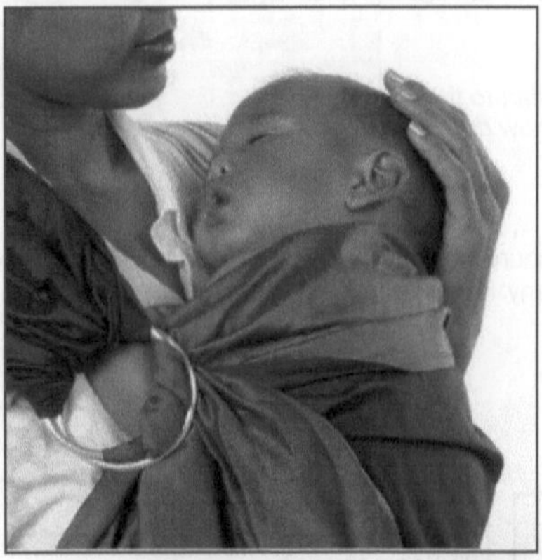

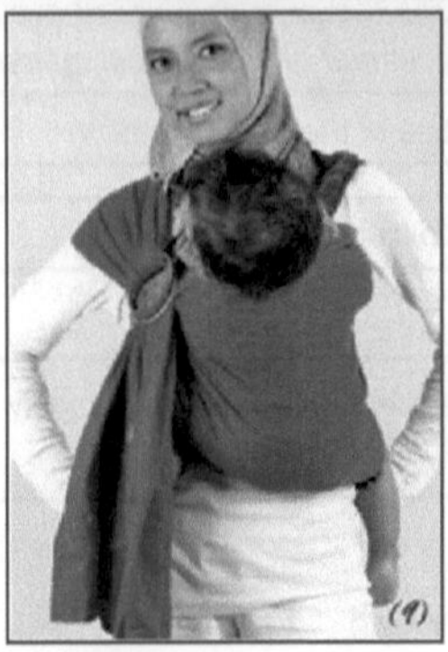

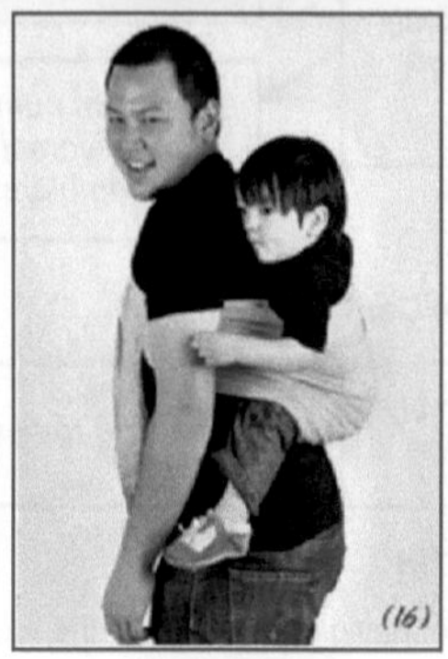

ZOOMING

Have the contradictions been removed? – Yes. – ~~No~~.

Super-effects:

1) lower row, center – the cape is used to breast-feed the child;

2) a lesson in inventiveness where it seemed everything had already been invented!

Negative effects: *none*

BRIEF DESCRIPTION

The solution is based on the use of a special construction of two rings inserted one into the other and held together by the stretched cloth which basically acts as the "baby-holding agent". Transformation models used: 24 Asymmetry and 35 Unite.

fig. 7.17,b

Such examples are an efficient means of teaching beginners as well as an efficient means of supporting the work aimed creating new solutions, particularly when there is an ongoing exchange of solutions to real-life problems.

Example 7.14. Invention of Conveyor.

This example was designed by student Oktay Tabak from Turkey. He was particularly interested in the invention of highly efficient managerial solutions: conveyorization, standardization, hierarchical management structures, etc. This is one of his works.

TREND

At first Ford plants automobiles were assembled from the start to the finish at the same work place called "station". PROBLEM: how do you accelerate car assembly to enable serial production?

Individual universal assembly stations

REDUCING

FIM: X-resource, together with available or modified resources, and without making the object more complex or introducing any negative properties, guarantees attainment of the following **IFR**:

[*fast assembly of numerous cars (serial production)*].

Standard Contradiction

Assembly station

(+) *Accelerate assembly of numerous cars* → *01 Productivity*

(-) *Insufficient number of universal assemblers* → *02 Universality, adaptability*

Factors: *01 Productivity*; *02 Universality, adaptability*

Navigators:
- *01 Change in the aggregate state*
- *03 Segmentation*
- *04 Replacement of mechanical matter*
- *27 Full use of thermal expansion*
- *08 Periodic action*

Radical Contradiction

Assembly station ⇒ *must be to assemble cars* & *must not be as there are not enough highly qualified universal assembly workers*

INVENTING

Group of navigators to model the idea for the conveyor invented by Henry Ford:

03 Segmentation: *a) disassemble an object into individual parts; c) raise the degree of disassembly (reduction into parts) of an object – divide the technological process into groups of operations so that they have the same duration;*

04 Replacement of mechanical matter*: c) replace static fields with dynamic fields, fixed fields with fields changing in time, unstructured fields with fields with a specific structure – move the cars, not workers, from station to station; cars can be moved by a conveyor or move by themselves.*

08 Periodic action *(added by the developer of the example): a) transition from a continuous function to a periodic one (impulse) – each station performs specialized operations over the duration of the conveyor's operating cycle (period).*

ZOOMING

Have the contradictions been removed? – Yes. – ~~No~~.

Super-effects: *the way to mass technologies!*

Negative effects: *monotonous work at specialized stations.*

BRIEF DESCRIPTION

In 1908 Henry Ford launched the world's first conveyor to assemble the famous T-model. Over the next 20 years the company manufactured 15 million cars of this model – only in black color! In 1924 those cars accounted for 50% of all cars in the world. From the conveyor idea, one can easily extract models 03 Segmentation, 04 Replacement of mechanical matter and 08 Periodic action.

fig. 7.18

Example 7.15. Breakfast of the Clever *Common Blackbird.*

This year (2015), on the 4th of July, we drank morning coffee with my dear friend, Heinrich Kochs, at his house near Hannover. A small bird flew to the table and began to twirl near it (1).

I photographed this restless creature and drew attention of all those present to her behavior. It seemed as if she wants to say or show something to us.

Heinrich looked (2) at this bird and said that it's a thrush-girl, in German *Amsel weiblich* (in English *Common Blackbird*). The thrush-girl has a more brown color than thrush-boy. But what Heinrich said and did next was like a morning fairy tale!

He said that the bird was asking him to turn on the water for watering the grass in the area and that the bird always does the same thing when she wants to... eat! We were surprised, but my friend moved the sprinkler and turned on the water (3). The bird immediately began to run on the wet ground, lowering his beak to the ground.

And indeed (4), she soon had her "breakfast" of rain-worms!

That is, the bird sees similarities with the situation when there is rain, i.e., the rain-worms crawl out of the soil on the ground during and after the rain. Clearly, the bird noticed that the rain-worms appear after the watering! And she found that exactly this tall man initiates the "rain" here!

And now she comes flying and asks him to turn on the water so that she can eat!!

What else is there to say?! So, does Nature think?

ASSIGNMENT. I ask the reader to perform the extracting and reinventing that this bird invented as a "technology of procuring food" for herself with a help of the rain, which is turned with a help of this "tall God."

Example 7.16. Humpback Whale Invents the "Air Fishing Net".

This line is created by a humpback whale. He swam below and aside from the school of fish and swims around this school. He produces especially small air bubbles that do not dissolve quickly in water. The bubbles rise up as a loud and white “wall”!

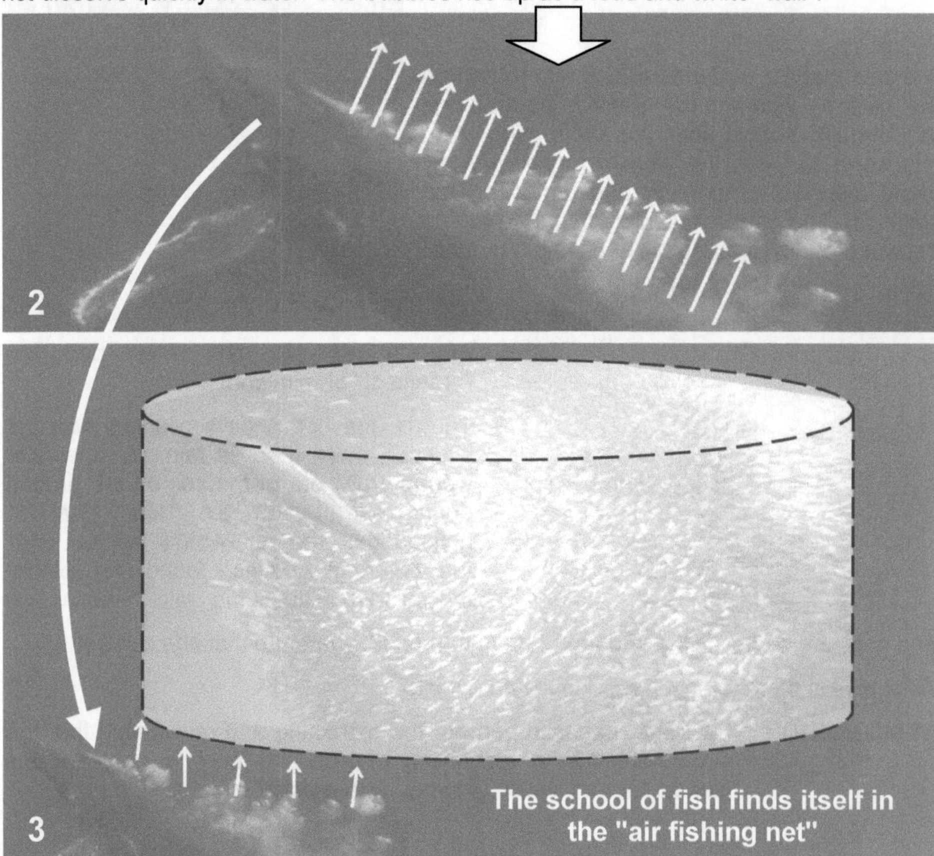

The fish try to escape upward!

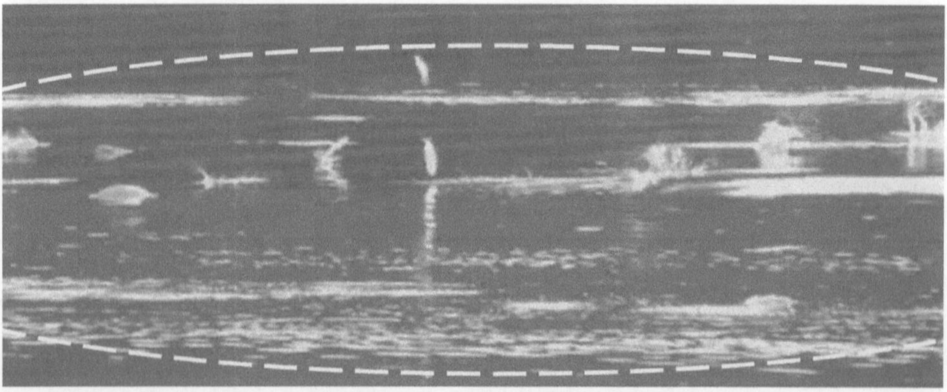

Then the companions of the first whale fly up to the ocean surface through the school of fish with open mouths to scoop them up! The whales feast!

Then another whale "works" as the “fish-beater” so his companions can feast.

It is incredibly surprising that the negotiation for such a hunt has realized by the whales from different families that met by chance in this place of the ocean! **How do they do it?**

They do understand that the noisy and colorful bubbles scare fish and bunch them it into the “air cylinder.” They understand that they must swim underneath and close the ring in order that the rising bubbles could form a cylinder, inside which will be a school of fish. And then the companions can feast on this school of fish.

How is it even possible? **There is no doubt—They think!**

ASSIGNMENT: I recommend the reader to independently make the extracting and reinventing for this unbelievable example!

You can compare your modeling for this and previous examples with the masters at www.modern-triz-academy.com.

Example 7.17. Answer to the task P4 "Magic Faucet"

Naturally, this assignment must be treated as a kind of joke. We will refrain from demonstrating the "real" invention—with or without TRIZ. But we will perform the familiar reinventing procedure according to TRIZ.

For that purpose, we will assume that the operating principle of this "magic faucet" is known and that required models and contradictions have already been obtained by extracting.

TREND

Here is the assignment: a "water jet" is topped by a huge faucet producing a non-stop water flow, and there is no feeding pipe – the faucet apparently hangs in midair! We know from TRIZ that if a phenomenon exists, all requisite resources are available! QUESTION: how such construction could have been invented using TRIZ methods?

REDUCING

FIM: X-resource, together with available or modified resources, and without making the object more complex or introducing any negative properties, guarantees attainment of the following **IFR**:

[*water is continuously fed to the faucet to obtain a non-stop water jet*].

Standard Contradiction

Water jet

(+) *A lot of water* → *19 Volume of the moveable object*

(-) *Large height* → *15 Length of the moveable object*

01 Change in the aggregate state
03 Segmentation
04 Replacement of mechanical matter
34 Matryoshka (nested doll)

Radical Contradiction

Water jet ⇒ *must be because… it is already there!* & *must not be because the feeding pipe is not there! (not visible!)*

INVENTING

Key model – 34 Matryoshka (nested doll): *the general idea is simple – the pipe is hidden inside the water jet!*

ZOOMING

Have the contradictions been removed? – Yes. – ~~No~~.

Super-effects: *an excellent aesthetically pleasing attraction!* Negative effects: *none*

BRIEF DESCRIPTION

The faucet rests on a solid glass pipe. The water passes through the pipe (inside it, naturally!) and gets to the faucet. Then it flows out of the faucet and falls down on the outer side of the pipe so that the air jet "hides" the pipe inside it. Principle implemented: 34 Matryoshka (nested doll).

fig. 7.19

Send us the reinventings of similarly curious objects!

Example 7.18. Reinventing for Examples 5.2, 5.3 and 6.3. Ice for a Drink.

TREND

To keep the drink (juice, cocktail, etc.) cool for a longer period of time, you can add pieces of ice. However, with the ice melting the taste of the drink is changing, as the relative content of water in the dwindling amount of the drink is increasing. How do you prepare an iced drink so that its taste remains unchanged in the course of consumption?

REDUCING

FIM: X-resource, without producing the inadmissible negative effects, provides together with other existing resources obtaining **IFR**:

[***taste of the drink, which has been cooled (with ice), must remain unchanged***].

Standard Contradiction

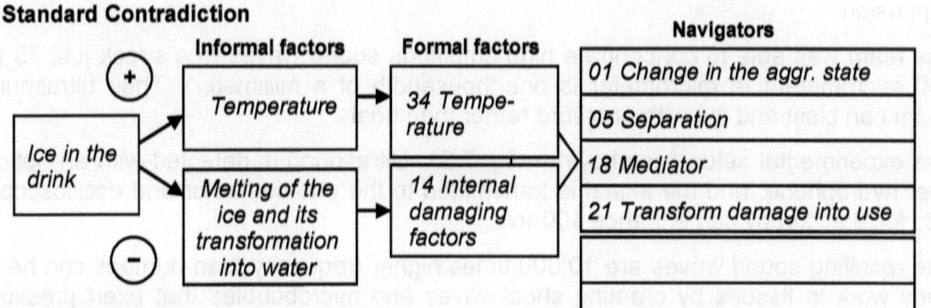

Radical Contradiction

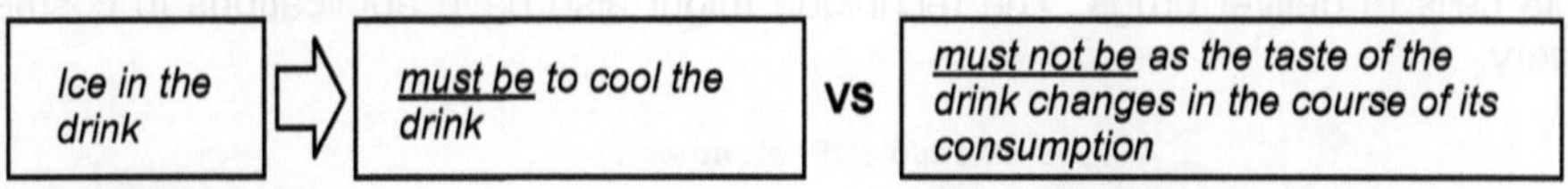

INVENTING

***Key model: 18 Mediator** – a) Use another object to transfer or transmit an action.*

***Key idea:** encapsulate the ice – the drink becomes cooler, and no water gets into the drink.*

Additional solution option based on the Radical Contradiction – on the (material) segmentation navigator and specialized navigator 38 Homogeneity: interacting objects should be made from the same materials – ice figurines should be made from the same drink (which is possible not for all drinks).

ZOOMING

Have the contradictions been removed? – Yes. – ~~No~~.

Super-effects: *it becomes possible to decorate the drink with nicely shaped figurines.*

Negative effects: -

BRIEF DESCRIPTION

Use hermetically encapsulated ice to cool drinks according to navigator 18 Mediator. Ice figurines can be made from the same drink in line with model 38 Homogeneity.

fig. 7.20

Example 7.19. Special very important case of finding "by your extracting" the NEW TRANSFORMATION MODEL

A carbon nanotube–coated (CNT) lens[54] that converts light to sound can focus high-pressure sound waves to finer points than ever before. The University of Michigan (UM) engineering researchers, who developed the new therapeutic ultrasound approach, say it could lead to an invisible knife for noninvasive surgery.

The UM researchers' system is unique because it performs three functions: it converts light to sound; it focuses to a tiny spot; and it amplifies the sound waves. To achieve the amplification, the researchers coated their lens with a layer of carbon nanotubes and a layer of a rubbery material called "polydimethylsiloxane." The carbon nanotube layer absorbs the light and generates heat from it. Then the rubbery layer, which expands when exposed to heat, drastically boosts the signal by the rapid thermal expansion.

The team was able to concentrate high-amplitude sound waves to a speck just 75 by 400 micrometers (a micrometer is one-thousandth of a millimeter). Their ultrasound beam can blast and cut with pressure rather than heat.

The experimental setup[55] is shown in fig. 7.21. Ultrasound is detected with an optical fiber hydrophone, and the signal is transmitted to the photodetector and oscilloscope. The focal accuracy can enhance 100-fold.

The resulting sound waves are 10,000 times higher frequency than humans can hear. They work in tissues by creating shockwaves and microbubbles that exert pressure toward the target, which could be tiny cancerous tumors, artery-clogging plaques, or single cells to deliver drugs. The technique might also have applications in cosmetic surgery.

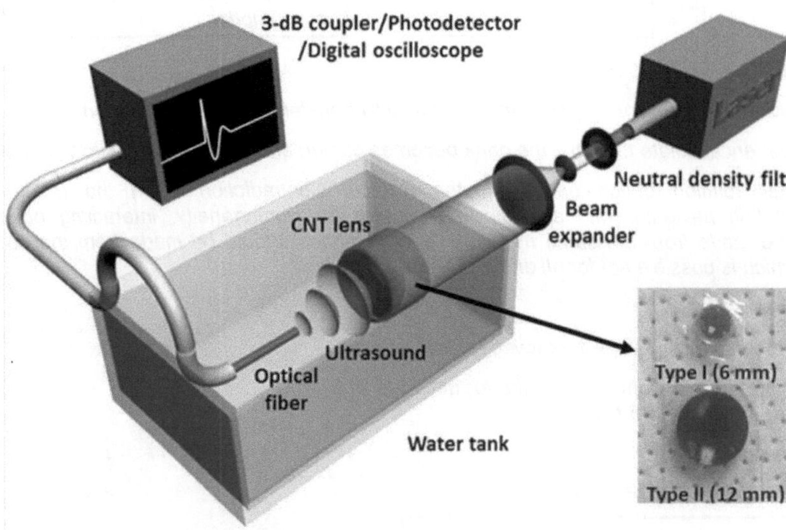

fig. 7.21. Experimental setup for testing the laser-generated focused ultrasound (LGFU)

---

54 www.ns.umich.edu/new/releases/21044-super-fine-sound-beam-could-one-day-be-an-invisible-scalpel

55 www.nature.com/srep/2012/121218/srep00989/full/srep00989.html

EXTRACTING

The result of the extracting with three dominant models is shown in the table (fig. 7.22).

Extracting-1			
LS	**No.**	**Navigator**	**Reason**
++	-	**FOCUSING**	Concentrating high-amplitude sound waves to a speck just 75 by 400 micrometers is necessary to enhance the focal accuracy 100-fold
++	18	mediator	Transforming the laser light into ultrasound is necessary to use the optic–acoustic lens
++	17	use of composite materials	Increasing the efficiency of the lens with achieving amplification is necessary to cover the lens with two special layers of carbon nanotubes and polydimethylsiloxane

fig. 7.22. Extracting-1 (in brief) for LGFU

ATTENTION. No one classical model can adequately represent the main transformation of this experiment and device of this achievement to increase the pressure and to decrease a size of sound speck! The key model should be named as "focusing."

Let us give an experimental definition to the model with conditional number 41:

41. focusing	a) concentration of action, effect, or influence on some object or several objects selected; b) concentration of influencing energy on an area of small (reduced, scaled-down) size instead of the usual influence of this energy on an area of large size.

REINVENTING

It is possible in such a case to add a "new" model in an extra field of the reinventing form and give an explanation of your idea to suggest the new model (fig. 7.23).

Focusing

INVENTION

Dominating model – suggestion: Focusing *– with preliminary definition as a) concentration of influencing energy on area smaller (reduced, scaled-down) in size instead of usual influence of this energy on bigger area in size.* ***Key idea:*** *the focused ultrasound is generated by using an optoacoustic transmitter (use of model* ***18 Mediator****), made with carbon-nanotube (CNT)-polymer composites (according to model* ***17 Use of composite materials****), which is formed on a concave surface (lens) that directly enables acoustic focusing.*

fig. 7.23. Including a suggestion with new model in Reinventing (fragment)

Example 7.20. How does this "submarine" move without screw-propellers?

TREND

Known muscle-powered submersibles have a serious drawback – they require application of considerable efforts to ensure prolonged navigation at relatively high speeds. The reasons is that all propulsion units used by such submersibles to date, including screw propellers, have relatively low efficiency ratios, and low efficiency ratios translate into high loss of power.

On the whole the power required (2.5-5.0 kW) is by about an order of magnitude higher than the normal power (0.2-0.4 kW) that can be sustained by a human being over a period of 2-4 hours.

Human power cannot be increased. What can we do?

REDUCING

FIM: The X-resource in the form of particles of substance or energy is located inside the operating zone and ensures, together with the other available resources, the obtaining of the following **IFR**:

[reduction of environmental resistance to motion and production of additional power].

Formulas and solutions (in general terms) of standard contradictions (SCs):

Submersible propulsion ► 01 Productivity **VS** 36 Power = 01, 02, 40
Submersible propulsion ► 01 Productivity **VS** 39 Loss of energy = 01, 02, 04, 14

Radical Contradiction:

Submersible propulsion ► must have increased efficiency **VS** must not have increased efficiency at existing propulsion devices, for example, screw-propellers

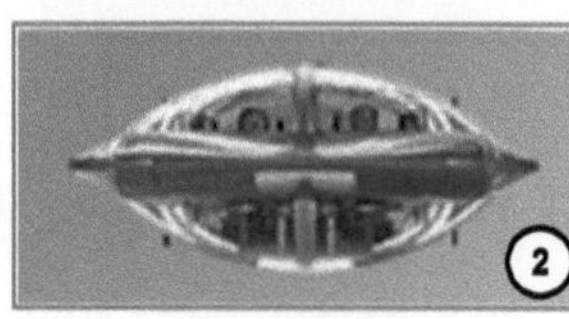

INVENTING

Dominant models: 01 Change in the aggregate state of the object and 04 Replacement of mechanical matter.

The solution is based on resource of energy and application of a physical/technical effect: it is suggested that the new submersible use a water propulsion unit. Outside water is sucked in by a muscle-powered pump and ejected in two places (fig. 3): (1) fore, so that the water "clings" to the sides of the submersible in accordance with the Coandă effect thereby rarefying the medium and reducing resistance to motion, and (2) aft, passing through water ducts lining the body of the submersible. As a result, the Coandă effect increases power efficiency of both propulsion units.

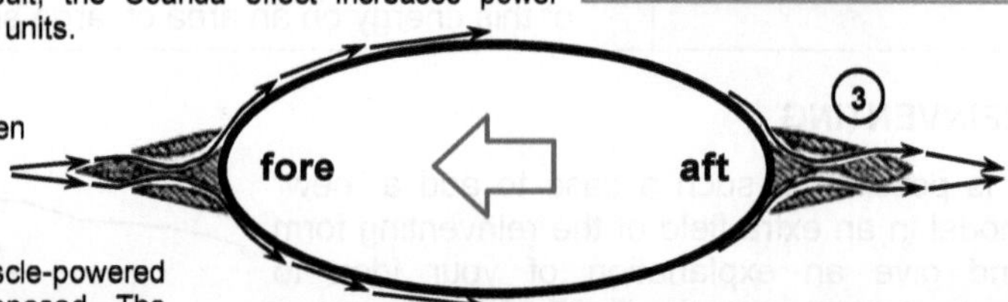

ZOOMING

Have the contradictions been removed? – Yes!

BRIEF DESCRIPTION

A new high-speed muscle-powered submersible has been proposed. The craft's propulsion unit is based on the use of the **Coandă effect.**

Dominant navigators: 01 and 04. The energetical resource is intensified.

Muscle-powered submersible Blues Space – project developed by the Marine Innovative Technologies company, Saint-Petersburg, General Director: V. Taradonov: 1) with opened entry/exit hatches; 2) front view.

Pressure profiles (3): thrust production by propelling force (aft) and additional thrust production by rarefaction (fore and aft) due to the pressure difference generated by the Coandă effect.

Submersible of the future: length – 3.5 m, height – 1.2 m, width – 2 m, weight (in the air) – about 2 tons. Diving depth – up to 30 m, underwater speed – up to 5 knots (up to 10 km/n), number of passengers – 2 or more. Source: **www.bluespace.ru**

fig. 7.24. Reinventing of the *Blue Space* submersible

Example 7.21. Special important case of mismatch of A-matrix recommendations with your extracting models

You have chosen suitable plus- and minus-factors in the A-matrix, but the relevant matrix cell does not contain the navigators that you selected at the extracting stage,

YOU MUST USE ONLY NAVIGATORS SELECTED BY YOU AT EXTRACTING!

You can describe the navigators and their influence in the "inventing" field of the "paper form" (the windows "idea" in software EASyTRIZ). These navigators could also be recorded in the additional fields under the set of field "navigators."

ATTENTION: The recommendation above is extremely important! If you stick to "your" navigators as you identified them at the extracting stage, you will avoid a "FICTITIOUS" or even "RASCALLY" representation of data!

Once more: Do not use inaccurately defined plus- and minus-factors from some A-matrix cell containing the "required" navigator! In other words, relying exclusively on Extracting-1 and Extracting-2 results, you will model the entire process correctly and objectively using formal factors and models from the A-matrix only when they truly correspond to the factors and models that were selected by you at extracting!

Example 7.22. Special important case of mismatch of plus- and/or minus-factors from A-matrix with your extracting

It is absolutely important first **TO IDENTIFY THE INFORMAL PLUS- AND MINUS-FACTORS WITHOUT RESORTING TO** factors of the A-matrix.

This will help you to hone your skills and develop a habit of fully and accurately defining the nature (content, meaning, essence) of the initial problem.

If you have a doubt when selecting some factor, you can try two or three other variants. Maybe you can get different navigators from different cells of the A-matrix. If some navigators can open you other ideas, put these reinventings on separate forms.

Example 7.23. On the role of the prototype in reinventing

One of the typical mistakes that students make in the course of reinventing is in the formulation of contradictions… for the invention (result-artifact). **This is totally wrong!**

ATTENTION: The CONTRADICTIONS identified at extracting and reinventing are formulated for the PROTOTYPE only !!!

In the result-artifact (i.e., new, invented), the contradictions and the problem are REMOVED with the help of transformation models identified by extracting!

LAST BUT NOT LEAST: All fields in the forms and program windows must be completed as concisely as possible.

Despite its brevity, the text that you input into the forms must be written in a language that can be understood by, say, senior school children. In particular, it is necessary to avoid using obscure specialist terminology, abbreviations, and slang words and expressions. Of course it is sometimes difficult to model without special terms as in examples 7.19 and 7.20, but then this is justified by indicating the effects.

7.4. WORKSHOP FOR CHAPTER 7

Instead of a workshop, let us consider some additional recommendations of the author.

It is advisable that you train your extracting and reinventing skills on an ongoing basis. This must become a habit and a professional need.

This is routinely done, for example, by chess players: When they are alone, they often do chess problems or analyze well-known chess games to pinpoint novel ideas. Similar training (and **THIS IS TRAINING**—what else would you call it?) is done by musicians and, naturally, sportsmen.

All these are steps in the long and winding road of self-perfection.

Modern TRIZ opens up new opportunities for self-perfection by providing a technology for exchanging efficient examples. This is supported by standardization of the information-presentation process **using the format Meta-Algorithm of Invention T-R-I-Z**.

We have considered several illustrative reinventing examples developed by various users of Modern TRIZ. In this part, several extracting and reinventing forms were applied. These examples mainly are very simple. After all, this is our ultimate objective: We want to give universal examples that are as simple as possible and that do not require special knowledge.[56]

These graduates have completed their training and have been certified (in most cases, within the framework of remote e-courses) to use the EASyTRIZ software package. Some of them were taught by the author at the Technical University of Berlin where he delivered a Modern TRIZ course for future Masters of Science in Global Production Engineering.

More complex professional formats have been implemented in the EASyTRIZ™ software package. Examples in that program's format are designed to store and exchange completed examples.

The purpose of this chapter was to show you that all readers of this book can master Modern TRIZ and create dozens of similar examples using both artifacts that surround them in their daily life or other items making up their professional environment. The purpose of this chapter was also to help readers who have embarked on a study TRIZ-basics **to quickly acquire practical skills required to use fundamental TRIZ notions and models**.

It is not enough to just read this book or other similar books.

To learn to invent with TRIZ, you must continuously train in extracting and reinventing.

Let's constantly improve ourselves!

---

[56] Some examples and forms are available for download at www.modern-triz-academy.com

8. Answers to Tests

The "super-task" (in Stanislavsky's parlance) of all solutions described in this book does NOT seek to provide readers with feelings of wonder, amazement, rapture, or any other state of heightened pleasant excitement.

The thing is, the "appearance of the miracle" is quickly followed by the "disappearance of the miracle." This very real psychological "law" is emphasized in section *4.1. Contradictions*.

However, as the author of this compilation—as well as many reinventings in it—I hope that the examples included in the book will not leave you indifferent and that the time you spend perusing those examples will not only prove useful but also give you moments of pleasure and help you to experience the joy of making many small discoveries.

The first practical objective of this section is to repeatedly demonstrate how solutions can be built by modeling the contradictions and applying the transformation models.

The second practical objective is to help readers to strengthen their problem-solving skills by studying similar (standard) methodological schemes used in the examples. All reinventings are built on the Meta-Algorithm of Invention T-R-I-Z (see *3. TRIZ Algorithms of Invention*). That is why all solutions are presented in either one of the standard reinventing formats or in a similar simplified format.

To prevent certain examples from appearing "ravenously serious,"[57] the author embellished some of them with stories, which he tried to make as funny as possible because of his modest literary abilities. Accordingly, you should treat those examples with some humor or, as they say, *cum grano salis* (with a pinch of salt). After all, the title of this books' series is *ABC-TRIZ* because of any ABC-book by definition has to be accessible, informal, and not boring to the extent that this is possible.

Another caveat: Clearly, it was not possible to write all examples so that their style and content are to everybody's liking. That is why we heartily hail "RE-reinventings" as may be performed and sent to the author by students or by any other readers of this book.

ATTENTION: If you have not studied the basic material presented in this book up to now, **please go back** and do it without reading the next sections!

This recommendation is exclusively for your benefit!

CHAPTERs 1 – 3

The answers to these chapters are in the other chapters of this book except for Problem P3. The answer to this problem is shown in the next part, ***Book II: How to become a genius***.

57 Translation of a part of the German expression "Ohne tierischen Ernst" attributed to the Danish physicist, Nobel Prize in Physics in 1922, Niels Henrik David Bohr (1885-1962).

M.A. Orloff, *ABC-TRIZ*, DOI 10.1007/978-3-319-29436-0_8

CHAPTER 4. MODELING THE PROBLEM

Reinventing for the Example 4.6. Swimmer.

Now we would like to show you the well-known brilliant solution of the original problem situation for the "Swimmer" assignment.

TREND

Open water bodies are used to train long-distance swimmers. In bad weather, training can become impossible. In a usual 50-meter pool the swimmer inevitably reaches the edge and has to turn around and push off to continue swimming, which impairs his technique and disrupts his rhythm. A pool with a complex circular (round, ellipsoid or 8-form) shape can remove the problem, but such pool will get too complex shape.

REDUCING

FIM: X-resource, together with available or modified resources, and without making the object more complex or introducing any negative properties, guarantees attainment of the following **IFR**:

[*adequate training of the long-distance swimmer in a pool*].

Standard Contradiction

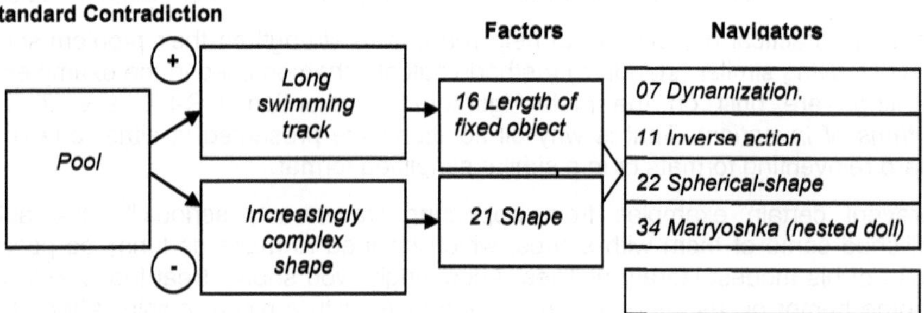

Radical Contradiction

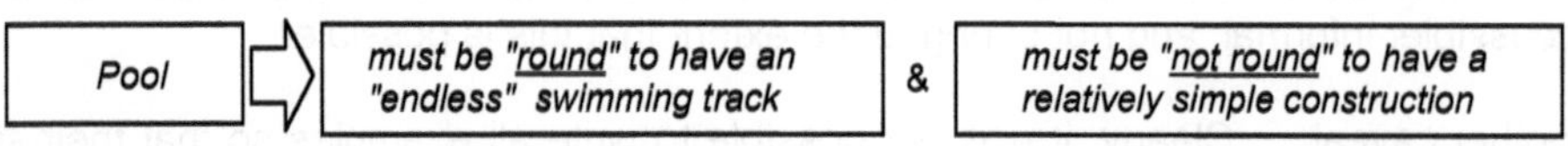

INVENTING

***Key model: 11 Inverse action** – b) make a moveable part of an object or the environment fixed or a fixed part moveable. **Key idea:** make the water moveable. Also: **07 Dynamization**. Take in account the **22 Spherical-shape** – proposed a round pool!*

ZOOMING

Have the contradictions been removed? – Yes. – ~~No~~.

Super-effects: *it becomes possible to manage training parameters (movement speed and phases), and adjust the swimmer's swimming technique "on the go" as the coach is located in the immediate vicinity of the swimmer.*

Negative effects: -

BRIEF DESCRIPTION

To provide long-distance swimmers with adequate training possibilities and simplify pool design, the water in the pool is made moveable in accordance with navigator 11 Inverse action.

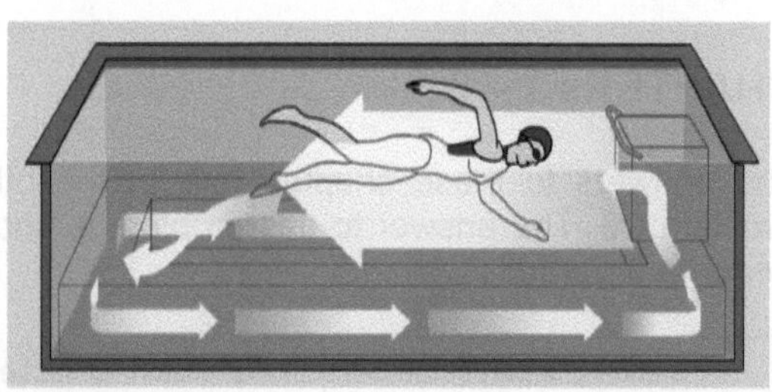

fig. 8.1

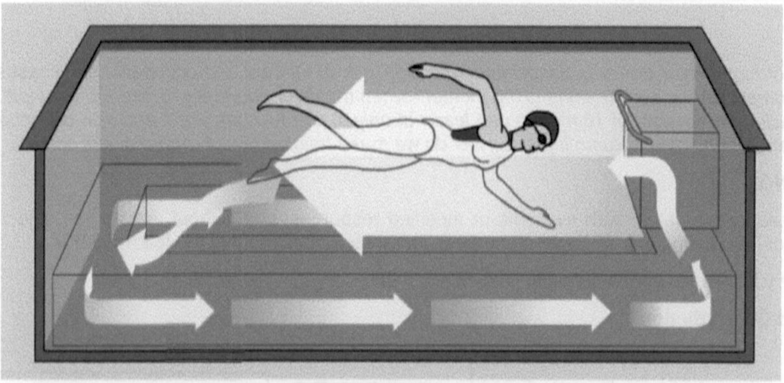

a) Water-flow swimming simulator operating scheme

b) Example of an "endless" water-track pool (Conrays)

c) Example of an "endless" water-track pool (Endless Pools)

fig. 8.2. Swimming simulators

Reinventing for the Example 4.7. Diver.

TREND

Training of high-board divers is associated with high risk of injuries and occupational diseases. The critical moment during such training is the moment when the diver enters the water. For example, a fall on the back from a height of 10 meters can lead not only to painful contusions and skin disruption, but also to dangerous spinal column injuries. How do we make training of high-board divers safer?

REDUCING

FIM: X-resource, together with available or modified resources, and without making the object more complex or introducing any negative properties, guarantees attainment of the following **IFR**:

[*safe entry into water from any height*].

Standard Contradiction

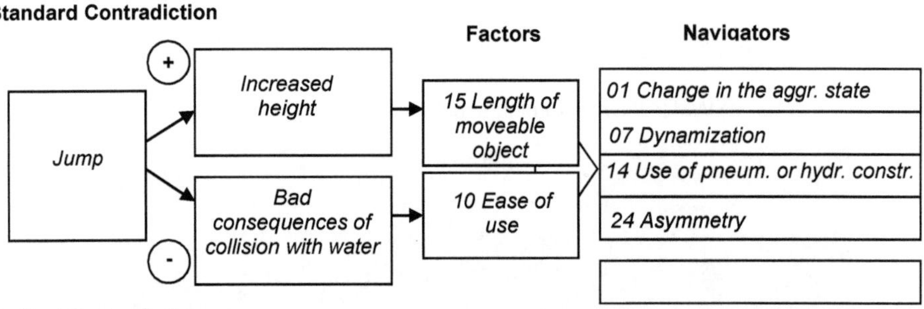

Radical Contradiction

Water	⇨	*must be "soft" so as not to cause injuries*	&	*must be "hard" in accordance with its natural properties*

INVENTING

Key models: 01 Change in the aggregate state – *a) transitions into "pseudo-states"; b) changes in concentration;*

07 Dynamization – *c) make an object moveable that is otherwise fixed;*

14 Use of pneumatic or hydraulic constructions – *use gaseous or fluid parts instead of fixed parts in an object: inflatable…, air cushion…*

Key idea: *inject compressed air at the time of the jump, make an "air cushion" on the surface of the water.*

ZOOMING

Have the contradictions been removed? – Yes. – ~~No~~.

Super-effects: *safe training of children*

Negative effects: *increased complexity of construction*

BRIEF DESCRIPTION

To make high-board diver training safe, a powerful jet of compressed air is injected into the water during the jump, creating an "air cushion" on the surface of the water. Navigators used: 01 Change in the aggregate state, 07 Dynamization and 14 Use of pneumatic or hydraulic constructions.

fig. 8.3

fig. 8.4. High-board diving pools fitted with "air cushions"

a) "Air cushion" from Natare

b) "Air cushion" from Myrthapools

According to the information provided by the Natare Corporation, the "air cushion" was used for the first time in 1971 in Point Claire, Quebec, Canada, and was later introduced in many pools in Canada, the USA, and Europe and then virtually in all Olympic-class pools.

fig. 8.5. Injection of compressed air (Natare)

Reinventing for Example 4.8. Invention of the Bicycle.

Despite the fact that the story of Comte de Sivrac is a bunch of lies (at least according to some experts), and he himself is a fictitious character, nobody and nothing can stop us from reinventing the object that was invented by *somebody* at the time mentioned in the legend.

To spice things up, we will preface the reinventing with a hypothetical story about how that ingenious object could have been invented by Comte de Sivrac. We should also mention that this made-up story replaces formal extracting.

We will resort to the same method to introduce the reinventings of two more clever contraptions that have considerably improved the "bicycle as we know it."

The Legend of Comte de Sivrac

So the count is sitting on the bank of the river and fishing. Suddenly he sees the wheel of a cart rolling by. It is rolling without falling. So he thinks to himself: Wouldn't be nice to take a ride on that wheel! But therein lies a dilemma: How does one ride a wheel?

So he quits fishing and goes back home to his castle, but now he has that idée fixe about how he wants to ride a wheel. So he goes to the stable, takes a wheel, straddles it with his feet planted on the shaft left and right, and then grabs a pole and just stands there, as he is afraid to let go and start riding. Then he picks up his courage, pushes himself off away from the pole and... BANG! He goes down head first! Luckily for him, there was a small haystack between the wooden floor and his forehead. Yeah.

So he thinks, this must have been the shortest ride in history because you cannot really ride a wheel for a long time. Besides, a cart has more wheels, and when you ride a cart you can SIT! So why not take two wheels and connect them – not on the same axle, like in a hansom, but one after the other. And there is a bonus – you can place a seat in between.

No sooner invented than done.

So the count makes himself a two-wheeled "horse" and rides around happy as a lark. He is having the time of his life riding that thing, and people around are having the time of their lives laughing their heads off. He pushes off with his legs, you know, to have his "horse" run faster. Then he raises his legs and rolls on while the rolling is good, somehow managing to stay on the wheel. Now, if that is not a miracle, I do not know what is. A man riding a wheel and not falling down! Soon his workers start making such "horses" for their own kids, and people start coming from other counties to look at the count's invention and build likenesses of it for their enjoyment.

As for the count, well, he soon gets bored with rocking on and falling off his "horse" because the "horse" stubbornly refuses to turn where he wants to go. So he buries it in the attic and goes back fishing.

It is a shame the count does not know TRIZ system evolution laws, or he could have invented tons of interesting things like a scooter, a skateboard, or even a monocycle plus a bunch of steering gadgets!

Also, he could have written down his inventing method presenting it as a special creative technique and giving it a fancy name, like "copying," "unite," or even "local property." And then he could have taught his descendants to use those.

Who knows how much farther and faster our civilization would have progressed!

Invention of the Bicycle

TREND

PROBLEM: how can we accelerate the man's movement along land surface using only his muscular strength?

If we take a wheel and stand on its shaft, time in motion will be too small! And it is DANGEROUS! And how are we going to sit while riding?

A cart – too heavy to keep pushing it and then jumping onto it.

So, we are stuck with the wheel. But what can we do to increase the duration of the trip?!

REDUCING

FIM: X-resource, together with available or modified resources, and without making the object more complex or introducing any negative properties, guarantees attainment of the following **IFR:**

[*man rapidly moving along land surface… on a wheel*].

Standard Contradiction

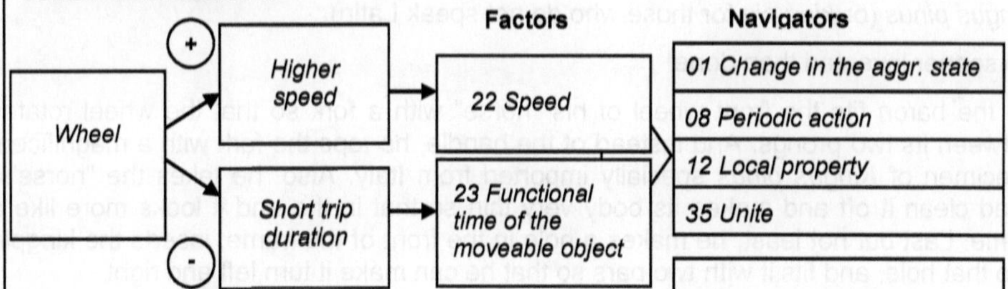

Radical Contradiction

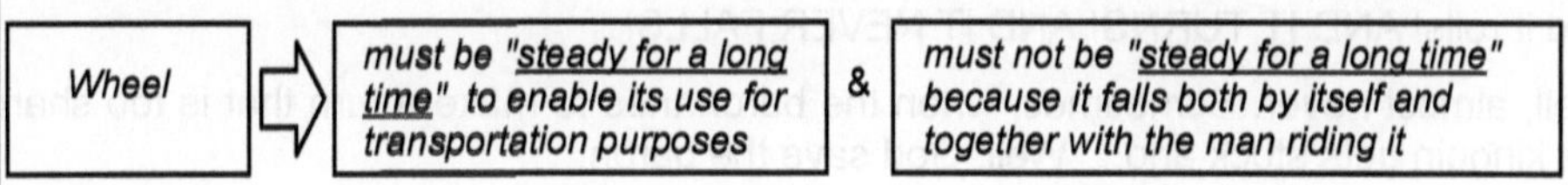

INVENTING

***Key model: 35 Unite** – a) Unite similar objects or objects for neighboring operations. **Key idea:** unite two wheels into an integral structure. Addition based on **Model12 Local property**: a seat between the wheels to sit on, and a handle in the front to hold onto while in motion.*

ZOOMING

Have the contradictions been removed? – Yes. – ~~No~~.

Super-effects: *we can MOVE! That is a feat in itself!*

Negative effects: *it is hard to turn! And it does not roll by itself for a long time anyway!*

The wheels in the drawing are open – the sidewall has been removed.

BRIEF DESCRIPTION

To enable continuous steady movement using wheels and muscular strength, two wheels have been united into an integral structure in accordance with the principle 35 Unite. The construction is fitted with a seat and a handle to hold onto while in motion.

fig. 8.6

Reinventing for Example 4.9. Invention of the Handle Bar and the Steering Front Wheel.

The Legend of Baron von Drais

So Baron von Drais is riding a "horse" he got from Comte de Sivrac. There is an audience, too: ladies, friends, kids, valets... And then this ram comes out of nowhere and plants itself right in the baron's way! The baron starts squirming and contorting his body to make the "horse" turn away, but it won't! And the baron comes down in a heap! Everybody starts laughing, but he hurts all over and does not think it funny, no sir!

Anyway, this baron is stubborn—and well-educated! So he starts thinking, how come he has to keep falling and almost killing himself almost every time he rides his two-wheeled "horse"?

And he comes up with this idea: When I ride my coach, he says to himself, I turn in a different way: Its front wheels (well, yeah, there are TWO front wheels, so what?) can turn to the side moving around a vertical shaft. And that shaft has a scientific name: *Kingus pinus* (or *king pin* for those who do not speak Latin).

No sooner invented than done!

So the baron fits the front wheel of his "horse" with a fork so that the wheel rotates between its two prongs. And instead of the handle, he tops the fork with a magnificent specimen of *Kingus pinus* specially imported from Italy. Also, he takes the "horse's" head clean it off and makes its body very thin so that in the end it looks more like a frame. Last but not least, he makes a hole in the front of the frame, inserts the kingpin into that hole, and fits it with two bars so that he can make it turn left and right.

When he finishes, few people can understand what exactly he has done, but many agree that it is a true miracle! So the baron decides to test the fruit of his endeavors—and it rolls! AND IT TURNS! AND IT NEVER FALLS!

Well, almost never. Sometimes, when the baron tries to make a turn that is too sharp, the kingpin gets stuck and... Well, God save the baron...

Anyway, he is very proud of his invention! So proud that he wants to show it off to the whole world and surprise all France first of all IN ADDITION to all Germany! So he runs/rides his new "horse" with a rotating king pin instead of a head from Berlin to Paris! How does it go, you ask? Well, like they say, "there's a whole lot o' shakin' goin' on!" There are rocks in the road, there are suicidal woodchucks, and there are envious ill-wishers plotting against him... But he sticks to his guns, or rather hangs onto his handle bars! Well, he is a BARON, for crying out loud! And barons are TOUGH!

Anyway, he does not have the first idea about TRIZ, either. That is why he fails to call his new method "dynamization" and write it down for posterity.

He is also totally ignorant of TRIZ system evolution laws, the immediate lamentable result of which is that must prove the worth of his invention by bruising his soft and not-so-soft tissues and body parts all the way from Germany to France. And then maybe back from France to Germany.

He probably should have fitted his "horse" with springs, too, to make the riding more comfortable. They did know how to make springs back then!

Invention of the Handle Bar and the Steering Front Wheel

TREND

Comte de Sivrac's "race horse" has one major drawback – it is hard to steer. It is difficult to negotiate turns, to "adapt" to turns, if you will. This is due to design flaws which make this "race horse" difficult to use on "real" roads.

PROBLEM: how can we increase the ease of steering of Comte de Sivrac's "race horse"?

REDUCING

FIM: X-resource, together with available or modified resources, and without making the object more complex or introducing any negative properties, guarantees attainment of the following **IFR:**

[*steady and comfortable negotiation of turns*].

Standard Contradiction

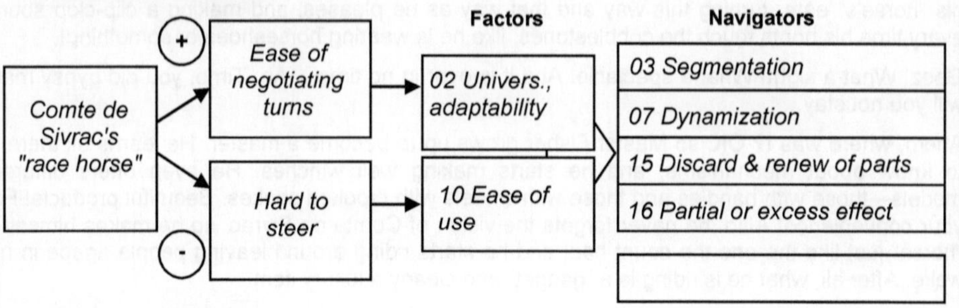

Radical Contradiction

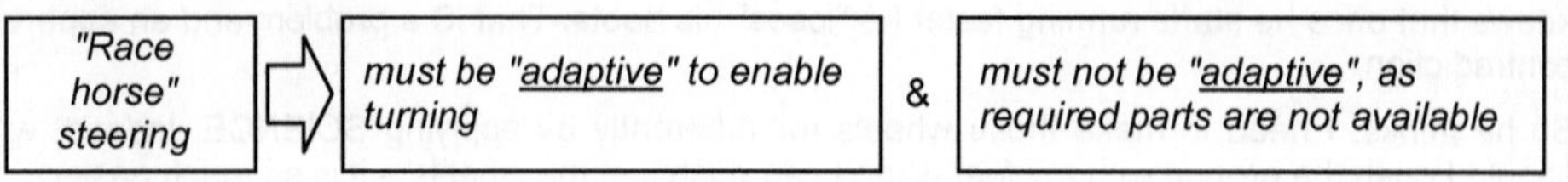

INVENTING

Key models: 03 Segmentation *– Divide the object into parts, and* ***07 Dynamization*** *– b) Divide the object into parts that can move with respect to each other.*

Key idea: *separate the front wheel from the integral structure, and connect it to a handle to make it possible for that wheel to be turned.*

ZOOMING

Have the contradictions been removed? – Yes. – ~~No~~.

Super-effects: - . Negative effects: *none*

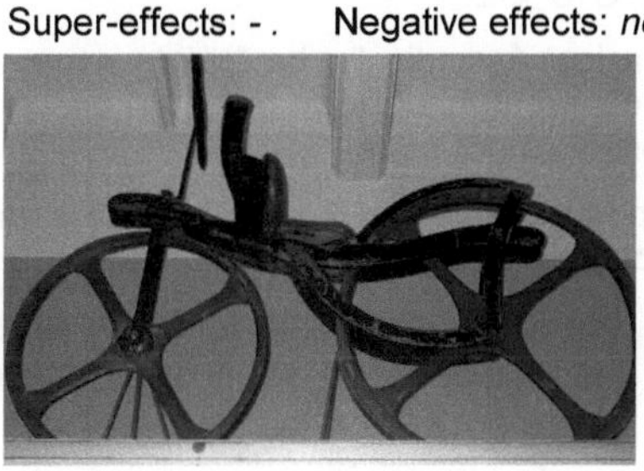

BRIEF DESCRIPTION

To enhance the ease of steering when negotiating turns, the front wheel of Comte de Sivrac's "race horse" is separated from the integral structure and fitted with a steering handle. Navigators used: 03 Segmentation and 07 Dynamization.

fig. 8.7

Reinventing for Example 4.10. Invention of the Pedals.

Pedals! Ah, forget pedals, the entire bicycle has been invented and reinvented hundreds of times by hundreds of talented people including Russian bonded locksmith Artamonov, German Master Fischer, and Baron von Drais (see above) to mention but a few. I have already told you about Comte de Sevrac, and I shudder to think what some people may say if I mention Leonardo da Vinci with his "bike," even though there is "evidence" that he is the one who invented it. I even had my photo taken near the "Da Vinci Bike" in the Museum of Inventions at the Invention Education Center in South Korea (you can see it at the end of the book after the Table of Contents). But who knows the truth? Still, I believe I must tell you another story—the one about Master Fischer—because it at least has a ring of truth to it.

The Legend of Master Fischer

When Master Fischer was a small boy, Baron von Drais rode past him as he was sitting on the porch of his parents' house! I can see it in my mind's eye: The baron is rushing forward twirling his "horse's" ears, turning this way and that way as he pleases, and making a clip-clop sound every time his boots touch the cobblestones, like he is wearing horseshoes or something!

Geez! What a sight! What a spectacle! And it is over in no time... Ah, Time, you old gypsy man, will you not stay...

Ahem, where was I? OK, so Master Fisher grows up to become a master. He learns all there is to know about mechanisms, and he starts making well winches. He even offers different models—those with handles and those with wheels with crooked spokes. Beautiful products! For your convenience! Also, he never forgets the vision of Comte de Sivrac, so he makes himself a "horse" just like the one the count had, and he starts riding around leaving people agape in his wake. After all, what he is riding is a "gadget" and clearly a luxury item.

So one day he is riding pounding the ground with his boots, and he sees some impudent girls laughing at him! Naturally, he takes it as a slight because he expects adulation! But, yeah, he knows that once he starts running faster he "loses" his boots. That IS a problem and an apparent contradiction.

So he thinks: I need to make those wheels roll differently by applying SCIENCE. Indeed, why should I push the ground with my feet when I can push just the wheels – it is so much easier!

No sooner invented than done.

So he takes a couple of well-winch handles and slaps them onto the front wheel shaft from left and right at a conic tip. And, ladies and gentlemen, we have a WINNER! Ah, my bad, I mean, we have PEDALS! Anyway, Master Fischer thinks to himself: NOW I'll show those girls. They'll know better than laugh at me! The boots stay on! I am pushing the pedals, the wheels are rolling, and everybody goes "Ah! Oh! WHO IS THAT MAN?"

And finally, the *coup de grace*: The pedals are fitted with square tips of the wheel shaft! (Well, in reality he does it to avoid accidents: Once the pedals with conic fastening come off, and he nearly dies!) In a word, he is CO-O-O-OL!

Again, it is a shame he is not familiar with TRIZ basics. Accordingly, he fails to name his inventing methods "mediator" or at least "copying" (well, not without "local property" and even "separation"). If not for his ignorance, he could have done so much more!

Also, he does not know TRIZ system evolution laws, and he has no idea how to make fast riding safer. All he can come up with is a huge front wheel. Some brainstorming!

Today we know for a fact that it is brainstorming that we must "thank" for the insanely large front wheel, which caused many injuries (especially head injuries) and even deaths.

Invention of the Pedals

TREND

Baron von Drais' "race horse" has a major drawback – the discomfort associated with the need to continuously use one's legs to push off the ground in order to accelerate. To increase trip duration, it is necessary to increase the pushing effort, which leads to higher energy expenditure. How can we enhance the ease of moving on Baron von Drais' "race horse" and reduce energy expenditure?

REDUCING

FIM: X-resource, together with available or modified resources, and without making the object more complex or introducing any negative properties, guarantees attainment of the following **IFR:**

[***ease of acceleration with minimum energy expenditure***].

Standard Contradiction

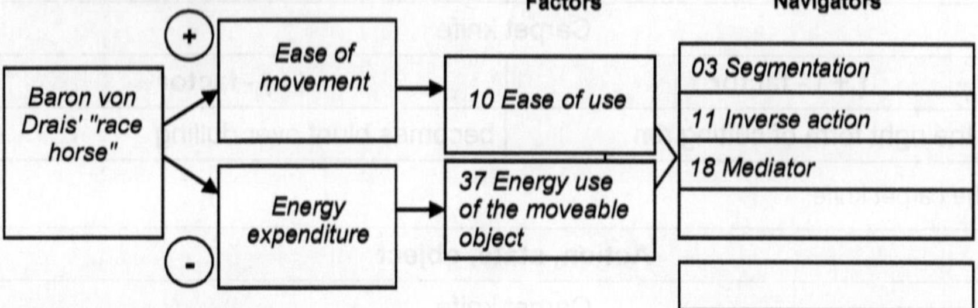

Radical Contradiction

Acceleration	⇒	*must be "comfortable" to spend less energy and increase trip duration*	&	*must be "uncomfortable", as the existing design provides no tools for that*

INVENTING

Idea inspired by navigator 11 Inverse action – a) instead of the action prescribed by the assignment, complete a reverse action: instead of pushing the ground so that it accelerates wheel rotation, push the wheel so that it rolls along the ground.

Idea inspired by navigator 18 Mediator – a) use another object to transfer or transmit the action.

The same navigators can be efficiently interpreted within the framework of the "Separation in structure" model used to resolve radical contradiction.

***Key idea for the solution:** To install mediating levers (similar to water well winch handles) to transfer the power from the legs directly to the front wheel.*

ZOOMING

Have the contradictions been removed? – Yes. – ~~No~~.
Super-effects: -. Negative effects: -

BRIEF DESCRIPTION

To improve ease of movement and save energy, we have used special levers, similar to water well winch handles, to transfer the effort from the legs directly to the front wheel. Navigator used: 18 Mediator.

In picture: later counterpart – "spider bicycle"; source: www.daociyou.com/BicycleEvolution7.htm.

fig. 8.8

WORKSHOP FOR CHAPTER 4

Exercise 1. Table of right answers to this exercise:

#	Exercise 4.3	#	Exercise 4.4	#	Exercise 4.5
1	RC	3	SC	5	RC
2	SC	4	RC	6	SC

Exercise 2. Wall Paper and Carpet Knife.

SC for the carpet knife:

Action, state, object	
Carpet knife	
(+) - factor	**(-) - factor**
needs the right form of cutting rim	becomes blunt over dulling

RC for the carpet knife:

Action, state, object	
Carpet knife	
+ Z	**- Z**
must be sharp to cut	must be blunt over dulling

Solution. It has been suggested to protect the cars with the help of a device which fills it with dense thick milky-white smoke upon detecting unauthorized entry. The device starts its work after the alarm system activation. The smoke has a special formula and does not damage the car interior but cannot eliminate for relatively long time (e.g. dozens of minutes).

(See Navigator *09 Change in color* in As-Catalog)

Exercise 3. Protection from Car Theft.

SC for the car:

Action, state, object	
Car	
(+) - factor	**(-) - factor**
Must be equipped with an antitheft alarm system	The criminal has a certain amount of time to disappear with a car from the scene of the crime

Solution. It has been suggested to protect cars with the help of a device that fills them with dense thick milky-white smoke on detecting unauthorized entry. The device begins working after the alarm system has been activated. The smoke has a special formula and does not damage the car interior, but it cannot be eliminated for quite some time (e.g. dozens of minutes).

(See Navigator *09 Change in color* in As-Catalog)

Exercise 4. Protection of Motorbike Rider in the Event of a Crash.

RC for the motorbike rider's protective clothing:

Action, state, object	
Motorbike rider's clothing	
+ Z	**- Z**
Must be thick to protect arms, legs, and other body parts in the event of a fall	Must be thin so as not to restrain the motorbike rider's movements

Solution. It has been suggested that motorbike riders wear inflatable life vests, which are activated only in the event of a fall.

(See Navigator *14 Use of pneumatic or hydraulic constructions* in As-Catalog.)

Exercise 5. Aircraft Carrier.

SC for the aircraft carrier:

Action, state, object	
Aircraft carrier	
(+) - factor	**(-) - factor**
Must be limited in size	Must hold many airplanes

SC for the airplane:

Action, state, object	
Airplane	
(+) - factor	**(-) - factor**
Must have wide wings	Must have much holding capacity

RC for the aircraft carrier:

Action, state, object	
Aircraft carrier	
+ Z	**- Z**
Must make many airplanes operate	Must have fewer airplanes be over the limited holding volume

RC for the wing:

Action, state, object	
Wing	
+ Z	**- Z**
Must be wide so the airplane can fly	Must be narrow enough to place the plane in the hold

CHAPTER 5. EXTRACTING

5.2. ANSWERS TO LESSON 2 (tests and validity tables)

Task 1. The Scrunchie "Operation". Prompt: dominating models – *02 Preliminary action*, *18 Mediator* and *34 Matryoshka (nested doll).*

Task 2. Structural Organization of Our Clothes. Prompt: dominating model – *34 Matryoshka (nested doll).*

Task 3. Live Flower Carpet. Prompt: dominating models – *02 Preliminary action, 17 Use of composite materials, 31 Use of porous materials* and *34 Matryoshka (nested doll).*

Task 4. "Mobile Traffic Light" Sample:

LC*	No.	Navigator	Substantiation for the Extracting
++	05	Separation	Single out (include) the required function
+	07	Dynamization	Make the object mobile
+	10	Copying	Copy the functions performed by an ordinary traffic light
++	12	Local property	Located in the spot where it is needed the most
+	19	Transition into another dimension	Ability to be located within a 2D plane rather than only in a unique 1D spot
+	29	Self-servicing	Independently replenishes its energy store using solar batteries

fig. 8.9. Result of Extracting-1 for Mobile Traffic Light (fig. 5.8)

* LC – Level of Compliance: (++) – completely efficient navigator;
(+) – partially efficient navigator

Task 5. "Fan-Cap."

The following TRIZ models can be identified in this solution (if the usual "plain vanilla" sun-cap is selected as the prototype):

LC*	No.	Navigator	Substantiation for the Extracting
++	05	Separation	Single out (include) the required function
+	07	Dynamization	Make the elements of the object mobile
+	10	Copying	Copy the functions performed by an ordinary fan
++	12	Local property	Located in the spot where it is needed the most
+	19	Transition into another dimension	Ability to be located within a 3D-space rather than within a 2D-plane only.
-	29	Self-servicing	None

fig. 8.10. Result of Extracting-1 for Fan-Cap (fig. 5.9)

Task 6. Compare your results for Task 4 and Task 5.

It is easy to see that almost all TRIZ models in these two completely different artifacts are the same! And, of course, that is also true for both dominating models – 05 Separation and 12 Local property.

This testifies to universality of TRIZ creative models. It also means that as you accumulate modeling experience, you will be able to transfer such experience and use it to create ideas to improve any object. All you have to do it apply your associative thinking and creative fantasizing abilities. Extracting, in turn, helps develop these abilities.

Figures 8.11,a through 8.11,d provide additional examples of implementation of dominating navigators 05 and 12, which were used to improve the two artifacts described previously. Incidentally, in these examples, the inventors have eliminated the last difference between the caps and the stoplight (in terms of creating thinking, naturally!) by using sun-powered accumulator batteries. Well, the show (evolution of caps through combination and integration) must go on!

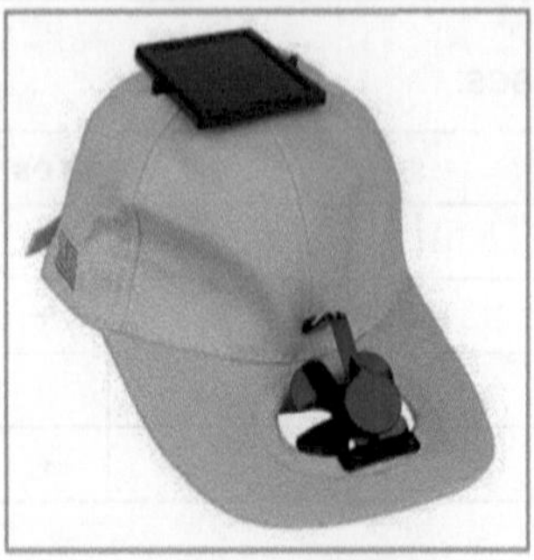

a) Cap with a sun-powered accumulator battery

d) Cap with penlights used for reading – or lighting the way!

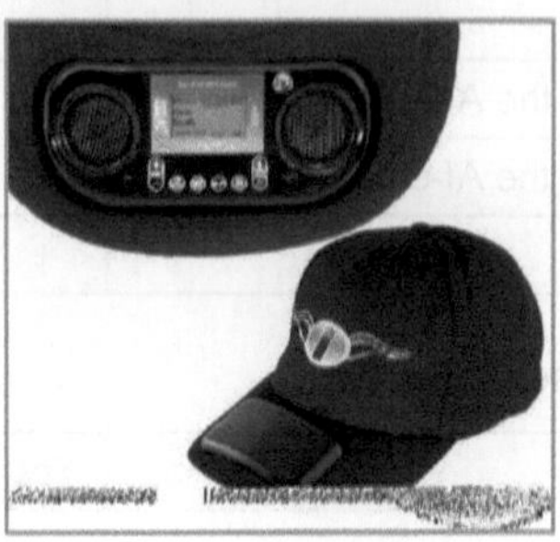

c) Cap with an built-in radio

b) Cap with a detachable fan!

fig. 8.11. Variants of implementation of dominating models 05 and 12 (and maybe others in your opinion!)

Task 7. "Ultrasonic whistles." Prompt: dominating models – *02 Preliminary action,* and *06 Use of mechanical oscillations.*

Task 8. "Avalanche rescue airbag." Prompt: dominating models – *25 Use of flexible covers and thin films,* and *14 Use of pneumatic or hydraulic constructions.*

Task 9. "Magic packing." Prompt: dominating model – *27 Use of thermal expansion.*

CHAPTER 6. INVENTING

WORKSHOP FOR CHAPTER 6

Exercise 1. Enter the correct answers (several correct answers are possible) about MAI T-R-I-Z and SMART.

Stage **Trend** is intended for the following purposes:

#	Purpose	Yes	No
01	To structure the initial problem situation	+	
02	Assign person responsible for solving the problem		-
03	Identify informal contradictions	+	
04	Determine the direction (trend) of search of solutions	+	

Stage **Reducing** is intended for the following purposes:

#	Purpose	Yes	No
01	Identify suitable formal models of contradictions	+	
02	Select the model of transformation of the A-matrix	+	
03	Determine the technology of the future solutions		-
04	Identify the IFR and FIM	+	

Stage **Inventing** is intended for the following purposes:

#	Purpose	Yes	No
01	Calculate the economic efficacy of a solution		-
02	Suggest ideas for solutions based on models from the As-catalogue	+	
03	Suggest ideas for solutions based on models from the Af-catalog	+	
04	Suggest ideas for solutions based on resources	+	

Stage **Zooming** is intended for the following purposes:

#	Purpose	Yes	No
01	Check whether the conflict is eliminated	+	
02	Develop technical documentation for the project		-
03	Check how could it be possible to develop a solution	+	
04	Check whether there are positive and negative effects	+	

Book II

How to Become a Genius

The author gave the subtitle to the second "book-part" according to the title of last outstanding book[58] (above) by Genrikh Altshuller in memory of the TRIZ founder and to emphasize the continuity of the TRIZ ideas and the **Theory of Developing the Creative Personality (TDCP)** initiated by G. Altshuller after TRIZ.

---

58 G.S. Altshuller, I.M. Vertkin *How to Become a Genius: Life Strategy for a Creative Personality*. – Minsk, Belarus, 1994 (in Russian)

To Seek and Not to Yield

Author's preface to Part 2

This emotional appeal is the leitmotif of the book. And I would like to tell my dear readers a brief story of my acquaintance with the title above.

Since my teenage years, a life motto for me has been (and still is) a wonderful phrase from a novel by[59] Veniamin Kaverin, *Two Captains,* which in English sounds like this:

To strive, to seek, to find and not to yield!

Much later I learned that this expression is the final line of short poem,[60] *"Ulysses,"* by Alfred Tennyson. And I understand that this line should be right interpreted only with the preceding two lines:

One equal temper of heroic hearts,
Made weak by time and fate, but strong in will
To strive, to seek, to find and not to yield!

A little later after that, I learned that the original line by Tennyson became the epitaph on the cross-monument to the outstanding British explorer, Robert Scott, who died after reaching the South Pole in 1912. He could not overcome the *last 11 miles* to the initial base in Antarctica where his life could have been saved.

This poetic line reflects the way that Scott and his companions, with amazing accuracy as the moving imperative of many other pioneers and discoverers—many other people who are courageous—were persistent and loyal to both their goal and their timeless noble moral values.

While working on this book (in 2012) and having a long history of understanding the original line of Tennyson, I *extracted*—using the term from my own book—two key aspects from it and recorded a concentrated and abridged version of the original:

To seek and not to yield.

As I see it, this is the essence of the complete line by Tennyson.

I think that the "super-task" of this book lies in such an extremely compressed poetic line to show you the way to achieve your goal.

I wish you good luck and courage in achieving your goals. But the fact is that you must be prepared to a certain degree to protect your own destiny, to defend your interests and to show great persistence during the difficult periods, which can sometimes last for days, months, or even years.

I am writing this because many thousands of inventors do not withstand a struggle against circumstances and do not reach their target both on the way and at their "last 11 miles."

---

59 Veniamin Alexandrovich Kaverin (Russian: Вениамин Александрович Каверин; real family name – Зильбер, or Zilber; April 19, 1902, Pskov – May 2, 1989, Moscow) – Soviet Russian writer; his best-known novel *"The Two Captains"* (1938–44) was completed during WWII (hereinafter this page – quoting *Wikipedia*)

60 Alfred Tennyson, 1st Baron Tennyson, FRS (6 August 1809 – 6 October 1892) – one of the most popular British poets, *Poet Laureate* of Great Britain and Ireland during much of Queen Victoria's reign; *"Ulysses"* (in Greek, *Odysseus*) is a poem in blank verse written in 1833 and published in 1842

In his latest book,[61] Genrikh Altshuller wrote about this: "The struggle for recognition of appeared idea – the inevitability, regular occurrence; the innovator must be able to prove the necessity of his invention... Nobody will do it instead him... The inventor, not prepared to fight for his invention, is as weak and impaired, as a surgeon who is afraid of the sight of blood, or an astronaut, which cannot withstand overload, or mountaineer, accustomed to walk only on the parquet..."

Genrikh Altshuller came to the conclusion that the next stage after TRIZ should be ***Theory of Development of Creative Personality (TDCP).*** The book ***"How to become a genius"*** was created as a result of studies of more than 1000 biographies of famous people who selflessly fought for their beliefs and ideas, who won, and sometimes died, but did not surrender. And the authors[62] proposed a "chess game" for the creative person against "life circumstances."

The TDCP concept seems to me to be very promising. It is necessary for its development to continue practical work on the analysis of the biographies of creative people and the search for constructive ideas to support the activities of such persons. I chose for this small book several biographies covering a long historical period from the beginning of the 19th century to the present day. My observations are consistent with *the idea of Genrikh Altshuller that many, if not all the makings of a future genius are rooted in childhood, in children's experiences and events.* This idea will be explored further to find practical recommendations for the nurture of talented individuals.

Here I will only speak to my readers about the idea to be attentive to oneself and one's children to be able to understand and assess *the moment of origin* of a particular interest in creative activities, hobby dreams, and useful concepts, which could become the goal of all life.

One of the most remarkable lives for is recorded in the biography of a German entrepreneur, engineer, and researcher Werner von Siemens. His great fortitude amazes me and provides a reliable example of how to follow life's goal.

In addition to the above, I would also cite the remarkable words of Werner von Siemens, which he wrote back in 1852. At the time Werner was 36 years old, but by that time he had already passed through many difficult trials of life.

But he was not hardened: He wisely and courageously defined his understanding of life and his life's particular path as it is expressed by these words. He persistently followed this understanding through all his life.

> Our aim and objective must be always higher than our strength can achieve, because only then will we be able to exert the latter to the fullest...
>
> Personal accomplishments of any kind should only be judged according to the benefits they create for others. Such deeds deserve to be honored only when they have made a contribution to the public good.[63]
>
> *Werner von Siemens*

---

61 G.S. Altshuller, I.M. Vertkin How to Become a Genius: Life Strategy for a Creative Personality. – Minsk, Belarus, 1994 (in Russian)

62 co-author, famous TRIZ-specialist Igor Vertkin (living in Great Britain now)

63 (translation by author of this book – M.O.; see references in section 3.2)

I spoke about Werner von Siemens in sections 2.3.1 and 3.2. The story of his life as well as the life stories of other prominent personalities, which are told in this book, will help you to overcome the trials you may encounter in your life and to achieve the goals to which you aspire.

In conclusion of this, the author's appeal, I invite you to become familiar with both the author and his TRIZ colleague,[64] Valery Mikhailovich Tsourikov (on the right [2003]). We met in 1972 when I was appointed to work as an assistant at the Department of Electronic Computers in Minsk Radio-Technical Institute (MRTI) and Valery transferred in MRTI to complete higher education.

We both had a great interest in TRIZ, but who could have imagined that TRIZ would become our profession and our life path? More precisely, two very different ways converged in one devoted goal—The development of creative tools for systematic engineering design.

By middle of the 1980s, Valery had created the world's first intelligent software,[65] called *Invention Machine*, for solving problems on the basis of TRIZ.

I started from the very beginning to study the origins of TRIZ and to establish accurate models and patterns of TRIZ. And the idea of simplifying TRIZ, I right took over from Valery, who once (at the end of 1980s) said that *TRIZ needs a "democratization" to become available not only to the elite, but for everyone*.

Today, the simple and reliable methods of the Modern TRIZ Academy have been used for many years to correctly and quickly train thousands of students and professionals.

Good luck!

Michael Orloff

Berlin, Germany,

2012 – October 2015

---

64 The author, like Valery Tsourikov, is a graduate of the Minsk Radio-Technical Institute (now Belarus State Information Science and Radio-Electronics University).

65 Dr. Valery Tsourikov, the designer of trailblazing intelligent software *Invention Machine* (from mid 1980s) based on TRIZ-modeling, pioneering in the world, and the founder (1992) of the world-renowned company *Invention Machine Corp.*, Boston, USA.

TRIZonal Concept

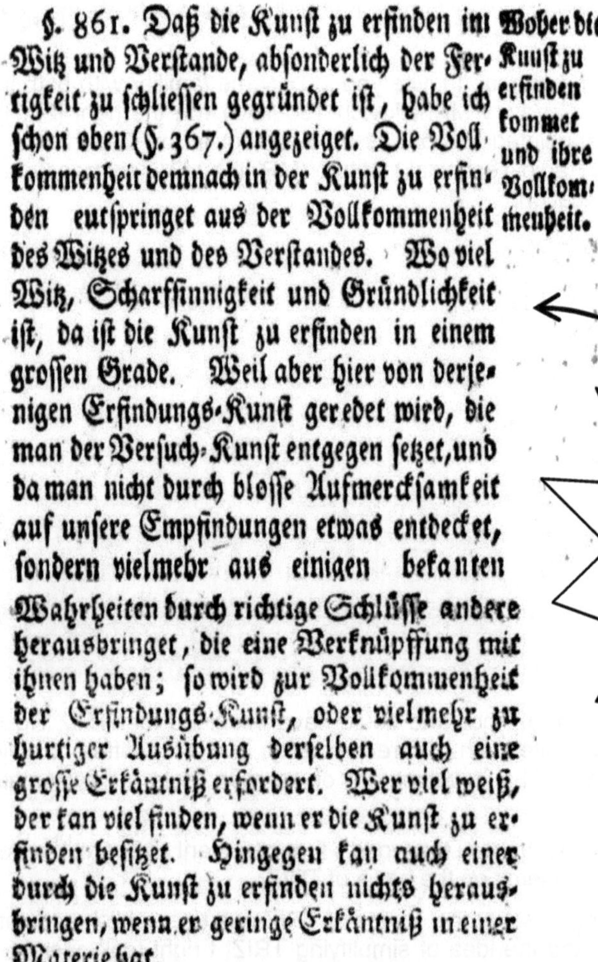

§. 861. Daß die Kunſt zu erfinden im Witz und Verſtande, abſonderlich der Fertigkeit zu ſchlieſſen gegründet iſt, habe ich ſchon oben (§. 367.) angezeiget. Die Vollkommenheit demnach in der Kunſt zu erfinden entſpringet aus der Vollkommenheit des Witzes und des Verſtandes. Wo viel Witz, Scharffſinnigkeit und Gründlichkeit iſt, da iſt die Kunſt zu erfinden in einem groſſen Grade. Weil aber hier von derjenigen Erfindungs-Kunſt geredet wird, die man der Verſuch-Kunſt entgegen ſetzet, und da man nicht durch bloſſe Aufmerckſamkeit auf unſere Empfindungen etwas entdecket, ſondern vielmehr aus einigen bekanten Wahrheiten durch richtige Schlüſſe andere herausbringet, die eine Verknüpffung mit ihnen haben; ſo wird zur Vollkommenheit der Erfindungs-Kunſt, oder vielmehr zu hurtiger Ausübung derſelben auch eine groſſe Erkäntniß erfordert. Wer viel weiß, der kan viel finden, wenn er die Kunſt zu erfinden beſitzet. Hingegen kan auch einer durch die Kunſt zu erfinden nichts herausbringen, wenn er geringe Erkäntniß in einer Materie hat.

Woher die Kunſt zu erfinden kommet und ihre Vollkommenheit.

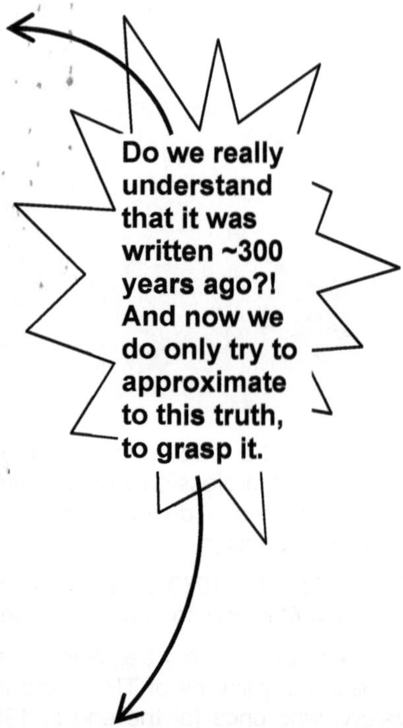

The art of invention is based on the quick wits and reason, and especially on the skill to make findings... Perfection in the art of invention comes from the perfection of the quick-wittedness and reasonableness. There, where a lot of quick-wittedness, reason and thoroughness are, the art of invention appears in a greater degree.

And since here we'll discuss the art of invention, which is opposed to the art of trials ... then perfection of the art of invention, or even more, its early implementation, requires knowledge. Who knows a lot, can achieve a lot, if one possesses the art of invention. But one, who does not have sufficient knowledge of the subject, can create nothing even with the art of invention. [66]

Christian von Wolff

[66] *Christian Freiherr von Wolff* (24 January 1679 – 9 April 1754) – German philosopher of the peak of Enlightenment rationality in Germany; translated by author (M.O.) from §861 of *"Vernünftige Gedanken von Gott, der Welt und des Menschen, auch allen Dingen überhaupt"* (1719), Magdeburg, 1752

With the TRIZonal[67] concept we briefly present three major zones of knowledge, art and… emotions, which can be extracted from the artifacts. These three zones are objectively realized and, consequently, are present in each and every artifact.

Definition of the term **"TRIZonal"**	**TRIZonal** is the totality of knowledge accumulated and materialized in an artifact as three interdependent zones (spheres) of knowledge (experience, skills).

At a minimum, the ideas are generated with the participation of three "mental" zones (fig.): 1) the applied (design) zone, 2) the creative zone, and 3) the psychological zone.

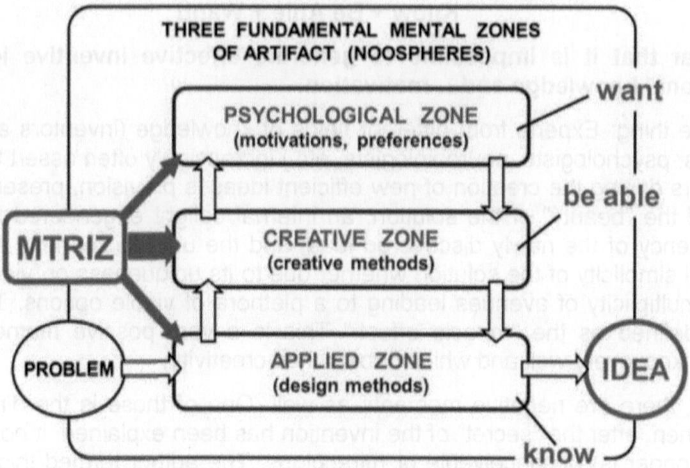

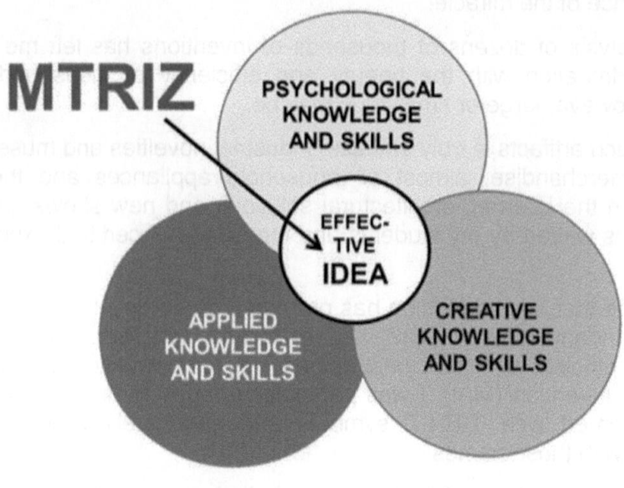

The theory and art of constructive design thinking are not taught in modern education. But maybe that is good because there is no creative methodology or tools that are comparable with TRIZ. Now **is the right time to start teaching TRIZ.**

---

[67] read as ['tri:z(ə)nl]

Instead, students are force-fed various cure-alls such as brainstorming and certain general creativity enhancement methods. This is not enough—not by a long shot!

The nucleus of new innovative thinking must coalesce, at a minimum, around applying knowledge, psychology, and TRIZ fundamentals presented in a palatable form and illustrated with convincing and easy-to-understand examples. And, most important, **TRIZ and Modern TRIZ penetrate into all three contiguous zones. So, this is just the ground on which to call this scheme "TRIZonal."**

Then the process of creating efficient ideas will use all three "specialized" zones: professional (applied) and creative knowledge and skills as well as commitment and tenacity in the attainment of formulated objectives (emotions, psychology), that is:

Know + Be Able + Want!

It is clear that it is impossible to generate effective inventive ideas without professional knowledge and... motivation.

One more thing: Experts from different fields of knowledge (inventors and engineers, designers, psychologists, neurobiologists, etc.) increasingly often assert that one of the key factors driving the creation of new efficient ideas is prevision, presentiment, and a feeling of the "beauty" of the solution, an internal delight engendered by the novelty and efficiency of the newly discovered idea, and the unexpectedness and oftentimes wonderful simplicity of the solution whether due to its uniqueness or, vice versa, to the sudden multiplicity of avenues leading to a plethora of viable options. These feelings can be defined as the "miracle effect." This is a very positive moment that many inventors know very well and which boosts their creativity.

However, there are negative moments as well. One of those is the directly opposite feeling when, after the "secret" of the invention has been explained, it no longer seems (and no longer is) unconceivable or miraculous. The author termed this phenomenon "disappearance of the miracle."

Still, my analysis of dozens of thousands of inventions has left me with a feeling of incessant admiration with the beauty and efficiency of ideas underlying all those inventions however large or small they may be.

The list of such artifacts is truly endless. Industrial novelties and museum exhibits, toys and store merchandise, almost all household appliances and the latest gadgets advertised on the Internet, architectural solutions and new shows, artistic and literary works, papers written by my students and thousands of certified experts. The list goes on and on...

It may well be that this admiration has prompted me to develop simple training tools to teach the fundamentals of TRIZ, i.e., extracting and reinventing. My penchant for simple and efficient solutions has helped me to formulate the idea for the Meta-Algorithm of Invention (MAI). I was particularly happy to discover that the MAI steps could be named with T-R-I-Z symbols and adequately filled with substance in accordance with those names.

In my opinion, in the immediate future psychological support methods will evolve even more intensively. In particular, such support has been used for several years on an experimental basis in our EASyTRIZ™ methods and software in the form of motivators based on color stimulation.

We will try to broaden these zones in following case study.

9 Operative Zone

The work is creative if its result is a creation of a novelty. And the very same process of creating the new can be deeply conscious and planned.[68]

Genrikh Altshuller

9.1 Reinventing of the Operative Zone (OZ)

9.1.1 Definition of the OZ

This section continues the study of the primary models of transformation introduced in the first section of the book. Now we will consider the structure of interaction of conflicting elements (constructs) in more detail.

We assume that we solve the design problem of improvements of some **Technical System (TS).** Design problems are caused by the interaction of the TS with the **World (W)** and the requirement to enhance the efficiency of TS functioning. The solution to the design problem requires changing the TS. Formulating and solving the problems are performed by the designer, i.e., the **Problem-Solver (PS).**

To solve the problem, the PS identifies a problem area in the TS, for example, a subsystem does not operate efficiently enough. Very simplistically, we define this subsystem as the **Operative Zone (OZ).**

Very conventionally (fig. 9.1), the OZ has been taken out of the TS, but the functional connectivity of the TS and the OZ is further illustrated by their belonging to the common area, which is encircled by a dotted line (edge of the hypergraph). One of the extreme cases is the full overlapping of both the OZ and the TS.

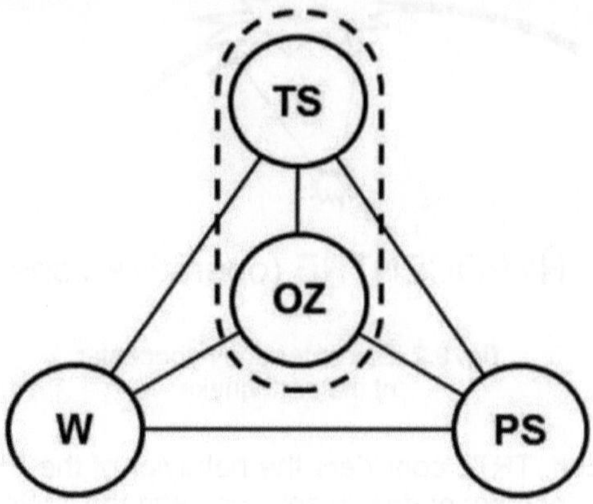

fig. 9.1. OZ as central area of TS to be changed by problem solving

The OZ should be modernized in solving this problem. Typically modernization consists of increasing the specific qualities of the subsystem and therefore the whole system. For example, the main qualities include productivity, reliability, and low power consumption.

The OZ usually contains a small amount of active participants, called the **"actor."** One or more actors, as well as the character of their interaction, will be changed to achieve the goal of transformation. And the purpose of the transformation is the elimination of contradictions in order to achieve the desired **Main Positive Function (MPF)** of the OZ.

[68] G. S. Altshuller *How to Learn to Invent.* – Tambov: Tambov Book Publishers, 1961 (in Russian)

M.A. Orloff, *ABC-TRIZ*, DOI 10.1007/978-3-319-29436-0_9

Usually changes are prevented by some causes, such as *the deterioration of one or more important qualities* in the trial project, while attempting to make changes. This situation is represented by **Standard Contradiction (SC)**. You also studied the sharper conflicts modeled by the **Radical Contradiction (RC)**, for example, when one and the same quality must satisfy two directly opposed requirements to change (or to maintain two opposed states for one and the same quality).

The solution is only possible with the elimination of contradictions.

If one has no idea how to eliminate contradictions, then the solution to the problem requires creative design thinking.

The advantages of TRIZ are fully disclosed here. First, the ***instrumental TRIZ*** gives the tools (models) for changing the OZ. At higher levels, TRIZ offers models for selection of the OZ and the realization of directed development of the TS.

Figuratively speaking, the OZ is the *hypocenter of the problem*, unlike the *epicenter of the problem*, which refers to external manifestations of the problem represented, in particular, in the form of contradictions. This metaphorical definition can be explained with the following simple scheme (fig. 9.2).

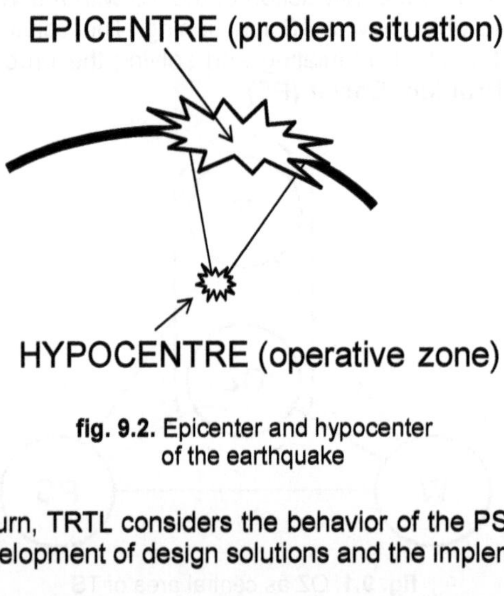

fig. 9.2. Epicenter and hypocenter of the earthquake

An earthquake starts in the *hypocenter* because of shifts or other changes that occur to the Earth's crust at some depth.

The *epicenter* is defined as the area where those subsurface changes manifest on the surface of the Earth.

In other words, "invisible" processes evolving in the hypocenter are the cause of the "visible" observable problems (destruction and other catastrophic consequences) that we see in the epicenter.

In turn, TRTL considers the behavior of the PS during problem solving and also in the development of design solutions and the implementation of created ideas into practice.

The Operative Zone (OZ) is one of the fundamental TRIZ notions.

When analyzing a problem situation within the TRIZ framework, we formulate models of the problem (or multiple problems) in the form of contradictions. Problems (regardless of the form of their presentation), conflicts, contradictions: All of these lie on the surface, i.e., in the "epicenter."

But what causes those contradictions? What subsurface processes create problems, conflicts, contradictions? What elements act as initiators or participants of those subsurface processes?

The common answer to all those questions can be located within the operating zone, i.e., the "hypocenter", of the problem situation.

Definition of the **Operative Zone**	The operative zone (OZ) is the totality of system components and, in some cases, system environment elements that are directly related to, and lead to the emergence of, the contradiction.
"Occam's razor" addition to the OZ definition	The OZ consists of elements that are necessary and sufficient to fully describe the processes (as well as causes and consequences) leading to emergence of the contradictions.

The formal definition of the OZ gives only a marginal idea of this notion. To truly master it, you must accumulate ample practical experience. Still, we will use examples provided in the following sections to review situations where the identification of the OZ is a relatively straightforward and intuitive process.

9.1.2 Reinventing with MAI T-R-I-Z

The origins of TRIZ are in the idea of extracting the models of transformation from the known inventions and in the idea of modeling the process of the invention as a whole (fig. 9.3).

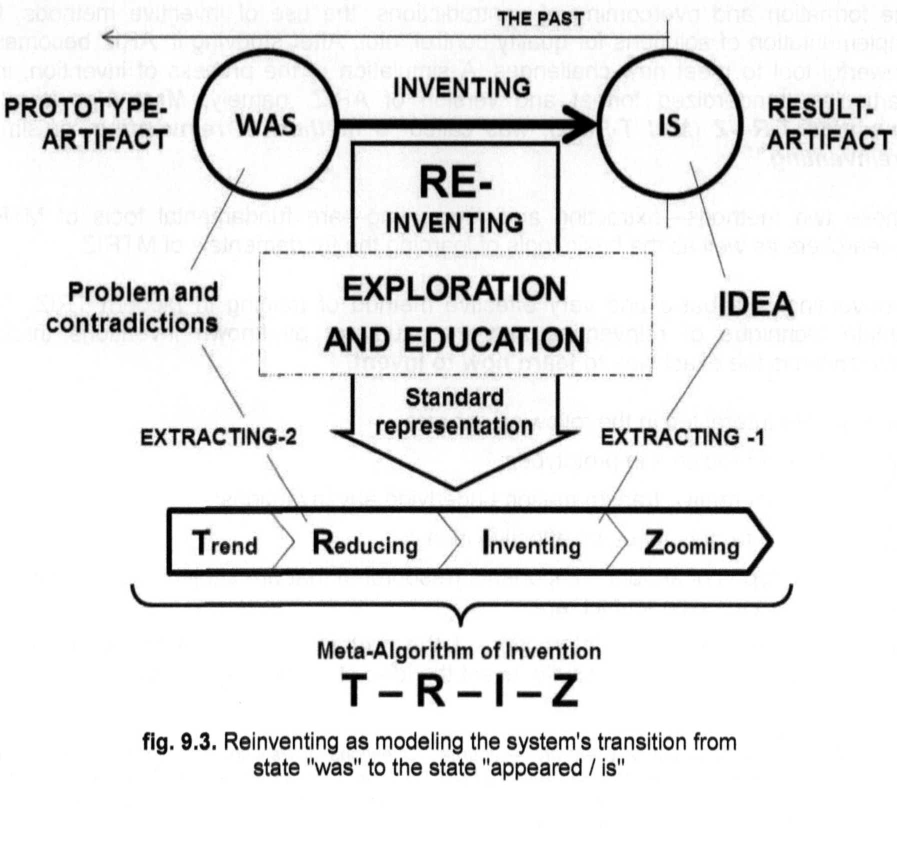

fig. 9.3. Reinventing as modeling the system's transition from state "was" to the state "appeared / is"

If we have some artifact (result-artifact) that is interesting to us, then it usually had a prototype, an artifact of the same purpose but with the known deficiencies that were eliminated in the result-artifact. Therefore, the invention of new construction is always a way to get from the prototype-artifact to the result-artifact.

TRIZ began from the studies of already-solved problems and from already-existing inventions. MTRIZ is applicable to any artifact, which always contains the creative idea. MTRIZ as a system both teaches and learns from the experience of civilization.

In fact, both of these ideas were implemented by Genrikh Altshuller throughout the development of TRIZ, since the very beginning, namely[69]:

1) Thousands of inventions were examined to *discover and identify the most common models of transformation (inventive methods),* thus providing a decisive change in the design of the prototype-artifact for obtaining the result-artifact with the desired properties and for eliminating the contradictions of the prototype. As a result, all model classes of different levels were identified in classical TRIZ. This exploratory approach, presented here in a modern structured and standardized form, was called "***a method of extracting***" or simply ***"extracting."***

2) Variants of the multistep methodological scheme were proposed and developed based on the fundamental concept of TRIZ (see chapter 3 in *ABC-TRIZ. Book I: How to learn to invent*)—the so-called "***Algorithm of Inventive Problem Solving (ARIZ)—*** which allowed to reproduce (simulate) the process of creating any invention to show the formation and overcoming of contradictions, the use of inventive methods, the implementation of solutions for quality control, etc.. After studying it, ARIZ becomes a powerful tool to meet new challenges. A simulation of the process of invention, in a particular standardized format and version of ARIZ, namely, ***Meta-Algorithm of Invention T-R-I-Z (MAI T-R-I-Z),*** was called "***a method of reinventing***" or simply "***reinventing***."

These two methods—extracting and reinventing—are fundamental tools of MTRIZ researchers as well as the basic tools of learning the fundamentals of MTRIZ.

Reinventing is a basic and very effective method of training in Modern TRIZ. This simple technique of reinventing and the study of all known inventions through reinventing is the exact way **to learn how to invent.**

First, we are interested in the following aspects:

> kinds of problems in prototypes;
>
> kinds of creative transformation underlying any inventions;
>
> types of targets to invent effective ideas;
>
> those types of artifact components (resources) that are subject to real change to appear as a new artifact; and
>
> the composition and interaction of the prototype components, which lead to problems and to necessity to invent the idea of eliminating the problems.

---

[69] *The Psychology of Inventive Creativity. – Journal "Questions of Psychology", No. 6, Moscow, 1956; coauthors: Genrikh Saulowitsch Altshuller (1926-1998) and Rafail Borisovich Shapiro (1926-1993);* explicit definition and explanation of the extracting and reinventing are missing in this and in all subsequent works on TRIZ by founder of TRIZ, as well as all other authors

Reinventing is fascinating process. Reinventing combines playing interest with exploration search. What could be more fun?!

I have used a four-stage description of the invention process based on MAI T-R-I-Z in the study of many examples, and you have made for them a fairly complete picture of the purpose and structure of the information at each stage of the process of invention in this scheme.

In this section, we discuss in more detail the elements of reinventing.

A simplified method of reinventing based on MAI T-R-I-Z:

1. For any result-artifact received to the study, the closest prototype-artifact preceding the result-artifact can be matched in the historical development of both the objects having the same purpose. Then the research and extraction of relevant information for each of the four stages of MAI T-R-I-Z is realized in reverse order (i.e., Z-I-R-T):

1) **Zooming:** identifying the benefits of the result-artifact. For this purpose, the results are studied in a different scale (zoom), i.e., at the levels of item, system, super-system, society, environment, etc. It is like zooming in with a digital camera or camcorder.

2) **Inventing:** Extraction 1 is to extract specialized and fundamental transformations from the result-artifact.

3) **Reducing:** Extraction 2 is to extract contradictions from the prototype-artifact.

4) **Trend:** The identification of problems and directions (trends) for their solution in the prototype-artifact.

2. To reconstruct (simulate) the complete process of invention in order to create a result-artifact, it is necessary to fill in a special standardized form of reinventing. At the same time, if necessary, the additional modeling elements are added: a description of the operative zone (OZ) and operative time (OT), a description of the ideal final result (IFR) and the functional ideal model (FIM), and other descriptions.

This allows us to represent the relationship between the processes of invention (inventing) and reinventing in the following form (fig. 9.4).

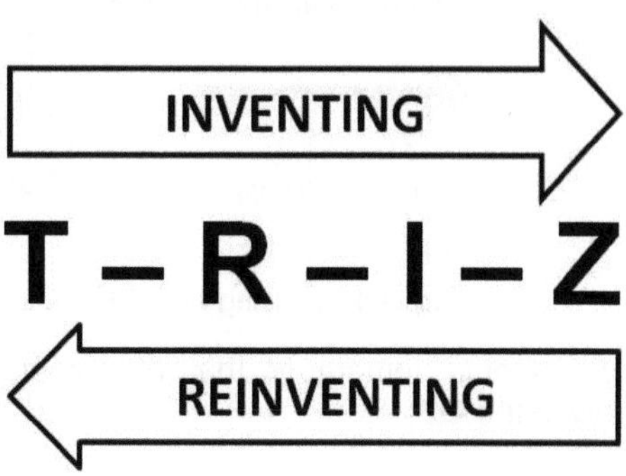

fig. 9.4. Inventing and reinventing based on MAI T-R-I-Z

Without a doubt it should be very useful for the reader to study attentively next two figures (9.5 and 9.6) ***to understand clearly the structural generality and the operative (executive) differences*** of applying MAI T-R-I-Z by education and invention.

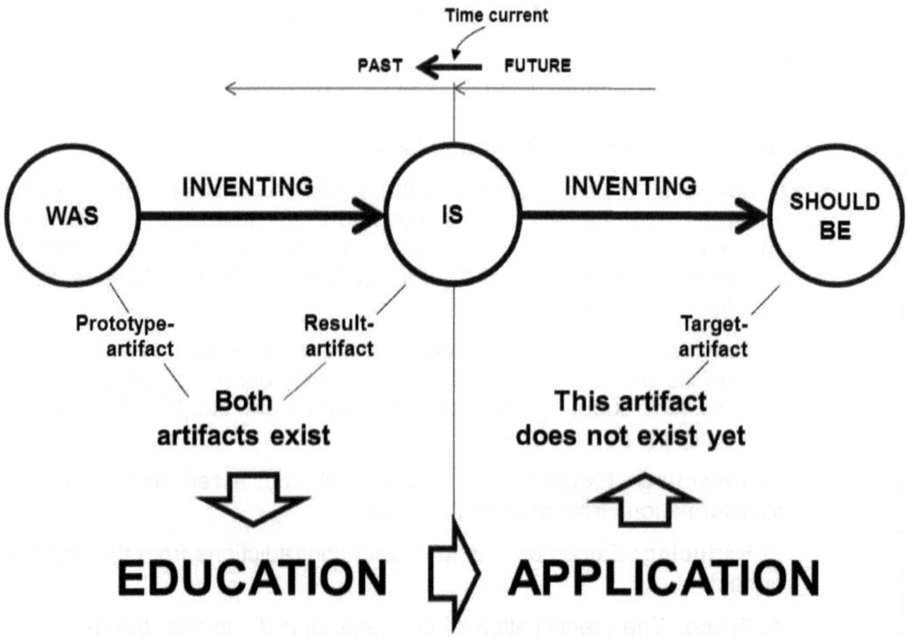

fig. 9.5. Inventing as transition between states "was" and "is" in the past (used for reinventing) and as transition between states "is" and "should be" to invent a new artifact (target-artifact)

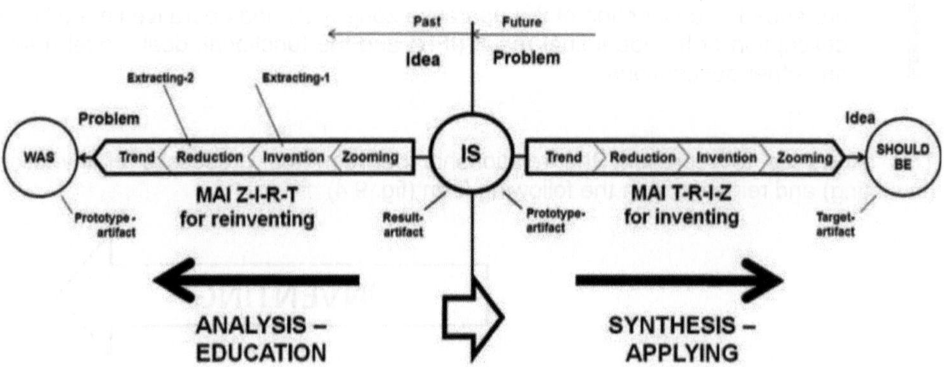

fig. 9.6. Applying MAI T-R-I-Z for education and new inventing

And we should remember that all transformations begins and are implemented primarily in the OZ!

9.2 Transformation of the OZ

9.2.1 Actors of the OZ

We will discuss several additional definitions that we will need in order to understand the nature of the OZ and to study the components acting in the OZ (fig. 9.7). You will find a summary description of this scheme just... before section 9.4 WORKSHOP FOR CHAPTER 9 ☺.

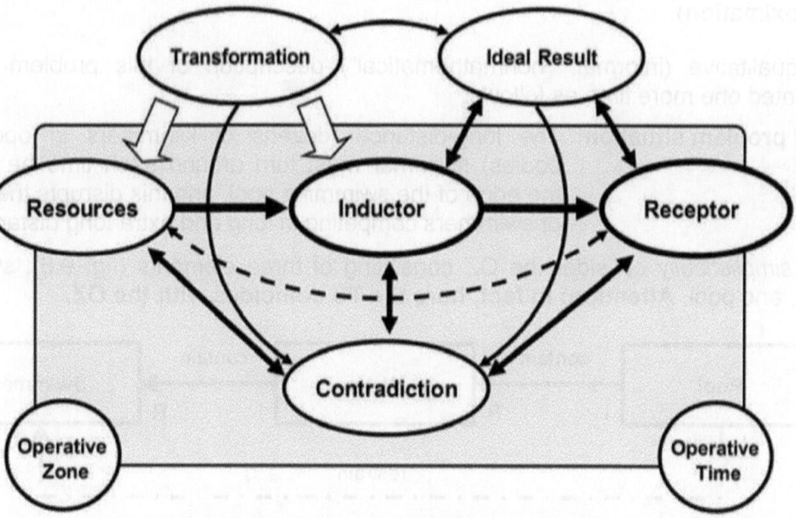

fig. 9.7. Main components of OZ at a glance

First and foremost is the determination of **"actors,"** i.e., OZ elements that actively participate in the resolution of problem situations and act as carriers of relevant processes and their results. Such elements can be conventionally divided into two categories: effect inductors and effect receptors.

Definition of **"actors"**	**Actors** are core OZ elements that actively participate in the problem situation and act as carriers of processes originating in the OZ and leading to the emergence of specific contradictive properties.
Definition of **"inductor"**	**The inductor** is an actor that creates an effect impacting another actor (receptor) through the transfer of energy, information, or material and initiates a change or action of the receptor.
Definition of **"receptor"**	**The receptor** is an actor that receives the effect generated by the inductor and changes, or is spurred into action, under the influence of such effect.
Definition of **"main receptor"**	**The main receptor is the product of the OZ.** If the OZ represents the entire system, the main receptor is the product of such system.
Definition of **"main inductor"**	**NB: Generally the main inductor directly affects the main receptor and should surely be transformed in order to invent a solution!**

The structure of the OZ can be illustrated by a number of graphic schemes. The simplest is the functional-structural scheme reflecting the OZ elements and the functional relations between such elements (informational or physical effects of diverse nature, for example, transfer of materials, mechanical or electromagnetic effect, etc.).

Let us consider the functional-structural model of the problem in **Example 4.6. Training of Long-Distance Swimmers (the "Swimmer" problem)** from the previous textbook "ABC-TRIZ. Book I: How to learn to invent."

Example 9.1. Functional-Structural Modeling of the OZ for the "Swimmer" (First Approximation).

The qualitative (informal, "nonmathematical") description of this problem can be presented one more time as follows.

Initial problem situation: The long-distance (dozens of kilometers in open-water bodies) swimmer must turn around each time he reaches the edge of the swimming pool, and this disrupts the training of swimmers competing in long and extra-long distances.

Let's *simplistically* consider the OZ consisting of three elements (fig. 9.8): swimmer, water, and pool. **Attention: in fact, here the TS coincides with the OZ.**

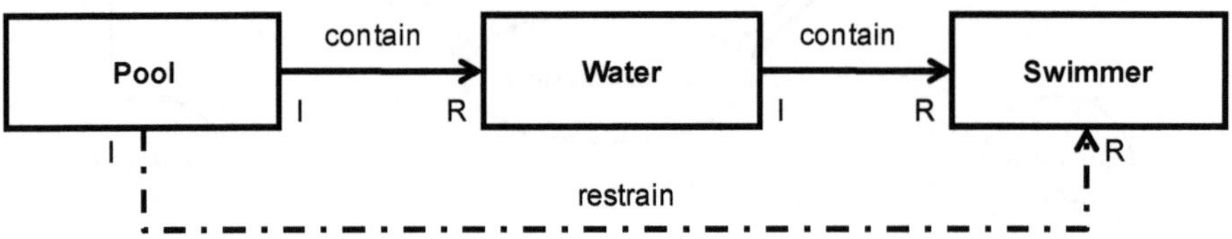

fig. 9.8. Operating zone model for the "Swimmer" problem

NOTE 1. Beginners usually consider the problematic situation in this manner, i.e., with no details, but considering the entire system as a whole: Aha! That's good, but this is bad! And they immediately start to "redo" the whole system! To suggest a huge round pool! Or, pool similar to 8! And so on... too complex!

OK, let's follow them! The water is "holding" the swimmer, and the swimmer is "pushing it off" to swim. The swimmer is the main receptor. The pool contains the water, and the water together with the pool can be regarded as a system that has both a positive and negative effect on the swimmer in the initial situation.

The **main positive function (MPF)** of the pool is to train the swimmer. However, the edges of the pool create a negative effect on the swimmer by restraining his movement and coercing him to turn around. These results in loss of energy by the swimmer and loss of time required to cover the distance.

The effect that the elements have on each other are designated with functional actions (contain, restrain).

NOTE 2. To eliminate the psychological negative fixing on the previous construction, all of the functional actions are formulated with infinitive verbs.

The "roles" of the elements are designated with the initial letters of the words "inductor" (I) and "receptor" (R). If necessary, we can also introduce I and R indexing.

So, we can see only the one actor acting negatively to "stop" swimmer at the edge. And we will change something in this TS (OZ) to get an "endless" water edge!

The standard contradiction was formulated as follows:

Pool ► endless swimming track VS complex shape

The short solution on the basis of the SC is shown in fig. 9.9. Full reinventing for this case is given in "ABC-TRIZ. Book I: How to learn to invent" (answers for Chapter 4. Modeling the Problem in Part 8 and Answers to Tests, fig. 8.1).

TREND

Open water bodies are used to train long-distance swimmers. In bad weather, training can become impossible. In a usual 50-meter pool the swimmer inevitably reaches the edge and has to turn around and push off to continue swimming, which impairs his technique and disrupts his rhythm. A pool with a complex circular (round, ellipsoid or 8-form) shape can remove the problem, but such pool will get too complex shape.

REDUCING

FIM: X-resource, together with available or modified resources, and without making the object more complex or introducing any negative properties, guarantees attainment of the following **IFR**:

[*adequate training of the long-distance swimmer in a pool*].

Standard Contradiction

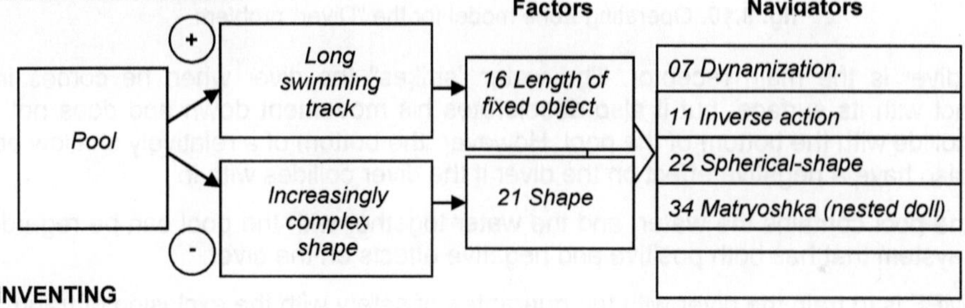

INVENTING

Key model: 11 Inverse action *– b) make a moveable part of an object or the environment fixed or a fixed part moveable.* ***Key idea:*** *make the water moveable. Also:* ***07 Dynamization****. Take in account the* ***22 Spherical-shape*** *– proposed a round pool!*

ZOOMING

Have the contradictions been removed? – Yes. – ~~No~~.

Super-effects: *it becomes possible to manage training parameters (movement speed and phases), and adjust the swimmer's swimming technique "on the go" as the coach is located in the immediate vicinity of the swimmer.*

Negative effects: -

BRIEF DESCRIPTION

To provide long-distance swimmers with adequate training possibilities and simplify pool design, the water in the pool is made moveable in accordance with navigator 11 Inverse action.

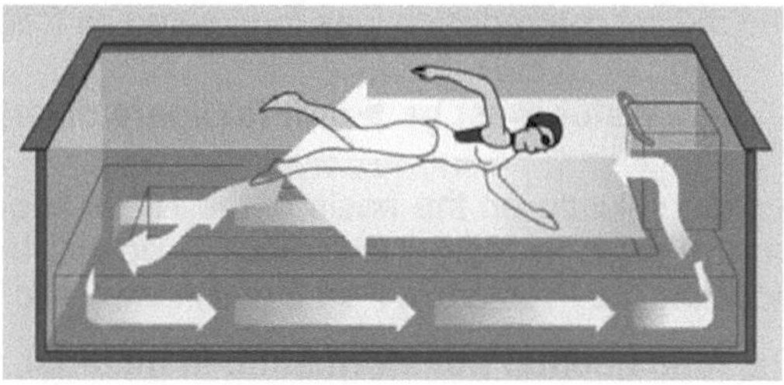

fig. 9.9. Short version of reinventing for solving to problem "Swimmer"

Let's undertake a similar review of the functional-structural model for the problem in **Example 4.7. Training of High-Board Divers (the "Diver")** from "ABC-TRIZ. Book I: How to learn to invent."

Example 1.2. Development of the Functional-Structural Model of the OZ for the "Diver" Problem.

Initial problem situation: The diver can sustain injuries at the time he enters the water when jumping from a diving tower or high-board. The probability of injuries increases if the diver enters the water at a wrong angle.

The OZ also consists of three elements (fig. 9.10): diver, water, and pool.

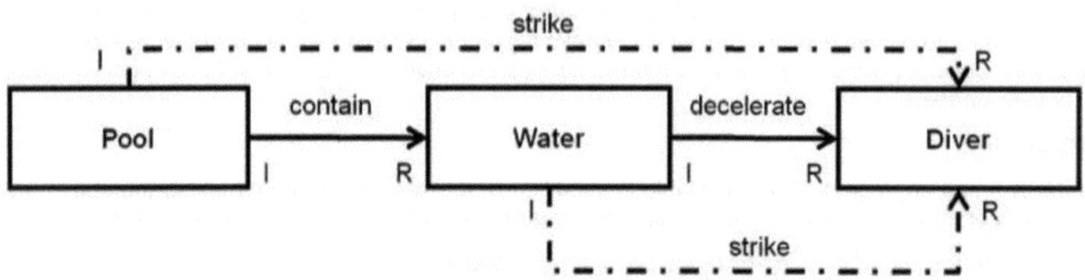

fig. 9.10. Operating Zone model for the "Diver" problem

The diver is the main receptor. The water "strikes" the diver when he comes into contact with its surface, but it also decelerates his movement down and does not let him collide with the bottom of the pool. However, the bottom of a relatively shallow pool can also have a negative effect on the diver if the diver collides with it.

So, the pool contains the water, and the water together with the pool can be regarded as a system that has both positive and negative effects on the diver.

The MPF is to train the diver with the guarantee of safety with the exclusion of trauma.

In fact, here the OZ also coincides with the TS, except for the diving tower, which really does not affect the future solution; This is important for definition of the OZ in this case.

Which actor could be changed in this OZ? Both the pool and the water potentially could be transformed to get MPF. And it looks simpler to start from analysis of water properties.

It was done in reality.

The radical contradiction was formulated as following:

Water must be ►soft (to spare diver) VS hard (physical property)

A short solution on the basis of the RC is shown in fig. 9.11. Full reinventing for this case is given in "ABC-TRIZ. Book I: How to learn to invent" (answers for Chapter 4. Modeling the Problem in Part 8 and Answers to Tests, fig. 8.3).

The **fundamental transformation in material,** supported by the specialized models of *01 Change in the aggregate state* and *14 Use of pneumatic or hydraulic constructions,* can be found in section **S25. Afs-catalogue.**

TREND

Training of high-board divers is associated with high risk of injuries and occupational diseases. The critical moment during such training is the moment when the diver enters the water. For example, a fall on the back from a height of 10 meters can lead not only to painful contusions and skin disruption, but also to dangerous spinal column injuries.

How do we make training of high-board divers safer?

REDUCING

FIM: X-resource, together with available or modified resources, and without producing any negative properties, guarantees attainment of the following **IFR**:

[*safe entry into water from any height*].

Radical Contradiction (two forms):

a) Water has to be ▶soft (to spare diver) VS hard (physical property)

b) | *Water* ⇨ | *must be* ***"soft"*** *so as not to cause injuries* | & | *must be* ***"hard"*** *in accordance with its natural properties* |

INVENTING

Key specialized models from Afs-catalogue of 4 fundamental transformations: 01 Change in the aggregate state – *a) transitions into "pseudo-states"; b) changes in concentration;* ***14 Use of pneumatic or hydraulic constructions*** – *use gaseous or fluid parts instead of fixed parts in an object: inflatable…, air cushion…*

Key idea: *inject compressed air at the time of the jump, make an* ***"preliminary installed air cushion"*** *on the surface of the water.*

ZOOMING

Have the contradictions been removed? – Yes. – ~~No~~.

Super-effects: 1) *safe training of children; 2) inviting the visitors for active recreation in diving pool by jumping from the big height!*

Negative effects: *increased complexity of construction.*

BRIEF DESCRIPTION

To make high-board diver training safe, a powerful jet of compressed air is injected into the water during the jump, creating an "air cushion" on the surface of the water. Navigators used: 01 Change in the aggregate state and 14 Use of pneumatic or hydraulic constructions to realize fundamental transformation in material / energy.

fig. 9.11. Short version of reinventing for solving to the "Diver" problem

This and similar problems are resolved by changing (transforming) the resources of the OZ. Incidentally, in many cases ideas for solutions are generated when describing and analyzing such resources.

Accordingly, we proceed to analyze, and select resources for the resolution at least one of the two problems presented in the "Swimmer" problem.

9.2.2 Resources of the OZ

Definitions of resources and their classification are provided in the tables below.

<table>
<tr><td>Definition
of "resources"</td><td>Resources are the totality of material and nonmaterial (model) organizational components of the artifact[70] as an integral systemic object.</td></tr>
<tr><td>Systemic-technical resources
(addition 1 to the definition of "resources")</td><td>Systemic-technical resources are the totality of system-defining, -forming, and -realizing components of the artifact.</td></tr>
<tr><td colspan="2">System-defining components are models that determine the purpose and effectiveness of the artifact.
System-forming components are all of the components comprising the systemic organization of the artifact: functional composition, process-related conceptualizations, structure-related conceptualizations, and, naturally, their underpinning material and nonmaterial constituent parts such as construction and knowledge (information), in particular, ideas, especially conceptual ideas.
System-realizing components include, first and foremost, the integral construction of the artifact determining all of its processes and functions and shaping its actual interaction with surrounding artifacts.</td></tr>
<tr><td>Physical–technical resources
(addition 2 to the definition of "resources")</td><td>Physical–technical resources are the totality of system-forming and -realizing components of the artifact residing in its construction.</td></tr>
<tr><td colspan="2">Construction defines the physical realization of the artifact in space–time–material and all processes involving its interaction with materials, energy, and information occurring both within the construction and in the course of its exchange with surrounding artifacts.</td></tr>
<tr><td>OZ resources
(addition 3 to the definition of "resources")</td><td>OZ resources (operative resources) are the totality of system-defining, -forming, and -realizing components of the artifact residing in the OZ.</td></tr>
</table>

The well-known classification table (fig. 9.12) divides resources into eight types, which are then lumped into two large groups: systemic-technical resources and physical-technical resources. It is always preferable to select, from among the resources listed in the table, one with the leftmost value of any kind.

> One of the key TRIZ statements is the reminder (relevant for the purposes of both analysis and synthesis) that when any resource of the artifact is changed, all other resources of such artifact, and possibly the resources of the surrounding artifacts, are also changed.

Successful resolution of the problem relies on the resources involved in analyzing and modifying the OZ and synthesizing an idea for the solution. Accordingly, the size of the

[70] Here the term "artifact" is used in its broadest meaning and refers to any integrated systemic formation (whether material or nonmaterial) including any construction, information, process, phenomenon, work of art of any type, etc.

OZ (number and composition of its elements and relations) is directly dependent on the resources that are taken into consideration.

SYSTEMIC-TECHNICAL RESOURCES			
SYSTEMIC	**INFORMATIONAL**	**FUNCTIONAL**	**STRUCTURAL**
Related to general system properties	Related to transmission of information-bearing messages	Related to creation of functions	Related to the composition of the object
Purpose of the system, its efficiency, productivity, reliability, safety, survivability, durability, etc.	Data integrity, accuracy, validity, interference immunity, methods and efficiency of measurement, management, encoding, etc.	Main useful function compliant with the purpose of the system, auxiliary functions, negative functions, description of the operating principle (functional model)	List of component–intercomponent relations, types of structures (linear, branching, parallel, closed, etc.)
PHYSICAL–TECHNICAL RESOURCES			
SPATIAL	**TEMPORAL**	**MATERIAL**	**ENERGETIC**
Related to geometric properties	Related to temporal evaluations	Related to properties of materials	Related to properties of energy and its manifestations
Shape of the object, dimensions of the object (length, width, height, diameter, etc.), shape features (presence of cavities, projections, etc).	Frequency of events, duration of time intervals, duration of time lags/leads. **Operating time (OT):** Interval of time of the existence of the problem situation	Chemical composition, physical properties, special engineering properties	Types of applied and measured energy including mechanical, gravitational, thermal, electromagnetic, and other forces; methods of energy utilization, etc.

Properties of the resource	**Value:** free → inexpensive → expensive
	Quality: harmful → neutral → useful
	Quantity: unlimited → sufficient → insufficient
	Readiness for application: ready → change in progress → creation in progress

fig. 9.12. Classification of OZ resources and value scale used to select resources required to resolve

When the problem is successfully resolved through the transformation of resources, development of contradiction models is required only if we must adjust the formulation of the problem or simplify the description (reducing) of the initial problem situation. Some important aspects of operating with resources of various types are specified in the table in fig. 9.12. In this section, we will limit ourselves to reviewing practical examples and comments from the viewpoint of transformations of resources involved in solution-finding.

Example 9.3. OZ Resources for the "Swimmer" Problem

In its most general form, the problem was reduced to the following question: What do we must change to make sure that the swimmer does not collide with the pool's edges?

Despite the generality of the question, it contains a requirement that determines the main practical purpose of the future change. This change can be expressed in different ways, but – as we saw before even such wording of the question prompts some potentially fruitful ideas such as building a round or figure 8–shaped pool with an "endless" swimming track. With such track, the swimmer would be able to swim for an unlimited period of time without running into the pool's edges because such edges (in the usual sense) would simply not be there.

Inasmuch as we were not happy with those ideas, and their implementation appeared excessively complex, the original purpose was replaced with the following: The pool must have an endless swimming track so that the swimmer never reaches pool's edges, and it must have a simple construction. Such dual requirement follows from the definition and analysis of the standard contradiction inherent in this problem. It should also be noted that a more elaborate description of requirements applicable to the "image" of the future solution is provided in the following section **9.2.3. Ideal Target Modeling**.

To change something, we must determine what changes are possible and expedient in the first place.

Now, analysis of the initial situation established that the OZ consists of three elements: swimmer, water, and pool. These elements possess numerous properties, and each such property is related to a certain type of resource.

Let us draw an inventory of resources available in the initial problem situation (fig. 9.13). Descriptions can be provided by direct reference to specific available or required resources or by formulating requirements applicable to certain unknown components that presumably belong to a certain type of resource.

The entirety of properties and requirements presented above makes up an initial "composite sketch" of the problem situation. Some of those properties are in direct opposition to the target requirements. Thus, in a usual rectangular pool, structural relation 2 directly conflicts with structural relation 3 because in the progress of the swimmer from one side of the pool to the other, he eventually comes into contact with the edge and must turn around.

In a round pool, there are no such contacts or turnarounds, but in this solution there emerges a conflict between the properties related to the spatial resource: *unlimited* track and *limited* pool (or, better yet, a marginally smaller pool).

If we keep "shrinking" the pool subject to a certain limit, we can determine the minimal size of the pool in which the swimmer can theoretically keep swimming. Such size can be comparable with human body dimensions; for example, it can be limited by the double height of the tallest man. Let us assume that the length of the pool is 5 meters. How can a man keep swimming in such pool for an unlimited time without touching its edges?

Systemic-technical	**Systemic**	1. Systemic result of the training process: Swim a certain distance in specified or minimal time. 2. Inexpensive pool.
	Informational	1. The length of the distance covered by the swimmer is measured by the number of swimming track lengths that he swims. 2. The swimming time is measured directly by using some timepiece.
	Functional	1. The water is holding the swimmer (on the surface). 2. The swimmer is pushing the water away to keep swimming. 3. The swimmer is moving along the surface of the water (swimming).
	Structural	Structural relation 1: The swimmer is moving relative to the water. Structural relation 2: The swimmer is moving relative to the pool. Structural relation 3 (objective): The swimmer is not able to touch the edge of the pool.
Physical-technical	**Spatial**	1. Unlimited swimming track. 2. Small or even very small pool.
	Temporal	OT: Training interval. It is during this time interval that there emerges the conflict related to the construction of the pool.
	Material	Round or figure 8–shaped pools require a lot of building materials.
	Energetic	Large energy expenditure to keep the water warm in a large pool.

fig. 9.13. Description of OZ resources for the "Swimmer" problem

Hypothetically, the answer could be this: If the swimmer were motionless relative to the pool, he would not be able to reach the edge of the pool. This, however, gives rise to the question: How would then the swimmer be able to swim, i.e., keep moving relative to the water?

Still, the hypothetical answer, as formulated above, contains an idea (direction) that can, in principle, help resolve the problem in its general structural form using the structural resource. With no unequivocal practical recommendations following from such idea, we can only suggest a number of potentially realizable solutions.

Thus, there is some practical value in the following two specific ideas:

SOLUTION 1: Fasten the swimmer, for example, in the middle of a small pool with an elastic (i.e. extensible, for example, made of rubber; under certain conditions, though, it can be nonelastic) cable attached to his belt (see fig. 9.14);

SOLUTION 2: Make the water moveable (see fig. 9.15).

The first solution has one significant advantage and one significant disadvantage. The advantage is its exceptional simplicity. The disadvantages include the following: It is difficult to measure the distance and speed with which the swimmer is moving (relative to the water) in certain time intervals and cables may prevent free arm movement.

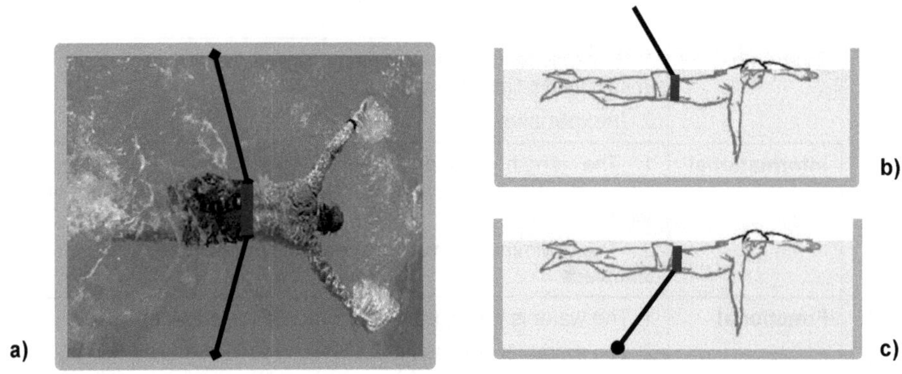

fig. 9.14. Possible solutions of the "Swimmer" problem: Swimmer fastened with a cable

Because of the evident new problems, we do not provide the adjusted resource analysis table for this solution.

The second solution (fig. 9.15 and 9.16) corresponds to the control solution shown in fig. 9.9.

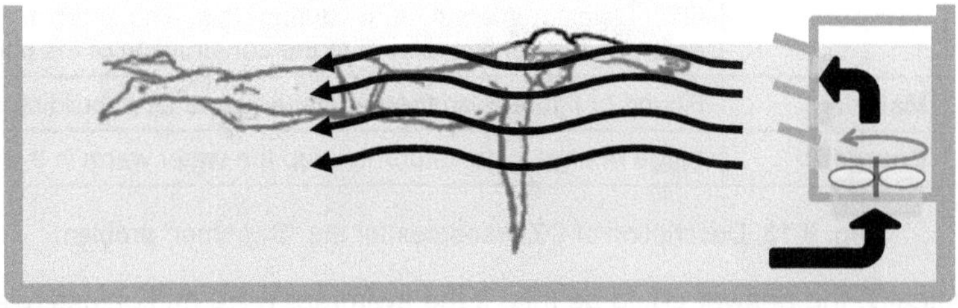

fig. 9.15. Swimmer problem: Water-flow solution

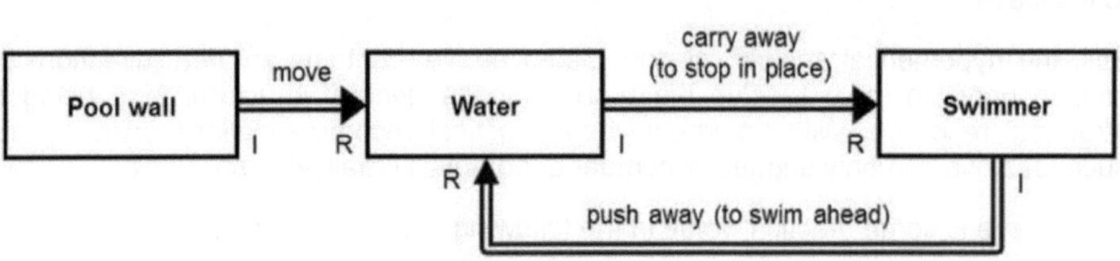

fig. 9.16. "Swimmer" problem: Solution model

The idea underlying this solution is to make the water move inside the pool. To do that, it is necessary to install, at one edge of the pool, a sufficiently powerful regulated pump forcefully ejecting water through special slits. The slits are designed to ensure uniform water flow of a certain "height." The speed of the water flow can be regulated by changing the speed of rotation of the pump's blades.

The resource analysis table for this solution is presented below (fig. 9.17):

LC	Resource	Conformity
	Systemic	1. Systemic result of the training process: Swim a certain distance in a specified or minimal time. 2. Inexpensive pool.
	Informational	1. The length of the distance covered by the swimmer is *calculated based on the known water flow speed.* 2. The swimming time is measured directly by using some timepiece. *A more accurate time measurement method is required to account for water-flow speed changes.* *3. It is necessary to monitor changes of water-flow speed.*
*	**Functional**	1. The swimmer is pushing the water away to keep swimming AND is moving along the surface of the water (swimming). 2. The water is holding the swimmer (on the surface) AND **produces an increasing and controllable resistance of motion.**
**	**Structural** **Key idea**	Structural relation 1: The swimmer is moving relative to the water. *Structural relation 2: The water is moving relative to the pool. It is necessary to complement the system with a new element, i.e., a pump feeding water in the direction opposed to that in which the swimmer is moving as well as other structural elements to manage the water flow.* *Structural relation 3: The swimmer is not moving relative to the pool and accordingly is not able to touch the edge of the pool.*
	Spatial	1. Unlimited swimming track. 2. Small or even very small pool.
	Temporal	OT: Time to reach the edge of the pool (when moving forward or back).
	Material	Construction of the pumping system.
	Energetic	*Operation of the pump and other new elements will entail additional energy consumption.*

fig. 9.17. Description of OZ resources for the "Swimmer" problem: Water flow

Three stars mark the dominating resource that must be changed to realize the key idea underlying the solution. The white arrow points to the key change in the dominating resource. The black arrow points to the emergence of a problematic property (disadvantage).

Example 9.4. Solution Zooming: "Swimmer" Problem

Critical Review of the Solution

When the swimmer is swimming against the water flow, he can swim for an unlimited period of time even in a small pool. However, there remains a possibility that under certain conditions the swimmer may reach an edge of the pool.

If the swimmer applies relatively smaller physical exertions, the water can pull him downstream, and he may touch the back edge of the pool with his legs. If the swimmer applies relatively larger exertions, he can overcome water resistance and reach the front edge of the pool with his arms or his head. The OT that remains until one of these situations occurs is very short.

There arises the "small-pool" problem. The OZ functional/structural scheme for this problem can be presented as in fig. 9.18. The problem can be solved through enhancement of the informational resource, i.e., better management of water-flow parameters.

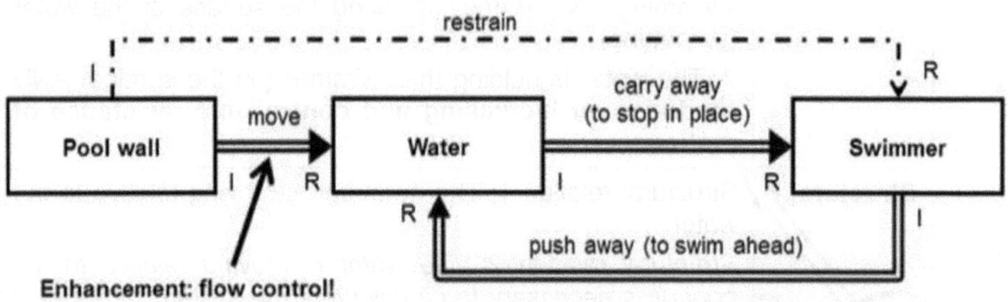

fig. 9.18. OZ model for the "small pool" problem

Enhancement of the Solution

You must pay attention to two things. First, the description of the material resource contains no special requirements to the new construction! Second, a new element (pump) must be introduced into the system. Again, no additional requirements are related, among other things, to the place of installation of the pump (space) or its material.

Hence, you are "free" to implement the water flow idea even in an existing pool. Indeed, what can stop us from submerging a pump at one end of the pool during the training of a marathon swimmer? We can also embed a pump into the wall of an existing pool and activate it during training sessions.

9.2.3 Ideal Target Modeling

The ideal final result is like the rope that the alpinist is holding when climbing up a steep slope. The rope is not pulling him up, but it supports him and does not let him fall. As soon as he lets go of the rope, the fall becomes inevitable.[71]

Genrikh Altshuller

Without going into details, we introduce another important concept here: the **General Contradiction (GC).**

This concept is closely linked to the MPF of the TS. By the way, we believe that the content of the MPF is clear intuitively and therefore did not impose a specific definition.

General contradiction	**The general contradiction** (in classical TRIZ this contradiction is called *administrative*) is a systemic requirement that merely reflects the general need to attain a certain property (or state) related to achievement of the MPF, such as the desired functioning of the system, or to remove an obstacle preventing the achievement of MPF.

NOTICE. For resolving, the **general contradiction** is usually reduced to the standard or radical contradiction.

In this particular case, the GC may coincide with the MPF, but usually the GC reflects some additional requirement that must be implemented to achieve the goal (target) of the MPF in its entirety.

We show this in the following examples.

Example 9.5. Swimmer.

MPF: Properly training ultra long–distance swimmers in an ordinary pool.

Problem: The small size of the pool, and the need to turn over at the edges, which prevents the correct movement.

GC: Swim without having to turns over, but it is not known how to do this.

In itself, the GP has only one part of the form of contradictions—the targeting part. The goal is to achieve a certain result or eliminate the deficiency. The second part means only one thing: the inability to do so in a situation in the here-and-now, that is, quickly and efficiently.

In this example, we once considered the design of a round pool, but this is difficult and takes a long time to build! So this solution is not of the here or the now!

This is the GC!

---

71 G. S. Altshuller *Creation as an Exact Science* (1979; later editions are available).

Example 9.6. Diver.

MPF: Proper training jump from a big height.

Problem: Injuries by reason of impact with the water.

GC: To eliminate injury at training. And the second part is not pointed out here: Initially it was not known how to do it!

Example 9.7. Kremlin Stars – Look at *Problem P3. How Do You Design the Ruby Stars of the Moscow Kremlin?* in "ABC-TRIZ. Book I: How to learn to invent".

MPF: Big and bright star at high altitude.

Problem: Danger of the star falling in high winds.

GC: Design of robust stars in any wind. But when the constructors started to design, they did not know how to do it!

The metaphor in the epigraph helps to further our understanding of the ***ideal final result (IFR)*** notion. Indeed, not only does the rope keep the mountaineer from falling, it also both points and creates the way to the top!

Then again, somebody must have fixed the rope higher up the slope, and that somebody must have been a master—a pioneer who saw the aim and the way leading to it—who had to climb up relying only on himself (in our case, on his knowledge and experience) and who blazed the trail to the peak for those who would follow.

The difference is that we—those who look for new ways to secure the success of our companies—have to act as pioneers in our problem situations. We must learn to see the aim—as well as the way leading to it—and install durable and reliable "ropes" facilitating the ascent to the top, i.e., to the final ***ideal*** result or destination.

Definition of **"ideal final result"**	**The ideal final result (IFR)** is the *definition of the objective* of artifact enhancement presented as the required functional state or action that best (ideally) conforms to the designated purpose of the artifact.
Definition of **"functional ideal model"**	**The functional ideal model (FIM)** is the formally arranged *model-metaphorical description* of the way in which the artifact must function in order to attain the IFR.
Problem-solving motion **"target–meta-trend–paths"**	The process of the generation of ideas on the basis of TRIZ models and methods is an *objective-directed system* where the IFR and the FIM are defined through objectives and meta-trends (general directions and methods used to approach the objectives, i.e., "targets"), and the path to each objective is formed by TRIZ transformation models.

Beginners normally experience noticeable difficulties with the formulation of the IFR. The reason is psychological inertia, which draws the would-be (apology for a) inventor back to the prototype, whereas the objective is to obtain a new artifact with new properties, which must be stated in the form of the IFR.

To weaken psychological inertia during IFR formulation, TRIZ practitioners have developed the following two time-proven rules:

1) During the initial stages, you must not think how and with what (with the help of what resources) you will obtain your solution.
2) An unknown resource or action required to obtain the desired result can be temporarily replaced with a metaphorical symbol, for example, the ***X-resource***.

Let us remind our readers the basic FIMs.[72] To better remember these titles and their definitions, we have also proposed also new expanded titles of known functional ideal models.

> **1. Maxi-FIM** or *"FIM of the Operative Zone itself"*:
>
> The OZ ***by itself*** ensures obtaining the **IFR:**
>
> [required *functioning]*

> **2. Macro-FIM** or *"FIM of the enlarged OZ"*:
>
> The ***X-resource***, without making the system excessively complex and without causing inadmissible negative effects, ensures—together with other available resources—obtaining the **IFR:**
>
> [required *functioning*]

> **3. Micro-FIM** or *"FIM of the deepened OZ"*:
>
> The ***X-resource*** in the form of *particles of substance or energy* is located inside the OZ and ensures, together with other available resources, obtaining the **IFR:**
>
> [required *functioning]*

The repeated scheme in fig. 9.19 illustrates the interaction of the IFR, the FIM, and other notions known already to reader (I hope).

> The search for a solution starts with formulating the contradiction as a model of the problem. Then we establish the objective (IFR). Then we select one or more FIMs to identify the meta-trend leading to the IFR. Then we select transformation models, i.e., specific paths that will take us to our destination. Last, but not least, we walk those paths: We invent specific solutions and change resources of the prototype artifact so as to finally arrive at a new artifact consistent with the sought-after IFR.

It should be noted that oftentimes once you have correctly formulated the IFR and correctly identified the relevant FIM, they THEMSELVES will lead you to the idea for the solution.

[72] M. Orloff, *Inventive Thinking through TRIZ*, Part 8.2. *Functional Ideal Modeling*.

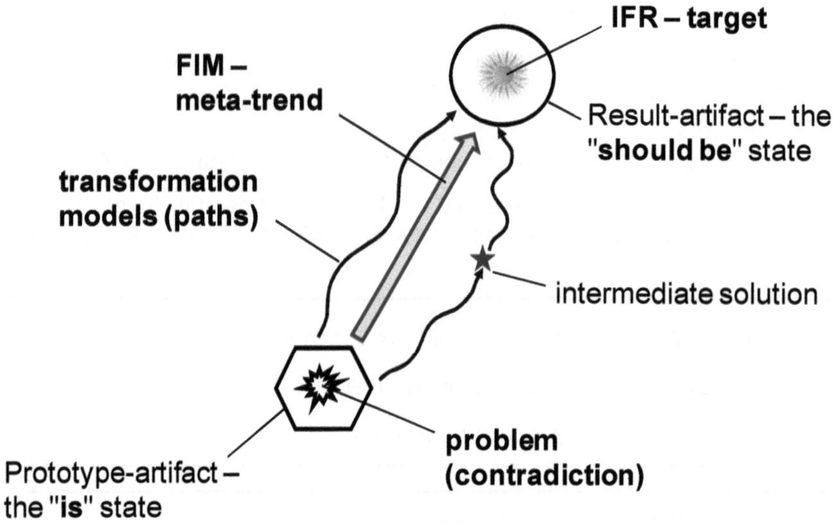

ATTENTION: You should know this scheme by heart!

fig. 9.19. Connection of primary notions of problem resolution with IFR and FIM

Example 9.8. The IFR and the FIM for the "Diver" Problem

In its most general form, the problem was reduced to the following question: What must we change to make sure that divers are not injured during their training sessions as a result of awkward water entry?

Now, analysis of the initial situation was established that the OZ consists of three elements: diver, water, and pool.

In this case, a review of available resources offers no apparent prompts that might facilitate the generation of an idea for the solution. Accordingly, we omit the resource table for this problem.

All we come up with is a notion that we must somehow make the water "soft"! Interestingly, the soft-water idea proves to be quite constructive and ultimately yields a solution, albeit from a somewhat different angle. Namely, the soft-water requirement can be regarded as an ***ideal final result*** for the future solution.

We need to realize some sort of "soft" water that would gently impact the diver at the time he penetrates the surface, and then decelerate his downward motion until he begins to come back up. Essentially, we are now but one step away from finalizing the IFR wording for this problem.

IFR: ***The water must be soft!***

OT: The IFR must exist for the duration of the jump performed by the diver including the moment when he enters the water.

Operative resources (without the summary table): water, air, atmospheric pressure, and gravitation.

And what does it mean: "The water must be soft"? It means that the water density must be considerably less than its natural value.

Micro-FIM or "FIM of the deepened OZ":

> ***The X-resource*** in the form of *particles of substance or energy* is located inside the OZ and ensures, together with the other available resources, obtaining the IFR
>
> ["soft" water = *low water density]*

And one cheerless digression now:

> I remember one episode that happened to me in Tenerife (Canary Islands). I swam for a long time over submerged rocks in the middle of the output of a small bay. There was a little choppiness from the ocean. I did not notice that the choppiness had increased (a mistake!), and soon strong waves rose above me. At this moment, after piercing an another wave, I saw a continuous white "boiling" environment under the water that deafened me with a hiss and in which was difficult to float upward. It was difficult to discern where was up, where was down, and during my descent and with a strained consciousness, I saw the black jagged cliffs rushing to meet me. The density of the water, which was richly saturated with air, was inadequate for me to swim to the surface.
>
> I had to think fast to determine how to get out of this turmoil or, "scientifically," this turbulence. And it was not immediately clear that swimming directly to the shore would not work because I always ended up drifting back to the "foam" of the counter-flow, which was probably formed by the shore. An orange flare launched by lifeguards rose above the shore to become, as I thought later, my visible object.
>
> Rising to the next wave, I saw that close by, approximately 100 meters away from me, surfers were standing on the long smooth waves rushing to the shore along one of the sides of the rocky bay. And some of them, lying on the surfboard, were glancing in my direction. Obviously, they noticed that I was in danger in this huge foamy area of water as well as continuous high waves. I realized that I should first swim along the shore toward the surfers, where there was no counter-flow, and then to turn toward the shore.
>
> So, let this story be a lesson for beginners, which I was then.

Ideal solution idea: Feed a powerful jet of compressed air to ensure that the water in the OZ is rapidly saturated with air!

The schematic representation of the solution is shown in fig. 9.20.

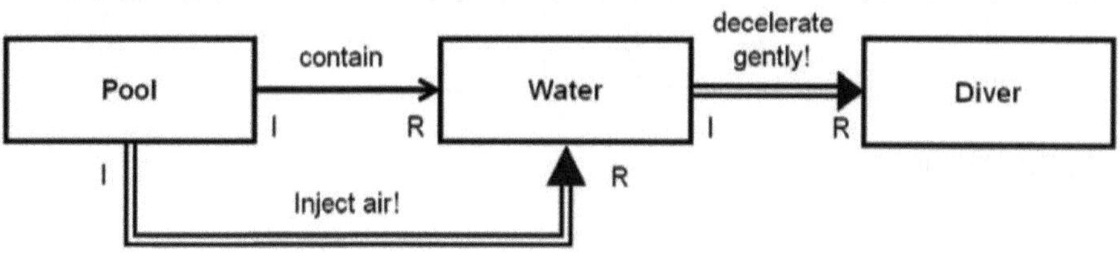

fig. 9.20. Solution model for the "diver" problem

Compressed air injected into the water upward from the bottom of the pool at the time when the diver is making the jump creates (in the hypocenter, which in this case coincides with the epicenter!) an "air cushion" on the surface, which considerably reduces the impact experienced by the diver.

Example 9.9. Solution Zooming: "Diver" Problem.

If the water is oversaturated with air, the diver may fall through the air–water mixture and hit the bottom of the pool.

Conversely, if the level of saturation is insufficient, the efficacy of the air cushion may be seriously impaired.

These extremes are illustrated by the functional–structural scheme of the OZ as shown in fig. 9.21, which models certain problems that may emerge if the air cushion malfunctions.

Zooming at the level of the OZ: The problem is eliminated or at least mitigated to a considerable extent.

Zooming at the level of the coach: 1) It becomes possible to try more complex jumps; and 2) it becomes possible to start training children at a younger age.

Zooming at the level of the pool: There emerges a commercially viable opportunity to enroll more visitors who will derive pleasure from safe high-board diving.

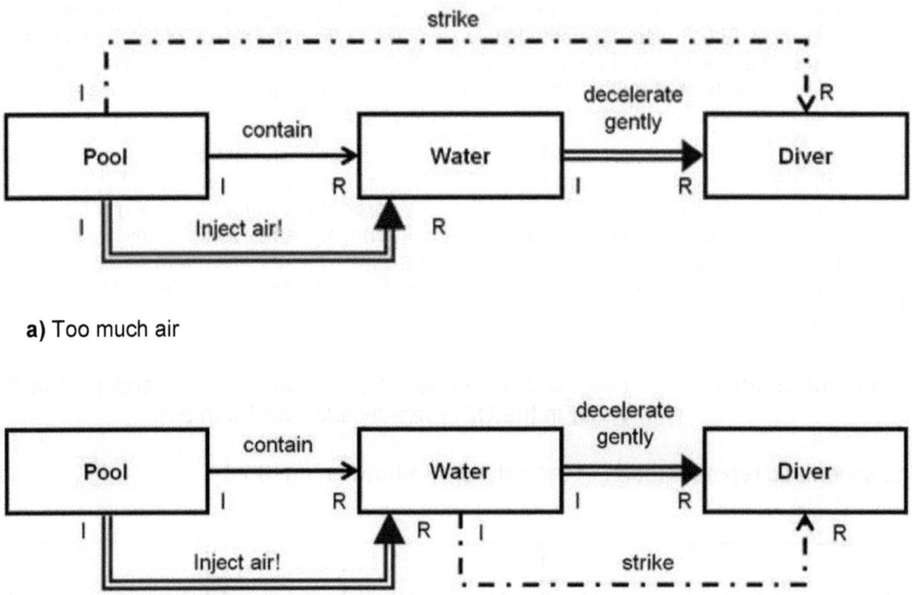

a) Too much air

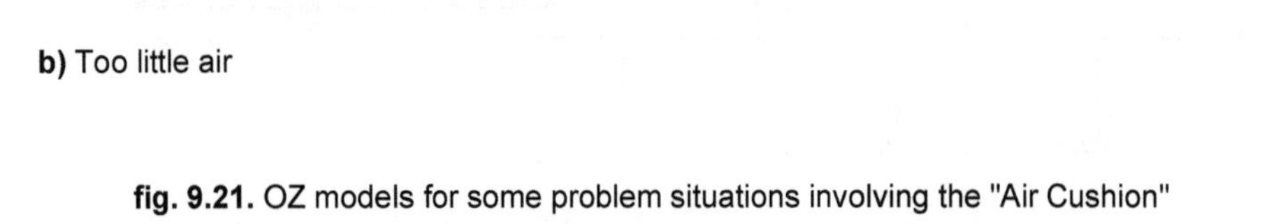

b) Too little air

fig. 9.21. OZ models for some problem situations involving the "Air Cushion"

9.3 The OZ makes this itself

9.3.1 Altshuller's Experiment-1: *Girl's Brilliant Solution!*

Here we will show a solution corresponding well to the most metaphorical FIM, namely, the Maxi-FIM.

> **According to this model, we assume that the OZ has everything necessary for a solution. It remains only to find hidden resources and reconstruct the OZ and with it the system as a whole.**

Now let us remember the resourceful kindergarten girl. Let's remember her inventive solution to learn from her victory an enlightening experience for the future.

Here is the story[73] once again according to the description of Genrikh Altshuller in his book "To Find an Idea." Then we will consider this example in four steps from the occurrence of the problem to assessment of the quality of its solution.

Example 9.10. Genrikh Altshuller's Experiment-1 (solution of Problem P1 from the textbook "ABC-TRIZ. Book I: How to learn to invent")

<u>**FIRST EXPERIMENT**</u>

Invention process

So, there are two ropes hanging down from the ceiling in the playroom (fig. 9.22). You must grab them with both your hands. But the ropes are too far apart—If you grab one of them, you cannot reach the other!

fig. 9.22. Original problem situation with two ropes

And then there was a girl who solved the problem. At first she acted too usually (brainstorming! MO): she grabbed a rope but could not reach to the other and threw the rope, grabbed another... And here she became thoughtful. She stopped hustling and started thinking!

---

[73] G. S. Altshuller, *To Find an Idea: Introduction to the Theory of Inventive Problem Solving*. – Novosibirsk, Science Publishers, 1986 (in Russian, later editions exist)

"I pull this rope," she said the educatress, "and you give me that rope, please."

The educatress said that she and the gentleman (Genrikh Altshuller) could not interfere with the game. And the girl went on thinking.

She was looking around the playroom for something.

Then she went to the window sill, rummaged in the toys, and pulled out a tattered doll. It was the necessary second "person" who could give her the rope! And the girl found this person in the form of a doll!

She tied the rope to the doll (fig. 9.23) and swung the resulting "pendulum," ran to the second rope and grabbed it, came back and caught the swinging doll...

So, a very big candy "Gulliver" was honestly earned for the creation of a genuine miracle!

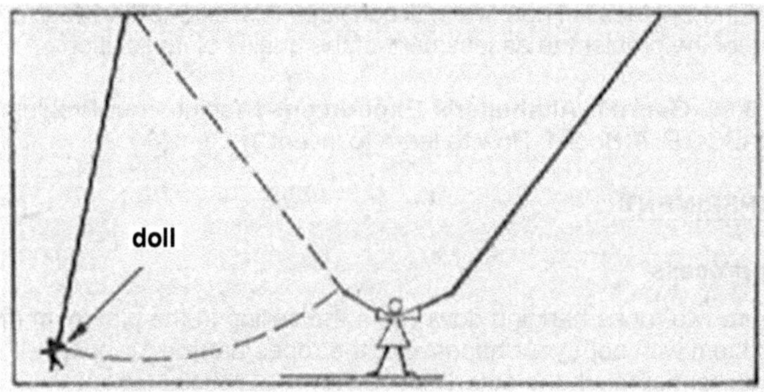

fig. 9.23. First girl's solution to Problem P1

Approximate modeling of the invention process

Stage 1. Trend

Diagnosis of the initial problem situation: It is impossible to independently hold one of the ropes and grab the other rope because the ropes are hung at a large distance from each other.

The girl asked for help But none could be given according to the conditions of the game. Failure was imminent the same as all her predecessors! But she persevered and continued to think.

General Contradiction (GC): *We must provide a solution to the problem, but we do not know how to do it (this wording is typical GC).* What can be done?

Stage 2. Reducing

Operative Zone (OZ): Objects are in the place of occurrence and (in most cases) solve the problem): two long thin ropes and a child.

Operative Time (OT): The time required to grasp the ropes.

Surroundings (room): Participants in the experiment, furniture, and toys.

Standard contradiction (SC): It is necessary that someone gives (pushed, pulled) the rope to the girl, but this is no visible person who can do it. We show this SC graphically (fig. 9.24).

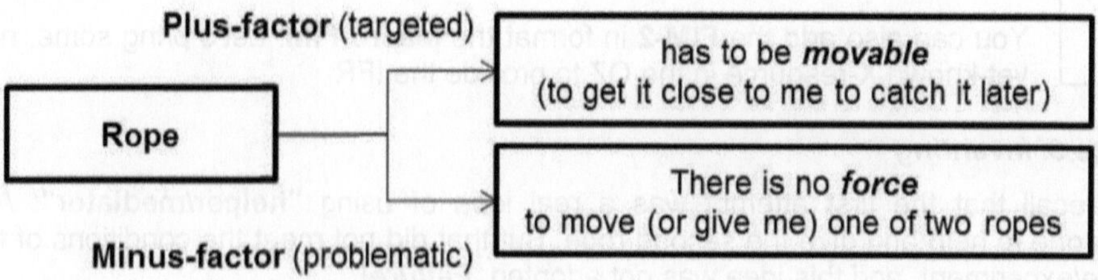

fig. 9.24. Standard contradiction as a model of the problem in a form of conflict of various properties of the object

Text variant: The rope must be *movable*, but there is *no physical force* to do this.

Fixing this text SC in the format of formula:

Rope ► movable **VS** force.

Radical contradiction (RC): The rope could be given by an assistant, for example, someone from the experimenters, but the experimenters *could not do* it under the terms of the game/experiment.

We show this RC graphically (fig. 9.25).

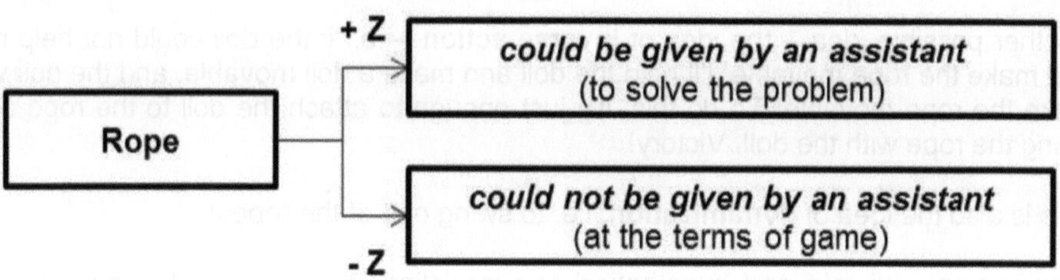

fig. 9.25. Radical contradiction as a model of the problem in a form of conflicting requirements for the same aspect

Fixing this text RC in the format of a formula:

Rope ► ***must be given*** *(by assistant)*

VS *must not be given* *(by assistant* because he is absent).

A special feature of this model is that it reflects the psychological basis of the first attempts to solve the problem by using an external resource.

THINK!

Ideal final result **(IFR):** The rope moves ***ITSELF*** to the right distance for me.

In fact, with such an IFR, we are ready to accept as a guiding model the **Maxi-FIM** as the **FIM-1**: The OZ by itself ensures obtaining the IFR pointed out above.

You can also add the **FIM-2** in format the **Macro-FIM**: Let's bring some, not-yet-known X-resource in the OZ to provide the IFR.

Stage 3. Inventing

We recall that the first attempt was a real idea of using **"helper/mediator"**: Ask someone to help and give the second rope. But that did not meet the conditions of the game/experiment, and this idea was not adopted. ***Failure!***

Assume that the girl continued to look for a person "who" could help her? And perhaps, in search of a helper, she remembered the doll: *Maybe the doll could help?!*

> Do not be too severe toward this "literal hypothesis." This is only a version for a coherent story about a possible development of the situation. And, in fact, as will be seen below, it is generally not so important that the inventor has subjectively "thought" at that time. And, importantly, *how could it be possible to reconstruct changes objectively present in the known solution with the objective models!*
>
> Only this will continue to be important and not the "literary–psychological" side of the description. So, let's wait until the end of this story and the consideration of other examples. In the meantime, we continue in the same style, which is in very close resembles a joke.

There is a new idea here—the idea of **copying**—i.e., using the doll-mediator as a copy of an experimenter-mediator! But the doll is not a man; it cannot move the rope; and therefore it cannot eliminate the contradictions! Failure again?!

Another possible idea— the idea of **inverse action**—i.e., if the doll could not help me and make the rope movable, I'll help the doll and make a doll movable, and the doll will make the rope movable! To do this, it's just enough to attach the doll to the rope and swing the rope with the doll. Victory!

This is also the idea of **dynamization,** i.e. to swing one of the ropes!

However approximate and hypothetical this modeling is, it contains the main result: *The mentioned models can be used to generate ideas of solutions.* These models are perfectly workable during the practical steps of a particular creative process, and they lead from the initial problem situation to a constructive solution.

Indeed, the X-resource is the doll-mediator, the doll-copy, doll that made a previously fixed rope moveable! And this is a real fact.

Stage 4. Zooming

Verification of solution: The goal is achieved, and *both contradictions are eliminated.*

Verification of adverse effects: None! The goal is achieved without unacceptable consequences for the project, the inventor, and the environment.

Verification of positive effects: Yes, but in a surrounding (in the "super-system"), for instance, as a positive example of creative and volitional behavior to other children.

Can we summarize the explored solutions and models to other situations? Yes, we can! For example, the very first and necessary model, ***dynamization,*** can be represented at least in this minimal version: *Make an object moveable that is otherwise fixed.*

Is it long? The relatively complex and unusual? At first glance, yes.

> But after solution and with very little experience in the analysis of only several problems, the following to this simulation scheme (of course without literary speculation and flattering), it becomes a matter of habit and therefore simple. This makes it possible to apply the acquired skills in solving the new challenges.

9.3.2 Altshuller's Experiment-2: *Boy's Brilliant Solution!*

Example 9.11. Genrikh Altshuller's Experiment-2 (second solution of Problem P1 from the textbook "ABC-TRIZ. Book I: How to learn to invent")

<u>**SECOND EXPERIMENT**</u>

The game–experiment was complicated: All the items were taken away from the room, and the experimenters themselves were out the room.

Invention process

After many confused and crying children, who were consoled with a large candy "Gulliver," the real inventor appeared in the room. Like everyone else, the boy first run from rope to rope, and after getting an unsatisfactory result, he stopped, but did not call for help or just cry. No! He clearly started to ponder something. Then he took off a sandal, tied it to a one rope, swung the rope, ran up and grabbed the second rope, quickly returned, and caught the first rope, which was still swinging like a pendulum.

He won!

Approximate modeling and analysis of the invention process

Stage 1. Trend

The first instinctive brainstorming-idea on the basis of analogy (imitation, copying): Swing the rope! But it is thin and light, so it quickly stops swinging. There is not enough time to run and grab the second rope and then catch first swinging rope because it was not swinging as necessary.

General Contradiction: How do we make a rope swing with a large deviation and a long-enough swing time?

Stage 2. Reducing

Operative zone: Two long thin ropes and a child.

Operative time (OT): The time needed to grasp the ropes.

Additional resources in the room: Only the child's clothing and footwear. There is nothing else in the OZ or in the environment (the room).

Standard contradiction (SC): We must get at least one long rope *swinging for a long-enough time*, but the ropes are *thin and light*, and therefore they swing with a small deviation and quickly stop (fig. 9.26).

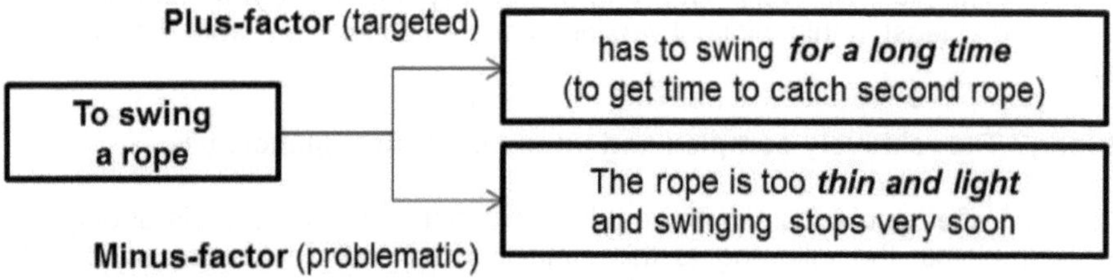

fig. 9.26. Standard contradiction

Fixing this text SC in the format of a formula:

a) Rope ▶ Swing for a long time **VS** thin and light.
b) Rope ▶ Swing for a long time **VS** internal harmful factors.

The variants are shown for you to pay attention to the fact that the models of TRIZ are still not arithmetic. It is possible to formulate different variants of the SC. This reflects a different understanding of the initial problem situations in terms of functionality, i.e., a different depth of description in the physics of processes and phenomena. This also reflects different experiences and different people's knowledge about problem solving.

Yet this description is "running" inside the OZ on a conflict interval (OT) and relates generally to the resources and components involved in the conflict interaction in the initial problem situation.

This means that these models represent the conflict system factors in a problem situation quite adequately.

This is important to continue operation with such SCs in order to select suitable transformation models to change a system that was already met in practice in situations modeled by similar plus- and minus-factors.

Radical contradiction (RC): The rope *should be swung* (to be able to swing one of the ropes and then catch it while holding the second rope), and it *should not be swung* because it is too light to swing for a long time. We show this RC graphically (fig. 9.27).

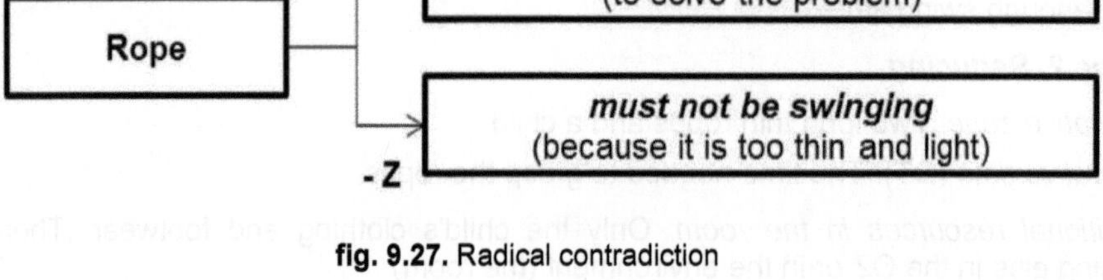

fig. 9.27. Radical contradiction

Fixing this text RC in the format of a formula:

Rope ► ***Must be swung*** (to solve the problem)

VS *must not be swung* (because it too light).

THINK!

Ideal final result **(IFR):** The rope should swing like a pendulum with a large deviation from the original position and for a long-enough time.

And here we are also ready to accept as a guiding model the **Maxi-FIM** as the **FIM-1**: The OZ by itself ensures obtaining the IFR pointed out above.

.

Stage 3. Inventing

Solution of SC: We can apply different approaches in according to several transformation models.

The first idea is ***dynamization***: Swing one of the ropes. But, as we know, this idea cannot be implemented due to the fact that the rope is light. This is described as the SC and the RC.

The second idea is ***Copying***: Make the rope swing (copy) a real pendulum with resources that are in the OZ.

The third idea is to eliminate the flaw in the SC and in RC: Make one of the ropes heavy (to increase the weight, that is, ***Change in the aggregate state of the object***) by adding the load (applying the ***mediator*** again) in the form of sandals (***copying***: in this case, simplified copying of the "plummet" of a pendulum).

The problem is solved!

Stage 4. Zooming

Verification of solution: The goal is achieved because *both contradictions are eliminated.*

Can we generalize the explored solutions and models to other situations? Yes, we can in the form of models that fit such names as *Change in the aggregate state of the object, Copying, and Mediator.*

Let us try to figure out what those models advise us to do. The composite sketch of a possible solution looks like this:

01 Change in the aggregate state of the object: b) Changes in concentration or consistency, degree of flexibility, temperature, etc. Alright, let us assume that we must **increase the weight** of the rope;

10 Copying: a) Use a simplified and inexpensive copy instead of an inaccessible, complicated, expensive, inappropriate, or fragile object. Once more: **Someone or something must help me** and to swing the rope!

18 Mediator: a) Use another object to transfer or transmit an action; or b) temporarily connect an object with another (easily separable) object. Well, it keeps getting better and better! We must **connect** some relatively **heavy object** to the rope! But the room is empty!

This is where inventing thinking kicks in: You can take off your footwear, say, a sandal, tie it to one of the ropes in advance (fig. 9.28,a) and… swing (9.28,b) that rope like a pendulum! Then you should run and grab the second rope (9.28,c), get back to the central position, and catch the first rope (9.28,d). That's it! The "comic strip" is complete! The mediator has worked!

Solution of the RC: Make a light rope heavier (use of material and energy resources) for some period of time (use of time resource).

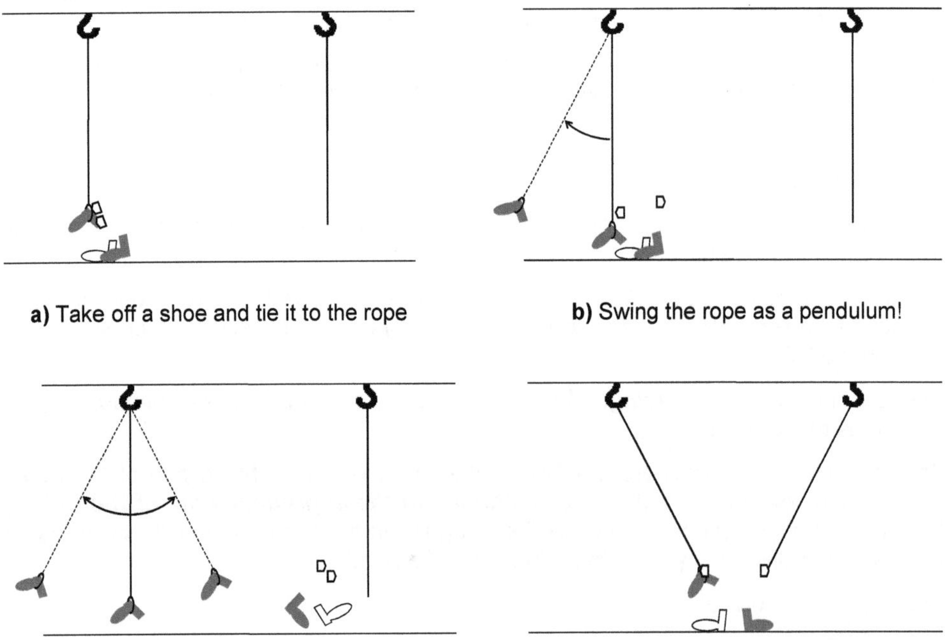

a) Take off a shoe and tie it to the rope

b) Swing the rope as a pendulum!

c) Run and grab the second rope!

d) Get back and catch the first rope

fig. 9.28. First boy's solution of "ropes problem"

Example 9.12. Special solution for Genrikh Altshuller's Experiment-2 (second solution of Problem P1 from the textbook "ABC-TRIZ. Book I: How to learn to invent")

A boy found out that he had, in his pocket, a spool of cord left over from the crafts lesson where he was making a kite. The boy tied the cord to one of the ropes, took the second rope into his hand and got as close to the first rope as he could. Then he used his free hand to pull the first rope up by the cord. Interestingly, he held the cord… in his teeth so that he could reach out with his free hand and pull in another arm's length. He kept doing that until he could grab the first rope with his "pulling" hand. There you have it, a "two-in-one": The invention of a solution and the "school of survival"!

SUPER-EFFECT ☺: Indeed, such a lesson in creativity can probably be regarded as training for survival in circumstances where it appears that there are **NO RESOURCES**, but if you think about it, **THE RESOURCES WILL BE FOUND**, and **THE SOLUTION WILL BE FOUND**!

REINVENTING: CHILDREN'S PROBLEM WITH TWO ROPES

TREND Two thin ropes were hanged in kindergarten, in a room where children play. Children were set the goal: you need at the same time to take the both ropes with your hands. However, the distance between the ropes is too large to take one rope in hand and to reach the second one!

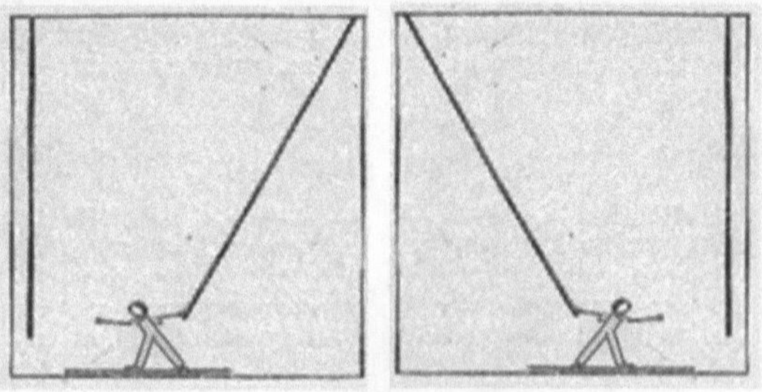

You can try to swing at least one rope, then to run back to the another rope and grab it, and then quickly move to the first swinging rope and catch it, and then to hold both ropes. But the rope is too light and stops soon. What could be done?

REDUCING **FIM-1:** The OZ by itself ensures the obtaining of the IFR

[one of the ropes itself approaches me]

SC (here reinventing made without A-matrix only after extracting):

Rope ► *long and thin* **VS** *losses energy* ► *models 01, 10, 18*

Several suitable variants of plus- and minus-factors do not include the dominant model(s) for this example. This means that the completion of the respective cells of A-matrix with new model(s) is possible.

RC: **Rope** ► *must swing long enough* to grab the other rope, and then to catch a swinging rope **VS** *must not swing for a long time,* since it is stopped quickly over the lightness.

INVENTING **Solving to SC and RC:** in order to give more weight (***model 01***) to the rope, two key models are applied – model ***18 Mediator*** (*use another object to transfer or transmit an action; temporarily connect an object with another (easily separable) object)* and ***10 Copying*** *(use a simplified and inexpensive copy):* tie to a rope toy or shoes!

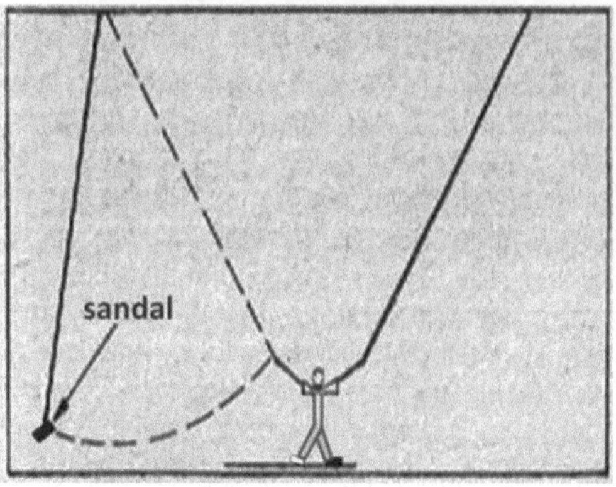

And then on the model ***11 Inverse action*** the rope **itself** will "fly" in the hand and it will not be necessary to try reaching it hopelessly!

ZOOMING The contradictions are eliminated.

fig. 9.29. Reinventing of boy's solution for second version of Problem P1

9.3.3 Moscow Kremlin Stars: *Turn Harm Into Good!*

fig. 9.30. Spasskaya Tower of the Moscow Kremlin (photo by the author – 19.04.2015)

Surprisingly, this solution is known to a very small number of Russians and even Muscovites, but of course it unknown to tourists.

Visiting Red Square in Moscow and seeing the stars on the towers, a few people notice a surprising feature of these stars, which, after discovering, surprises the observers.

In this case, we consider the solution as both ***direct analysis and inventing***, i.e., direct passage through ***MAI T-R-I-Z*** from raising issues before the invention of the idea.

So, in 1937, the five highest towers of the Moscow Kremlin got new stars with the three-layer structure of a special glass having a ruby color. Inside the stars were mounted powerful lamps so that the stars are visible from a great distance.

When installing the stars it was necessary to solve the problem of protecting them from strong wind. In addition, the stars should be raised to a 70-meter height (fig. 9.30).

The initial problem may be briefly expressed in the following form: The construction of the star should be strong enough to withstand hurricane winds. Obviously, this is a ***general contradiction***, the general framework of which is usually the question "What can be done to ...?" This implies that it is not known how to do it.

Indeed, at the time of the architectural concept, the designers of the Kremlin stars were not ready to answer this question.

In the methodological purposes, we will consider three options for the use of MAI T-R-I-Z in the version SMART with passage through the SC and the RC.

In the first variant, the route to solving the SC is the longest and should be realized in a twofold pass through SMART.

Then we show the possibility of solutions with a single SMART pass through the SC route.

And then we'll show a variant of single passage through the RC route.

At the same time, our task is to show that the practice may have different variants to achieve the best effective solution. But according to TRIZ, such variants will still be a small, and this speaks in favor of the application of TRIZ.

However, in all of these cases, we will rely on the same FIM, namely, the Maxi-FIM! That is, the solution will be achieved through a minimal restructuring of the OZ and with small changes in the surrounding objects.

Example 9.13. SOLUTION 1: FIRST SC-ROUTE for the Moscow Kremlin Stars (solution of Problem P3 from the textbook "ABC-TRIZ. Book I: How to learn to invent")

CYCLE 1

TREND

So, a strong wind has put a lot of pressure on the star. How do we protect the star from falling?

REDUCING

Let us construct an SC.

Let's start with the text option and write informal SC in the following form:

STAR Has *a large surface area* to be visible from afar,
VS *large surface area* leads to *large windage (sail area)* and *large pressure* from strong winds.

The formula variant: Star ► large surface area **VS** large pressure.

For clarity, we represent the SC in graphical form (fig. 9.31).

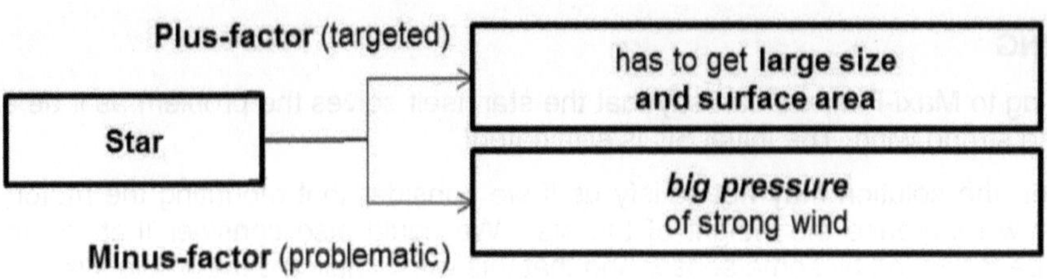

fig. 9.31. Standard contradiction for Solution 1

It is possible confidently to include into the OZ the entire construction of star so that the hypocenter and epicenter (in this simplified description) coincide. This OZ contains the following elements: the actual star and the support of star in the form of sharp peak of a tower (spire).

Simplistically we can say that the problem of the OZ is that much pressure is placed on the star when a strong frontal wind arises, and the star, the support, or the point where the support connects with the star can break.

Let's formulate the IFR and the FIM.

Maxi-FIM or "FIM of the Operative Zone Itself":

The OZ ***itself*** ensures obtaining the **IFR:**

[the star is resistant to strong wind]

INVENTING

To work around this problem, two factors of the SC are used in the A-matrix: The first factor corresponds to the positive purpose of solving the problem, namely, the star should have a large area, which corresponds to a plus-input of the A-matrix with number *18. Surface of the fixed object*; the second factor corresponds to the minus-input *31. Tension, pressure*, because the pressure of the wind creates a problem.

At the intersection of line 18 and column 31 in the A-matrix, there is the cell in which we see the recorded numbers of transformation models—02, 07, 26 and 27—which statistically are often encountered in practice by the solution to the contradiction represented by factors 18 and 31.

Then, the formal solution can be written as a formula:

Star ► 18 **VS** 31 ► 02, 07, 26, 27

To create an idea, it is necessary to look at the description of these models in the As-catalogue and try to interpret them in relation to this problem. The greatest creative potential in this case has a navigator of *07 Dynamization*: *c) make an object moveable that is otherwise fixed.*

The idea may be to rotate the star sideways to the wind!

Assume that a small motor is mounted within the star and that this motor, by instruction from a sensor that detects wind direction and force, rotates the star so that its wide front side is always set parallel to the wind's direction!

Clearly, in this case, the wind pressure on the star will be minimal!

ZOOMING

According to Maxi-FIM, we can say that the star itself solves the problem as it defends itself in a strong wind. The initial SC is eliminated!

However, the solution may not satisfy us if we consider that mounting the motor into the star will increase the weight of the star. We could also consider that the motor should be managed by some sensor and that this will complicate the whole system.

This means that we can first try to improve the obtained solution to avoid a complex construction. To do this, we will pass Cycle 2.

These solutions are given in the tables of reinventing (fig. 9.32 и 9.33).

CYCLE 1

TREND

The main Kremlin Stars are large – about 5 meters tall. Each star weighs 1 ton or more. All stars have large surface area and, consequently, large windage. As a result, in high winds there is a risk that they may be thrown off their towers.

GENERAL CONTRADICTION: how do we design the stars to make them reliable and resistant to strong wind?

REDUCING

Maxi-FIM, or "FIM of Operative Zone itself":

The OZ ***by itself*** ensures the obtaining of the **IFR:**

[*the star is reliably steady in strong wind*]

Standard Contradiction

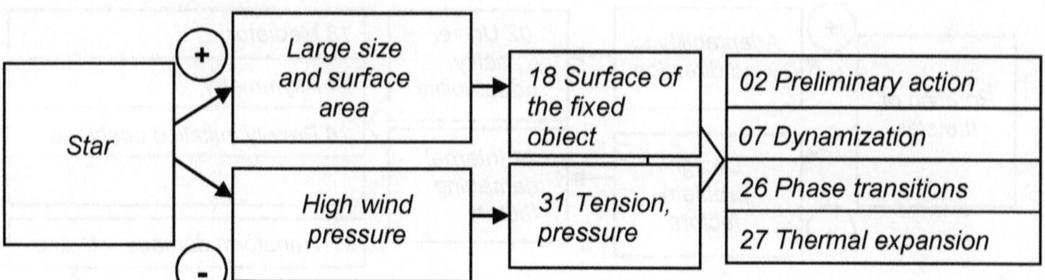

INVENTING

Key model: 07 Dynamization – *c) make an object moveable that is otherwise fixed.*

Key idea: *make the star mobile, rotate it.*

ZOOMING Have the contradictions been removed? – Yes. – ~~No~~.

Super-effects: *none.* Negative effects: *possibly more complex design*

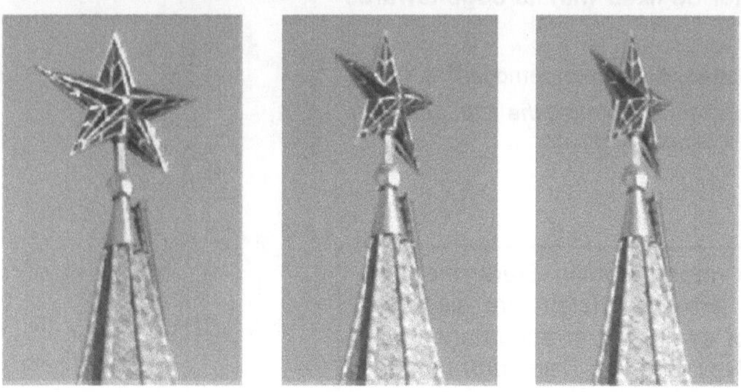

BRIEF DESCRIPTION

To ensure their reliability and sustainability in strong winds, Kremlin Stars are put in motion in accordance with model *07 Dynamization*.

fig. 9.32. Reinventing cycle 1 for solution 1

CYCLE 2

TREND Kremlin stars must turn so that, while adapting to the direction of the wind, they position themselves "sidewise", i.e. with their rays towards the wind. The "cross-section" area of the star is much smaller than its face area. And smaller area means lower windage and, consequently, lower pressure exerted by the wind on the star. However, due to its huge weight, the star is extremely sluggish, and a powerful motor is needed to accelerate its rotation. However, when the star turns faster, there emerges a problem related to the "fragility" or "rigidity" of its design. The elements of the star, especially the "ruby" veneers, can become dislodged, break down or fall out. In other words, the design has inherent flaws which may create problems if the star needs to be forcefully turned. *What can we do?*

REDUCING **Maxi-FIM.** The OZ ***by itself*** ensures the obtaining of the **IFR:**

[*the star itself responds to the direction and strength of the wind*]

Standard Contradiction **Factors** **Navigators**

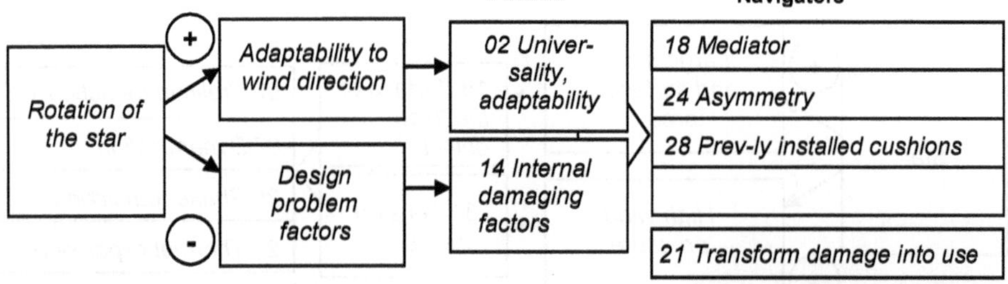

INVENTING

Key model: *24 Asymmetry – a) move from a symmetrical shape of an object to an asymmetrical one.*

Key idea: displace the axis of rotation of the star relative to its axis of symmetry, so that the star is transformed into a "wind vane". This idea uses also model *21 Transform damage into use:* the stronger the wind, the more securely will the star be fixed with its edge towards the wind!

ZOOMING Have the contradictions been removed? – Yes.

Super-effects: *no energy is spent on rotating the star, and there is no need to install complex motors or drives.*

Negative effects: *none*

BRIEF DESCRIPTION

To use wind power to rotate Kremlin Stars, they are designed so that their axes of rotation are displaced relative to their axes of symmetry in accordance with model *24 Asymmetry*. Navigator *21 Transform damage into use* is interpreted here very spectacular.

fig. 9.33. Reinventing cycle 2 for solution 1

Example 9.14. SOLUTION 2: SECOND SC-ROUTE for Moscow Kremlin Stars (solution of Problem P3 from the textbook "ABC-TRIZ. Book I: How to learn to invent")

If you replace informal factors with the formal from the A-matrix, as shown in fig. 9.34, it becomes possible to apply to the solution of the original problem one (or sometimes more than one) of the four models shown in fig. 9.34 in the extreme right block.

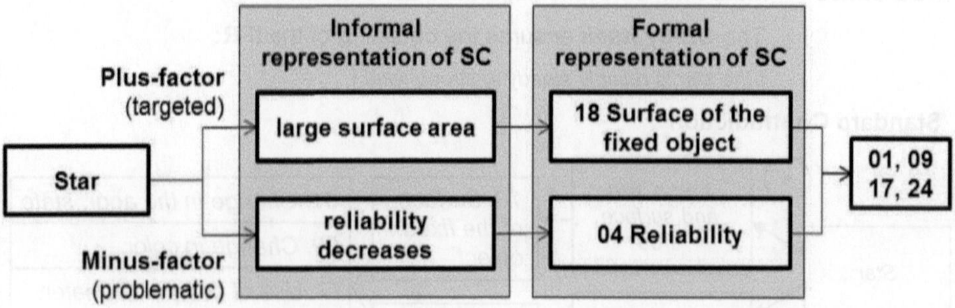

fig. 9.34. Selection formal plus- and minus-factors from the A-matrix

In this case, the selection of formal factors is obvious; however, it does not always work so successfully. Sometimes you must perform a bit of trial-and-error and go through several routes of MAI T-R-I-Z. Yet we must remember that we are in the OZ and we are working with a specific problem that is rather accurately represented by the informal SC.

> In this case of using the A-matrix, first we are working on the basis of managing and correcting methods; second, in any case, we disproportionately reduced the potential number of attempts (trials), e.g., which are particularly peculiar to brainstorming.

To avoid repetition, we note only that we already know that the most interesting choice here is model *24 Asymmetry* corresponding to the results of a full investigation on the basis of the Extraction-1C (see both cycles in Example 9.13) as well as "check solution."

Last, it is useful to note two other important facts related to the formulation of the SC and using BICO:

1) As a rule, difficult problems have several mutually conflicting factors, which means there exists the possibility of a few SC for the initial situation. In this case, usually only one of the SCs is chosen using certain bases (see explanations below), or pass several options of SC (as discussed briefly in Example 9.11), or some special methods for solving the composition of several SC could be applied.

2) Even if the conflicting factors are clearly visible and the SC is made quite easy, this does not mean that the problem will be easily solved; this could be, in particular, due to a lack of specific professional knowledge to create new solutions, a lack of resources, or a lack of creative knowledge and skills.

TREND The main Kremlin Stars are large – about 5 meters tall. Each star weighs 1 ton or more. All stars have large surface area and, consequently, large windage. As a result, in high winds there is a risk that they may be thrown off their towers.

GENERAL CONTRADICTION: how do we design the stars to make them reliable and resistant to strong wind?

REDUCING **Maxi-FIM,** or "FIM of Operative Zone itself":

The OZ ***by itself*** ensures the obtaining of the **IFR:**

[*the star is reliably steady in strong wind*]

Standard Contradiction

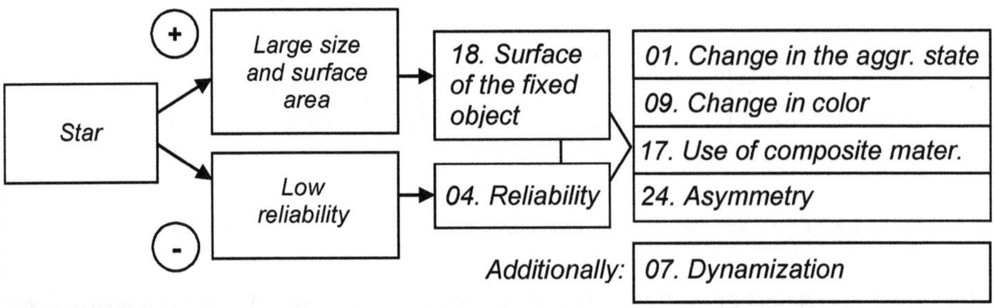

INVENTING

Key model: *24 Asymmetry – a) move from a symmetrical shape of an object to an asymmetrical one.* What does a "nonsymmetrical star" resemble? A wind vane! So turn the star into a wind vane! This is a clear case of application of model *07 Dynamization.*

ZOOMING Have the contradictions been removed? – Yes. – ~~No~~.

Super-effects: *the wind **ITSELF** aligns the star with its own direction.* The solution objectively implements model *21 Transform damage into use*, as the stronger the wind, the better it keeps the star in a position where its smallest cross section "cuts through" the air flow, while its largest cross section is streamline and wind flows around the star well. The design is simplified. No energy is consumed.

Negative effects: *none*

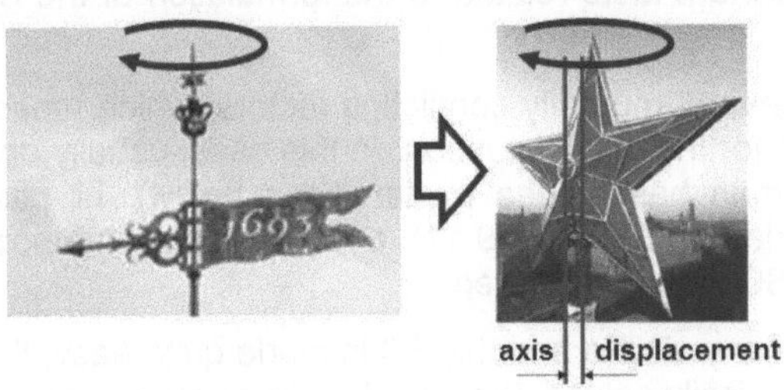

BRIEF DESCRIPTION

To simplify design and reduce energy expenditure, Moscow Kremlin Stars are designed so that their axes of rotation are displaced relative to their axes of symmetry in accordance with model *24 Asymmetry*; model *07 Dynamization* and 21 *Transform damage into use* are also realized here.

fig. 9.35. Reinventing for solution 2

Example 9.15. SOLUTION 3: RC-ROUTE for Moscow Kremlin Stars (solution of Problem P3 from the textbook "ABC-TRIZ. Book I: How to learn to invent")

The solution of the radical contradiction (RC) is based on the concept of resources, that is, on the real representation of the object in different "aspect spaces" in various descriptions such as spatial, temporal, structural, and functional description; energy (i.e., which fields and forces act in or on the object); and material (i.e., from what substance is the object constructed), etc.

In this textbook, using the shortened version of the Af-catalogue (section S26) involves working with four resources in accordance with the names of the fundamental transformations.

We continue the analysis of problem P3. Stars of the Moscow Kremlin.

Recall the RC first for the task in the **text variant**:

Star ***must be large*** (to have *a large surface area* to be visible from afar),

VS ***must be little*** (to have a *slight windage [sail area]* and to be *resistant* to strong wind).

Formula variant: Star ► large surface area **VS** small surface area.

For visualization, let's represent this RC in graphical form (fig. 9.36).

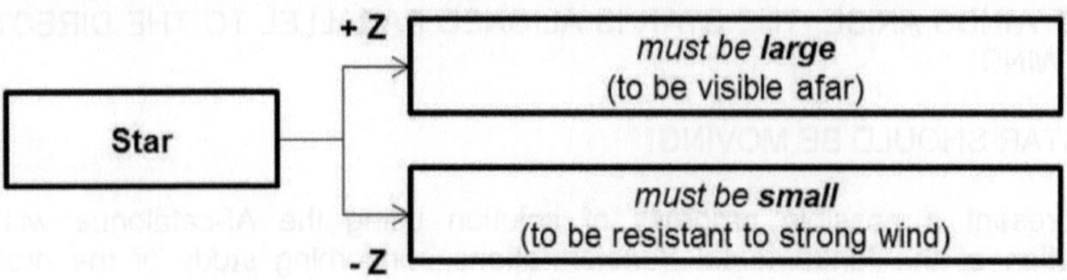

fig. 9.36. Radical contradiction for the solution 3

Ideal Final Result and Functional Ideal Model

In solving any problem, we must first try to use the Maxi-FIM. This is understandable because presumably the solution using the resources of the OZ should lead to minimal changes in the system. Although, this is true if these changes will be implemented only within the OZ.

Let us formulate the IFR and the FIM for this problem.

Maxi-FIM or "FIM of the Operative Zone itself":

The OZ ***by itself*** ensures obtaining the **IFR:**

Star ***must be maximally large*** (to have *a large surface area* to be visible from afar),

AND ***must be minimally little*** (to have a *slight windage (sail area)* and to be *resistant* to strong wind).

Formula variant: Star ► maximal surface area **AND** minimal (zero!) surface area.

At first glance, on initially considering about how the emerging problem situation will present itself in practice, the change of **"VS"** to **"AND"** in the RC is not possible!

But that's the paradoxical model of the effectiveness of TRIZ in that

the impossible becomes possible,

the incompatible becomes compatible,

and the insoluble becomes soluble!

Question 1: In what position relative to the wind does the star have maximum sail?

Answer 1: The front area of the star is maximally large, so the star windage is also maximal.

Question 2: In what position relative to the star is the minimum windage?

Answer 2: The vertical transverse section of star, made orthogonally to the frontal view, has a smaller surface area by many times and therefore a many-fold smaller windage. Furthermore, the transverse shape of the star with sharp beams on the edges and relatively small convex part in the middle ensures good streamlining airflow!

THE GENERAL IDEA OF A SOLUTION: WE MUST MAKE SURE THAT WHEN STRONG WINDS ARISE, THE STAR IS ALIGNED PARALLEL TO THE DIRECTION OF THE WIND!

HERE: STAR SHOULD BE MOVING!

Let's represent a possible progress of solution using the Af-catalogue with an interpretation of the fundamental transformations concerning study of the problem situation presented as the RC.

01. Separation of the conflicting properties in space.

The star as a spatial figure has different properties in terms of maintaining its stability in strong wind. Indeed, if the star is oriented frontally, that is, with its all of its surface to the wind, the pressure on the star will be maximal. A minimal pressure will exist if the star remains sideways with its beams to the wind. Of course, the star is clearly visible from afar with its frontal projection, but it will better withstand the wind when it is turned to the wind with its lateral projection. The maximal area of the cross-section is much less than the area of the frontal section.

02. Separation of the conflicting properties in time.

It is clear that in at strong wind (operative time), the star should be oriented sideways against the wind.

03. Separation of the conflicting properties in structure.

Hence, the idea follows that the additional elements should be introduced into the construction of the star to monitor the direction and strength of the wind and to cause rotation of the star in the most reliable position. It is easy to see that this could lead to a complex structure (i.e., the new standard contradiction).

And is it possible to simplify the design, i.e., do not go in the conventional direction of brainstorming?

04. Separation of the conflicting properties in material (energy).

And here is the time to bring to bear the two TRIZ rules.

One of them corresponds to a high-performance solution with a double-win in the specialized model *21 Transform damage into use*: A harmful factor not only does not harm, it also performs a useful function!

The second rule is associated with the concept of the "ideal solution" when the function (result) is accomplished, and no material costs are required.

Although there is an obviously fantastic "claim" to this rule, it simply leads to unexpected and highly effective ideas.

> A path (trend, tendency) to such decisions is indicated by the Maxi-FIM: Let the star be set in the best position against strong winds!
>
> Here we have almost everything we need! Actually, the wind as "harmful factor" has a large and harmful energy, but the star will itself be installed (to rotate!) with its beams against the wind.
>
> Now, connect these factors together: Let the wind set the beams of the star to meet itself!
>
> ***How to do it? Answer: by analogy with the weathervane!***
>
> ***A slight shift of the bearing axis relative to the axis of the symmetry of the star turns the star into a "weathervane"!***

Brief *Zooming*:

1) Everything is fine at the system level! There are no costs! Only the position of the axis of bearing is shifted a little bit.

2) Yes, there is a radical change at the subsystem level, that is, the real price paid for the result, i.e., the mounting of a special bearing support.

3) At the level of the super-system, there is great win with a super-effect—The Kremlin stars periodically may be completely visible from different directions!

The results of this inventing, in actuality the reinventing of an historic solution, is represented in condensed standard form in fig. 9.37. The demonstrated procedure has received in MTRIZ the method name **RICO (Radical In Cluster Out).**

However, proof requires a windy day. The star will be rotated... by the wind itself! In contrast, if there is no wind, there is no danger of the star falling!

This shows the result and action of technical ideas realized in the construction of the large Kremlin stars to ensure their safety, reliability, and resistance to strong winds.

The unknown is not only a technical solution (it's really little known by experts!), *it is also a functional manifestation, i.e., mobility and rotation*. But any careful observer from *Red Square* could see how this solution works after looking at the stars on the *Spasskaya* and *Nikolskaya* towers. A similar phenomenon can be seen even on the three highest towers in the Kremlin wall surrounding the Kremlin.

But ... only few notice it!

Let's talk now about the technical and creative solutions.

The Kremlin stars are constructed on the principle of a weathervane! They are mounted on the rotation axis with a small offset relative to the axis of symmetry of the star (fig. 9.37). Therefore, the wind produces greater pressure on the part of the star that has a large area. And the star is rotated to set the beams of the "lesser part" to the wind! So, on the principle of action, the star is a weathervane!

Extracting-1S of specialized transformations

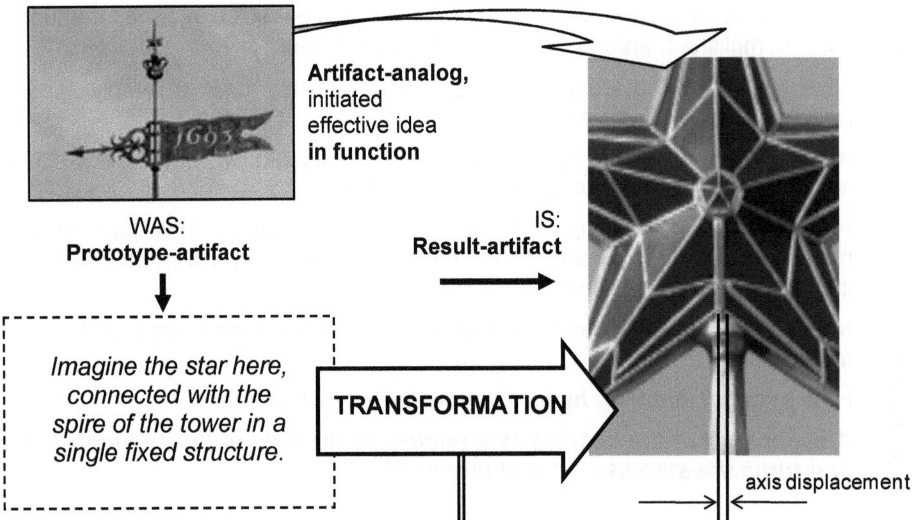

Extracting-1S			
LC	**No.**	**Navigator**	**Substantiation for the Extracting**
+	**07**	Dynamization	Make the star movable
+	**10**	Copying	weather vane
	11	Inverse action	Not to resist and succumb to the wind
	21	Transform damage into use	"Harmful" wind itself sets the star in a safe position (i.e., turned sideways to the wind)
+	**24**	Asymmetry	By way of displacement of the vane axis of rotation from the axis of symmetry

fig. 9.37. Results (briefly) of Extracting-1S for the Kremlin stars

Extracting-1F of fundamental transformations

Extracting the fundamental transformation requires much more skill than extracting the specialized navigators.

Modeling for the example is presented in a table assuming that the interpretation of the extracted patterns are more obvious (fig. 9.38).

It is clear that for the acquisition of reliable modeling skills, you first need to study the textbook from cover to cover and second to solve dozens of similar problems. So, go ahead! This game is worth burning the candle!

Extracting-1F			
LC	**No.**	**Navigator**	**Substantiation for the Extracting**
+	**02**	Separation in time	Outside the operative time (absence of wind), the system is stationary, and at the interval OT (in high winds) it is movable
+	**03**	Separation in structure	Each part of the system (the individual star and the supporting axis) is symmetric, but the system as a whole (a star on the bearing axis) is asymmetric

fig. 9.38. Results (briefly) of Extracting-1F for the Kremlin stars

In conclusion of this section, we present the following two specially designed pages. They are designed to ensure that students and teachers can make a copy/memo of the SC and the RC for permanent use.

These pages contain a short definition to represent the SC (fig. 9.39) and the RC (fig. 9.40) with each on one spread, both left and right, as well as examples for the formulation of the SC and the RC on the basis of the problem Kremlin stars.

These definitions of the SC and the RC should definitely know by heart!

Note also that in the resolution of the considered problem situations (such different ones), the model (thinking navigator) *07 Dynamization* is dominant.

This generalization is the essence of TRIZ and the real discovery of Genrikh Altshuller.

<u>**Conclusion of section "9.3.3 Moscow Kremlin Stars: Turn Harm into Good!"**</u>

For this reinventing, like almost for all of the examples for training purposes, we have presented two ways to solve the initial problem, i.e.. through the SC and through the RC.

In practice, the solution in SMART T-R-I-Z usually goes first through the SC, and if a solution is not achieved, it is possible to traverse the "inner cycle", i.e., to pass route through the RC. But if the first full cycle of SMART T-R-I-Z has not yielded a final idea, it is necessary to perform a second "outer cycle" taking into account the experience gained in previous versions. And so on until victory.

Note: We can correct the problem-solving process: We can improve the contradictions, clarify the factors, consider more appropriate model transformations, etc.

But this is the real **DESIGN OF IDEAS**, the synthesis of solutions!

This is a targeted and controlled synthesis of a new design concept, a new organization of the process!

This is TRIZ!

And this is not all that TRIZ possesses! It is only the origins, the most basic model of TRIZ.

So, the great prospect for your self-improvement is in the near future!

Good luck to you!

STANDARD CONTRADICTION

<u>Basic definition</u>

Standard contradiction (1)	***Standard contradiction*** (in classic TRIZ *technical contradiction*) *is a* binary (two-factor) model that reflects the incompatible demands ***in two different functional properties*** of the construct (or different constructs).

<u>Basic examples of SC representation</u>

<u>Text version:</u>

Star must have a ***large surface area*** to be visible afar,

BUT (VS) this leads ***to large windage (sail area)*** and ***low reliability*** in strong wind.

<u>Formula version:</u>

Star ► large surface area VS low reliability

First (plus-) factor Second (minus-) factor

<u>Graphic version:</u>

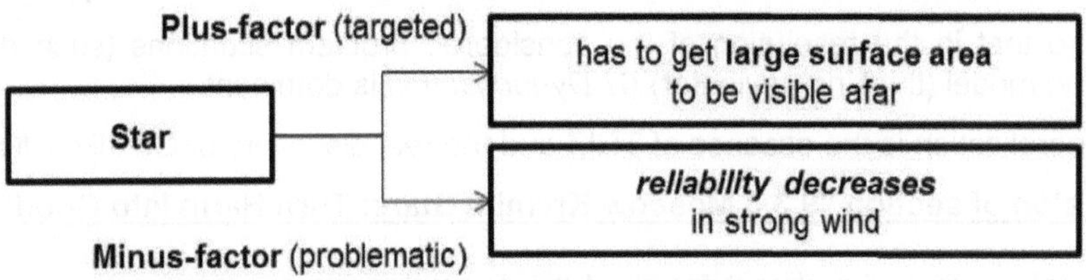

<u>A generalized addition to a graphic definition of SC:</u>

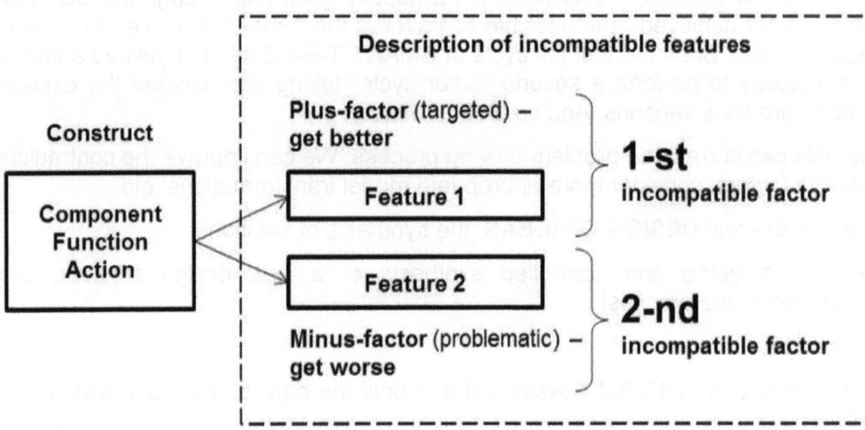

fig. 9.39. Definition of the SC

RADICAL CONTRADICTION

Basic definition

Radical contradiction **(1)**	**Radical contradiction** (in classic TRIZ *physical contradiction*) is a combination of ***two oppositely directed*** and ***mutually exclusive requirements for one and the same system property***.

Basic examples of RC representation

Text version:

Star ***must be large*** (to have *a large surface area* to be visible from afar),
BUT (VS) ***must be little*** (to have a *slight windage (sail area)* and to be *resistant* to strong wind).

Formula version:

Star ► LARGE surface area (+Z) VS SMALL surface area (-Z)

One and the same factor (Z)

Graphic version:

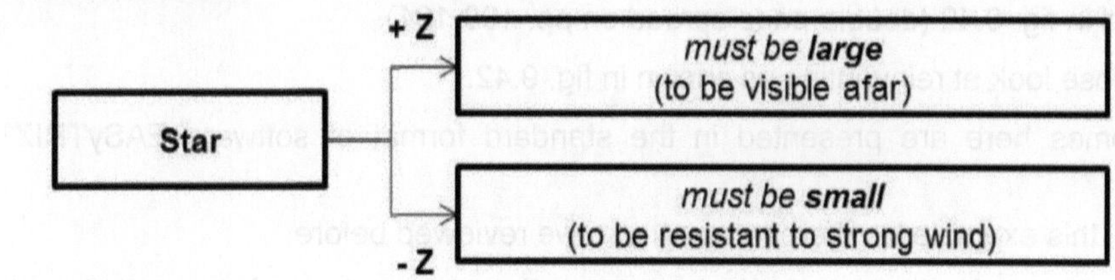

A generalized addition to a graphic definition of RC:

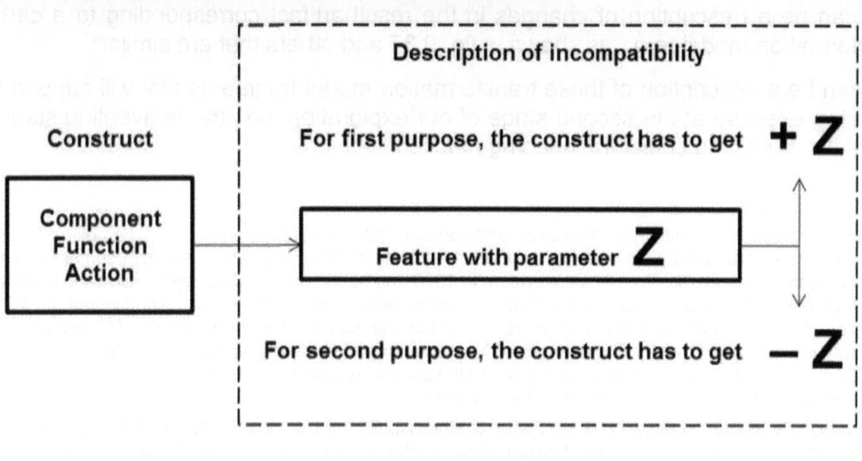

fig. 9.40. Definition of RC

9.3.4 Memories of the Future: *This Is da Vinci!!*

Modern TRIZ accumulates past experience and acts as a "time machine" allowing you to navigate through the past and apply the experience that was extracted from the past in the future. Just think—In fact, each thing that civilization invented in the past was once "the future" for its inventor, wasn't it?![74]

Modern TRIZ allows remembering how the future was created to continue the advancement of modern civilization into "the tomorrow."

Using a simile coined by the head of one of German Inventor Clubs, reinventing can be metaphorically described as a "time machine." Let us use this machine to move back in time and peep into the workshop of the great Leonardo da Vinci.[75]

Example 9.16. Leonardo da Vinci's Bridge

This bridge was invented by Leonardo da Vinci and presented in a sketch alongside with designs of several other bridges of all possible types, including floating, rotating (fig. 9.41), and portable bridges, which were collapsible and could be quickly reassembled. Those bridges were developed first and foremost for military uses: Their task was to afford passage to friendly armies (as necessary) and deny it to enemy armies (by keeping the parts separated).

For now we must focus not on the technical aspects of these inventions but rather on their creative aspects. We want to identify models—i.e., da Vinci's workshop tools—and compare them with TRIZ catalogues. Extracting and reinventing outcomes are presented in fig. 9.42 (double-page spread on pp. 190-191).

Take a close look at reinventing as shown in fig. 9.42.

All outcomes here are presented in the standard format of software EASyTRIZ™ Junior™.

However, this example is different from those we reviewed before.

In the table for Extracting 1, the program offers two options for completing the substantiation line:

1) It can be a description of changes in the result-artifact corresponding to a certain transformation model (e.g., as shown in fig. 9.37 and others that are similar).

2) It can be a description of those transformation-model fragments that will support the modeling exercise at the second stage of our exploration, i.e., the reinventing stage. If that is the case, you select the heading function.

---

[74] the title reproduces the film title *"Memories of the Future"* (German: *"Erinnerungen an die Zukunft"*; in the English film distribution as *"Chariots of the Gods"*) – West Germany' popular-scientific documentary film made by the Austrian director Harald Raynlem in 1970 and based on the books by the Swiss journalist and ufologist Erich von Däniken *"Chariots of the Gods?"* and *"Return to the stars"*. The filmmakers put forward their own versions of the responses to questions about who built the pyramids of Egypt, whom portray the statues of Easter Island, who and for whom someone drew giant figures on the Nazca Plateau in South America, suggesting that it is related to the possibility visiting Earth by aliens; de.wikipedia.org/wiki/Erinnerungen_an_die_Zukunft

[75] *Leonardo da Vinci* (It.: *Leonardo di ser Piero da Vinci*, April 15, 1452, near Florence, Italy – May 2, 1519, Clos Lucé, Amboise, France) – great Italian artist (painter, sculptor, architect) and scientist (anatomist, naturalist), inventor, writer, one of the most notable representatives of Cinquecento art; en.wikipedia.org/wiki/Leonardo_da_Vinci

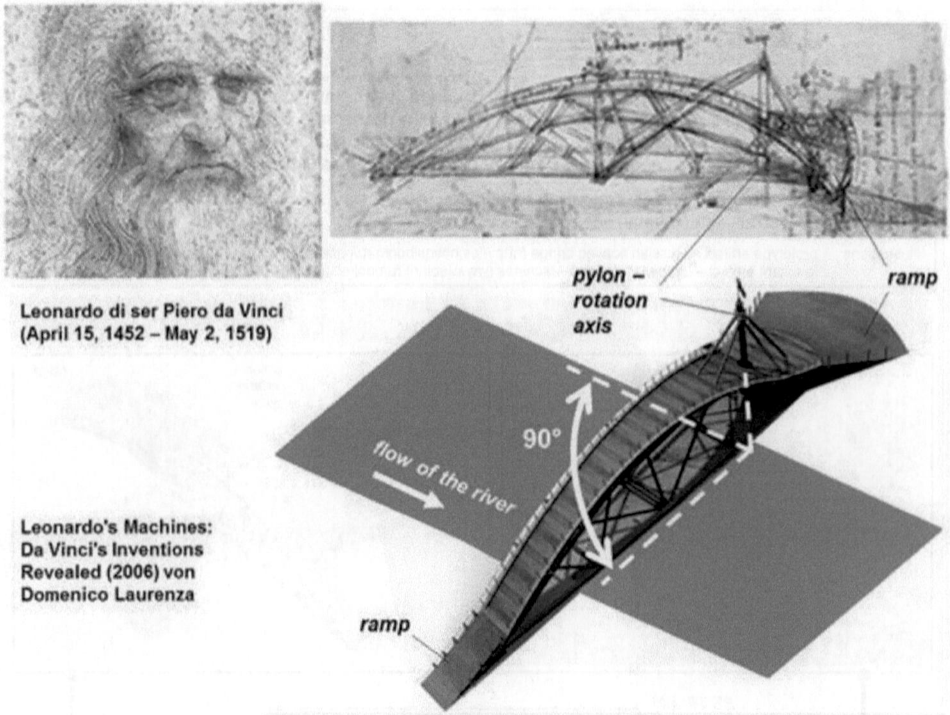

fig. 9.41. Rotating bridge by da Vinci

The possibility to create two descriptions is useful because at training sessions, extracting is often the only objective—and then we need the substantiation form.

In contrast, if we go for full reinventing, it is good to specify at the extracting stage which of the functional analogs embodied in the transformation model will be interpreted in the inventing section; accordingly, the user specifies the purpose of this part of the extracting table line.

Next, the functional ideal model is selected from the three options considered in section **9.2.3. Ideal Target Modeling**.

Formulation of the IFR is the pivotal element of the FIM structure.

The general recommendation is this: It should be a brief description fixing the target that must be met to design a new solution. It should be a concise answer to the question: WHAT do YOU want in the end?

The FIM at reinventing should reflect the most real changes in the OZ, namely, by the variants: 1) at the level of the OZ itself; 2) with extending the OZ and using a resource that is not in the OZ; and 3) at the micro-level of OZ, i.e., changes in material and/or energy.

To expand your understanding of these things, look at the next examples.

ARTIFACT — Leonardo da Vinci's Bridge

Description Leonardo da Vinci invented a rotating bridge for military use. Own troops could use the bridge joining the banks of a river to safely cross it. When necessary, the bridge could be "retracted" (turned away) to prevent enemy troops from quickly crossing over. Besides, the bridge could be used to cross rivers in peacetime, and enabled the passage of river-going vessels.

The bridge was rotated by cables around an axial pylon. It was also supported by cables and could use boats or empty barrels as additional supports.

Illustrations: prototype artifact – Russian floating bridge (http://les.novosibdom.ru/node/468); resultant artifact – *Leonardo da Vinci Machines* (www.labirint.ru/books/121061).

EXTRACTING

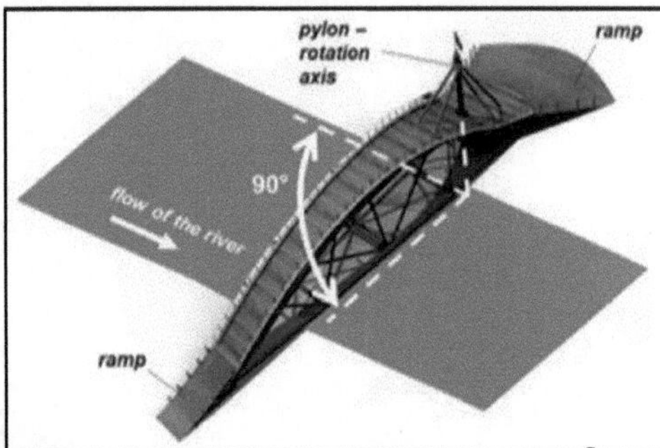

LC	No.	Navigator	Function
++	07	Dynamization	c) make an object moveable that is otherwise fixed
	12	Local property	c) every object should exist under conditions that correspond best to its functions
++	19	Transition into another dimension	a) an object is shaped so that it can move or is placed not only in a linear fashion, but also in two dimensions, meaning on a surface
	32	Counter-weight	a) compensate for the weight of an object with its connection to another object with lifting power

REINVENTING

TREND Known bridges (floating and prefabricated bridges) cannot be used in situations where it may be necessary to promptly provide a crossing and then quickly "retract" the bridge to allow the passage of vessels. Construction of a draw bridge across a relatively wide river presents numerous technical challenges. In such cases the bridge is usually built with multiple spans, whereupon each span can be turned into a small draw bridge. However, such bridges are stationary, and cannot be easily dismantled.

A different type of bridge is needed where a crossing has to be made at an arbitrary spot, provided, of course, that the width of the river is not too large, say, under 100 meters. What can you suggest?

REDUCING **Macro-FIM:** The X-resource, without making the system excessively complex and without causing inadmissible negative effects, ensures, together with other available resources, the obtaining of the

IFR: [a crossing that can be quickly provided and quickly removed]

STANDARD CONTRADICTION (SC) – Extracting-2S

Bridge	(+) *A crossing that can be quickly provided and quickly removed*	**From Extracting-1**
	(–) *Immobility, complexity, large weight of the structure*	07 Dynamization 12 Local property 19 Transition into another dimension 32 Counter-weight

RADICAL CONTRADICTION (RC) – Extracting-2R

Bridge ⇨ *must be to provide crossing* **&** *must not be to allow passage to awaiting vessels*

INVENTING

To attain Objective 1 (acceleration of provision and removal of the crossing), the bridge is made moveable (capable of horizontal rotation) – dominant models *07 Dynamization* and *19 Transition into Another Dimension.*

To attain Objective 2 (realization of two bridge positions – across the river and alongside the bank), the bridge is rotated by pull cables around an axial pylon – model *12 Local Property.*

To attain Objective 3 (bridge weight balancing), a counterweight is fixed under a moveable ramp in the shorter section – model 32 *Counter-Weight.*

Fundamental models: *Temporal* – the bridge performs its two main functions during different time intervals; *Structural* – certain elements are introduced to enable the bridge to rotate.

Scientific Effects: Application: rotary pulleys (for example, by analogy with windmills); cable-pulling hoists; balance scales; internal enforcement ribs (beams, girders) to make the structure sturdier.

ZOOMING

Contradictions Removed:	Yes.
Super-Effects:	Can be used for military purposes (deny crossing to enemy troops).
Negative Effects:	Relative complexity of construction.
Development Trend:	Automation of the crossing construction/removal operation.
Changes in Ambient Systems:	Bridge abutments and ramps must be aligned with the banks.
Beauty of Solution: 100	Highest grade. Substantiation: 1) such structure was not known before, there are no direct counterparts; 2) functionality is changed very fast.

fig. 9.42. Extracting and Reinventing the Leonardo da Vinci's bridge

Example 9.17. Rotating bridge in Vihtakanta, Finland

Bridge da Vinci has one major drawback: There is a need to balance the weight with a heavy counterweight at the shore, that is, a ramp on the side of the pylon is a "parasitic" load created by the rotation of the bridge.

However, it is possible to build a self-balanced bridge (fig. 9.43) if a small island is in the middle of the riverbed. If the island does not exist, sometimes it is possible to install an artificial support as has been realized in Vihtakanta's bridge in Finland. The rotating part is marked in fig. 9.43,a with a brace, and the axis of rotation is indicated by vertical dashed line. The rotation prop is shown in fig. 9.43,b.

The bridge in a state of 90° rotation is shown in fig. 9.43,c. A huge cargo ship freely runs parallel to the turned bridge span (fig. 9.43,d), and then the bridge is rotated back to its initial position (fig. 9.43,e).

The advantage of this bridge is that there is no limit to the height of the passing ships.

fig. 9.43. Rotating bridge in Vihtakanta (photo and video by author – 03-04.08.2014)

Here we will build the contradictions for the da Vinci's bridge as well as record the IFR and the FIM for Vihtakanta's solution shown in fig. 9.43, and put all this into MAI T-R-I-Z form (fig. 9.44) for extracting and reinventing.

TREND Rotating da Vinci's bridge has one major drawback: the need to balance the weight with a heavy counterweight at the shore. That is, a ramp on the side of the pylon is a "parasitic" load by the rotation of the bridge! It's OK for temporarily bridge but not good for permanent functioning. QUESTION: how such construction could have been changed using TRIZ methods?

REDUCING **FIM:** The OZ by itself ensures the obtaining of the **IFR**:

[light and movable bridge construction].

Standard Contradiction

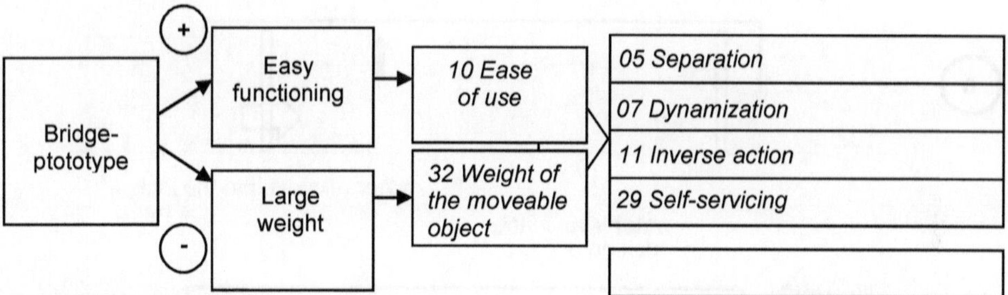

Radical Contradiction

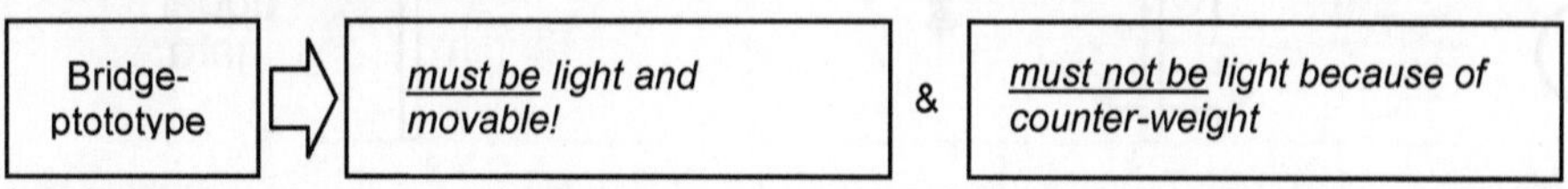

INVENTING **Key model – 29 Self-servicing:** the general idea is simple – the rotating bridge uses the weight of its both wings balanced at central supporting pillar built in the center of riverbed! Excessive counterweight is no longer needed.

ZOOMING Have the contradictions been removed? – Yes. – ~~No~~.
Super-effects: possibility to construct long and light bridge using the cable-stayed structure with central pylon by analogy with da Vinci's bridge! Negative effect: necessity to build a big supporting pillar at the center of riverbed.

BRIEF DESCRIPTION Rotating bridge has two wings self-balanced and equal in weight, and a central supporting pillar built at center of riverbed to make functioning easier in accordance with model 29 Self-servicing.

fig. 9.44. Extracting and reinventing Vihtakanta's bridge

Example 9.18. Leonardo da Vinci's Lock

First, we recall the principle of the lock.

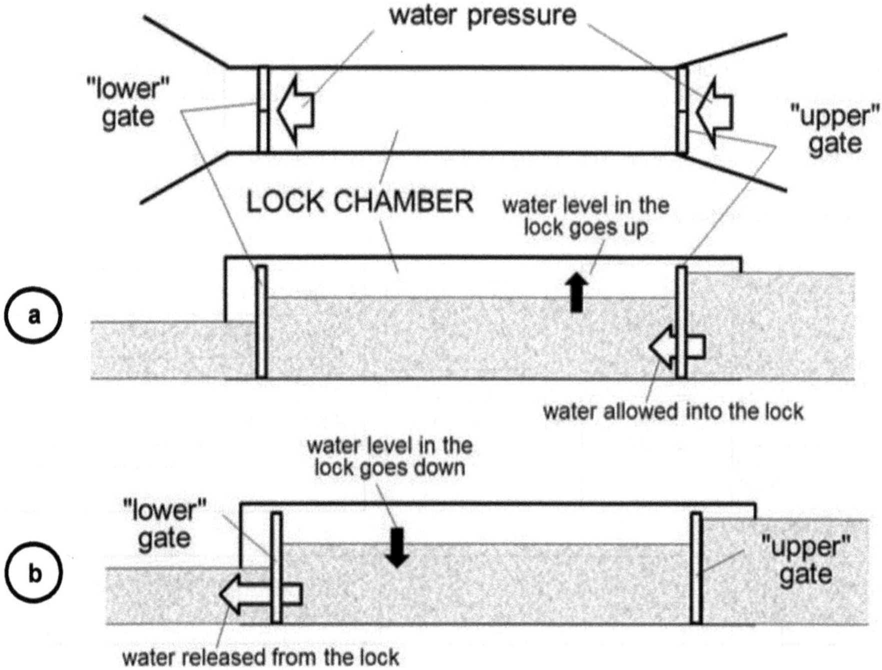

fig. 9.45. General principle of the lock

If a vessel is moving downstream, the chamber of the lock (fig. 9.45,a) is filled. When the water levels in the chamber and on the high side of the water are equalized, the upper gate is opened, and the vessel enters the chamber. The upper gate is then closed.

The water is drained from the chamber through the door in the lower gate (fig. 9.45,b). When water levels in the chamber and on the low side of the water are equalized, the lower gate is opened, and the vessel exits the chamber and proceeds downstream. This process is shown in detail in fig. 9.46.

If a vessel is moving upstream (fig. 9.47), it enters the chamber, and the lower gate is closed. The chamber is then filled with water flowing in through the door in the upper gate. When water levels in the chamber and on the high side of the water are equalized, the upper gate is opened, and the vessel exits the chamber and proceeds upstream.

The process is repeated as necessary. The upper gate may close or remain open if a downstream moving vessel is expected to pass through the lock.

> Old lock gate(s) had many problems:
>
> - They leaked water because their wings were not tight when closed.
>
> - It was hard to open the wings slightly to let in the water in the lock chamber from the upstream section or let it out of the lock chamber into the downstream section.

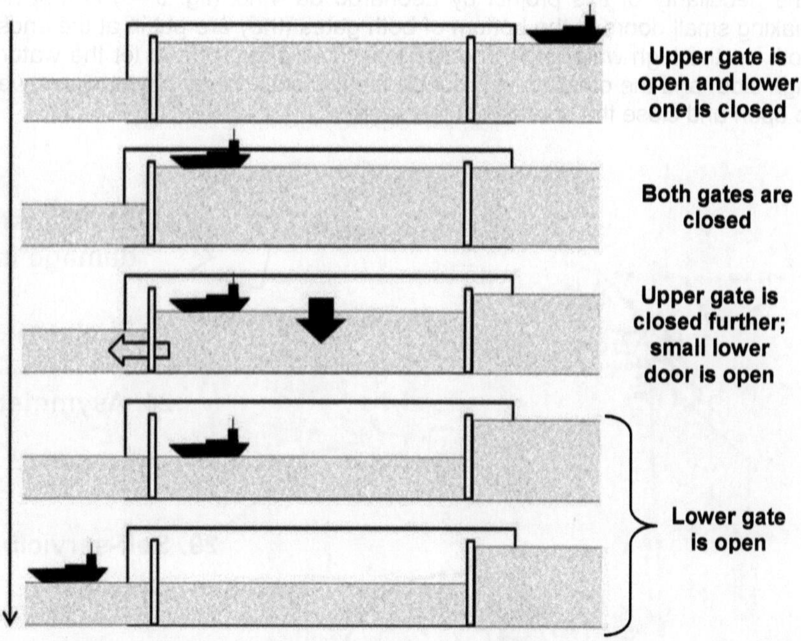

fig. 9.46. The vessel going from high water downward

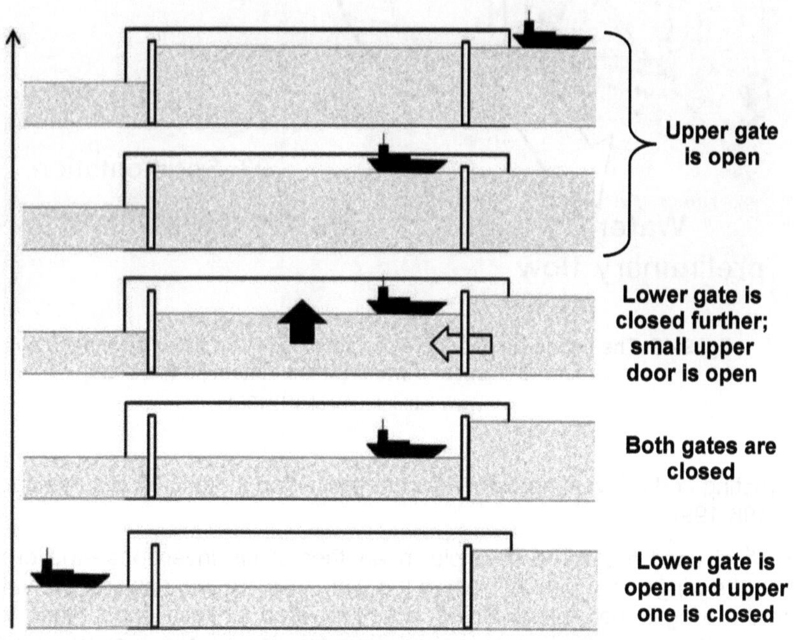

fig. 9.47. The vessel going from low water upward

The peculiarity of this project by Leonardo da Vinci (fig. 9.48) is that he suggested making small doors at the bottom of both gates (they are put in at the ends of the lock, both on the high-water side and the low-water side): one to let the water in from the high side, and the other to let it out on the low side. Lever mechanisms were designed to open and close the doors from the shore.

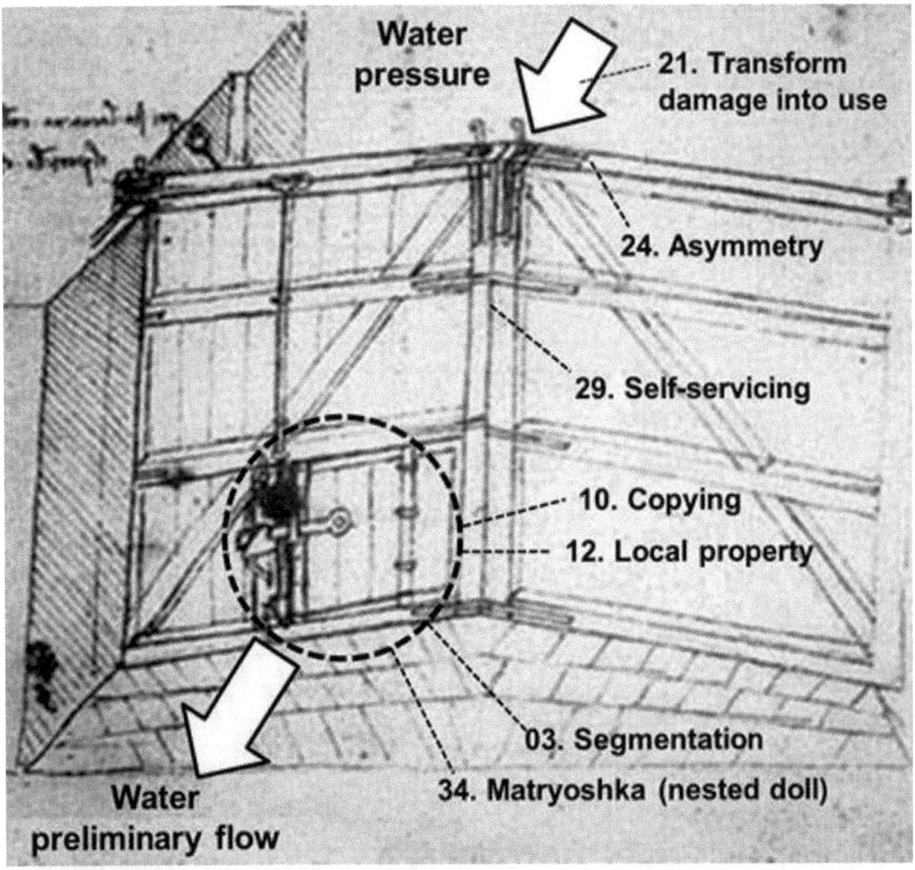

fig. 9.48. The Leonardo da Vinci's project of lock with gates having the built-in door for the release of water (at the bottom [in the circle])
www.labirint.ru/books/121061

Extracting and reinventing outcomes are presented in fig. 9.52 (double-page spread on pp. 198-199).

Da Vinci used the same principle in another of his inventions—the ornithopter (fig. 9.49): Its wings had "valves." When the wing was moving upward, the valves opened, and the wing turned into a lattice, which lowered air resistance; when the wing was moving downward, the valves closed, thus forming a solid supporting (when gliding–bearing) surface.

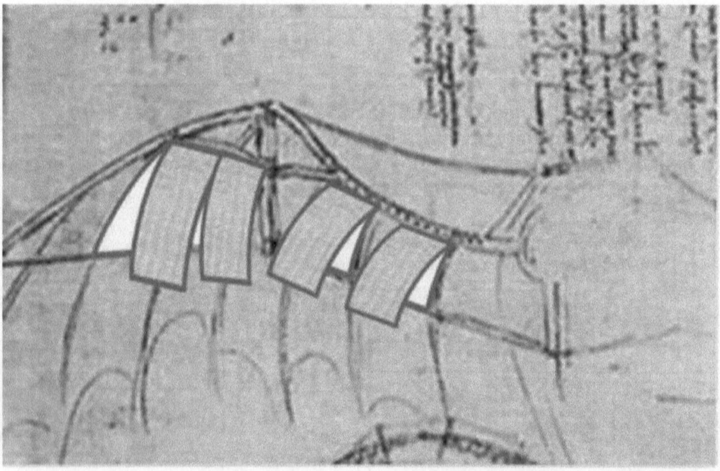

fig. 9.49. Da Vinci's wing had get such a shape and construction – "developed" by the author (M.O.)

And the ideas of the great master live on!

Similar "doors-within-doors" are shown in fig. 9.50 and fig. 9.51.

Invention of the door-within-door for cats is attributed to Sir Isaac Newton,[76] who allegedly wanted to make it possible for his cat to carry her kittens through the door so that he would not have to open and close it himself because it distracted the great physicist from his darkroom light experiments.

fig. 9.50. Built-in small door for cats
www.mindhobby.com

fig. 9.51. "Matryoshka" of the doors
www.prohandmade.ru

[76] *Sir Isaac Newton* (January 4, 1643 – March 31, 1727) – great English physicist, mathematician and astronomer, one of the founding fathers of classical physics.

ARTIFACT — Leonardo da Vinci's Lock

<u>Description</u> Leonardo da Vinci proposed to make input (in the upper gate) and output (in the lower gate) special preliminary openings in the form of small doors built-in the lock gate which could be opened with lever mechanisms operated from the shore before opening the main gate. The gates jutted slightly out towards "high water".

Illustrations: prototype artifact – old lock (www.tourism.ru/docs/report/cycle/54/111/614/img/norway-08_006.jpg); resultant artifact – www.newepoch.ru/journals/18/zubov.html.

EXTRACTING

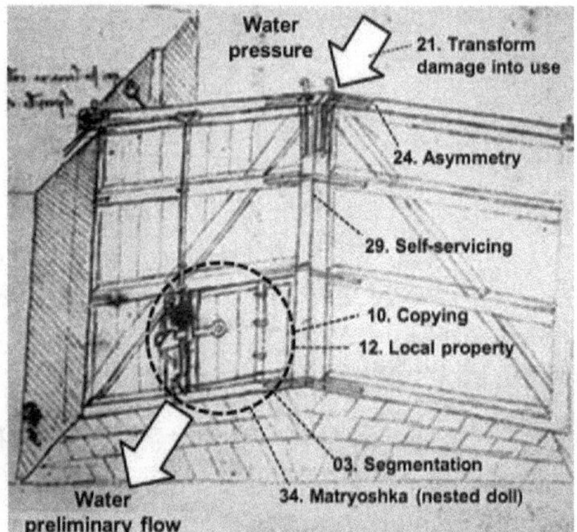

WAS: Artifact-Prototype — PROBLEM → Trend > Reducing > Inventing > Zooming → IDEA — IS: Artifact-Result

EXTRACTING-1

LC	No.	Navigator	Function
++	03	Segmentation	a) disassemble an object into individual parts b) make it possible to disassemble an object
++	21	Transform damage into use	a) use damaging factors, especially damaging influences from the environment to achieve a useful effect
+	24	Asymmetry	b) increase the degree if the object is already asymmetrical.
++	34	Matryoshka	a) an object is inside another object that is also inside another, etc.

REINVENTING

TREND Old lock gate(s) had a lot of problems:

- they leaked water because their wings were not tight when closed;
- it was hard to open the wings to let the water in the lock chamber from the upstream section or let it out of the lock chamber into the downstream section.

How can these deficiencies be removed?

REDUCING **Maxi-FIM:** The OZ by itself ensures the obtaining of the

IFR: [simple and fast water ingress and egress]

STANDARD CONTRADICTION (SC) – Extracting-2S

From Extracting-1

Lock gates	(+)	*Fast water ingress/egress, ease of operation*	03 Segmentation 21 Transform damage into use 24 Asymmetry 34 Matryoshka
	(–)	*Complexity of operation due to high water pressure*	

RADICAL CONTRADICTION (RC) – Extracting-2R

Lock gates	*must be tightly closed and prevent the flow of water from the higher level to the lower level*	&	*must not be tightly closed to allow water to fill the lock chamber*

INVENTING To attain Objective 1 (improved ease of operation and faster water ingress/egress), an additional door is inserted in the lower part of each gate that can be opened and closed from the shore with a special lever mechanism – dominant models *03 Segmentation, 05 Separation* and *34 Matryoshka (Nested Doll).*

To attain Objective 2 (gate tightening), the gates are jut slightly out towards "high water" so that water pressure squeezes them closer together – in accordance with models *21 Transform Damage into Use*, *24 Asymmetry* and *29 Self-Servicing.*

Fundamental models: *Spatial* – part of the large gate wing can open, whereas the rest is solid; *Structural* – division of wing gates into parts.

Scientific Effects: Hydraulic effects; lever mechanisms.

ZOOMING

Contradictions Removed: Yes.

Super-Effects: Possible application in dams to regulate water flow during both droughts and floods (in his engineering pursuits, Leonardo da Vinci attached very high importance to finding a solution to this problem).

Negative Effects: Relative complexity of construction.

Development Trend: Automated operation of the additional door.

Changes in Ambient Systems: The more reliable and predictable management of water flows in locks and dams makes it possible, in principle, to simplify ambient systems.

Extended Uses: Da Vinci used the same principle in another of his inventions – the ornithopter: its wings had "valves" – when the wing was moving up, the valves opened, and the wing turned into a lattice, which lowered air resistance; when the wing was moving down, the valves closed, forming a solid supporting (when gliding – bearing) surface.

Beauty of Solution: 100 Highest grade. Substantiation:
1) such structure was not known before, there are no direct counterparts; 2) functionality is realized very fast.

Рис. 9.52. Extracting and reinventing Leonardo da Vinci's lock

9.3.5 *Does Sir Norman Foster also like da Vinci?*

Example 9.19. The Dome of the Reichstag. You can now imagine yourself into the role of the head architect Sir Norman Foster,[77] who had the idea for the reconstruction of the Reichstag (Parliament building) in Berlin. The number one idea, both architecturally–technically as well as in its symbolic character, is to see the glass dome (fig. 9.53 and 9.54) as an element in the system of natural illumination of the main chamber and thus make a sight worth seeing in Berlin, just like the Eiffel tower in Paris or Big Ben and Westminster Abbey in London.

Let's examine the first assignment with the dome. There is a ramp on the inner side of the dome that allows visitors to reach the upper platform. How could the ramp be built so that the streams of visitors from above and below don't meet?? If the ramp were built like in fig. 9.55, the visitors would get in each other's way. In such a project, there are always strong technical contradictions: The ramp has a certain shape that causes the visitors to run into each other when ascending and descending. This leads to loss of time and difficulties. A more optimal shape is needed for the ramp.

fig. 9.53. Reichstag from a bird's-eye view
Basic picture's fragment from:
http://www.morgenpost.de/berlin/article104723496/Berlin-aus-der-Luft.html

fig. 9.54. Dome of the Reichstag
(author's picture of May 8, 2004)

Now we conclude the orientation examples where initial models for technical contradictions were developed like is normally performed in the *trend* stage. Directed solving of problems begins with making the models precise and concrete in the *reducing* stage. This continues with the resolution of contradictions in the *inventing* stage, and it ends with the *zooming* stage. Try to find a solution yourself and compare your ideas with the answers that you will find below.

Are you now ready to reconstruct the thought processes of the architect Sir Norman Foster? If so, let's try it. If not, read this textbook again from the beginning!

Here we'll combine solutions based on technical and physical contradictions, especially because we want to move to an investigation of transformations on the basis of physical contradictions after this example:

Ideal result: Streams of visitors do not encounter each other!

[77] *Norman Robert Foster*, Baron Foster of Thames Bank (born June 1, 1935) is world-famous British architect whose company, Foster + Partners, maintains an international design for high-tech architecture

Standard contradiction: Ramp ▶ 21 Shape **VS** 25 Loss of time **▶ 02, 15, 19, 22**

Radical contradiction: There must be streams of visitors moving in opposite directions because they go to and from the upper platform, and there cannot be streams of visitors moving in opposite directions so that they don't disturb each other.

Leading resource: space.

The A-matrix recommends navigators *02 Preliminary action, 15 Discard and renewal of parts, 19 Transition to another dimension,* and *22 Spherical shape.*

The Afs-catalogue (simplified, section S25) recommends these navigators, which seem extremely promising: *05 Separation, 10 Copying, 19 Transition into another dimension, 22 Spherical shape, 34 Matryoshka.*

Of course, we have all the recommendations and their interpretations after extracting, which makes our modeling clear and easy:

05 Separation: Remove the disruptive part (for example, the visitors leaving), and separate the part needed (analogy).

10 Copying: Use copies (install a second ramp!).

19 Transition into another dimension: Shape a construction with several floors (somehow place a ramp over the other one).

22 Spherical shape: Use spirals (they already are in use here!).

34 Matryoshka: Put objects into each other; let an object move through the hollow space of another (the ramps must be installed within each other!?).

Fig. 9.56 shows a simple and effective solution: The second ramp is turned 180º in its installation, and its curves spiral between the curves of the first ramp. The ramps are identical, i.e., they are copies of each other.

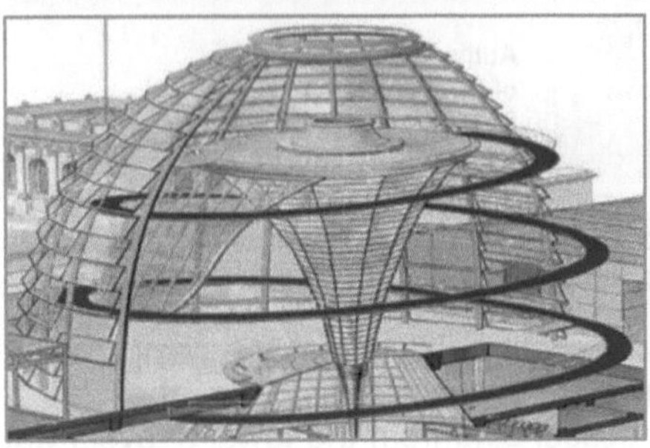

fig. 9.55. Hypothetical construction of single ramp

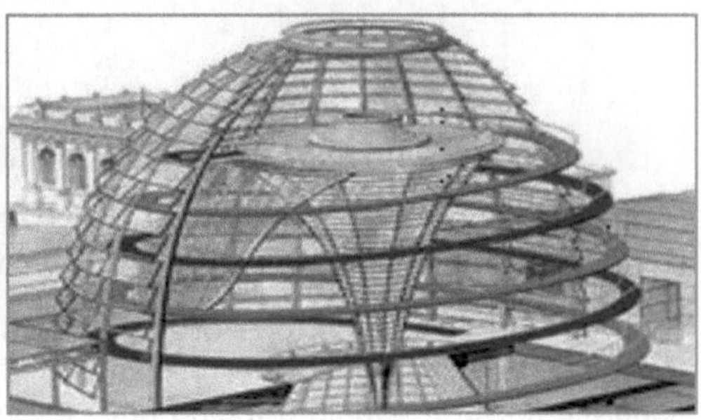

fig. 9.56. Real double spiral ramp – separately ascent and descent

Fig. 9.57 shows separated entrance and exit for a double-helix ramp construction as well as also the directions of motion on both ramps.

A great idea continues its life in the new designs. Sir Norman Forest applied this idea in the construction of ramps (fig. 9.58) at a top-level outstanding architectural and cultural edifice—*Palace of Peace and Reconciliation*—in Astana, the new capital of Kazakhstan.

fig. 9.57. White arrows – movement along the ascending ramp; black arrows – for descending (basic picture's fragment from: https://c1.staticflickr.com/9/8003/6987727706_fae0c9fb56_b.jpg)

Author's pictures of September 13, 2014, Astana

fig. 9.58. Double-spiral separate ramps—One for ascent (white arrow) and one for descent (black arrow)—in the Palace of Peace and Reconciliation, Astana, Kazakhstan

And now we can reveal one of the secrets of creating the dome and ramps in the Reichstag's building. For the first time in the history of our civilization, the idea to separate flows of people climbing and those descending from the tower's levels was invented by Leonardo da Vinci (fig. 9.59, a). Among the many sketches made by hand of da Vinci, there is also a project with four staircases in volute formation around the central part of the structure (fig. 9.59, b).

The staircase in Chambord castle (fig. 9.59, c) is one of the most outstanding implementations of the double-helix structure made during the construction of the castle in 1519 to 1547 on the basis, again!, of da Vinci's sketches after his death.

There is no doubt that the masterworks of the great da Vinci were familiar and likened to those of Sir Norman Foster, who reproduced the brilliant idea of a double-helix ramp in his own architectural masterpieces!

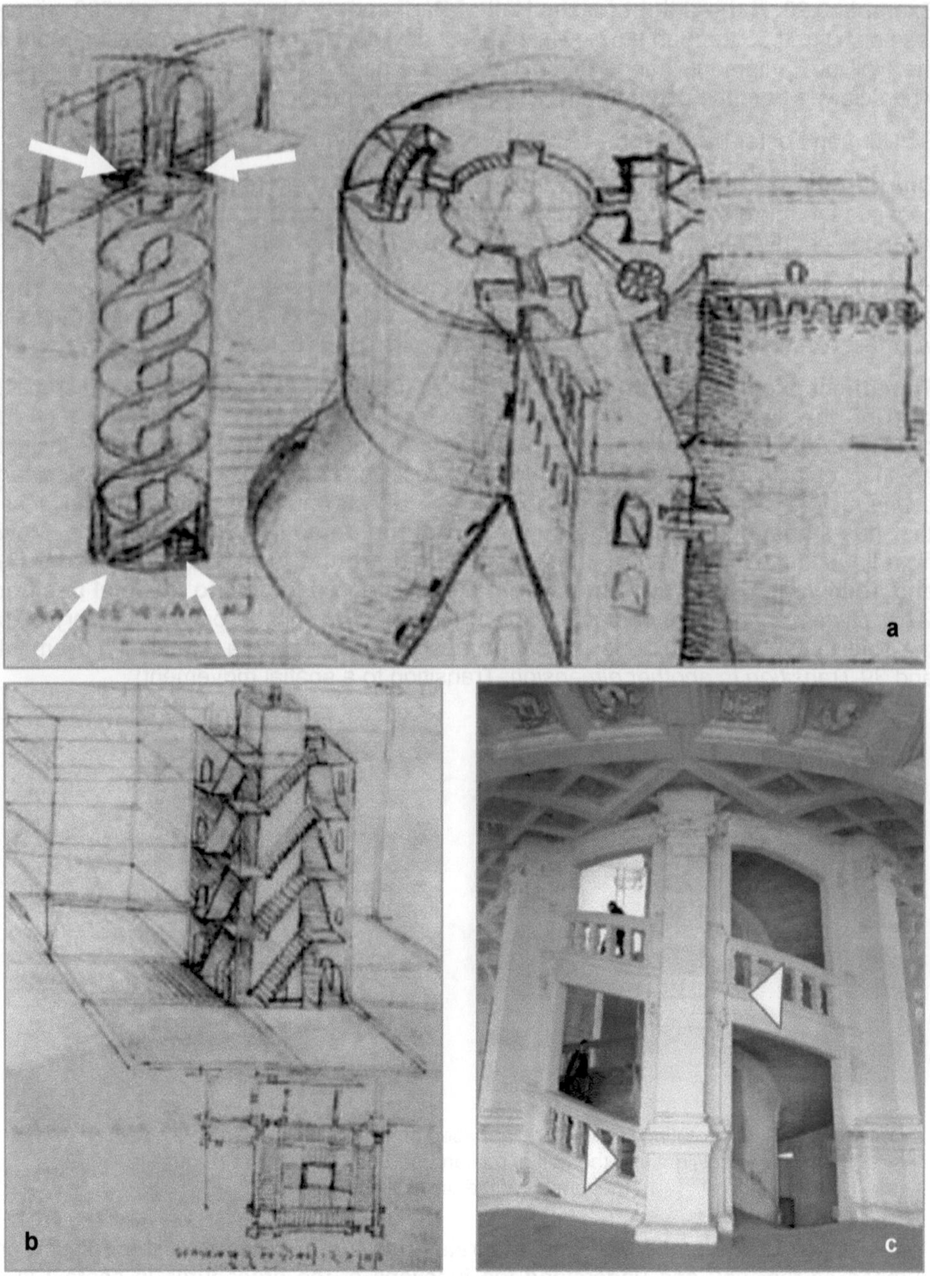

fig. 9.59. Da Vinci's sketches of a) fortress double-helix staircase (1488) and b) fourfold stairs (ab. 1490). Double-helix staircase (c) in the castle Château de Chambord at Chambord, Loir-et-Cher, France, presumably based on the concept by da Vinci

– basic pictures from magazine *"Наука и жизнь"*, №9, 2014 (Science and Life, Russian: *Nauka i Zhizn*)

Example 9.20. Natural light for the Hall of Parliament. A large cone with 360 mirrors (see also fig. 9.55 through 9.57), which reflect daylight directly into the meeting room in the Hall of Parliament, hangs under the central point of the dome of the Reichstag when viewed from the center of the observation platform (fig. 9.60).

Sir Norman Foster had to solve a number of contradictions.

One of them is the following radical contradiction: Daylight is constantly present because the mirrors are stationary **VS** the light should not be present on meeting days when the brightly shining sun blinds people in the meeting hall.

Here the spatial, temporal, and structural resources are clearly dominant. We can find an entire series of appropriate Af-procedures in section S25 and S26 including effects we must examine in our description of the control solution below.

To redirect excess sunlight from the mirrors (*05 Separation*: separate the disruptive part of the light; *12 Local property*: every part must function under optimal conditions–the mirrors), **we need a previously installed shade** (*18 Mediator*: Include another object temporarily; *28 Previously installed cushions* and *39 Preliminary counteraction*: Disaster measures and inverse effects must be planned in advance) **that has a shape similar to the upper part of the dome** (*22 Spherical-shape*: Move from flat surfaces to spherical ones), **is installed around the cone with mirrors (3), and follows the sun's movement from the starting position 1 to the final position 2** (*07 Dynamization*: The characteristics of the object should be optimal in every step; the object should be moveable; *22 Spherical shape*: Change to a rotational movement; and *39 Transition to another dimension*: Transition to a spatial movement).

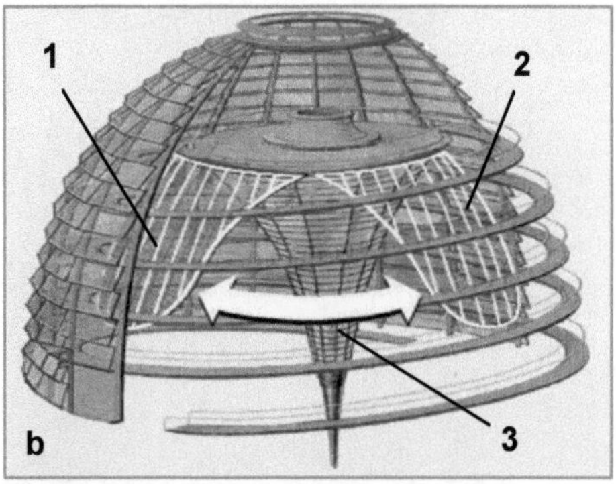

fig. 9.60. Rotating "sun umbrella" based on the "principle of a sunflower."
a) Path of natural light into the hall of parliament (author picture of August 12, 2009);
b) movement of the protective sun visor

A description of the procedure was integrated into the description of the solution on purpose so that you can understand the functions of the procedures in context in a detailed way. You should read the description carefully and think about the fragments in bold as many times as is needed until the entire description is understandable as a whole. And, as usual, you are invited to perform extracting and reinventing!

Sun energy is involved in the movement of the shade with implementation of the Maxi-FIM approach: The OZ itself resolves its problem!

9.3.6 Water from Air: *"Magic" Particles in the OZ*

Example 9.21. The WATERCONE®

This very impressive device[78] (fig. 9.61) was invented and developed by engineer Stephan Augustin from Munich, Germany. It is also important to note that his activity is mainly related to charitable-movement "donationware." The pilot project in Yemen was realized with the financial support of the foundation *Hans-Sauer-Stiftung*, Munich.

One look at our planet shows that more than 70% of the global surface is covered by water. A large percentage f Earths total water reserves, 97.2 %, are covered by oceanic saltwater and therefore are mostly inappropriate for human consumption and use. The remaining 2.7% of freshwater is equally unavailable to man because it is confined to the ice of the polar caps and mountain glaciers. Only ground water (0.625 % of Earth's total water reserves) and the water from rivers and lakes (0.017%) remain easily accessible to man.

For ecological, economical, geographical, and/or political reasons, 40% (2.5 billion people) of the world's population has no access to clean water!

Currently most people in poor and sunny regions must use dirty water for daily use. Seawater is not a viable source because most desalination units are cumbersome, need constant technical maintenance and support, and, most importantly, are very expensive.

Artifact-invention

The WATERCONE is a product that enables anyone, in a most simple fashion, to operate an independent, cheap, and mobile solar object that facilitates potable-water generation from sea water or brackish water on the basis of condensation by a solar still. Technically speaking, it is a solar still.

This invention is represented by a conical, self-supporting, and stackable unit made from transparent, thermo-formable polycarbonate (same as water dispensers) outfitted with a screw cap spout at the tip and an inward circular collecting trough at the base.

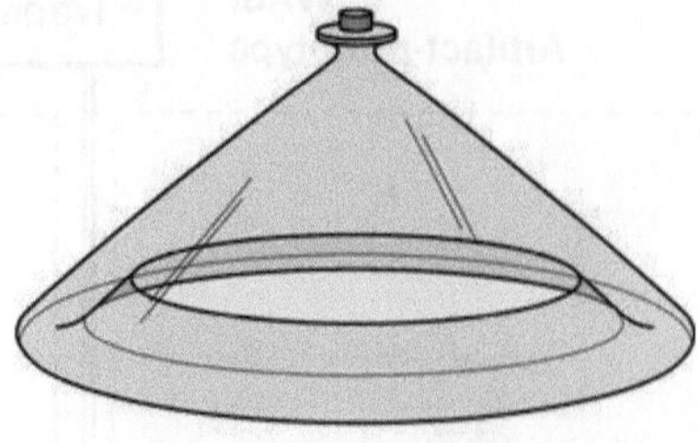

fig. 9.61. The cone part of the WATERCONE®

Pour salty/brackish water into a pan (fig. 9.62). Then float the WATERCONE on top. The black pan absorbs the sunlight and heats up the water to support evaporation. The cone can be set directly on the ground.

fig. 9.62. Installing the device on a pan (a) in mud (b) and in a swamp (c)

[78] The text here is based on citation from the following Web sources: www.augustin.net; www.watercone.com and http://www.kabeleins.de/tv/abenteuer-leben/videos/we-are-made-in-germany-2-clip

The evaporated water condenses in the form of droplets on the inner wall of the cone. These droplets trickle down the inner wall into a circular trough at the inner base of the cone (fig. 9.63,a). By unscrewing the cap at the tip of the cone and turning the cone upside down, one can empty the potable water gathered in the trough directly into a drinking device (fig. 9.63,b). The WATERCONE (with a base diameter of 60 to 80 cm) yields between 1.0 to 1.7 liters of condensed water per day (in 24 hours).

fig. 9.63. Condensed water on the pan wall (a) and pouring the potable water into the bottle (b)

what is the most important in this example (fig. 9.64 and 9.65) is using a Micro-FIM because of getting the ideal final result from just... air, from the OZ ***in the form of water droplets***!

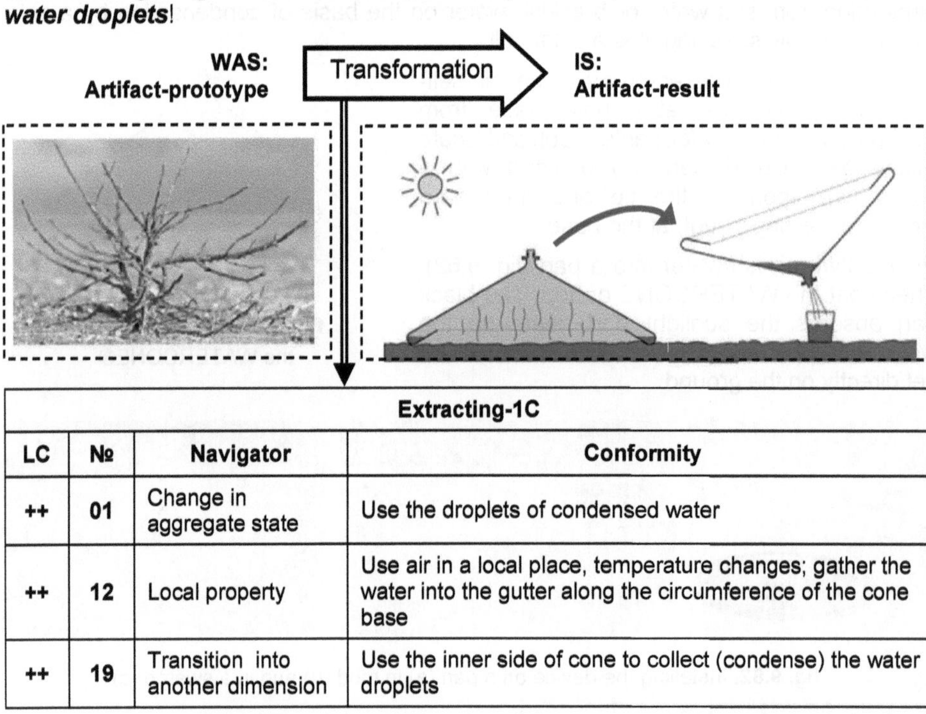

Extracting-1C			
LC	**№**	**Navigator**	**Conformity**
++	01	Change in aggregate state	Use the droplets of condensed water
++	12	Local property	Use air in a local place, temperature changes; gather the water into the gutter along the circumference of the cone base
++	19	Transition into another dimension	Use the inner side of cone to collect (condense) the water droplets

fig. 9.64. Results (briefly) of Extracting-1C for WATERCONE

REINVENTING: ZELTEC – WATERCONE

TREND Vast areas on the planet are deprived of sufficient water.

There are even more arid places where water is scarce in the hot season. Climate warming steadily exacerbates the lack of water. Water desalination by traditional methods on an industrial scale requires a huge amount of energy, which is also lacking.

Moreover, the supply of such water is complicated in the drylands because of the low population density, so the water has to be delivered a small number of people over large areas, so the cost of water is too large. What could be done?

REDUCING Let's formulate a fantastic requirement:

Micro-FIM, or *"FIM of deepened OZ"*:

The ***X-resource*** in the form of *particles of substance or energy* is located inside the OZ and ensures, together with other available resources, the obtaining of the **IFR:**

[*water in sufficient quantities* ***ITSELF*** *comes to my house every day*]

SC: **1) Water ▶** *quantity of material* **VS** *temperature* **▶ 26 VS 34 ▶** *12, 19, 23*

2) Water ▶ *productivity* **VS** *temperature* **▶ 01 VS 34 ▶** *01, 02, 04, 33*

RC: **Water ▶** *must be* (to drink) **VS** *must not be (due to heat)*.

INVENTING With a view to extract water from the air through condensation when temperature changes, such as day and night, the following models are applied: ***01 Change in the aggregate state of an object, 02 Preliminary action, 12 Local property and 19 Transition into another dimension.*** It is proposed a device for condensation and collection of water on the inner walls of the transparent cone, standing on a protective (and for other goals) disk, from cooled air, such as at night! The condensate flows down into the gutter along the circumference of the cone base, and pours out through an upper mouth of the inverted cone.

ZOOMING Contradictions eliminated.

fig. 9.65. Reinventing of solution for invention of the WATERCONE®

And now I would like to go back to the diagram of the OZ shown in fig. 9.7.

My resume for this part of the book should be done as a brief description of actions in the OZ during the process of problem-solving and idea-generating as follows.

The OZ transformation process according to fig. 9.7 has the following actions:

1) Selection of the inductor and receptor within the OZ;

2) definition of the contradictions that prevent acquisition of required properties by the receptor or that reflect other problems existing in the OZ;

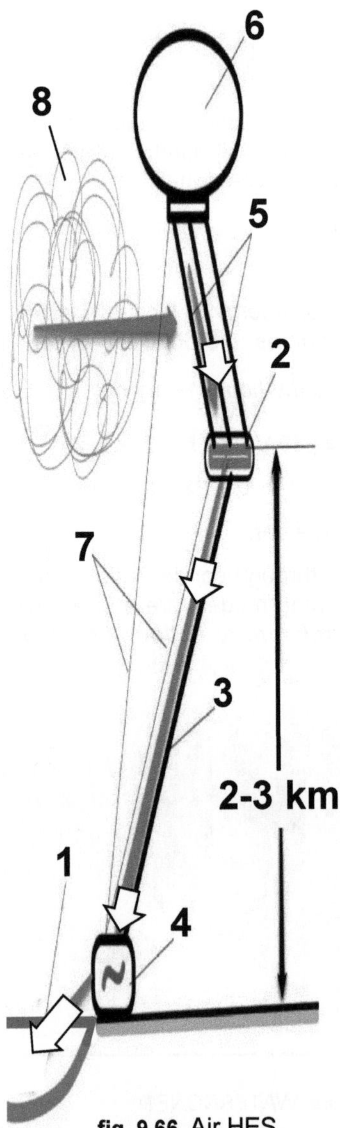

fig. 9.66. Air HES
http://airhes.com

3) determination of resources, both those already existing in the OZ and those that are required additionally;

4) formulation of the IFR and the FIMs for the OZ;

5) selection of transformation models subject to the following:

- transformations are usually applied to the inductor, its own resources and other resources available in, or introduced into, the OZ;

- realization of transformations is directed toward replacement of contradictions with the ideal final result or the complete removal of contradictions.

The method of selection of transformations is determined by the type of contradiction, by the type of resource, or by the nature of the IFR, which also leads to the selection of dominant resources that are most actively "involved" in the transformation. We have seen examples of such selections in this and previous sections.

At this juncture, we draw the readers' attention to the very important role of correct selection of the OZ's actors and resources, definition of the ideal final result, and, naturally, selection of transformation models to solve the problem with the highest degree of efficiency.

Example 9.22. Air HES by Andrey Kasantsev

The Air Hydroelectric Station (HES) is an invention of Russian engineer Andrey Kazantsev (patent no. WO2013157991A1).

The Air HES (fig. 9.66; the texts below are from http://airhes.com) provides a downstream (water's outflow) 1, upstream (water's inflow) 2, conduit (pipe, penstock) 3, turbo generator 4, mesh, fabric or film surfaces 5 to collect water from cloud 8, airship (air balloon) 6, and fastening ropes (tethers, lines) 7.

The airship (6) lifts the surfaces (5) at the height of near or above the dew point (condensation level of clouds) for current atmospheric conditions (typically 2 to 3 km). There super-cooled atmospheric moisture begins to condense from clouds on the surfaces (5). The drainage system on the surfaces (5) collects the water in a small reservoir (upstream 2) where water, under pressure from the whole hydraulic head (2 to 3 km), flows through the penstock or conduit (3) to the downstream (1) on the ground, thus producing electricity in the turbo generator (4).

The Air HES can be easily mounted in any convenient place for the consumer of electricity and water simply by lifting and moving it entirely using the same airship (6).

If at this point winds are blowing steadily, or if it is a portable unit (for example, for tourists or military), one can do without the airship (6) and use the surfaces (5) like a paragliding wing for self-containment of the assembly in the air (as occurs when you fly a kite).

Also, the surfaces (5) can be maintained with full or partial metallization (for example, by weaving metal wire). This will increase the structural strength and reduce solar heat in order to increase the condensation of water vapor through the filing of an electric field (for example, as has been experimented with this corona discharge), as well as the need to decrease the amount of ice due to current supply.

Extracting-1C	
Navigator	**Conformity**
01. Change in aggregate state	Use droplets of condensed water
02. Local property	Use water from cloud with dew point at right height above sea level; gather water with fabric and transport it down to generate electricity
19. Transition into another dimension	Use positioning at high altitude to collect water from cloud
20. Universality	Produce water and electricity
29. Self-servicing	Water cycle is an endless source of energy for mankind
32. Counter-weight	Raise fabric or metal membrane with dirigible or kite

fig. 9.67. Extracting of solution for invention of Air HES by Andrey Kazantsev

The principal difference of Air HES is the condensation of moisture from the air that at first glance it seems an odd and impractical curiosity. Nevertheless, there is in fact nothing unusual. There are several such working systems, known as fog collectors, in the world. For example, a device for collecting drinking water in Chile was tested in 1987.

Interestingly, already having obtained this solution, I began to search the Internet for key words for such ideas. I unexpectedly found in an article from 1915 that the genius Nikola Tesla[79] was nearly half a step toward realization of this idea: He was correct in his principle estimation of the required resource, but he had not found a scheme to implement it even though the necessary technology was available hundreds of years ago. It's a pity. If he had pursued the idea, then we would live in a different world—clean, ecological, abundant, without wars over oil, etc. Unfortunately, humanity has lost a hundred years due to this omission!

It is known that the solar energy reaching to the planet is approximately 10,000 times greater than the needs of humanity. Approximately a quarter of it goes to the evaporation of water and virtually always more or less evenly accumulates in the atmosphere at any point around the world. Because the annual precipitation is approximately 1 m of rainfall with an average height of 5 km, this gives a potential capacity of approximately 810 TW, which is more than 60 times greater than all the current needs of humanity (13 to 16 TW).

Standard hydroelectricity can only use a small fraction of this energy because all precipitation loses most of its potential energy on the way to the ground to (1) overcome the resistance of the air and (2) actually hit the ground. In order to use this potential energy more cost-conscious, it is necessary to collect water at the altitude where it condenses and uses all possible vertical hydraulic heads. This is what constitutes the essence of this invention.

9.4 WORKSHOP FOR CHAPTER 9

It is advisable that you train your extracting and reinventing skills on an ongoing basis. This must become both a habit and a professional requirement.

This is routinely performed, for example, by chess players: When they are alone, they often do chess problems or analyze well-known chess games to pinpoint novel ideas. Similar training (and this IS training or what else would you call it?) is done by musicians and, naturally, sportsmen.

All of these are steps in the long and winding road of self-perfection.

Modern TRIZ opens up new opportunities for self-perfection by providing a technology for exchanging efficient examples. This is supported by standardization of the information presentation process using the format of Meta-Algorithm of Invention T-R-I-Z.

You could start your self-training right here!

Exercise 1. Is the OZ usually a local part (area, space) of the TS?

Exercise 2. Could the TS in its entirety sometimes be considered as the OZ?

Exercise 3. Is the OZ an epicenter of a problem?

---

79 Nikola Tesla (Serbian Cyrillic: Никола Тесла; 10 July 1856, village of Smiljan, Austrian Empire, modern-day Croatia – 7 January 1943, New York) was a Serbian American, great inventor, electrical and mechanical engineer, physicist, and futurist best known for his contributions to the design of the modern alternating current (AC) electricity supply system

Exercise 4. Is the reinventing studying the transition between the system's states between "was" and "is"?

Exercise 5. Is inventing a transition between the system's states between "is" and "should be"?

Exercise 6. What is a main inductor—the pool or the water—in the OZ for the "Swimmer" problem?

Exercise 7. What is a main inductor—the pool or the water—in the OZ for the "Diver" problem?

Exercise 8. Could you repeat mentally and/or retell and explain to listeners the scheme of TRIZ-functioning (fig. 9.19)?

Exercise 9. Is it possible to change the GC for example 9.5 with the radical contradiction formulated as follows:

"The pool wall *must be* (to hold in the water) and
must not be (so that the swimmer cannot touch the wall)"

Exercise 10. Is it possible to change the GC for example 9.5 with the radical contradiction formulated as follows:

"The swimmer *must reach* the pool wall (because he is swimming) and
must not reach the pool wall (so that he cannot touch it)"

Exercise 11. Can we formulate the following the Maxi-FIM for example 9.5:

The operating zone ***itself*** ensures obtaining the IFR
[*the pool wall itself does not allow the swimmer to approach it*].

Exercise 12. Can we formulate the following the Maxi-FIM for example 9.5:

The operating zone itself ensures obtaining the IFR
[*the water does not allow the swimmer to approach the pool wall*].

Exercise 13. Specify the possible physical actions of the inductor "pool wall" required to attain the FIM formulated in Exercise 3 above: a) push off the swimmer, b) move away from the swimmer, c) restrain the swimmer.

Exercise 14. Specify the possible physical actions of the inductor "water" required to attain the FIM formulated in Exercise 3 above: a) push off the swimmer, b) carry away the swimmer, c) restrain the swimmer.

Task 1. Concert by Andrea Bocelli on May 23, 2009, in Berlin German Opera[80].

When walking onto the stage, the illustrious tenor is accompanied by an assistant because, alas, the singer is blind. Born with poor eyesight, Andrea lost any hope to improve his vision when he had a soccer accident at the age of 12. The ball hit the boy in the face, and he had to undergo more than two dozen eye surgeries.

Andrea Bocelli leaves the stage together with his assistant after having performed several arias. Suddenly he comes back onstage alone without his escort! Bocelli unhurriedly walks along the edge of the stage to the conductor's podium, and the shock that I experience watching him do that briefly distracts me from his performance. At the end of my set, I see him walking amid thunderous applause, retracing his steps between the edge of the stage and the orchestra pit and disappearing into the wings.

Creative technical question: What made it possible for the singer to move like that?

80 Die Deutsche Oper Berlin, Bismarckstraße 34–37.

Use the following small prompt: The singer controlled his movements by himself.

Task 2. Advertising Clouds in Clear Skies on a Sunny Day.

This event happened in Long Island, NY, on a sunny day in July 2002. Huge letters suddenly began to materialize one after another in the sky above Garden City. They "drew themselves" element by element in a rapid left-to-right succession like in a PowerPoint animation.

It was so unexpected and overwhelming that one could only admire the result without even trying to find out how exactly an image of the brand of a well-known brewery came into existence out of nowhere. One could see, though, that vertical elements of the letters filled themselves out from left to right along their entire height at once.

After a short break, the same brand made a second and then a third appearance. The sky was adorned with the name of the famous Dutch beer manufacturer – HEINEKEN.

In all that excitement, the author remembered that he had a small digital camera (which he carries around just for such occasions), and he had enough time to take a couple of shots before the wind dispersed the letters (see fig. 9.68).

fig. 9.68. "Cloudy" advertisement in clear blue skies

Problem 1: Using sufficient available information and the results of TRIZ analysis of the situation and of the operating zone, devise a method of making such letters.

Provide a general answer including key spatial and structural principle, major inductors (technical objects!), and description of the process of realization. Physical and chemical details can be omitted.

Small prompt: This method can be used to create in the sky not only words made of "rectangular" letters but also any "curvaceous" images.

Problem 2: Extract TRIZ models from this wonderful technical and aesthetically pleasing solution.

And remember the **phenomenon of "disappearance of the miracle"** (see 4.4. Contradiction as an Attribute of Development in "ABC-TRIZ. Book I: How to learn to invent") according to which the inventive idea:

1. is not obvious until it is discovered;
2. is perceived at the time of its emergence as an unusual, outstanding phenomenon, i.e., a "miracle"; and
3. is perceived as a usual, simple phenomenon after it has been explained.

Do not forget *the miraculous in the mundane!*

It may well be *the quintessence of creativity.*

Task 3. Nonfalling toothbrush. Perform extracting for a toothbrush capable of holding itself up on a smooth even surface (fig. 9.69). To make it possible, the base of the toothbrush is made in the form of an elastic suction cup (marked with a white ellipse). The toothbrush is pressed to the smooth surface, the air is squeezed out from under the elastic base, and the toothbrush sticks to the surface. The toothbrush can grip not only the surface of the sink or bathroom table but also to glazed tiles and wall-hung mirrors.

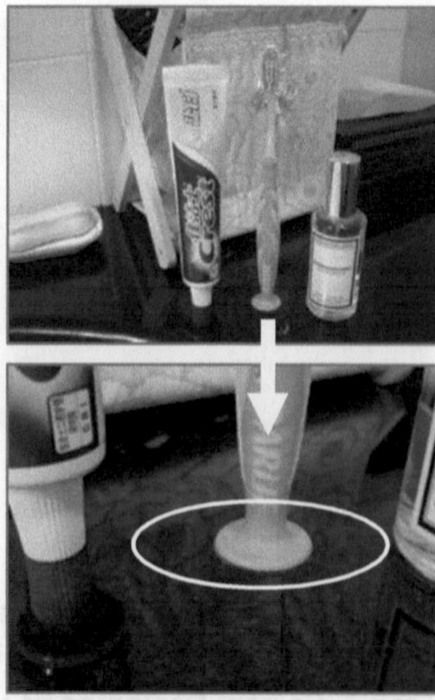

fig. 9.69. Non-falling toothbrush

Task 4. Visualization of driver actions during parking. *Nissan* came up with a system helping to visualize the maneuvers performed by the driver while he or she is parking the car. The system consists of four miniature video cameras providing a panoramic view of the cars environs. The on-board computer processes the information received from the cameras and plots an overhead view of the car and the adjacent area (see photo from www.mikeslist.com in fig. 9.70).

fig. 9.70. View of the car on the driver's monitor during parking

The resultant picture is similar to the view that could have been obtained from one video camera raised approximately 15 meters above the car. Clearly, with such visual support, the driver's maneuvers will be much more accurate and safe.

PROMPT: The prototype artifact is a car without such parking-management system. When parking such car, the driver looks through the front and side windows, uses side and rearview mirrors, and often turns back to look through the rear window. Sometimes the driver must get out of the car in order to take a closer look at the objects surrounding the car so that he/she can avoid damaging such objects, and the car, during the course of parking.

Task 5. Bookmark/book-holder. To make it easier to hold an opened book, the reader can use a special device in the form of a thumb-mounted plastic plate. When the book is closed, the plate doubles as a bookmark (see photo from www.thumbthing.com in fig. 9.71).

fig. 9.71. Bookmark (book-holder)

PROMPT: The prototype artifact is a book that we have to hold using two hands, which is not always convenient.

fig. 9.72. Golf practice field (pictured by the author in 2006 at a golf club in South Korea)

Task 6. Golf club practice grounds. Golf players frequenting golf clubs practice using artificial playing fields surrounded by tall mesh-wire fences (fig. 9.72). The field has a strong tilt toward the ball-hitting position. This is done to make the balls roll back by themselves towards the place where golf players can reuse them for practice. It also makes it much easier to collect the balls.

PROMPT: The prototype artifact is a golf field where the player or a golf club employee must walk around and collect the balls manually. This takes much time and effort. It also makes it virtually impossible for several dozen players to practice at the same time.

Exercise 15. Make Extracting and Reinventing for the artifacts shown in fig. 9.49, 9.50, 9.51, 9.58, 9.59, and 9.60,a and 9.60,b.

Task 7. Correct Designing of the Junction of Vihtakanta bridge and both ends of a road.

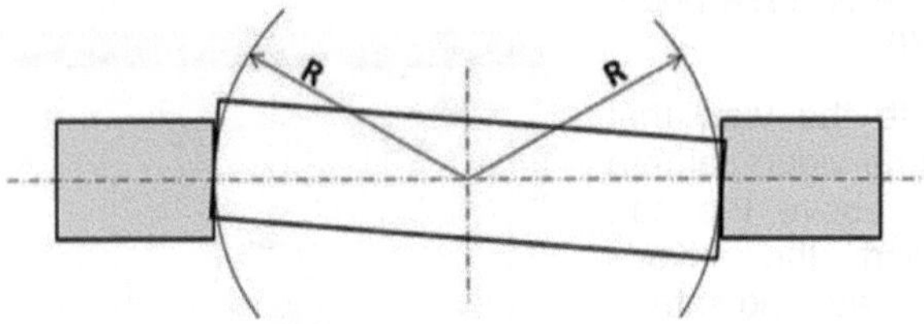

fig. 9.73. "Impossibility" of connection and disconnection of the bridge and a road because of right-angled shape of the ends of bridge and the both butts of the road

This simple example is a little bit prankish!

Imagine that the edges of the bridge have a right-angled shape (fig. 9.73).

At the beginning of rotation it cannot get out of contact with the road, and when returning it cannot be directly docked to the road – see fig. 9.73.

And here's a problem for you: How do we dock the bridge to the road on both sides at the end of the rotation? I hope you found this as an "unsolvable" general contradiction!

Try, please, to formulate your own contradictions and IFR, then select the FIM and invent a possible solution. After getting the idea, perform the extracting and the reinventing, perhaps a model(s) may appear that is not already in the A-matrix.

You can find a clear tip for the solution in fig. 9.44,a.

Task 8. How to connect reliably the Vihtakanta bridge with the ends of a road.

How to connect securely a bridge in Vihtakanta (see. fig. 9.43,e) with the road on both sides of the bridge at the end of the rotation?

You can see a problem if you pay attention to the levels at which the actors of the OZ (i.e., the bridge and the road edges) are before docking (fig. 9.72, in white circle). Really it is very difficult to get both the bridge and ends of a road exactly on one level! But the bridge and road have to on exactly the same level to facilitate transport motion!

Tip 1: Be sure to use the Maxi-FIM in the construction of the bridge!

Tip 2: Look please at some docking assembly inside the white circle in fig. 1.44,b! What does it do in your opinion?!

Try independently to formulate the contradictions and the IFR, then select the FIM and invent a possible solution.

After the solution it is very useful to perform the extracting and reinventing, and perhaps there should be some model(s) that will not be taken from the A-matrix.

fig. 9.74. Video (by author – 04.08.2014) of docking the bridge and a road in Vihtakanta is used here only to exaggerate the speculative situation for this example

Task 9. The amazing thing—To catch fish by "trolling"!

Last summer (2014), we had a little vacation in Finland at our friends Galina and Yuri Naryshkin.

It was exactly 50 years since I became friends with Yuri. Our friendship began in the boxing section in Minsk. He was in the heavyweight division, and I was a junior middleweight, although we were often put to work together in a mixed pair: speed against heavy impact (and vice versa).

Who are not the fishermen cannot live in the house by the lake! And this was the first time I fished "for real"!

We set 10 fishing rods on the holder (fig. 9.75, a), equipped them with planers (behind our backs in fig. 9.75,b), and moved slowly along the shore watching the planers and looking at the bottom profile on the sonar screen.

The scheme of trolling (c) is set up as follows: Each fishing line is equipped with a separate planer (d), and on the left side of the boat should be the left-sided planers (marked with L), and on the right the right-sided (marked with R) planars. The planers have bevels, by which they are pulled apart to reveal the fishing lines at the motion of boat (acting similar to a kite in the air). Planers ensure the absence of crossing the fishing lines so that the lines are not tangled. An ingenious invention!

Each planer has a small signal flag (e), which falls as soon as a fish takes the bait and becomes hooked. Then we gently pull the right fishing rod and in time raise a landing net for caught fish (it is "my" pike in a landing net in fig. f). Good fishing brings bigger fish for the master (pike of 8 kg in fig. g and salmon of 4 kg in fig. h). The author, as a rookie, was pleased on the next day even with a small zander (i).

And now the learning task: Realize two reinventings. The first one for the single planer and the second for the right placement of planers by releasing the fishing lines in different lengths. Main tip: Use only the Maxi-FIM!

Specific recommendation for reinventing: Explain the physical principle of the planer.

Oh, and visit the unique Finnish lakes to fish, hunt for mushrooms, and collect bilberries and cowberries. And once we even met with Finnish President Ms. Tarja Halonen, who was also enthusiastically picking cranberries!

To continue learning TRIZ, it really pays to carefully study this case!

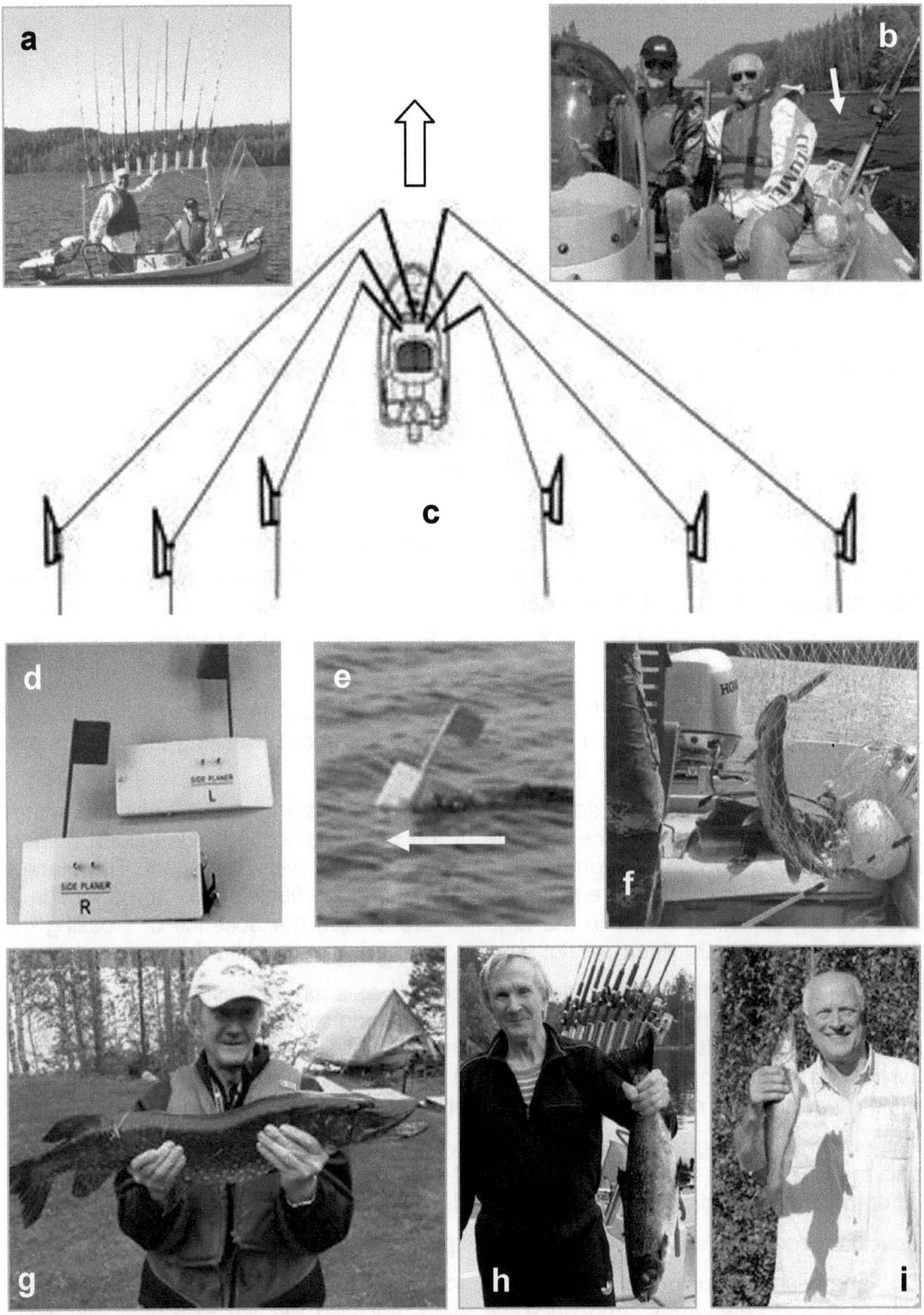

fig. 9.75. Fishing in Finnland (photo by author – 03-05.08.2014)

10 Directed Development of Systems

Throughout the history of humanity, the technics is continuously enriched with new and new methods of invention. If to formulate briefly, it was a way from random windfalls and godsends towards the conscious and systematic solving to new technical challenges.[81]

G. S. Altshuller

10.1 System Development and Evolution

10.1.1 TRIZ Laws

In this section, we will give a few ideas of TRIZ regarding the development and evolution of systems. Knowledge of patterns of development, as introduced and postulated in TRIZ, is necessary to control the process of creating the effective ideas and solutions. Then we can talk about tactics and strategy for solving to problems based on the "laws" of TRIZ.

Each artifact is the fruit of integration of diverse knowledge and creative talent. An important objective of modern science as a whole, and of TRIZ in particular, is to learn to extract that knowledge from the history of evolution of artifacts, technical systems, and civilization as a whole. Another objective is to learn to apply that invaluable experience to efficiently manage the subsequent evolution of our civilization.

Among technical characteristics, there is always one that dominates the system and has the greatest effect on efficiency. Sometimes that characteristic is assumed to be the equivalent of efficiency, but that is not always true.

The increase of each characteristic is constrained by the physical properties of the system or by the limited range of implementation of the ideas embedded in the system. As a result, parametrical evolutional progress can be represented in the form of a logistical curve (fig. 10.1).

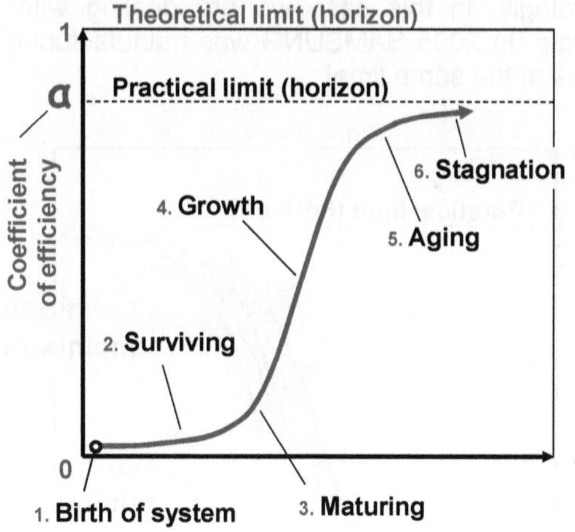

fig. 10.1. Logistical curve (S-curve) featuring evolution of the main parameter of a system type

As a rule, after developmental resources of a system of a given type have been exhausted, there emerges a new type. Systems of that type are then used for the same purposes, but they have superior technical characteristics and efficiency (fig. 10.2). Normally, such a transition is evolutional.

[81] translated by M. Orloff from G. S. Altshuller *How to Learn to Invent*. – Tambov, Tambov Book Publishers, 1961 (in Russian)

M.A. Orloff, *ABC-TRIZ*, DOI 10.1007/978-3-319-29436-0_10

For example, when propeller aircraft exhausted their capacity to increase their flight speed, they were replaced with jet aircraft. However, in terms of efficiency, both types of aircraft still have their "niches": Propeller aircraft are good for short-distance flights, whereas jet aircraft are mostly used for long-distance flights.

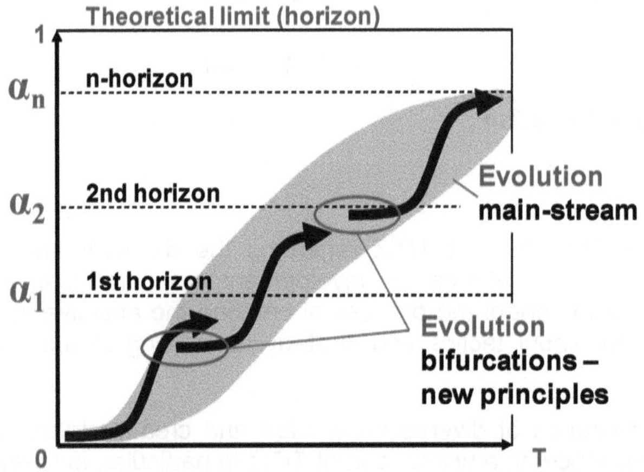

fig. 10.2. Family of evolutionary S-curves: Evolution as bifurcations—division into types

Commercial evolution of virtually any artifact is driven by its adaptation to the needs of different users. That is why there are so many types of cars, aircraft, mobile phones, television sets, computers, etc. Such transitions do not modify the type of the system. Accordingly, in this case we are dealing with "intra-typical evolution" (fig. 10.3). Example: In 2005 SAMSUNG was manufacturing more than… 164 models of mobile phones at the same time!

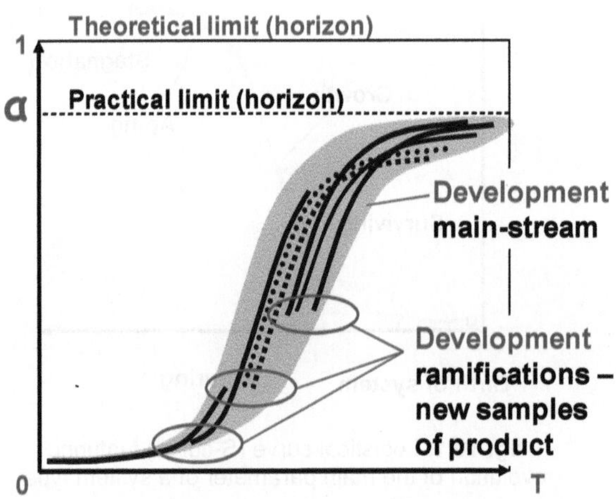

fig. 10.3. S-curves' family of development: development as ramifications—Division within the type

Each "point" of ramification or bifurcation is a conflict between the two main driving forces: growing needs (demand, "pull forces") and growing abilities (supply, "push forces). If abilities cannot meet growing needs, contradictions arise (fig. 10.4).

Those contradictions intensify when OZ resources prove to be deficient or depleted to the point of exhaustion and when transformation potential proves to be limited. In such situations, TRIZ comes in very handy.

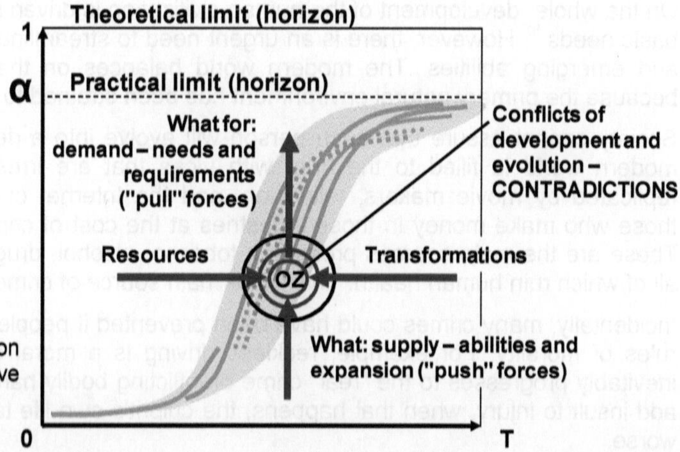

fig. 10.4. OZ representation on the S-curve

Fig. 10.5 shows the still rampant S-curve displaying the "might" of our civilization as the sum total of assimilated material, power, and information resources. Future growth will be driven by enhanced information content of artifacts. The dominant trend will be development of efficient evolution methods and management models. TRIZ will occupy the leading position among such methods.

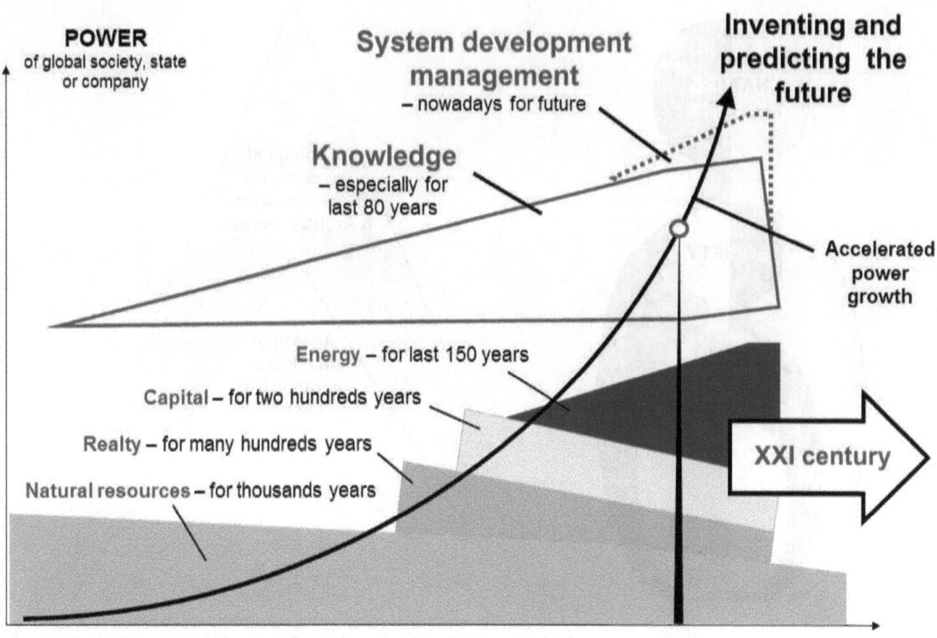

fig. 10.5. Dominating strategic platforms of development of the human civilization

On the whole, development of the human civilization is driven by qualitative changes in basic needs.[82] However, there is an urgent need to streamline the use of both existing and emerging abilities. The modern world balances on the brink of self-extinction because the primary natural environment has been strained to the limit of its capacity.

Society cannot assure that each person will evolve into a decent human being. The modern world is filled to the brim with vices that are irresponsibly and immorally replicated by movie makers, television, and the Internet or—to be more exact—by those who make money in those industries at the cost of crippling millions of people. These are the main sources promoting tobacco, alcohol, drug, and substance abuse, all of which ruin human health. This is the main source of crime.

Incidentally, many crimes could have been prevented if people had simply followed the rules of morality. For example, reckless driving is a moral crime because it almost inevitably progresses to the "real" crime of inflicting bodily harm or even death. And to add insult to injury, when that happens, the culprit's own life takes a sharp turn for the worse.

Self-actualization according to Maslow dictates the need to create a human being capable of ignoring or countering negative influences and being focused on self-development and maximum realization of his or her abilities and his or her higher intellectual and spiritual needs.

Apparently, large complex systems (fig. 10.6) consist of systems, and subsystems in turn consist of elements, whereas systems are incorporated into higher-ranked super-systems.

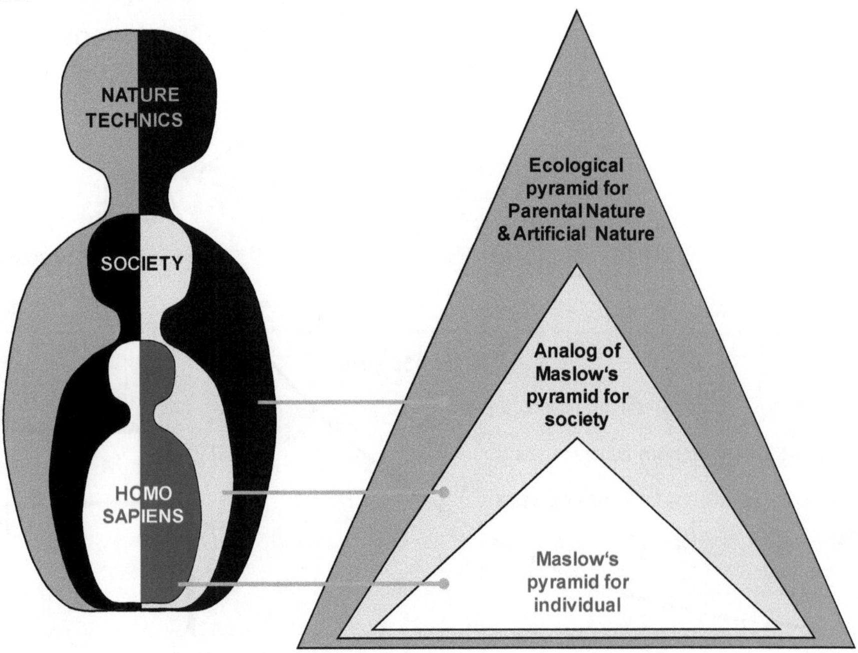

fig. 10.6. "Matryoshka" of civilization systems

82 See, in particular, the works of the founder of humanistic psychology Abraham Maslow (1908-1970); see also http://en.wikipedia.org/wiki/Abraham_Maslow.

System evolution can be presented (fig. 10.7) by a poly-model in the form of a nine-screen scheme[83].

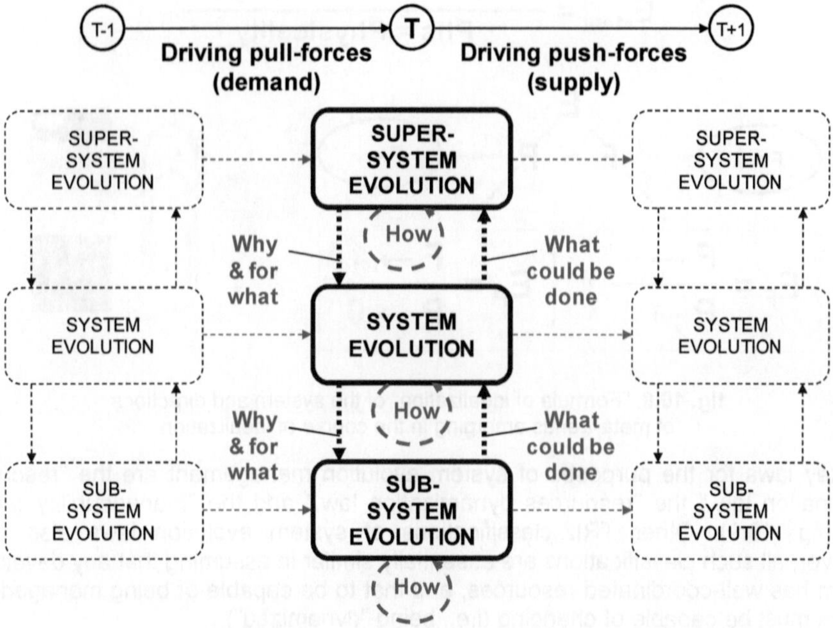

fig. 10.7. Nine-screen model of system evolution

Higher-ranked systems pose requirements (demand) to lower-ranked systems. In turn, evolution of "inferior" systems creates new abilities (supply) for meeting the needs of the more general "superior" systems. G. Altshuller held this model in high esteem[84]: "Purpose of TRIZ: relying on the study of objective laws of development of technical systems, formulate thinking organization rules within the poly-screen scheme."

These "thinking organization rules" incorporate certain notions, such as "idealization" (fig. 10.8) and laws for TRIZ system evolution (fig. 10.9).

Idealization (fig. 10.8,a), according to TRIZ, means that throughout their life cycles, all systems seek to improve the ratio of their functional abilities **F** to the costs of their realization **P**. There are two possible meta-trends (fig. 10.8,b), which may also form various combinations: According to the "functional expansion law," it is possible to increase functionality $\mathbf{F\to\infty}$ with limited investments $\mathbf{P\to 1}$ (any parameter can almost always be rebased to 1) or, according to the "physical compression law", it is possible to decrease costs $\mathbf{P\to 0}$ with limited functionality $\mathbf{F\to 1}$.

An example of the combined action of these meta-trends is shown in fig. 10.8,c where the ongoing increase of capacity of memory modules is accompanied by the ongoing reduction of their physical dimensions: Today (2009) an 8-GB solid-state memory module has the size of a paper clip. In the 1970s, magnetic discs were thousands of times larger and had capacities that were thousands of times smaller.

---

83 G. S. Altshuller *To Find an Idea. Introduction to the Theory of Inventive Problem Solving.*– Novosibirsk: Science Publishers, 1986.

84 Ibidem

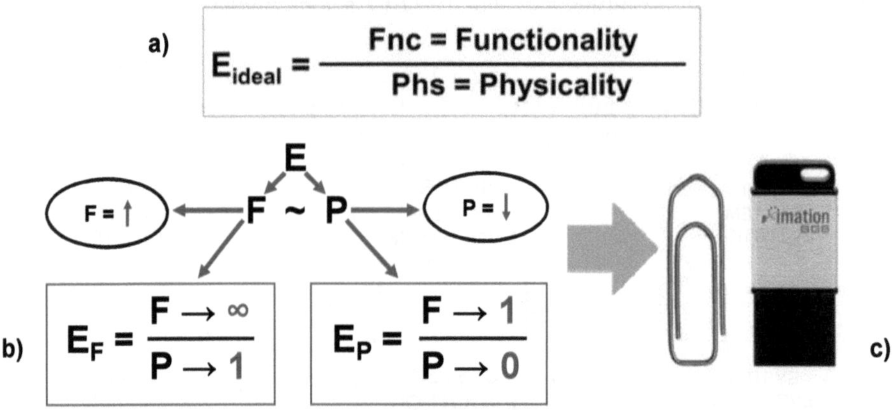

fig. 10.8. "Formula of idealization" of the system and directions of meta-trends emerging in the course of idealization

The key laws for the purposes of system evolution management are the "resources coordination law," the "resources dynamization law," and the "manageability growth law" (fig. 10.9). Other TRIZ-classifications of system evolution laws also exist. However, all such classifications are essentially similar in assuming that any developed system has well-coordinated resources, and that to be capable of being managed, the system must be capable of changing (i.e., being "dynamized").

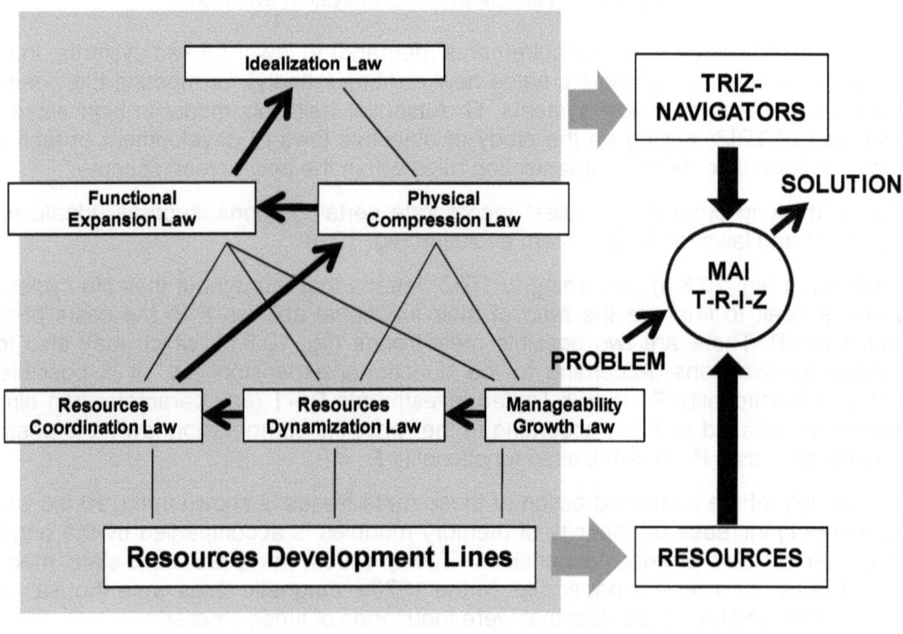

fig. 10.9. Aggregated scheme featuring TRIZ laws

Example 10.1. Mobile Evolution. Fig 10.10 shows a range[85] of mobile phones starting with the first 1.5-kg MOTOROLA manufactured in 1984 and ending with a small modern gadget.

fig. 10.10. "Mobile Evolution" exposition as a demonstration of the "smart artifact" trend. These mockups are empty and are assembled each into other like *Matryoshka*.

Functional abilities of "smart" modern mobile phones include photographing and video-recording, "keyboard" text input (fig. 10.11), screen projection of photo and video files (fig. 10.12), touch screens, GPS navigation, Internet data exchange, etc.

fig. 10.11. Ideal keyboard (in TRIZ terms): It is not there, but its functions are being performed

fig. 10.12. Projecting images from a mobile phone

However, in the late 20th to early 21st century, human civilization encountered in its evolution problems that require immediate remedial action by all people inhabiting planet Earth. Those problems include the following:

- global air and water pollution and scarcity of water in many countries, even those not classified as "poor";
- exhaustion of natural resources, primarily fuel resources;
- inability of transportation systems to deal with growing passenger and cargo flows in many countries, particularly in urban areas;
- overpopulation of many cities and ongoing urbanization;
- increasing incidence of almost all types of crime and drug abuse; and
- growth of terrorism, armaments, and military threats.

85 "Mobile Evolution" is presented by the British artist Kyle Bean in the style of a "Matryoshka" – cardboard mobile phones can be placed one inside the other. Source: www.membrana.ru.
Photo: www.kylebean.co.uk/portfolio.

Unfortunately, even available ideas whose implementation might help resolve some of those problems do not always receive requisite support due to existing political and economic discord and controversy between the most developed countries.

One such ideas involves the invention of an extraordinary transportation system. During their movement on the S-curve, systems pass through several stages of development, each of which may correspond to certain level of the invention.[86]

The method of selection of transformations is determined by the type of contradiction, by the type of resource, or by the nature of the IFR, which also leads to the selection of dominating resources most actively "involved" in the transformation. We saw examples of such selections in the previous sections. Other methods of selection and application of transformation models will be presented below.

At this juncture, we would like to draw the readers' attention to the very important role played by the selection of the OZ actors and resources, the definition of the ideal final result, and, naturally, the selection of transformation models (fig. 10.13). There are five levels of inventive solutions in TRIZ.[87] This complex multifaceted selection can generate solutions of very different levels.

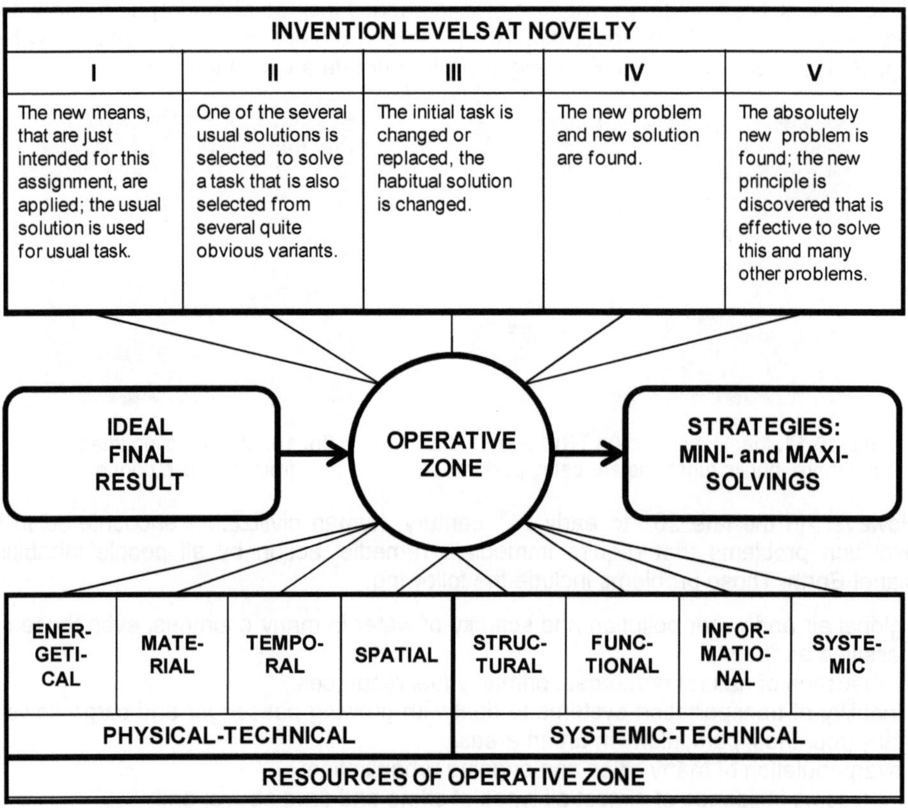

fig. 10.13. Connection between major systemic aspects of OZ transformation

86 Orloff, Michael *Modern TRIZ: A Practical Course with EASyTRIZ Technology*. – Springer, NY, 2012

87 G. S. Altshuller, I. M. Vertkin *How to Become a Genius. Life Strategy of a Creative Individual*. – Minsk: Belarus, 1994 (*italics* at the end of the quotation are mine – M.O.)

Clearly, level III+ solutions can radically change the object.[88] This means that in strategic terms, they will gravitate toward "maxi-solutions," which involve a change of the operating principle and, consequently, of the whole construction.

However, many clients are interested in inexpensive improvements of existing technologies and products and therefore prefer the "mini-solution" strategy, which tens to produce the required result with minimal changes to the system.

It should be noted that in many cases perception of the extent of "radicality" of changes depends on the overall systemic analysis of the solution and the identification of new implementation options.

Thus, in the course of solving the "Swimmer" problem, we managed to overcome an acute radical contradiction: To have and to not have an endless water track. This provoked a change of the pool design, i.e., implementation of a level III "maxi-solution"!

However, in the course of zooming, it was established that that solution could be downgraded to level II by generating a powerful water current in an ordinary pool, thereby eliminating the need to build a special pool, which clearly fits in the category of "mini-solution" strategy.

The solution to the "Diver" problem —despite the apparent "unusualness" of its implementation in existing pools, the breathtaking emotions produced by observing churning water, the high power costs (alas!), and the technical sophistication—can be classified (in terms of inventive novelty and complexity) as a level II "mini-solution."

10.1.2 String "Rail Sky Way"[89]

Question: Why do you think trains are not rushing back and forth at speeds of, say, 500 km/h? Answer: Because under the weight of the train, the rail sags so much that all advantages of resistance to rolling are eliminated, and acceleration becomes possible only at an immense loss of power.

One might say that the wheel moving along the sagging rail always must roll up!

Fundamental contradiction: With increasing weight, there is a persisting need for the wheel to work its way out of a... depression; as a result, power costs soar, and effective speed drops!

Comment: Any knowledgeable expert will say that soon we will discover magnetic levitation and start riding around in wheel-less magnetic cushion trains! Response: Wrong, because power costs will still remain high, plus the construction of such railroads (in several dimensions) is bound to be extremely expensive.

Comment: At this point many readers might say. “OK, we will be using our cars!” Response: Cars can hardly be an object of unrestrained admiration.

Let us turn to facts:

1) In the 21st century, global environmental and safety problems will become even more acute because transportation vehicles, due to their massive use, have become the most dangerous of all human inventions.

---

88 Orloff, Michael *Inventive Thinking through TRIZ: A Practical Guide*", parts *"3.2. Levels of Inventions"* and *"14.1. Development of Systems"*

89 additional information: http://rsw-systems.com/

EVERY YEAR CARS KILL

one and a half million—that's 1,500,000—people,

while approximately 50 million sustain injuries that make them cripples and invalids, which is unacceptable from the positions of humanism and sustainable development of our civilization.

WHO PROTESTS AGAINST THIS MASS MURDER?

2) The overall length of communications spanning the planet now stands at approximately 35 million kilometers with motor highways, railroads, and pipelines accounting for 32 million, 1.2 million, and 1 million kilometers, respectively. These communications are built on approximately 60 million hectares of land,

which is equal to the aggregate area of such countries as Germany and the United Kingdom.

This land does not breathe, nor does it produce oxygen, because it is devoid of plants and stripped of top soil with its humus, which was the product of million years of natural creation. Areas adjacent to major transportation routes are characterized by abnormal patterns of migration of large and small domestic and wild animals, with more than 1 billion of them ending up as road kill every year. Within the territory,

which is larger than the area specified by one order of magnitude

soil and all that grows and feeds on it are contaminated with carcinogenic and more than 100 other deleterious substances representing fuel-burning products, tire and road-top residues, deicing salt mixtures, etc.

NB The solution to this global problem was proposed and actively pursued by inventor and engineer Anatoly Yunitskiy.

Here is the gist of the problem:

The infrastructure of major transportation communications created in the 19th and 20th centuries rests directly on land surface.

Solution trend:

Create a second transportation level, i.e., raise the rail road above the earth!

Solution to the problem:

Create a perfectly straight rail road!

How is this possible?

The rail is made of perfectly tight strings!

What do we get?

Unprecedented safety!
Any speeds up to 1,000 km/h!
Incomparably fewer road construction costs!
Negligibly small land use!
Trillions of euros in annual fuel savings!

In terms of TRIZ, this is **transition into another dimension (!)** in all senses: technically, economically, environmentally, socially, morally, and aesthetically!

This is BEAUTIFUL!

Example 10.2. String Transportation System.

The trend governing the evolution of many modern artifacts is systemically based on networking patterns. This trend is not new, but it keeps manifesting in new ways. Historical examples include the development of pedestrian roads, then roads for horse-drawn carriages and motor vehicles; development of railroads; development of electric networks; development of telephone and then radio networks; and, most recently, development of computer networks (Internet) and mixed networks.

This trend is one of the ideal final results of the Yunitskiy transport system, which abandons expensive traditional railroads in favor of light high-speed networks capable of connecting a multitude of human settlements and making cultural riches stored in megalopolises and large cities readily available to people living elsewhere.

They used to say that ***states grow around roads***, i.e., the rate of growth of a country is linearly dependent on the rate of growth of its road network. And now I say that roads may grow *for* the state by coming to wherever people wish to live and by making it possible to rapidly travel to other settlements ***at an absolutely reasonable cost*** at least at distances ranging from 200 km to 500 km (and up to 1000 km).

Systems often evolve along the spiral of history. Indeed, from time immemorial people have settled at places that offered the best chances of survival and then joined those settlements with paths, ways, and roads. Quite recently—during the course of the last 200 years—states have grown around railroads. Small stations became settlements and cities. The application of Yunitskiy roads opens up new vistas and offers an alternative to the existing transportation arrangements—once again building and living decision-making can be informed by the notions of beauty and health—while roads, in the sense of Yunitskiy theory, can follow people to their new settlements to merge into the existing Yunitskiy transportation network and provide connections to the existing "traditional" traffic arteries. After all, it is too early to write them off.

For distances of less than 2,000 km, the use of ordinary airplanes, a time-intensive and environmentally hazardous means of transportation that relies heavily on a high-maintenance airport network, will probably, if not completely, be discontinued by 90 %.

Motor vehicles at such distances are radically inefficient, to say nothing of their inherently hazardous nature. Still, they have an advantage of being ubiquitous!

Clearly, the only reason for the continued existence of motor vehicles with their internal combustion engines is pernicious—and unthinkably tenacious—intellectual inertia, which is supported by a powerful market press working for the benefit of manufacturers of petroleum products. Meanwhile, there are more reasonable uses for oil than burning it in motor vehicles, locomotives, ships, and heat-/power-generating stations. Electric vehicles can become a significant improvement, although even they need electric power, which must be generated and stored.

The oil and gas industry produces more and more leaders capable of thinking in terms of global ecology and overall economic efficiency, people who support new ideas, including the idea that delivery of petroleum products can be effected not only by railroads and pipelines but also by Yunitskiy lines.

So let us consider Yunitskiy ideas about how to elevate the world onto a network of inexpensive high-speed roads, eliminate traffic jams in large cities, and save the lives of hundreds of thousands of people... per year! Clearly, this last argument outweighs everything else!

We will present Yunitskiy ideas in the form of a series of comparative analyses and reinventings.

First of all, let's look at the radical principle of the string Yunitskiy road.

What can be more familiar and even boring than a rail or the steel wheel of a railway car rolling along that rail? Well, maybe a motorcar wheel rolling along a concrete highway or a battered dusty country road.

Now let us consider the five (according to the author) wheel revolutions in the history of the human civilization:

> 1st **Revolution** (6000 years ago in the Sumerian state in Mesopotamia): invention of the wheel as such – first made of wood;
>
> 2nd **Revolution** (5000 years ago in Egypt; 4000 years ago in Europe): invention of the pair "wheel–manmade road" (road made of wood, stone, etc.);
>
> 3rd **Revolution** (500 years ago in ore mines in England, Ireland, Russia): invention of the pair "cast-iron wheel–cast iron rail";
>
> 4th **Revolution** (160 years ago by Robert Thomson[90]): invention of the pair "pneumatic wheel–road" (dirt road, stone road, concrete road, etc.); and
>
> 5th **Revolution** (35 years ago [late 1970s] by Anatoly Yunitskiy[91]): invention of the pair "steel wheel–steel string rail".

To be more exact, Anatoly Yunitskiy

INVENTED A COMPLETELY NEW... TYPE OF ROAD!!!

A manmade road of a new type using cast-iron rails did not have a prototype in nature! Anatoly Yunitskiy's string rail has prototypes both in technology *and* in nature. For example, Consider cables or spider-web threads. It is not identical to them because its structure and properties are wonderful and truly revolutionary!

---

[90] *Robert William Thomson* (1822-1873) – Scottish inventor, author of the idea for the pneumatic tire, which he patented in 1846 to 1847

[91] *Anatoly Eduardovich Yunitskiy* (b. 1949) – inventor of surface and circumplanetary string transportation systems; see www.yunitskiy.com and http://rsw-systems.com

Some people may say: "Well, it's not like you've invented a bicycle or a wheel! What's so special, so unbelievable about it?"

Example 10.3. New Civilization of Hybrid Transportation

Let us consider the key contradictions inherent in the pair "wheel–road" (fig. 10.14).

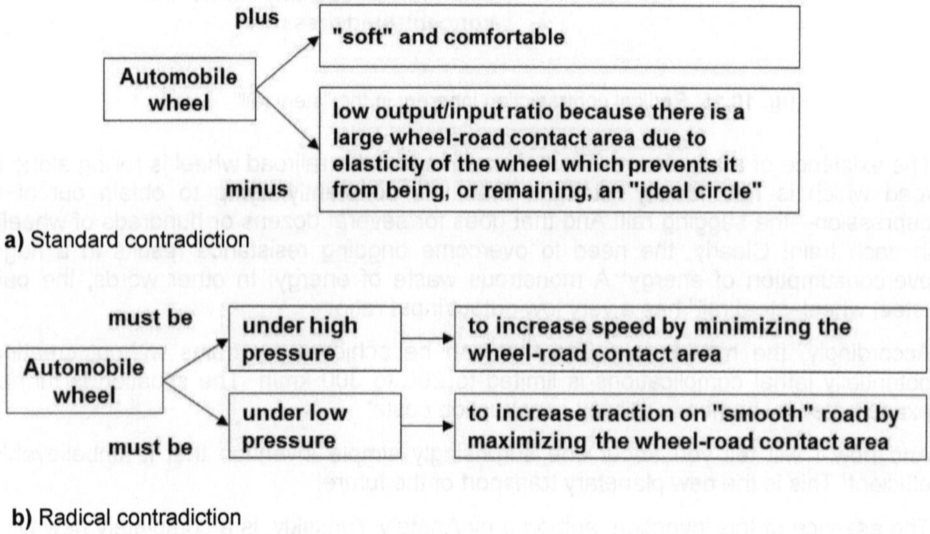

fig. 10.14. Major contradictions inherent in the pneumatic automobile wheel

Because of the road profile and the ensuing low efficiency of the pair "wheel–road" it is impossible to universally support movement even at speeds ranging from 150 to 200 km/h. If you are taking an autobahn from Berlin to Munich, you will move at an average speed of 120 km/h even if traffic conditions are favorable (e.g., there are no traffic jams, construction works, detours, etc.)! That is progress for you! It is a tall task to get from Berlin to Munich in 7 hours—It all depends on traffic jams, road repairs, and construction work!

In addition, the motor vehicle is the worst environmental pollutant and the worst "legitimate" murderer killing **1,500,000** people and maiming approximately **5,000,000** people throughout the world every year!

Conclusions:

1. Motor vehicles must not be used for mass-scale transportation of people and cargoes for distances exceeding several hundred kilometers!

2. In the future, the rules governing city traffic, as well as conditions and scale of parking in cities, must be radically changed!

And now let us consider the radical problem affecting the railroad pair "steel wheel–steel rail" (fig. 10.15).

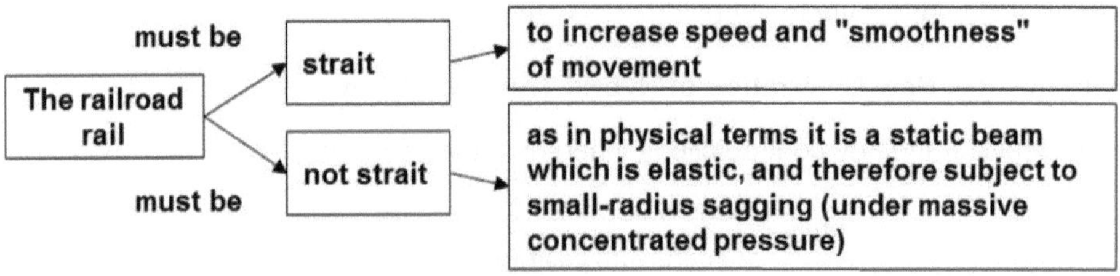

fig. 10.15. Radical contradiction inherent in the "steel rail"

The existence of a sag, however small, means that the railroad wheel is rolling along a road which is not "ideally" straight, that it is constantly trying to obtain out of a depression—the sagging rail! And that goes for several dozens or hundreds of wheels in each train! Clearly, the need to overcome ongoing resistance results in a huge overconsumption of energy! A monstrous waste of energy! In other words, the pair "steel wheel–steel rail" has a very low output/input ratio!

Accordingly, the maximum speed that can be achieved by trains without creating potentially lethal complications is limited to 200 to 300 km/h. The situation is further exacerbated by immense railway construction costs!

And now I will tell you about one surprisingly simple invention that is unbelievably efficient! This is the new planetary transport of the future!

The essence of this invention, authored by Anatoly Yunitskiy, is a completely new type rail construction (fig. 10.16).

fig. 10.16. String rail construction

Pursuant to the TRIZ model that we know as "matryoshka" (named after the famous Russian nested doll), the new rail is made of dozens and hundreds of wires. Each such wire is tight as a string, and all of them together make one powerful string which is completely straight!

In fig. 10.17 you can see the relative "sag" radii of motorway surface (Ram = 1, used as the base unit), high-speed railroad rail (Rrr = 3), and Yunitskiy string rail (Rsr = 10).

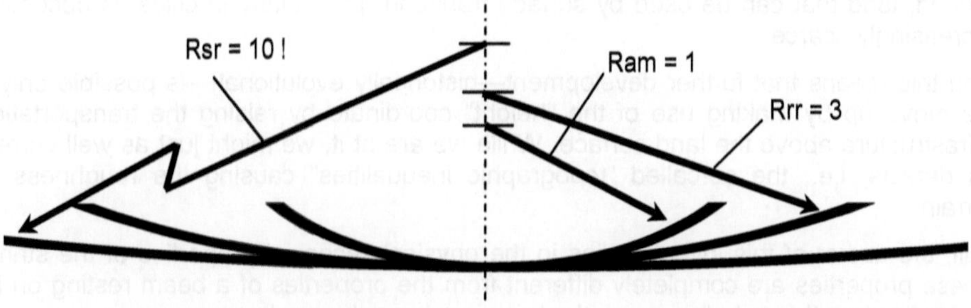

fig. 10.17. Rigidity and straightness of Yunitskiy's road is 10 times better than those of motorway road surface and 3 times better than those of a high-speed railway rail

The second dominant TRIZ model, which is *de facto* realized here, is "Transition into another dimension"—in all senses of the word!

Technically, this means that the entire transportation structure is raised above ground (fig. 10.18) to a certain height to minimize up and down sloping and to assure that the road is completely straight for dozens and hundreds of kilometers and that flight in *a unibus* while contemplating the beauty of nature gives pleasure to each and every passenger!

a) RSW station

b) High-speed transportation vehicle looks organic in any "landscape"

fig. 10.18. Examples from engineering projects designed by Anatoly Yunitskiy

Indeed, land that can be used by surface transport, particularly in cities, is becoming increasingly scarce.

And this means that further development—historically evolutional!—is possible only if we move up by making use of the "height" coordinate by raising the transportation infrastructure above the land surface. While we are at it, we might just as well correct its defects, i.e., the so-called "topographic inequalities" causing the roughness of terrain.

Still, the secret of this invention lies in the physical–technical properties of the string! These properties are completely different from the properties of a beam resting on its ends. All other things being equal, the beam sags much more!

As a result, the maximum speed that can be achieved by Yunitskiy transportation modules (*yunibuses* or *unibuses*) without complications is 200 to 500 km/h.

ATTENTION!

The string is an "intelligent" system that is capable of self-restoration and is exclusively—to the maximum extent possible (!) —straight and even along the way.

That is why Yunitskiy's invention is the last revolution in the development of the pair "wheel–rail": It cannot be surpassed or improved upon except in terms of the efficiency of materials used for its construction but NOT in terms of its operating principle!

In addition, like any other powerful invention, the string-based "Rail Sky Way" creates extraordinary super-effects:

DIRECT ADVANTAGES!

- energy (fuel) consumption is 20 times less compared with motor vehicles!
- energy consumption is 4 times less compared with railway transport!
- RSW construction costs are 20 to 30 times less compared with those of a motor highway of comparable length!
- RSW construction costs are 10 to 15 times less compared with those of a railroad of comparable length!
- passenger and cargo transportation costs are more than 10 times less!

And last but not least, STU has certain systemic properties that are extremely promising.

There is again a well-known saying: ***States grew along the roads.***

We will compactly present Yunitsky's ideas in the form of a series of comparative analyses and reinventings.

First, let's represent the advantages of Yunitskiy's string road compared with a conventional railway (figs. 10.19 to 10.21).

Example 10.4. Another Application. And we will consider another of the reinventings (fig. 10.22) of Yunitsky's inventions, many of which he has made on the basis of TRIZ or on the basis of creative systematic thinking, which he has repeatedly told me is very close to TRIZ.

ARTIFACT **STRING TRANSPORT BY YUNITSKIY**

SYSTEM "POLYSCREEN" ANALYSIS

OLD SYSTEM: expensive, resource-demanding, energy-intensive, consuming the earth, environmenttally harmful, rigid network

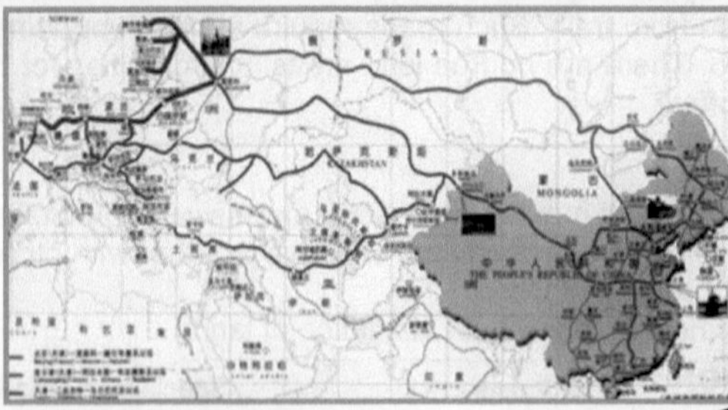

RSW projects **String Transport System** for **The New Silk Road "China-Europe"**

NEW SYSTEM: inexpensive, material-efficient, energy-economical, not consuming the land, environmentally-clean, flexible network

ROLLING STOCK: ponderous, highly concentrated load, relatively low speed, low efficiency

ROLLING STOCK: lightweight distributed load, speed (up to 200 km / h), high-speed (200-400 km / h) and super-high-speed (400-800 km / h) modes, high efficiency

RAIL: heavy, cast making sag and high rolling resistance, requires a strong roadbed

RAIL: light, stringed, assembled, with minimal rolling resistance, NOT required subgrade at all!

RAILWAY SUBGRADE: massive subgrade, consuming the earth, crosscutting the brooks, breaking the ways of land transport, as well as animal trails

RAILWAY SUBGRADE: light masts, NOT deprives the ground, NOT crosscuts the brooks and small rivers, NOT breaks the ways of land transport, as well as animal trails

TODAY **TOMORROW**

The future will be with the light, raised over the existing roads, stringed transport networks of high-speed automatic Yunitskiy's modules!

City-forming mission of string transportation systems will be the removal of residential centers in the beautiful natural landscapes away from the metropolises of 100-200 km, and avoiding the excessive use of cars (even electrical).

A radical redistribution of cargo and passenger traffic in the ground-air space over the multi-million metropolises, elimination of traffic jams and pollutions, development of a new integrated architecture.

fig. 10.19. Comparing the old and new railroads

REINVENTING 1: Transition into the light aboveground structure

TREND Roads, especially railways, inherited a fundamental flaw of a concept of large concentrated load on the road from train. Such roads require strong supporting roadbed. This leads to the alienation of land. There are no free land areas in large cities for the construction of new roads. What can we offer?

REDUCING

Macro-FIM: The X-resource, without making the system excessively complex and without causing inadmissible negative effects, ensures, together with other available resources, the obtaining of the:

IFR: [construction of new roads without alienation of land]

STANDARD CONTRADICTION

		Factors of A-matrix	Specialized models recommended by A-matrix
RAILWAY	(+) *High-speed*	***22** Speed*	*19 Transition into another dimension*
	(-) *Large losses of land*	***18** Surface of the fixed object*	*22 Spherical-shape*
			24 Asymmetry

RADICAL CONTRADICTION

RAILWAY	*should be broad and resource-demanding to ensure the evenness of the way*	**VS**	*should not be broad and resource-demanding, so as not to take away the ground and cut the artificial and natural way*

INVENTING In order to radically reduce land loss by the construction of railways and highways, the road may raise above the ground as development of "air-land" class of roads – the dominant principle corresponds to the model *19 Transition into another dimension: a) possible improvement in the transition from the movement in the plane to the space; b) to use multi-storey layout.*

The road should be realized at the level (height) that does not require land withdrawal and does not break existing transport communications, as well as the movement of streams and small rivers, the movement of animals.

ZOOMING **Contradictions Removed:** YES

Super-Effects: The possibility of construction high-speed railway because of straightness of the road realized without steep slopes and abrupt turns.

Development Trend: To relieve traffic flows in large cities. To unload the passenger roads by construction of separate cargo roads.

Increased use: The possibility of use within the city to relieve the problem of traffic jams.

Beauty of Solution: 100 — Highest grade. Substantiation: 1) such structure was not known before, there are no direct counterparts; 2) the idea has great system, social, functional, architectural and economic potential.

fig. 10.20. Reinventing of the idea of the "third dimension" in Yunitsky's string railway system

REINVENTING 2: Transition to the stringed rail

TREND Heavy cast rail has not changed **HALF A MILLENNIUM**! This rail has a big weight and, together with the supporting elements (numerous, heavy concrete sleepers), it requires a powerful supporting roadbed for the route smoothness. This rail is not suitable for "ground-air" road. Can the rails be improved and developed?

REDUCING **Maxi-FIM:** The OZ by itself ensures the obtaining of the IFR: lightweight, straight, smooth rail

STANDARD CONTRADICTION (SC)

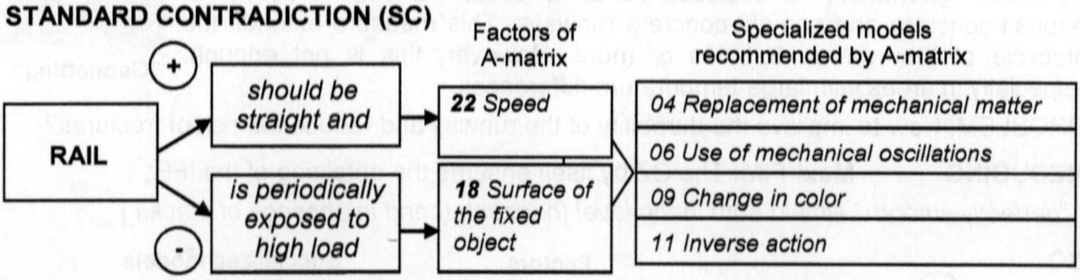

RADICAL CONTRADICTION (RC)

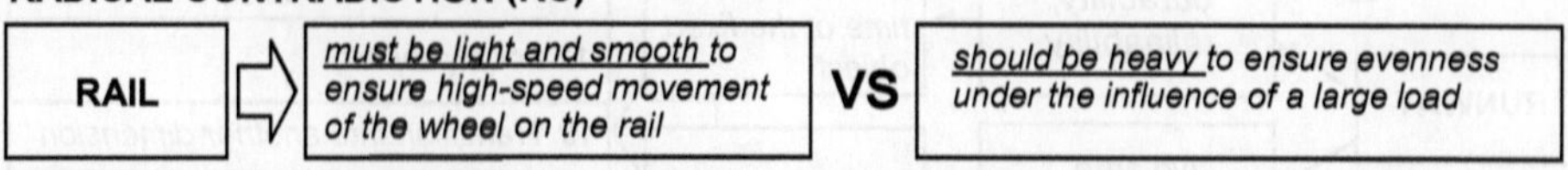

INVENTING **Solution of SC**. In order to radically reduce weight and ensure straightness and evenness of the road, the rail is made assembled and has a special structure, based on a strained "strings" (highly stressed steel wires) – the dominant principle is *11 Inverse action: a) instead of an action prescribed by the conditions of an assignment, complete a reverse action* (do not make rail heavier, and radically lighten it); *04 Replacement of mechanical matter: c) to move from unstructured fields to fields with a specific structure* (rail should be a composite structure with controlled parameters, in particular, evenness; www.yunitskiy.com).

Solution of RC. **01 In space:** load-bearing part of the rail (string structure) made high-strained, and the entire structure – with normal tension of materials; and rail as a whole system is becoming highly-even in string tension process by assembly directly on the masts of future road; **03 In structure:** rail is radically structured; **04 In material:** cheap wire instead of a expensive cast rail.

ZOOMING **Contradictions Removed:** YES

Super-Effects: The possibility to achieve extremely high speeds for a pair of "wheel - rail" (for example, up to 1000 km / h).

Development Trend: **The possibility to build the roads through forests, rivers, swamps, permafrost, etc**. The possibility of construction high-speed freight roads (for example, multi-ton units – oil, ore, products, fruits and vegetables, etc., drinking water! – that is particularly important – at a speed of 500 km / h) – **this goal is set for the first time in the history of land transport.**

Increased use: The possibility of building a highly smooth and durable airfields, highways, buildings, the ports remote in the sea, and others.

Beauty of Solution: 100 The idea of outstanding beauty and efficiency, meeting the TRIZ laws of systems development: inventive use of special physical-technical effects (the behavior of the string under static and dynamic loads), separation, structuring, controllability.

fig. 10.21. Reinventing of idea of the string railway by Yunitskiy

REINVENTING 3: YUNITSKIY LOAD BEARING STRUCTURES

"Geonetting"

TREND The cracks in the concrete structure, in both transverse and longitudinal direction, can develop during prolonged operation of runways at aerodromes. Then the runway needs of repair. The tendency of cracking can be figuratively represented as a process of forming the separate ice floes (even relatively large ones) on the frozen lake at the beginning of the thaw. Solid ice is able to take a big load, and individual ice floes – no. Therefore, so called "geonetting" is enclosed so as to cover the coating seams of asphalt-concrete and cement-concrete runways. This measure extends the lifecycle of the runway 3 times or more. However, this is not enough, especially in areas with large temperature differences.

PROBLEM: How to improve the durability of the runway and reduce the risk of fractures?

REDUCING **Maxi-FIM:** The OZ by itself ensures the obtaining of the IFR:

[*"perfectly smooth" airfield both in the level (horizontal) and in absence of cracks*].

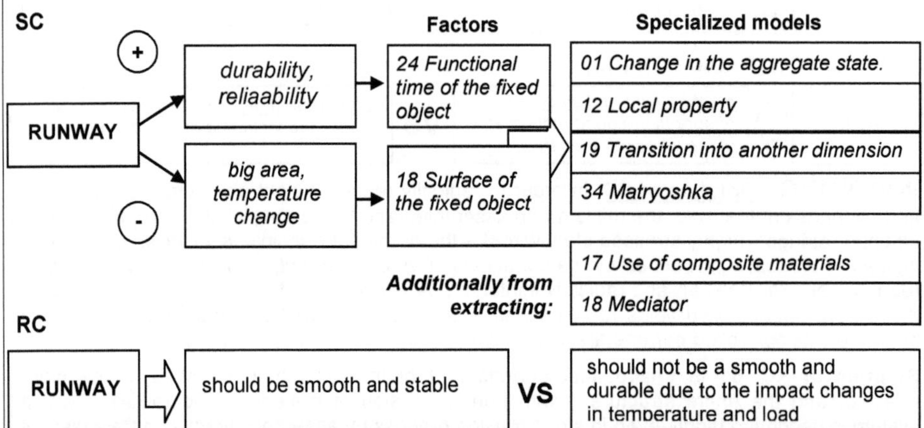

INVENTING In solving the problem, the four dominant navigators create a "portrait" of the future solution, and then present in the solution: **01, 17, 18** and **34** – cement-concrete cover of runway reinforced by strings of high tension according to the invention of Yunitskiy (www.yunitskiy.com).

ZOOMING **Contradictions Removed:** YES

Super-effects: an exceptional smoothness and durability of runways and airfield, the possibility of widespread use in the construction of high-rise buildings, bridges, roads, squares, etc.

BRIEF DESCRIPTION

To ensure high durability and evenness of the runway concrete slabs reinforced with high tension strings (inventor A.Yunitskiy). Dominant models: 01 Change in the aggregate state, 17 Use of composite materials, 18 Mediator and 34 Matryoshka.

fig. 10.22. Reinventing of runways and airfield with string technology by Yunitskiy

Today the overwhelming majority of Russian roads can be characterized by this well-known piece of "black humor" (fig. 10.23):

During the first 9 months of 2011, there have been more than 19 thousand fatalities on Russian roads, a four percent year-to-year increase.

fig. 10.23. It is true that even on a top-quality road, like in Germany where there are no speed limits on many Autobahns, moving at a speed shown in the poster to the right may produce the same results if there is fog, rain, snow, sleet, if you travel at night... **So let us change over to Yunitskiy lines to move at speeds from 200 km/h to 500 km/h!!**

fig. 10.24. Yunitskiy Modules: High-Speed, Speed, and Cargo Options
a – e) Passenger and cargo transportation modules; a, b, d, e, f, g) High-speed modules; f, g) Stations (they can also be built at the sides and on the roofs of high-rise buildings)
www.yunitskiy.com

Two Historical Prototypes and One Modern Challenger

Ex. 10.5. George Bennie: "Railplane"—integration of "railroad" and "aviation"

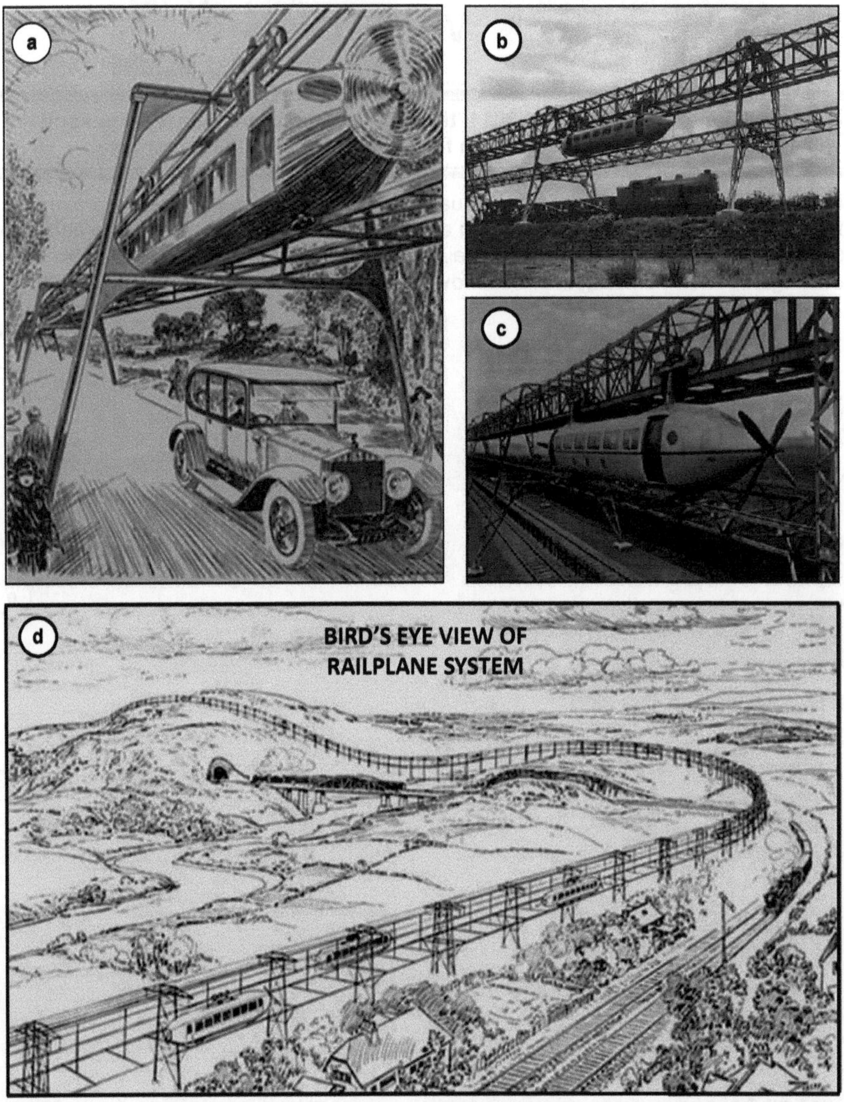

fig. 10.25. High-Speed Roads Built by George Bennie, UK Patent No. 191760
a, b, c) The roads could be laid over (next to) existing motorways and railroads; d) the roads could go uphill and downhill because propeller power was sufficient to negotiate any hill. See www.nas.gov.uk/downloads/DD17-117-2-1.pdf.

George Bennie (1891 to 1957) proposed, and in the 1930s built not far from Glasgow, Scotland, a 120-meter monorail road with modules driven by propellers. Unfortunately, it was next to impossible to attain meaningful speeds on that road because of its vertical and horizontal profiles and the extent of loaded rail depression. Other deficiencies included the hellish din produced by air engines and propellers…

Ex. 10.6. Konstantin Tsiolkovsky: Air Cushion Train—Over Chasms and Rivers!

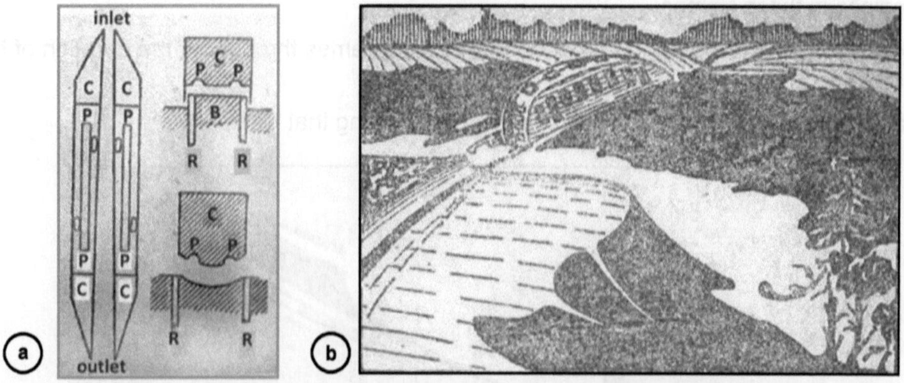

fig. 10.26. Air-Cushion Train Designed by Tsiolkovsky, 1927

a) C = car, P = pipes (inside the car!) pumped air into the space under the car. B = road bed, R = rails to assign direction; an air duct runs through the center of the car; it is used to suck the air in at the front (ingress) and discharge it at the back (egress) as a "jet" to create jet thrust; b) The train flies over (!?) a river!

Tsiolkovsky authored an extremely audacious idea[92]—that of a high-speed train running on an ***air cushion***! He believed such train would be able to accelerate to 1000 km/h and even fly over rivers and precipices. This would eliminate the need to build bridges in hard-to-reach areas!

But here is a question: How even and how straight must the railway be to assure safe movement at high speeds? With an ordinary rail, it is not enough to have "flanges" (see the lines in the top right-hand corner in fig. 10.26,a) that continuously maintain contact with the rail. Trying to keep the train inside an air "trough" (fig. 10.26,a, bottom right-hand corner) is problematic because we must exclude the possibility of cars derailing or rolling over.

Example 10.7. Tubular Rail Train

Halfway to … BARS (see fig. 10.84) and TRIZ!
Idealization: The outcome—the train—is there, but the road is not!
Models of segmentation, separation, and inverse action: The wheels keep still, but they are rolling, and the train accelerates using the wheels!

This idea is simply amazing in and of itself (fig. 10.27): You take the wheels off the train, then place and spin them within frames. Because of friction, the wheels accelerate the train to 200 km/h. The train always rests on several (minimum of two) frames.

Super-effect 1: No heavy road!

Super-effect 2: The train does not need an engine or a power system and therefore weighs less!

---

[92] N. A. Rynin, *K. E. Tsiolkovsky. His Life and Work* (in Russian: Н.А. Рынин, *К.Э.Циолковский. Его жизнь и работа.* – Ленинград, "Вестник знания", 1936)

But there is that same problem again: You must make the virtual "road" absolutely even and straight as well as obtain a large number of frames and also wheels and engines in these frames!

One more: There are hundreds of engines inside frames throughout the duration of the trip!

Last but not least: There is a problem with turns along that "road."

fig. 10.27. Tubular rail train with framed wheels (2006)
See www.tubularrail.com; http://www.membrana.ru/particle/3027

ATTACHMENT[93] TO SECTION 10.1.2 STRING "RAIL SKY WAY"

WATER

Humanity consumes 2 billion tons of petroleum products per year. Freshwater consumption is five times higher. One third of the Earth's population has no adequate access to clean, healthy water. Especially dramatic is the state in many countries, e.g., Saudi Arabia and Central Asia. Inhabitants must drink either dirty water or purified and artificially mineralized water. Russia possesses 80 % of the world's freshwater

[93] excerpts from Yunitsky's archive: *Economist and Systems Analyst S.A.Sibiryakov speaks about Yunitsky string systems*

resources. Those resources are renewable and can be supplied to the global market. However, there exist no means of transportation capable of delivering that unusual product to the market quickly and at reasonable cost.

Construction of string roads would solve that problem once and forever. Russian water could be carried alongside passengers and ordinary cargoes. Today the idea that Baikal water can be sold in Tokyo, New York, or Riyadh seems absurd because it is impossible to deliver that the product to the consumer. However, 100 years ago it was absurd to think that people would fly from Paris to Washington, and 15 years ago it was impossible to imagine a student calling any place from any place on the planet using a cell phone.

Today it is all real, and this reality has been made possible by technological development. Yunitsky's strings will open up an inexhaustible market, and we will be earning billions just on water alone.

ICE

It could be possible to use the permafrost zones of Russia as a "global refrigerator." What is the common state of water in Russia? *Ice, i.e., frozen water!* So here is a business idea: Supply water in the form ice briquettes. What for? To serve cold to the rest of the world.

Do you know that the USA spends three times more energy and resources on *air conditioning* than Russia does on *heating*? People in Asia are also ready to pay quite handsomely for ordinary ice.

Today (2012) the price of one ton of freshwater ice in the global market is USD$7000, whereas one ton of top-quality petrol costs merely USD$1000.

You do not believe that? A little more than a century ago nobody would have believed that Arab countries would become fabulously rich supplying oil to the West because in the 1880s there were no internal combustion engines. But now they exist.

So, like oil, water will become a profitable export product.

CARGOES

What is the planet's most burning problem today? Transportation and communication! European and Asian markets are locked in. American cargoes must be dragged along sea routes to Europe, with each trip lasting an interminable 200 to 300 hours, and it takes even longer to carry anything from Europe to Asia or from Asia to Europe.

Yes, sea-going vessels are the cheapest means of transportation, but their sluggishness means that huge resources are immobilized for extended periods of time, and business is slowed down. Aircraft cannot replace ships: They are simply too expensive.

Railroads? Not good enough, either. For example, railroads in Russia are not made for high speeds, and their state of repair and quality of service are clearly inferior. The idea of a trans-Eurasian railroad linking London to Tokyo through the expanses of Russia has been bandied about for a long time. However, it is virtually impossible to build railroads to the north of Baikal due to permafrost, snowdrifts, and "floating" soils.

What about the Northern Sea Route? It's difficult and slow. A motorway with a bridge across the Bering Strait? That's impossible—too costly.

Despite the information revolution, business is shackled by the slowness of cargo transfer. After all, there has been but a slight increase in the average speed of ships and trains since the 1930s.

String roads, in contrast, can pierce the entire country by opening heretofore impassable northern regions for development along "string ways" stretching out to the south, west, and east.

In other words, it is possible to inexpensively connect London to Tokyo and New York and Central Europe to southeastern Asia and India. It will take merely hours to move cargoes along those roads, which is dozens of times faster than trying to move them by sea or land using conventional means. And the cost of transportation will be less than USD$100 per one ton of cargo!

The usual barriers, i.e., snowdrifts, permafrost, rivers, and mountains, will simply cease to matter. String frosting will not be an issue, either because no ice can hold on to strings as transportation modules rush along on steel wheels. Last, but not least, such roads can be laid anywhere.

WORLD INTEGRATION

Central Asia is experiencing an acute shortage of freshwater and cold? No money can buy those things? Plans to turn northern rivers to Asia have been buried and forgotten?

Russia can supply Central Asia with clean and healthy water in the form of Chukotka's or Taymyr's ice, which it has in abundance, by creating string-based trading routes.

Excellent suppliers of fruits, vegetables, nuts, and cotton are locked in Central Asia and Kazakhstan? Russia can give them access to its own as well as European markets. The same benefits offered by the new type of transportation will become available to China, India, Iran, and Pakistan.

And Russia, wedged between Europe, America, and Asia, can become the planet's mega-transportation bridge.

fig. 10.28. Prototype of a future settlement

LIVING IN NATURE

I would like to live in a region with a pristine environment in order to be healthy, full of energy, and ready to move through continents along string roads (fig. 10.28).

People will be able to move throughout the Americas, Africa, and Eurasia freely and rapidly creating a sense of psychological unity for the modern civilization.

We do not mention such an excellent opportunity, ***as an exception to traffic "jams"*** and ***preservation of transportation links in any flood,*** including "civilized" Europe. It is extremely important!

Failsafe Idea from Yunitskiy—String Road:
A Fast, Safe, All-Weather Transport Conveyer!

10.2 Transfer of Experience

10.2.1 We learn by example: *Exemplis Discimus*

The purpose of this section is to show you that all readers of this book can master Modern TRIZ and create dozens of interesting and useful examples using both artifacts that surround them in their daily life and other items making up their professional environment.

And, naturally, the purpose of this section is to help the readers who have embarked on a study of TRIZ basics to quickly acquire practical skills required to use fundamental TRIZ notions and models.

Here you will see both complete and abridged examples of extracting and reinventing for very different artifacts. Presentation formats also vary to help you enjoy more freedom in presenting extracting and reinventing outcomes.

After all, your ultimate goal is not to always come up with a "long" description but rather ***to obtain a clear and compact model of the inventing process***.

The same freedom is granted to you by our software package EASyTRIZ™ Junior™. Accordingly, examples in this book also show you how to use that software tailored to satisfy the needs of the readers of this book.

Along with explication of methodological opportunities offered by MTRIZ-modeling, another "super-task" of our examples is to draw your attention to, and arouse your interest in, the following two aspects of exploration:

First, almost every artifact can produce exciting, sometimes extraordinary, esthetic effects whether it is the unexpected, paradoxical transformation from the state "was" to the state "is"; the surprising simplicity or, one might say, intellectual exquisiteness of many key ideas underpinning solutions to problems that have appeared unsolvable for many years; or the outstanding—or at least considerably enhanced—functional and economic efficiency of solutions, which ultimately brings them to fruition and practical implementation.

Second, many ideas directly reflect and encapsulate the fortunes and personalities of their authors who standing as paragons of moral fortitude, courage, and will to reach the final destination, to accomplish their dreams, to realize their ideals not just in spite of "organizational" difficulties (such as the need to cut through red tape) but by braving extremely dramatic circumstances and sometimes overcoming tragic personal blows administered by fate and—alas!—by society, which are often personified by specific individuals obstructing the ascension of ideas into the realm of praxis but sometimes occur simply due to general public inertia and aversion to novelty.

The only friendly request that I have for you before you read sections 10.2 and 10.3, and especially section 11, is this:

Seek the truth, but also try to find like-minded persons; bring together enthusiasts, people captivated by beautiful ideas, people who are honest, selfless, true to their word and dedicated to the work; people who are brave and naturally kind, friendly, and understanding.

I wish you success in mastering Modern TRIZ!

Example 10.8. To make the *invisible visible*

TREND *An aerodynamic plant generating artificial "wind" is used to study airflow behavior of motor vehicles. There are numerous instruments measuring, for example, pressure borne by various vehicle components. PROBLEM: how do you make air flows moving around the vehicle visible?*

REDUCTION **FIM:** ***X-resource***, without making the system excessively complex and without causing inadmissible negative effects, ensures, together with other available resources, the obtaining of the **IFR**: [*visualization of air flow*].

Standard Contradiction

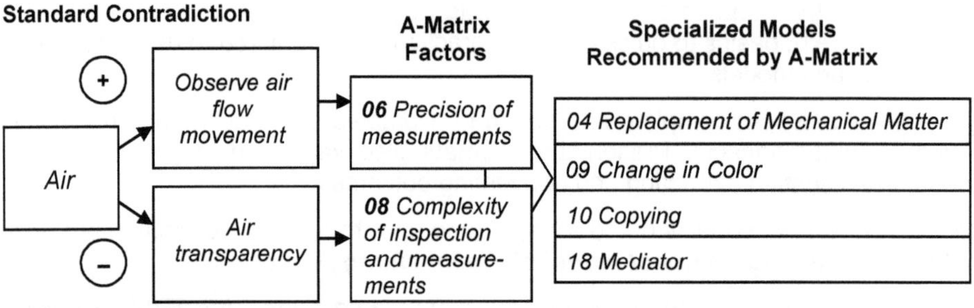

Radical Contradiction

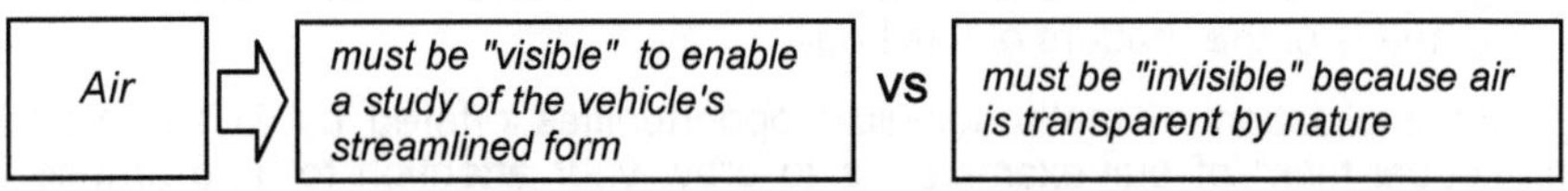

INVENTING When solving the problem, all navigators without exception combine to create a "portrait" of the future solution, and then are then incorporated into that solution: ***09 Change in color***: *c) use color supplements to observe objects or processes that are difficult to see;* and ***04 Replacement of mechanical matter***: *a) replace mechanical schemes with ... olfactory schemes;* ***10 Copying***: *b) replace an object or a system of objects with optical copies or their images;* and ***18 Mediator:*** *b) temporarily connect an object with another (easily separable) object.*

Solution: smoke! According to Model 04 – even with small, if necessary! But the main thing is air color, change in transparency, creation of color! Usually they use "dry ice" (pre-frozen carbon dioxide CO_2) which at room temperature is transformed into "white smoke" converting from solid form into gaseous form, while skipping the liquid form. To produce a large volume of "smoke", it is possible to use special generators which accelerate the rate of evaporation of "dry ice". Color can also be added to the "smoke".

ZOOMING

Have the contradictions been removed? – Yes.

Super-Effects: *possible use in cinema to create special effects.*

Negative Effects: -

BRIEF DESCRIPTION

"Smoke" obtained by evaporation of "dry ice" is used to visualize air flows around the vehicle.
Dominant models: 04 Replacement of mechanical matter, 09 Change in color, 10 Copying and 18 Mediator.

Illustration: fragment of a Volkswagen Wind Tunnel video clip at www.youtube.com

fig. 10.29. Volkswagen wind tunnel, Germany

Example 10.9. *Tie the pieces of experience and synthesize the new*

ARTIFACT **APPLE: on the Way to New iPOD and iPHONE**

EXTRACTING

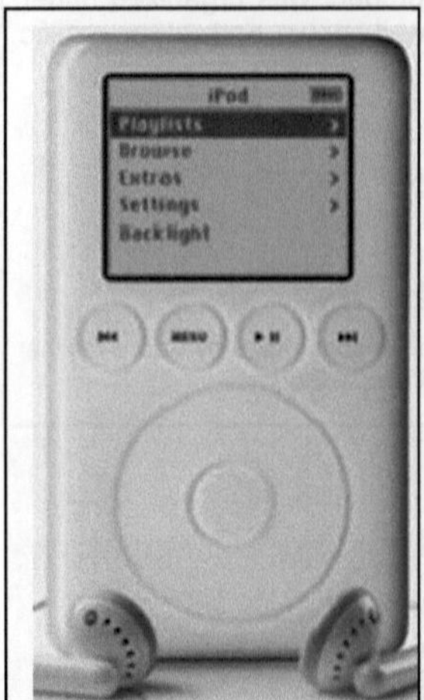

DESCRIPTION

"With iPOD, listening to music will never be the same!" – said Steve Jobs in October 2001 at the presentation of a new gadget which soon became the global sales leader among musical MP3 players. One of its functional innovations was the click wheel which could be turned to select menu items and browse through the list of recordings stored in the gadget's memory.

Still, several years later it was replaced with new generations of iPODs and iPHONEs. Their key innovations included popup menu lists and sensor scrolling (US patent 2007/0150830A1, 23.12.2005). Menu items could be selected, and lists could scrolled, by just moving the finger across the screen.

Illustrations: resultant artifact – "Apple Presents iPod", October 23, 2001, http://www.apple.com/pr/library/2001/10/23Apple-Presents-iPod.html

resultant artifact – www.google.com/patents

Application number: US 11/322,547
Publication number: US 2007/0150830A1
Filing date: Dec. 23, 2005

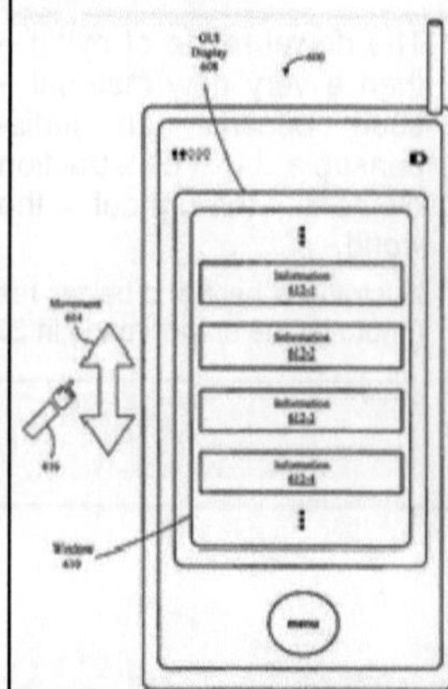

Проблема — Идея

WAS: **Prototype Artifact** → Тренд › Редукция › Изобретение › Зуминг → IS: **Resultant Artifact**

Экстрагирование-1

D	No.	Navigator	Function and Substantiation
+	04	Replacement of mechanical matter	Mechanical keys are discarded in favor of direct contact with the screen or, to be more exact, with the sensors underlying the screen (or other types of sensors)
	05	Separation	Critical controls are placed *on* the screen
+	10	Copying	Keys are replaced with their optical copies on the screen
	20	Universality	The screen performs multiple functions, including the movement sensor function

REINVENTING

TREND *Mechanical components are short-lived? What can we do?*

REDUCING **Maxi-FIM:** The operating zone by itself ensures the obtaining of the following result:
IFR: [*selecting menu items and list scrolling*]

SC: "Click wheel": 01 Productivity **VS** 23 Functional Time of the Moveable Object = 01, 02, **05,** 06

RC: "Click wheel" ► is needed (to select other menu items or selection positions in recording lists)
VS is not needed (if the wheel quickly breaks down)

INVENTING

SC Resolution Model ► *presented in extracting.*
RC Resolution Model ► *03 In Structure* and *04 In Material*: new elements and materials for the screen; new algorithms and new data structures for programmatic controls.

ZOOMING Contradictions Removed. Less time is needed to select menu items and scroll through lists.
Super-Effect: The screen can now be made larger (no need to waste space on the keys and the "click wheel").

fig. 10.30. Reinventing of iPod and iPhone screen with sliding popup menus

Example 10.10. To make the *unreliable* reliable

ARTIFACT **ARTUR AND KLAUS FISCHER DOWELS**

Description The dowel invented in 1958 by Professor Artur Fischer, founder of the Fischer Group, has become a truly global product. Mass production of this product was organized under the guidance of Professor Klaus Fischer (A. Fischer's son), also a well-known and rather prolific inventor, the current head of the Fischer Group (Germany).

The dowel made of nylon – then a very new material – soon became an indispensable construction element throughout the world.

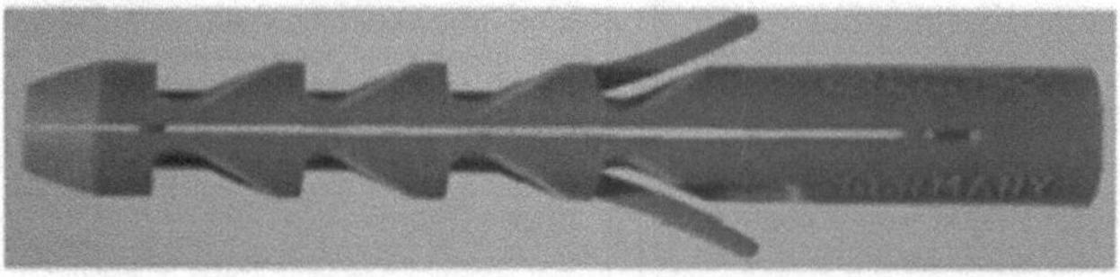

Illustrations here and below: resultant artifact and the stand dedicated to key inventions made by Arthur Fischer (photo by the author made in 2009 during his annual workshop at *Deutsches Museum Bonn*).

EXTRACTING

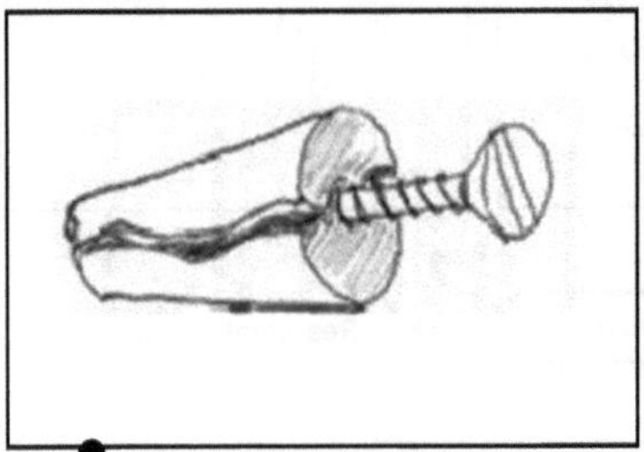

D	No	Navigator	Function / Substantiation
+	01	Change in the aggregate state	a) … use of elastic properties of solid objects: **the body of the dowel is made of elastic plastic**
+	03	Segmentation	a) disassemble an object into individual parts: **immobile and mobile parts, "wings" to fix the dowel inside the hole while the screw is being turned in**
+	04	Replacement of mechanical matter	c) replacement of static fields with **dynamic** ones, from those fixed to those changing **in operating time**
+	07	Dynamization	a) characteristics of an object … are changed to optimize every work procedure – **mobile body!**
+	F	01 In Space	A part of the dowel's body is made to be mobile (elastic)
+	F	03 In Structure	Cuts, crests, wings; elastic part – to assure screw movement
+	F	04 In Material	We need *elastic* and rather *durable* material.

Contradiction	Description
SC	A wooden dowel is simple, but it is easily destroyed, failing to "keep the shape"
RC	Dowel: "hard" (to hold the screw in) **VS** "soft" (to fill the hole)

REINVENTING

TREND *In the past, if we wanted to drive a screw into a brick or concrete wall, we made small wooden plugs and hammered them into pre-drilled holes. Unfortunately, sometimes the plugs crumbled and fell out of the wall. Generally speaking, it was not always possible to take the screw out of the wall and then put it in back again. Apparently, despite its apparent simplicity, plugs may break down – lose their shape, so to speak. What can we do?*

REDUCING **Macro-FIM:** X-resource, together with available or modified resources, and without making the object more complex or introducing any negative properties, guarantees attainment of the following **IFR**: [*the whole is tightly filled to hold the screw*]

STANDARD CONTRADICTION (SC) – Extracting-2S

Dowel ► 07 Complexity of construction **VS** 21 Shape = 04, 07, 11, 14

RADICAL CONTRADICTION (RC) – Extracting-2R

Dowel ► "hard" (to tightly fill the hole) **VS** "soft" (to let the screw in)

INVENTING **Dominant models used to resolve the SC:** 04 Replacement of mechanical matter, 07 Dynamization and, additionally, 01 Change in the aggregate state of an object. Invention of the dowel totally agrees with these navigators: the body of the dowel is made of elastic plastic (model 01) so that it expands, and its diameter increases, as the screw goes through it (models 04 and 07).

Dominant resources used to resolve the RC: structural resource (cuts, "wings", protrusions, etc.), spatial resource (new 3D shape), and material resource (use of elastic – but at the same time hard and durable – material) – when expanded, the dowel is tightly "wedged" in the hole.

Fundamental models: 01 Spatial – the parts are mobile and immobile, while the system as a whole is elastic; 03 Structural – cuts, crests, wings; 04 Material – the new plastic material (nylon) which is elastic, resilient and tensile.

Scientific effects: Application of a new material; elasticity; increased friction (inside the whole) due to the changing form of the gripping surface.

ZOOMING

Contradictions removed: Yes.

Super-effects: High accuracy and reliability, minimal waste.

Negative effects: Dowel is hard to extract, e.g., if it is necessary to change the diameter of the hole.

Development trend: Huge range of various dowels and wall ties.

Changes in ambient systems: Radical change in the method of installation of fastening components (screws, bolts, etc.).

Extended uses: Can be used with diverse materials of load-bearing structures (walls, ceilings, etc.).

Beauty of solution: 100 Highest grade: change of material *and* dynamization – an extremely powerful combination of transformation models!

fig. 10.31. Reinventing of dowel

Example 10.11. Make *a solid* liquid and vice versa

ARTIFACT	MAKING SILICON WAFERS FOR SOLAR PANELS

DESCRIPTION For many years, silicon wafers for solar panels were made by slicing solid ingots obtained from highly pure silicon melt. This method had a major disadvantage, though: up to 50% of the ingot got wasted in the process!

During the period from the 1970-es to the late 1990-es, numerous German concerns, institutes and universities (including *Siemens, Bayer AG, Brandl GmbH, Deutsche Solar, Fraunhofer Institute for Solar Energy Systems, University of Konstanz,* and others) suggested various technologies and constructions for the formation of wafers directly from the melt. In the 1990-es, a series of major successes were scored by scientists in Denmark and the USA. However, for a long time the new technologies remained incapable of assuring the level of quality required for mass production.

In 2010 the Massachusetts Institute of Technology (MIT) and the company *1366 Technologies* designed an industrial method **Ribbon Growth on Substrate (RGS)** and associated equipment under the guidance of Professor Dr. Emanuel Sachs, CTO, and Frank van Mierlo, CEO, MSc.

Illustration: resultant artifact – http://www.ecn.nl/docs/library/report/2002/rx02038.pdf (*Bayer AG* technology) and 2006_09_27_talk_g_hahn_pdf; inasmuch as the MIT technology is classified as "know-how", we review the reinventing of the original method for uninterrupted extraction of silicon wafers from the melt.

EXTRACTING

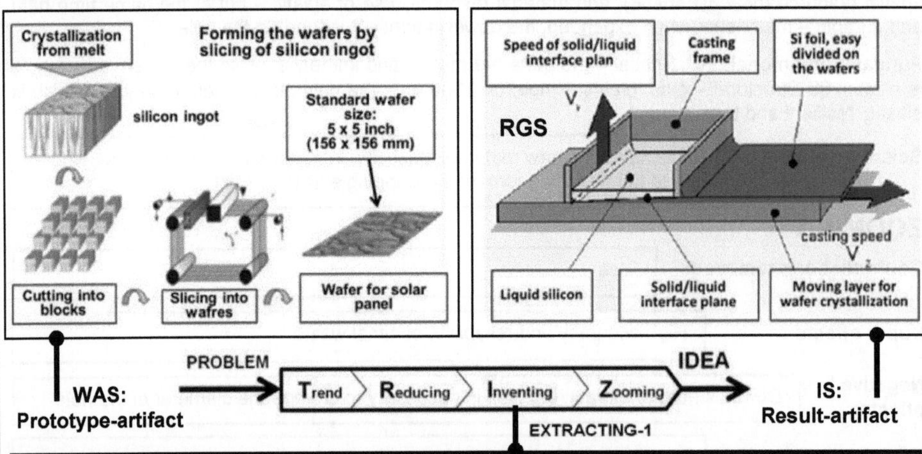

D	No	Navigator	Function / Substantiation
+	01	Change in the aggregate state	a) this includes transitions into "pseudo-states" ("pseudo-liquid") and into transitionary states such as the use of the elastic properties of solid objects as well as simple transitions such as from a solid to a liquid state
	02	Preliminary action	a) previous necessary (partial or complete) change of an object
+	11	Inverse action	a) instead of an action prescribed by the conditions of an assignment, complete a reverse action (heat an object instead of cooling it)
+	18	Mediator	a) use another object to transfer or transmit an action

REINVENTING

TREND *Formation of silicon wafers for solar panels by slicing the ingot produced a lot of waste. What can you do to remove this deficiency?*

REDUCING **Maxi-FIM:** The OZ by itself ensures the obtaining of the IFR:
[wafers of required dimensions without waste]

STANDARD CONTRADICTION (SC) – Extracting-2S

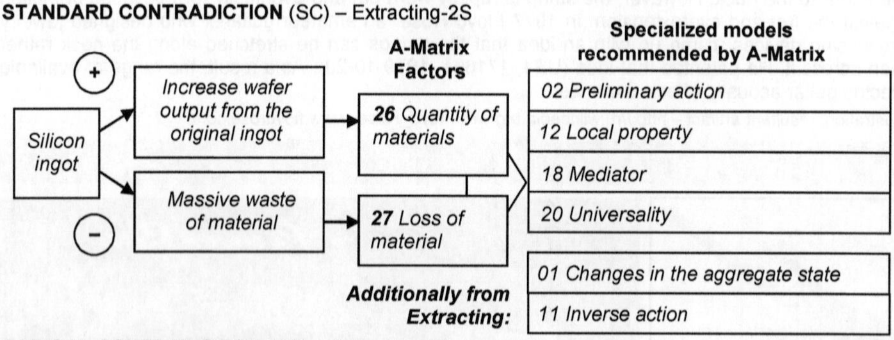

RADICAL CONTRADICTION (RC) – Extracting-2R

Silicon ingot ► must be to make the wafers **VS** must not be to avoid waste

INVENTING **To attain Objective 1 (waste reduction),** a zero-waste technology was proposed where the wafers are formed directly from the silicon melt – dominant models 01 Change in the aggregate state of an object, 02 Preliminary action and 11 Inverse action (instead of slicing the solid ingot or some semi-finished material, silicon is heated to melting temperature, and then a strip of film with required thickness is drawn out of the melt to be transformed into a waver when it cools down).

To attain Objective 2 (film extraction from silicon melt), the device is equipped with a moving base sheet; silicon melted in the lower section of the melting form sticks to it through adhesion, and is then drawn out in the form of thin film through a special slit to cool down while it keeps moving along lying on the base sheet – in accordance with model 18 Mediator.

Resolution of the Radical Contradiction using navigator 01 Separation in Space: some silicon in the form is solid (melt powder), and some is liquid (at the bottom of the form above the base sheet); 01 Separation in Material: when silicon cools down, it easily and BY ITSELF separates from the base sheet due to the fact that silicon and the material of which the base sheet is made have different thermal expansion coefficients..

Scientific effects: Change of the phase state of the matter; adhesion; variation in thermal expansion coefficients displayed by objects when their temperature changes, etc.

ZOOMING **Contradictions removed:** Yes.

Super-effects: High precision, continuity and productivity (in addition to absence of waste).

Extended uses: Possibility to make wafers with any dimensions; possibility to concurrently draw several films (wavers).

Beauty of solution: 100 — Highest grade. Substantiation: ideas based on the Inverse Action principle usually yield similar unexpected solutions. Here: do not slice anything; instead, form wafers when silicon is still in liquid stage, before the ingot is made (model 02 Preliminary Action and others).

fig. 10.32. Reinventing for making silicon wafers directly from the melt. Example prepared on the initiative and with the participation of Jesper Frausig (Denmark), a student of my (M.O.) class "Global Sustainable Innovention with Modern TRIZ" in the program Master of Science in Global Production Engineering at the Technical University of Berlin.

Example 10.12. *It's Floyd Rose! And you did not know it!*

ARTIFACT	FLOYD ROSE FOR AN ELECTRIC GUITAR

DESCRIPTION A technique called *bending* is used to create a smooth transition between notes (tremolo, vibrato). The string is moved transversely up and down at the place where it is pressed by the finger to the neck. However, the string is rapidly worn out and eventually torn because of friction against the fret and overextension. In 1977 Floyd Rose, an amateur guitarist who designed jewelry processing devices, came up with an idea that the strings can be stretched along the neck rather than across it. He patented that idea (US4, 171661, 1979-10-23). As a result, the range of available electric guitar acoustic effects expanded.

Illustration: resultant artifact – http://ru.wikipedia.org/wiki/Floyd_Rose, www.floydrose.com

EXTRACTING

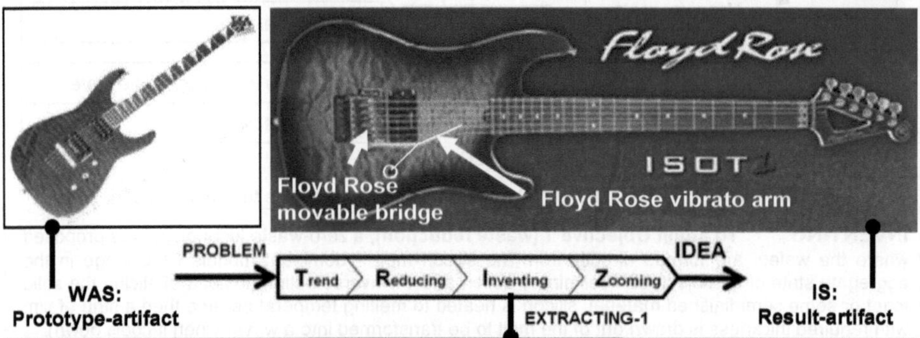

D	No	Navigator	Function / Substantiation
+	07	Dynamization	The bridge can move on the side where the strings are fixed to the soundboard so that the strings can be stretched by longitudinally.
+	11	Inverse action	The string is moved not transversely by the finger, but longitude-nally by a special mechanism; a right-hand Floyd Rose is used instead of a left-hand bend
	18	Mediator	String ends are fixed to the mobile Floyd Rose platform by special screws and clamps
+	19	Transition into another dimens.	The string is stretched longitudinally, and not transversely
+	32	Counter-weight	Springs equalize string tension, so that Floyd Rose can be used to both up and down the sound

REINVENTING

TREND *When doing a tremolo with an "old guitar", the string is rapidly moved by the finger along the fret, wears down and snaps! What can be done?*

REDUCING **Macro-FIM: *X-resource*** introduced into the system, together with available resources, guarantees attainment of **IFR**: [*tremolos which do not wear out strings*]

SC: String stretching ► 10 Ease of use **VS** 15 Length of the moveable object = 03, 11, 19, 37

RC: String stretching ► *must be* (to make a tremolo)
VS *must not be* (to prevent the string from wearing down against the fret)

INVENTING **SC Resolution Model** ► *presented in extracting.*

RC Resolution Model ► *01 In Space*: the string moves where it is fixed to the soundboard, and does not move where it is pressed to the neck (some details are not considered in this example).

ZOOMING Contradictions are removed.

fig. 10.33. Reinventing of Floyd Rose mechanism. From certification work by Vladimir Lysenko (Владимир Лысенко) Priazovskiy State Technical University, Mariupol, Ukraine

Example 10.13. To work at -200°C and at +20°C!

ARTIFACT **TAPS FOR CRYOGENIC DEVICES**

DESCRIPTION Taps used in cryogenic devices that work with liquid oxygen, nitrogen and other substances at temperatures below -120°C have a common problem: the freezing of the stem (a rod with the handwheel on one end and the disc stopping the liquid flow on the other end). A simple solution to this problem is to use an extended stem with a length calculated so that the upper part of the stem with the handwheel is operated at ambient (room) temperature.

Illustration: resultant artifact – http://www.asia.ru/ru/ProductInfo/1362693.html

EXTRACTING

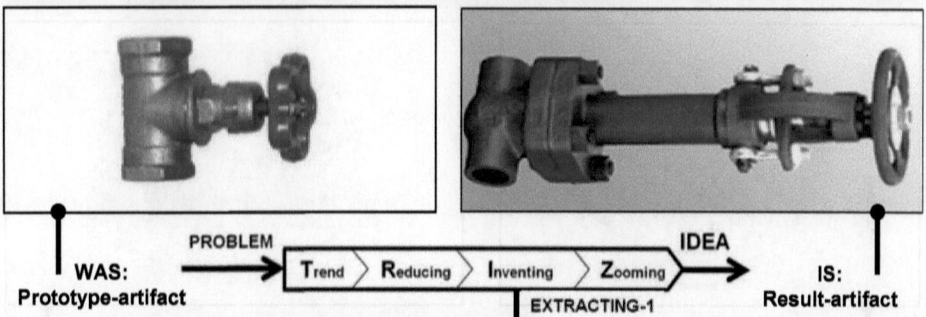

D	No	Navigator	Function / Substantiation
+	05	Separation	separate the "incompatible part" ("incompatible property") from the object or (turned completely around) include the only really necessary part (necessary property) into the object: the handwheel part of the stem must be placed in a "warm" environment", while the valve part must be inserted into the cryogenic zone
+	16	Partial or excess effect	when it is difficult to achieve the desired effect completely, we should try to achieve a bit less or a bit more. This can make the task much easier: make a 200% or 300% extension (as necessary to keep the handwheel out of the cold)
	18	Mediator	a) use another object to transfer or transmit an action: complement the stem with a similar "transitional, intermediate, supplementary" stem (and then, possibly, unit them into one)
	19	Transition into another dimension	a) it is possible to improve the transition from a surface to a three-dimensional space: the stem can be raised (third dimension) above the cryogenic zone (surface where the disc touches the valve seat).

REINVENTING

TREND *An ordinary tap does not work well with cryogenic devices – it just freezes. What can be done?*

REDUCING **Maxi-FIM:** The operating zone by itself ensures the obtaining of the following **IFR**: [*non-freezing tap stem*]

SC: Tap stem ► 02 Universality, Adaptability **VS** 34 Temperature = 01, 05

RC: Tap stem ► must be **"cold"** (as it operates in a cold environment)
VS must be **"warm"** (not to freeze up)

INVENTING

SC Resolution ► *05 Separation*: make the stem long so that the handwheel remains warm!

RC Resolution Model ► *01 In Space*: the stem is cold inside the tap, and it is warm outside the tap.

ZOOMING Contradictions are removed.

fig. 10.34. Reinventing of cryogenic valve. From certification work by Ekaterina Dontsova (Екатерина Донцова), Bauman Moscow State Technical University, Moscow, Russia

The following eight examples are based on the certification works of participants of Euro Project TEMPUS PROMENG MTRIZ-2, which was carried out by the Academy of Modern TRIZ.

Example 10.14. *Not every friendship leads to sharing... tea!*

ARTIFACT **TEAPOT FOR TWO DIFFERENT TEAS**

DESCRIPTION. Designer Ewa Sendecka, Krakow, Poland, has created a miracle tea for two. The glass teapot consists of two separate vessels. There are also two spouts, and now everyone can pour out into your cup the tea, which he likes. Author of teapot named her creation "Yin-Yang", i.e. different and even opposite aspects of the same whole, and here as different teas in one teapot.

EXTRACTING

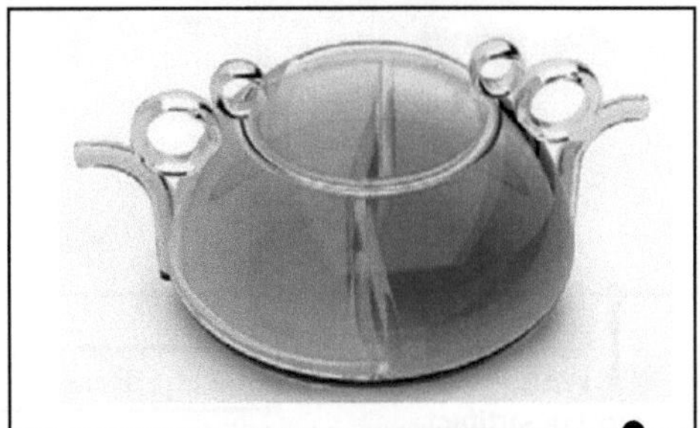

WAS: **Prototype-artifact** → PROBLEM → Trend > Reducing > Inventing > Zooming → IDEA → **IS:** **Result-artifact**

EXTRACTING-1

D	No	Navigator	Function / Substantiation
+	03	Segmentation	Separation of the teapot into two.
+	34	Matryoshka	Two vessels are located within a single teapot.

REINVENTING

TREND If family members like different teas, and there is only one teapot, often a dilemma appears, which tea will be made in a teapot, and which one in a cup. So, the tea party at the same time becomes a convenient and just impossible making the tea by turns. What to do?

REDUCING **Maxi-FIM:** The operating zone by itself ensures the obtaining the following **IFR**:
[*making two teas simultaneously*]

SC: Teapot ► 01 Productivity (making two teas) **VS** 10 Ease of use = 03, 04, 08, 34

RC: Teapot ► making two teas **VS** making one tea (because of only single teapot)

INVENTING

SC Resolution ► *look Extracting*

RC Resolution Model ► *01 Separation in space: the division of the teapot into two by a partition; making of two spouts.*

ZOOMING Contradictions are removed. It is a beautiful solution both aesthetically (with humor!) and technically!

fig. 10.35. An example from certification work by Olga Gladkova (Ольга Гладкова), absolute winner of the Euro Project TEMPUS PROMENG MTRIZ-2 (March 2013), Technical University of Zaporozhye, Ukraine

Example 10.15. *"Demon" maneuvers no worse than a UFO!*

ARTIFACT — WING WITHOUT PROTRUDING CONTROL ITEMS

DESCRIPTION. The company BAE Systems, UK, in 2010 introduced a model of unmanned turbojet aircraft DEMON. This machine has discarded the need of use to maneuver the ailerons, flaps and rudders. To achieve this, has become possible due to the fact that air injection takes place in a set of slits in the wing, that generates on its surface the desired pressure changes based on the Coanda effect and leads to rotation, descending or ascending. Such wing transformation leads to a simplification of the aircraft, loss of weight and increasing the reliability of the device

Illustration from www.membrana.ru/particle/3404

EXTRACTING

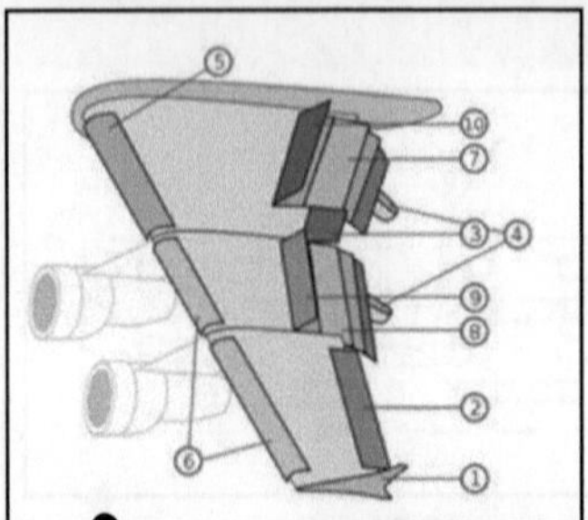

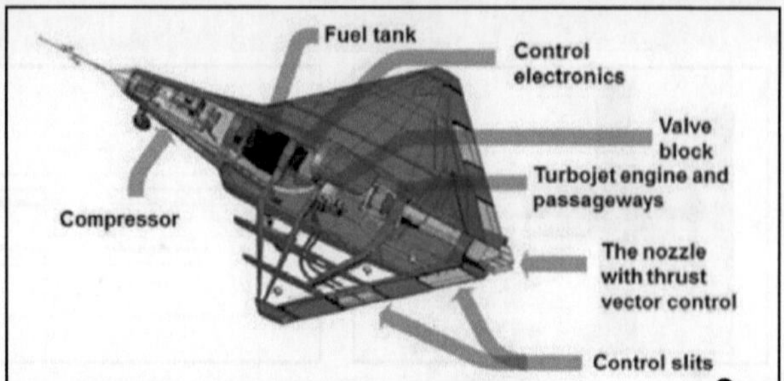

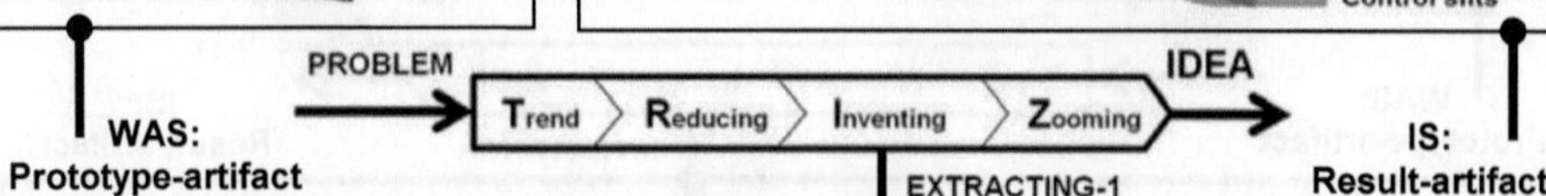

D	No	Navigator	Function / Substantiation
+	11	Inverse action	b) make a moveable part of an object or the environment fixed or a fixed part moveable: *abandonment of moving planes*
+	35	Unite	a) unite similar objects or objects for neighboring operations: *air escapes through the nozzle and through the control slits*

REINVENTING

TREND The plane with the traditional wing mechanization has the following disadvantages: 1) the increased complexity of the design; 2) the increased weight; 3) the complexity of the service, and 4) an increased chance of breakage. Is it possible to improve the wing?

REDUCING **Maxi-FIM:** The operating zone by itself ensures the obtaining the following **IFR**: [*no moving planes*]

SC: Wing ► 02 Universality, adaptability **VS** 04 Reliability = 01, 11, 18, 32

RC: Wing ► needed to maneuver **VS** impossible to maneuver (without moving elements)

INVENTING **SC resolution:** *based on model 11 Inverse action, the wing has a system of opening and closing slits instead of protruding elements!* **RC resolution:** *based on fundamental models 01Separation in space and 03 Separation in structure, and mainly the Coanda effect.*

ZOOMING A beautiful solution: no protruding elements and better reliability and maneuverability!

fig. 10.36. Example from certification work by Artem Muntjanu (Артём Мунтяну), "silver medalist" of Euro Project TEMPUS PROMENG MTRIZ-2 (April 2013), Admiral Makarov State University of Maritime and Inland Shipping, St. Petersburg, Russia

The following (also interlaced with other works further) are the works of the team members of the Baltic State Technical University "VOENMEH" after D. F. Ustinov, St. Petersburg, Russia, the absolute winner among university teams at Europroject TEMPUS PROMENG MTRIZ-2 (April 2013): Mark Ionin, Anastasiya Grigorjeva, and Aljona Protsjuk.

Example 10.16. *Armor is strong!*

ARTIFACT — **MULTILAYER ARMOR**

DESCRIPTION. Increasing the strength of a conventional armor realizes with increasing the thickness of the material. But this leads to an unjustified increase in the mass of armored vehicles and metal consumption. Often, even a greater thickness of armor does not protect against cumulative jet. Multi-layer armor is developed with enhanced reliability and reduced thickness.

Illustration from http://www.sciteclibrary.ru/rus/catalog/pages/7984.html

EXTRACTING

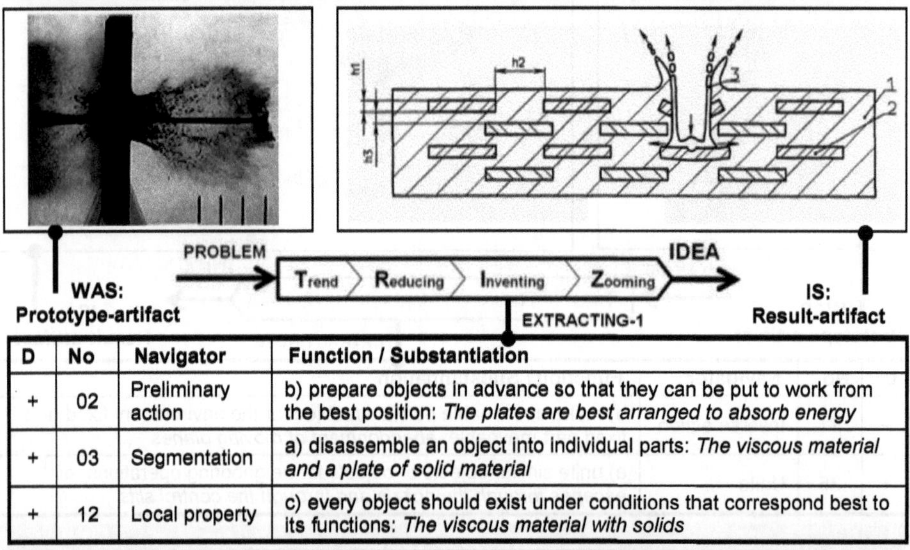

D	No	Navigator	Function / Substantiation
+	02	Preliminary action	b) prepare objects in advance so that they can be put to work from the best position: *The plates are best arranged to absorb energy*
+	03	Segmentation	a) disassemble an object into individual parts: *The viscous material and a plate of solid material*
+	12	Local property	c) every object should exist under conditions that correspond best to its functions: *The viscous material with solids*

REINVENTING

TREND Often, even a greater thickness of armor does not protect against cumulative jet. Other technologies are needed to confront different types of shells. What could be done?

REDUCING **Maxi-FIM:** The operating zone by itself ensures the obtaining the following **IFR**: [*durable and light armor that protects against cumulative and other missiles*]

SC: Armor ► 04 Reliability **VS** 29 Stabile structure of the object = 03, 05, 12, 18

RC: Armor ► should be thick (to withstand the cumulative jet)
VS must not be thick (as it becomes too severe)

INVENTING **SC resolution:** *it was decided to abandon the homogeneous armor. Armor makes inhomogeneous: the matrix of the viscous substance + hard plate (12 Local property). Due to such division into plates and the base material (03 Segmentation), significantly reduced the thickness of the armor, and hence mass. The plates are staggered to better extinguish the explosive energy (02 Preliminary action).* **RC resolution:** *rigid plates are not placed on the same line and staggered to provide more effective confrontation of the cumulative jet (Separation in space). Armor is not uniform as made of different materials with different properties (Separation in material and structure).*

ZOOMING A very effective solution with elimination of contradictions!

fig. 10.37. First example from certification work by Mark Ionin (Марк Ионин)

Example 10.17. *In case of danger, to become a... hang glider!*

ARTIFACT **PLANE WITH EXTENDABLE WINGS FOR EMERGENCIES**

DESCRIPTION. The idea is to equipped aircraft with retractable and extendable wings, which are utilized only in emergency situations. Wings' design is realized on the principle of "fantail": they are let out on both sides of fixed aircraft wing and expanded in the horizontal plane. So this plane turns into a glider, which will make planning and landing in case of engine failure or other emergency.

Illustration from http://www.sciteclibrary.ru

EXTRACTING

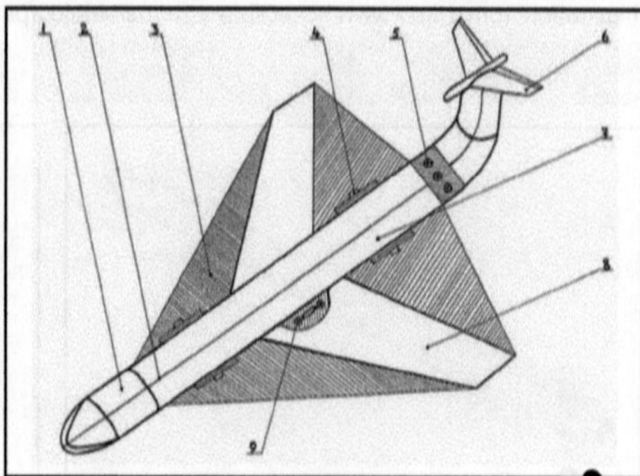

WAS: Prototype-artifact — PROBLEM → Trend > Reducing > Inventing > Zooming → IDEA — IS: Result-artifact

EXTRACTING-1

D	No	Navigator	Function / Substantiation
+	07	Dynamization	b) disassemble an object into parts that are moveable among each other: *Extendable consoles*
+	16	Partial or excess effect	to achieve a bit less or a bit more: *creating a very wide wings*
+	28	Previously install. cush.	increase the relatively low security of an object with safety measures in advance

REINVENTING

TREND When the aircraft engine failure, the plane has continuously to descend and land, but it is impossible because huge plane virtually impossible to plan, management of difficult. How to descend and also to control the plane?

REDUCING **Maxi-FIM:** The operating zone by itself ensures the obtaining the following **IFR**: [*aircraft capable of emergency planning and landing*]

SC: Plane ► 02 Universality, adaptability (to control)
VS 10 Ease of use (not enough load-bearing capacity of the wing) = 03, 07, 15, 16

RC: Plane ► should plan (to smoothly land)
VS should not plan (since the wings are narrow and does not allow to plan)

INVENTING **SC resolution:** *sliding additional wings are pre-placed in the plane for case of an emergency landing (28 Previously installed cushion). When emergency, consoles are expanded (07 Dynamization), and wing area is increased by almost 100%, that gives the ability to control the planning (16 Partial or excess effect).* **RC resolution:** *Separation in space and structure).*

ZOOMING Development: emergency surfaces could be flexible, as the fabric.

fig. 10.38. Second example from certification work by Mark Ionin (Марк Ионин)

Example 10.18. *A missile that could not be shot down!*

ARTIFACT **ROTATING GUIDED MISSILE**

DESCRIPTION. To improve the maneuverability, rotating guided missile is designed. The new missile has a body with a combat bay and jet. The housing includes a tail stabilizer, which consoles are chamfered at an angle to the longitudinal axis of the housing. The novelty is that the elements of stabilization of rocket rotation speed are formed as "destabilizers", installed at a small angle δ to the longitudinal axis of the housing.

Illustration from http://www.sciteclibrary.ru/rus/catalog/pages/7395.html

EXTRACTING

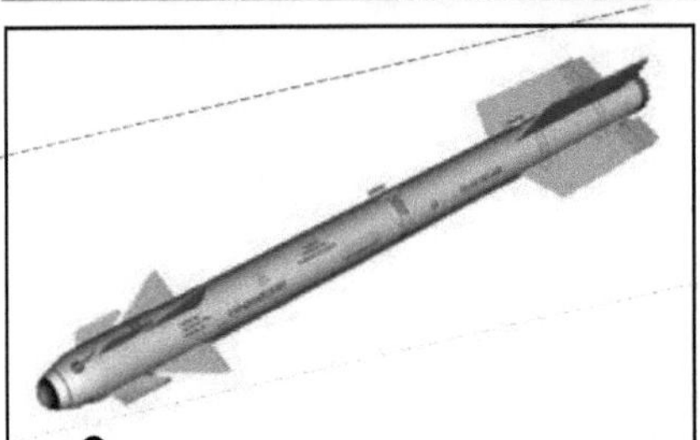

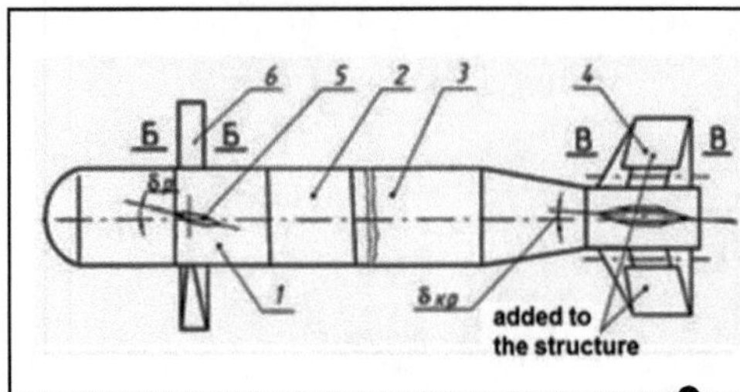

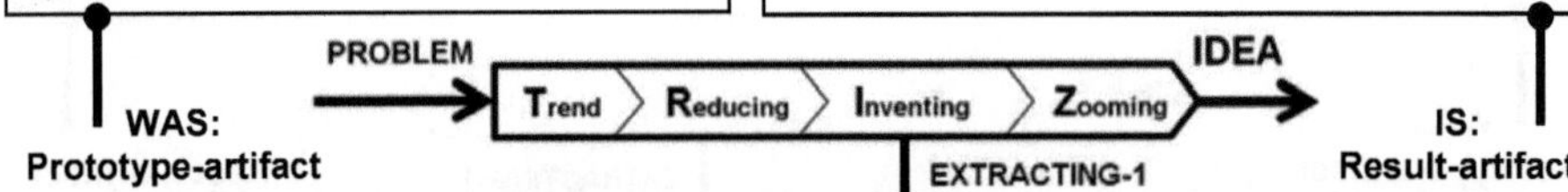

D	No	Navigator	Function / Substantiation
+	07	Dynamization	a) the characteristics of an object are changed to optimize every work procedure: *Ensuring good maneuverability with destabilizers*
+	16	Partial or excess effect	to achieve a bit less or a bit more: *enhanced dynamic performance and maneuverability*
+	22	Spherical-shape	c) change to turning movements: *improved dynamic characteristics due to the rotation*

REINVENTING

TREND Conventional missiles have a very bad dynamic performance and maneuverability, therefore they could be easy shot down. Anti-missile defense (AMD) finds them easily and quickly and destroys. It is necessary to increase the agility to successfully overcome AMD.

REDUCING **Maxi-FIM:** The operating zone by itself ensures the obtaining the following **IFR**: [*missile, which cannot bring down*]

SC: Missile ▶ 02 Universality, adaptability (missile should be maneuverable)
VS 10 Ease of use (excessive stabilization reduces maneuverability) = 03, 07, 15, 16

RC: Missile ▶ must overcome AMD
VS cannot overcome AMD (because of insufficient maneuverability))

INVENTING **SC resolution:** *Destabilizer missiles produces rotation of the missile (22 Spherical-shape), thereby its maneuverability is dramatically increased. (16 Partial or excess effect).* **RC resolution:** *accomplished by Separation in structure: one part has one property, and the object at whole has other property. Stabilizers make a missile rotating and, in some way, destabilize it, but this leads generally to a sharp increase in maneuverability.*

ZOOMING The combination of stabilizing and destabilizing components leads to an incredible increase in maneuverability..

fig. 10.39. Third example from certification work by Mark Ionin (Марк Ионин)

Example 10.19. ***How to optimize waste removal?***

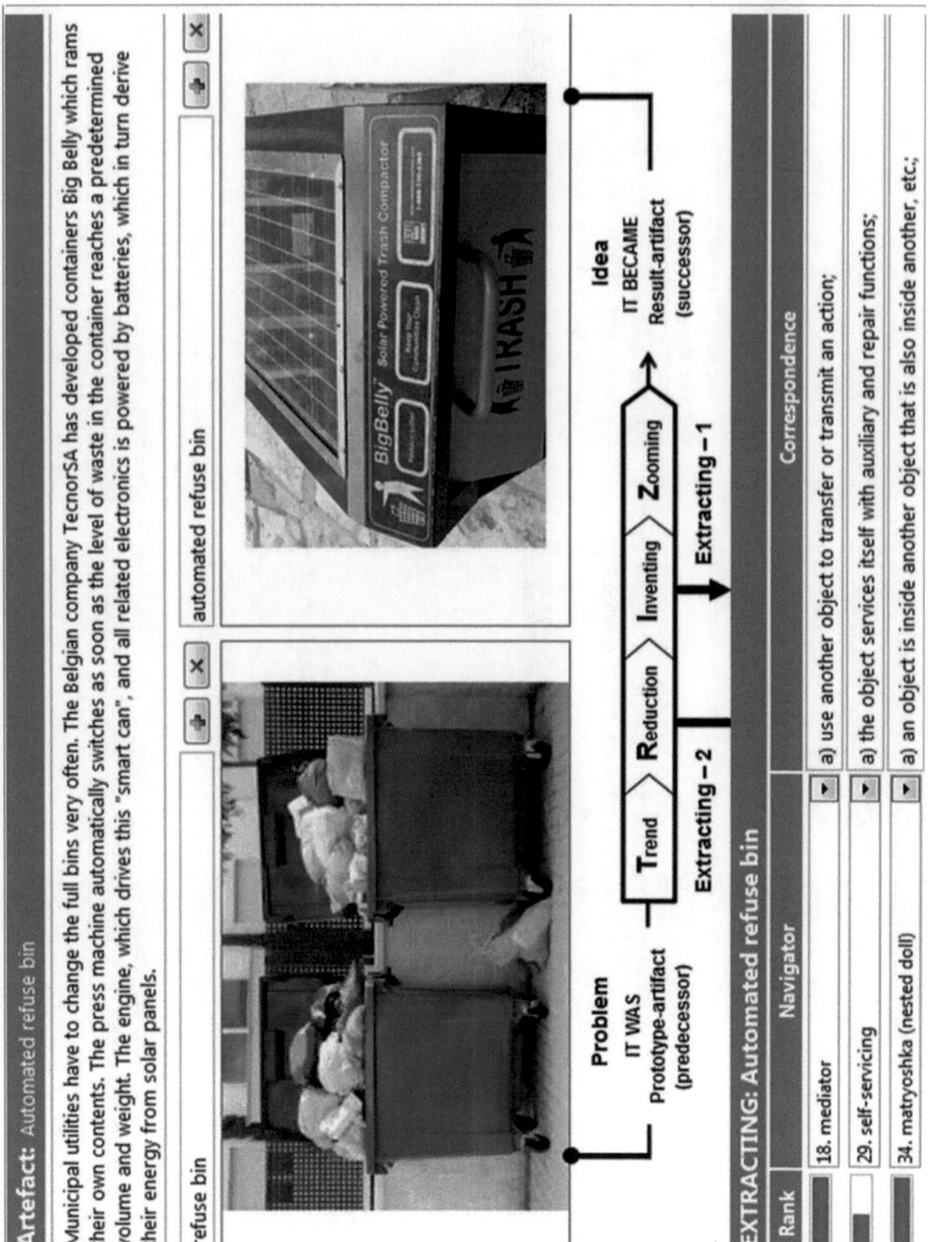

This example was developed by teacher Paulina Desewa from Foreign Language High School "Nikola Vaptsarov," Shumen, Bulgaria, at the UNESCO Institute of Informational Technologies in Education (IITE) project "Learning For the Future—Modern TRIZ."

fig. 10.40. Example from certification work by Paulina Desewa (beg.)

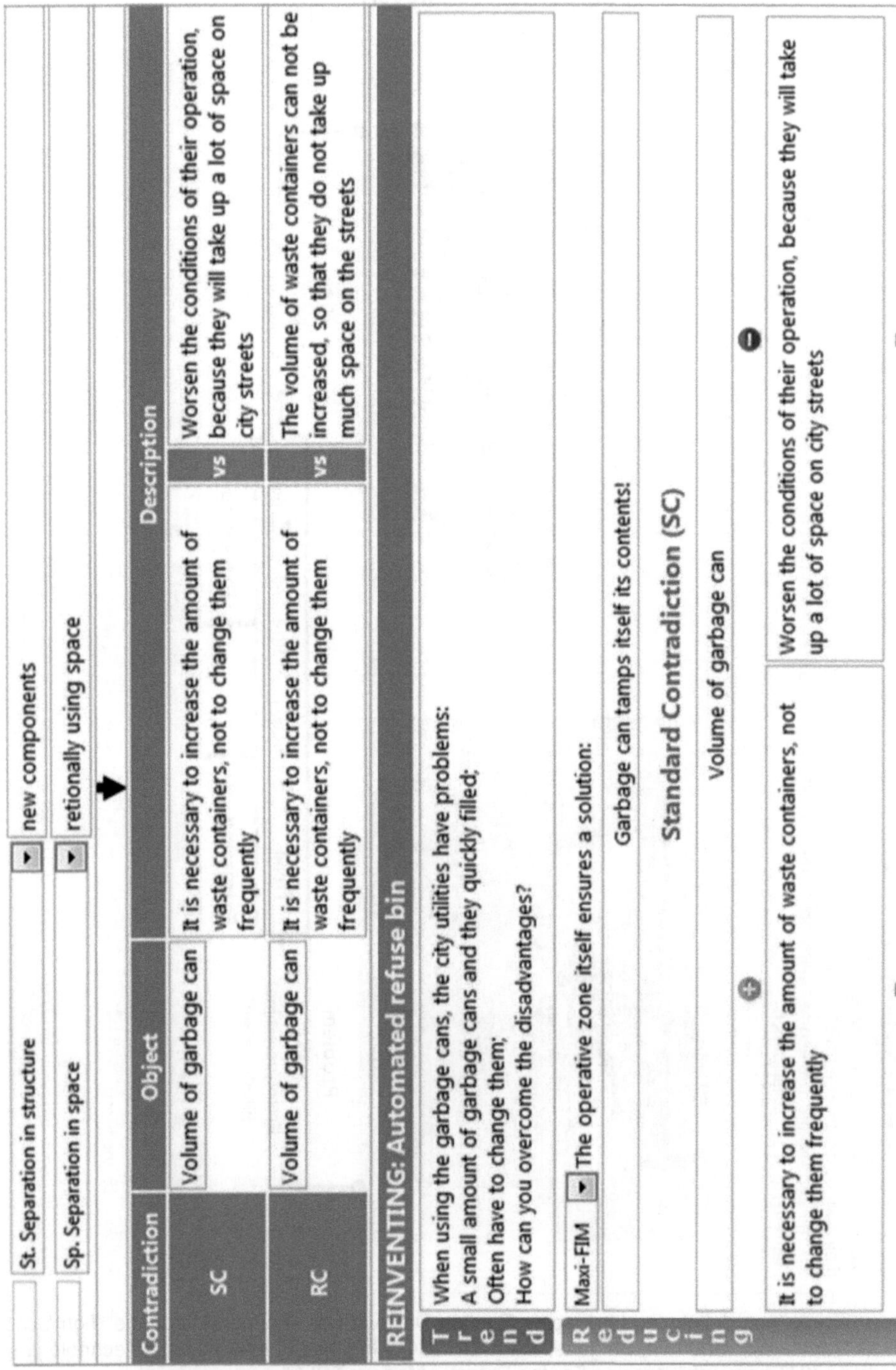

fig. 10.40. Example from certification work by Paulina Desewa (cont.)

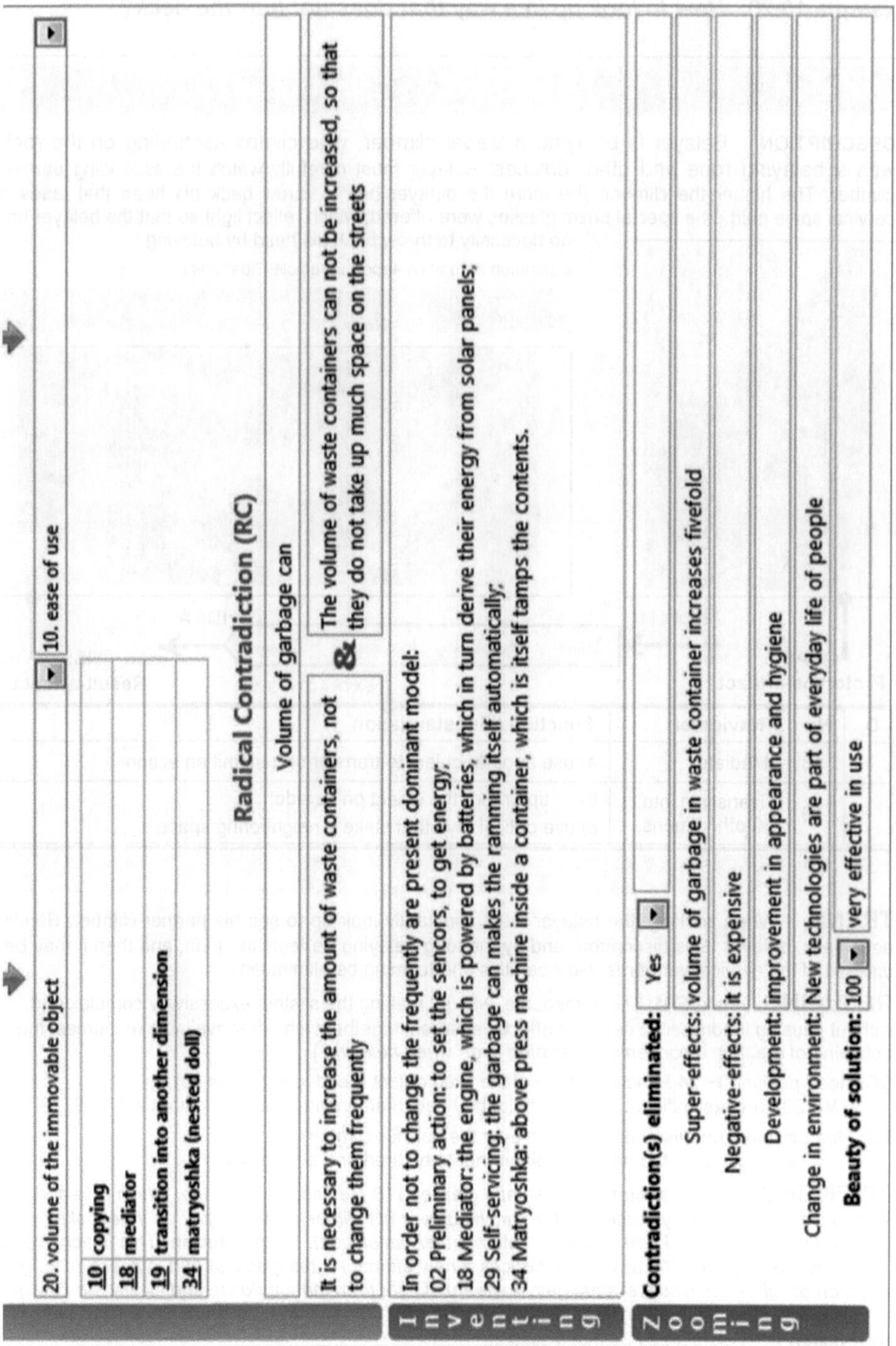

fig. 10.40. Example from certification work by Paulina Desewa (end). Screenshots from software EASyTRIZ Junior

Example 10.20. *How to look up in a way that does not hurt the neck?*

ARTIFACT **PRISMATIC GLASSES FOR CLIMBERS**

DESCRIPTION. Belayer is belaying a leader-climber, who climbs ascending on the rock, with a belaying rope and other devices. Belayer must carefully watch the ascending partner-climber. The higher the climber, the more the belayer has to throw back his head that leads to cervical spine pain. The special prism glasses were offered, which reflect light so that the belayer has no necessity to throw back his head by belaying.

Illustration from www.4sport.ua/articles?id=13541

EXTRACTING

WAS: Prototype-artifact — **PROBLEM** → Trend > Reducing > Inventing > Zooming → **IDEA** — **IS:** Result-artifact

EXTRACTING-1

D	No	Navigator	Function / Substantiation
+	18	Mediator	a) use another object to transfer or transmit an action
+	19	Transition into anoth. dimens.	b) … tip or turn the object on its side; c) use optical rays that strike a neighboring space…

REINVENTING

TREND When working, the belayer must constantly look up to see his partner-climber. Bending neck back, belayer feels discomfort, and by the long belaying he feels the pain, and then it may be a disease of the cervical vertebrae. How can this shortcoming be eliminated?

REDUCING **Maxi-FIM:** The X-resource, without making the system excessively complex and without causing inadmissible negative effects, ensures, together with other available resources, the obtaining of the IFR: [*convenient watching over when belaying*]

SC: Head position ▶ 24 Functional time of the fixed object (need long to look up)
VS 30 Force (significant efforts to hold the head in an uncomfortable position) = 17, 19, 39

RC: Head ▶ must be raised up (to watch over the partner-climber)
VS must be in the normal position (not to be tired and aching neck)

INVENTING **SC resolution:** *prismatic glasses (18 Mediator) was made, which refract light in such a way that the belayer sees and is watching over his climbing partner just as if he looked u and his head was thrown back (19 Transition into another dimension).* **RC resolution:** *This is achieved by Separation in structure (the system introduces a new element – the glasses) and in space (change of direction of light – the whole system gets a stream of light from the top down, and locally the flow goes in front of the face to the eyes of the belayer).*

ZOOMING Original and beautiful solution.

fig. 10.41. Second example from certification work by Aljona Protsjuk (Алёна Процюк)

Example 10.21. ***How to optimize a light bulb?***

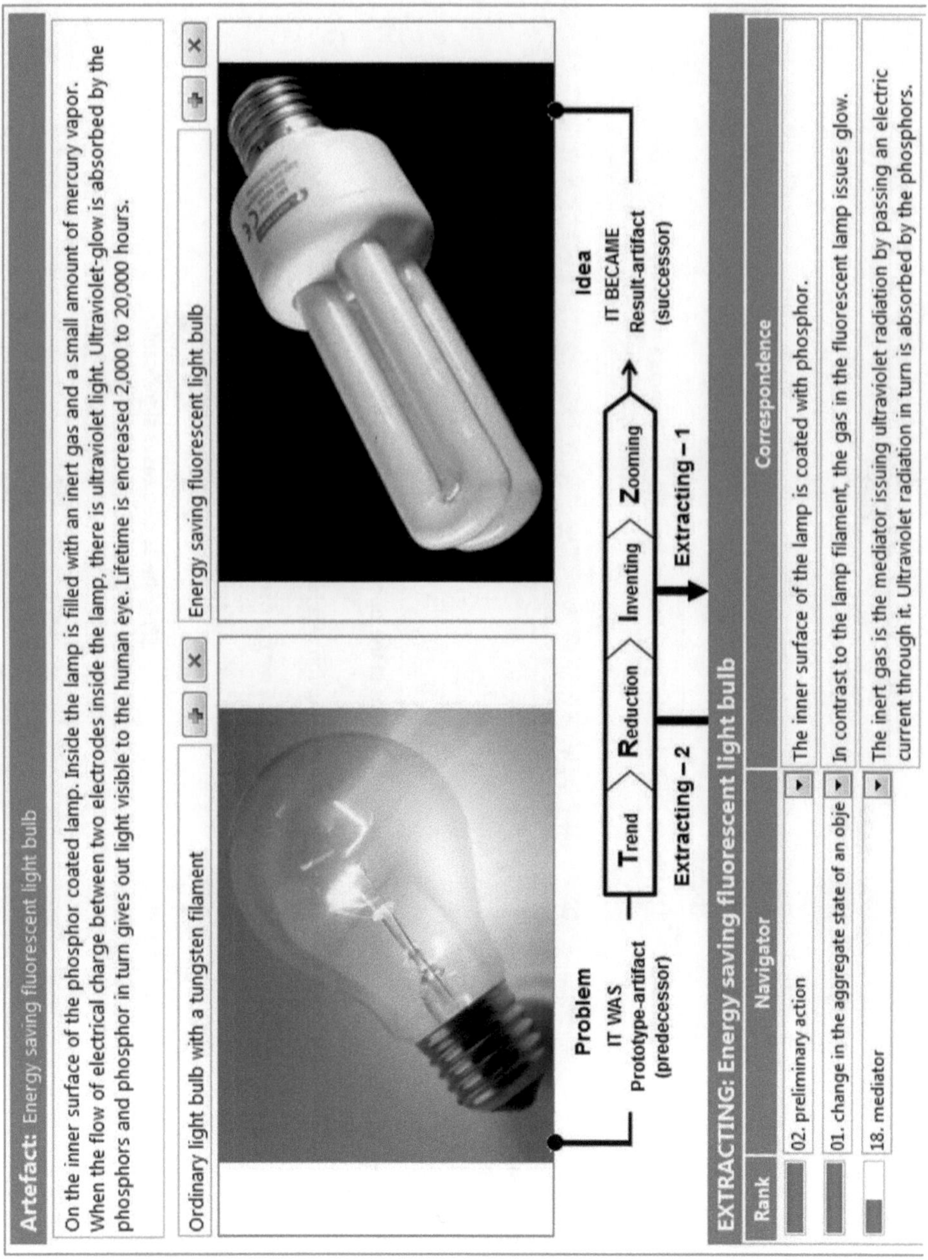

This example was developed by Nikita Panchenko (class 11) from High School "BEST," Almaty, Kazakhstan, absolute winner at UNESCO IITE project "Learning For the Future—Modern TRIZ."

fig. 10.42. Example from certification work by Nikita Panchenko (beg.)

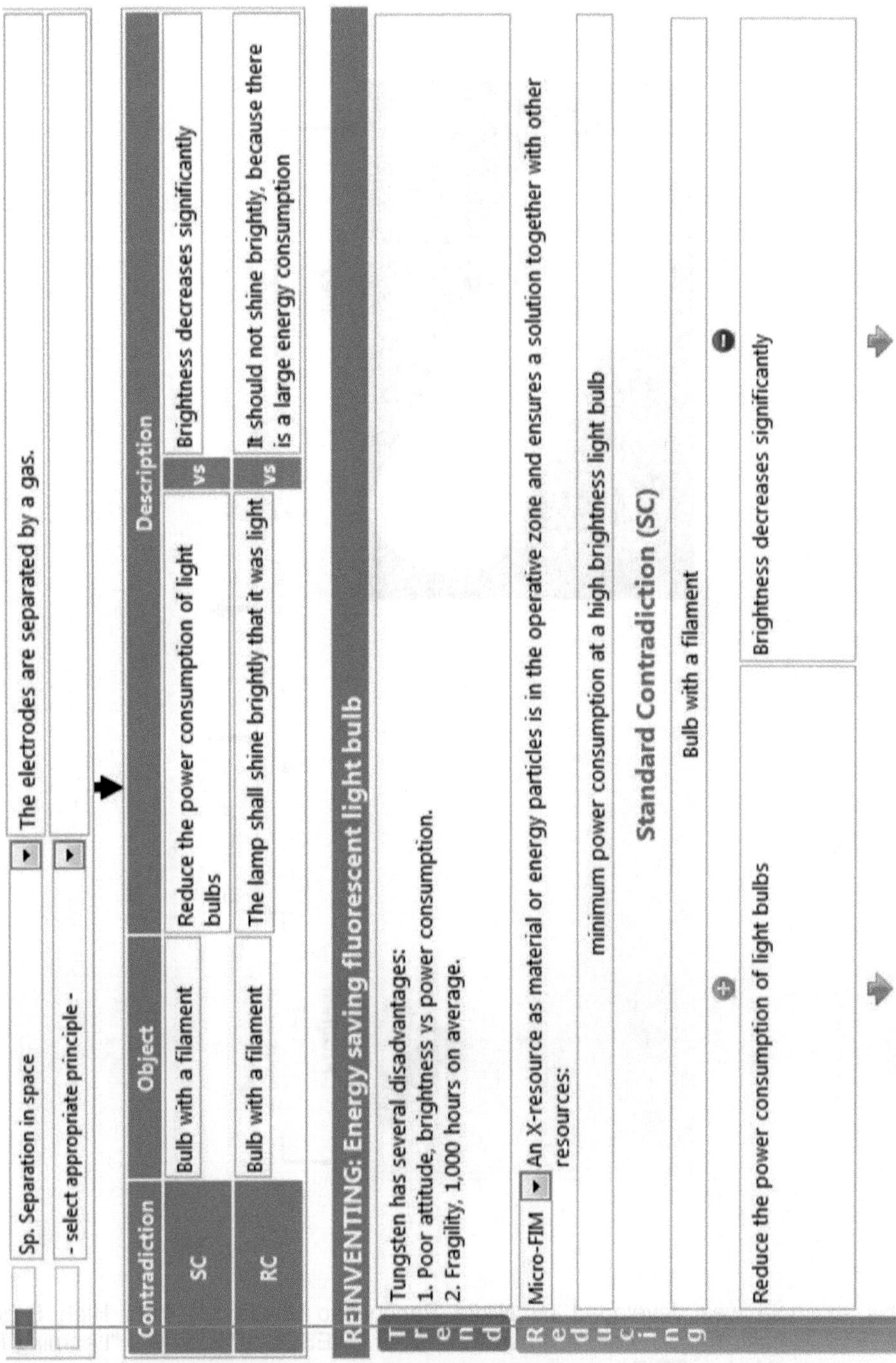

fig. 10.42. Example from certification work by Nikita Panchenko (cont.)

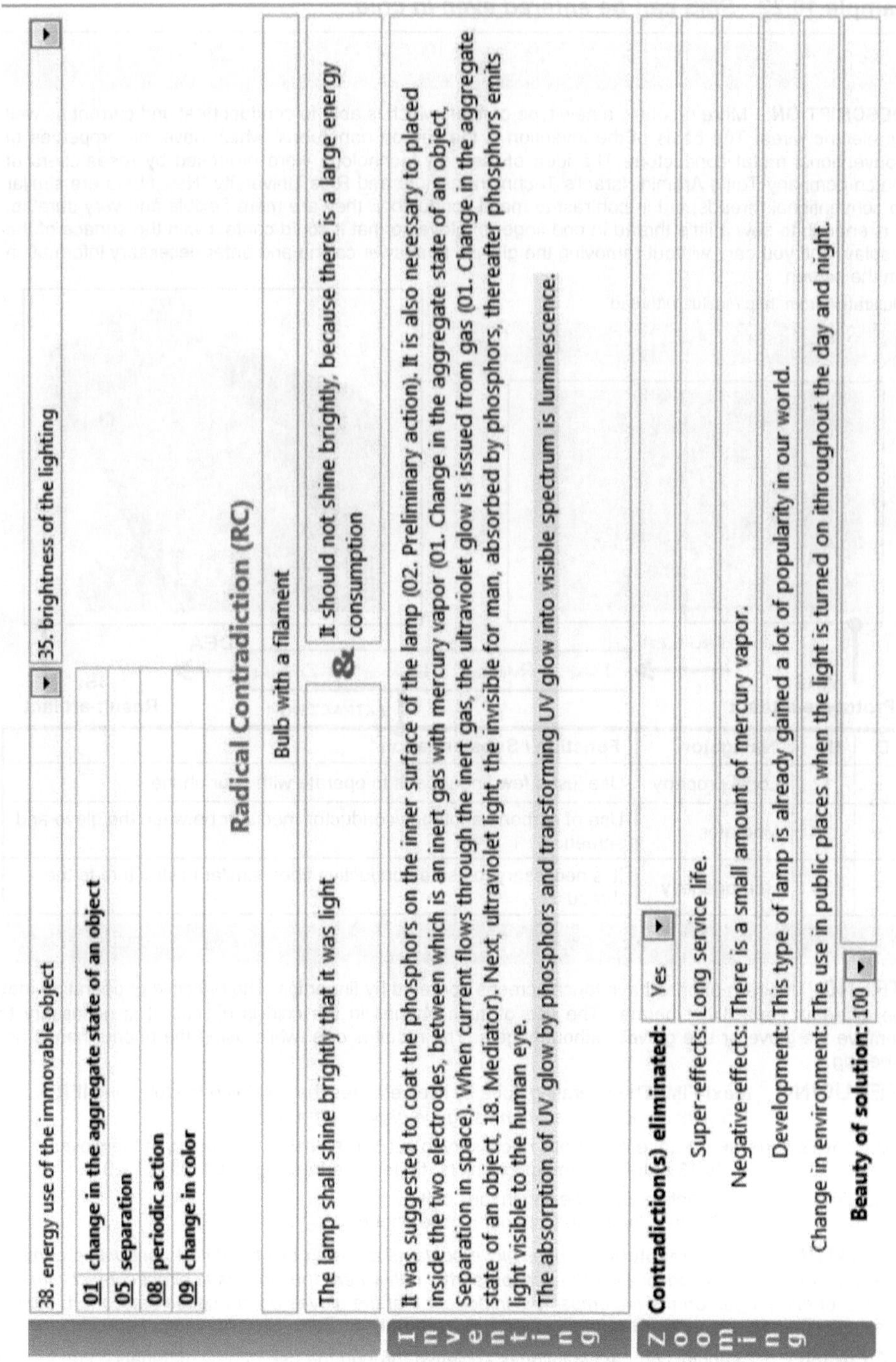

fig. 10.42. Example from certification work by Nikita Panchenko (end).
Screenshots from software EASyTRIZ Junior

Example 10.22. *SMS can be entered even in cold.*

ARTIFACT **ROTATING GUIDED MISSILE**

DESCRIPTION. More recently, a new type of fiber, which is able to conduct heat and current as well as electric wires. The basis of the invention is the carbon nanotubes, which have the properties of conventional metal conductors. The idea of the new technology were proposed by researchers of Dutch company Teijin Aramid, Israel's Technion Institute and Rice University. New fibers are similar to conventional threads, but in contrast to metal conductors, they are more flexible and very durable. It is enough to sew a little thread in one finger of glove so that it could contact with the surface of the display, and you can, without removing the gloves, to answer calling and enter necessary information on the screen.

Illustration from http://jmitut.ru/thread

EXTRACTING

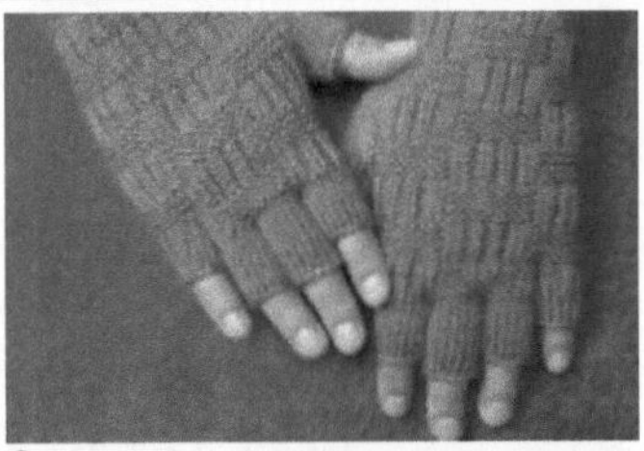

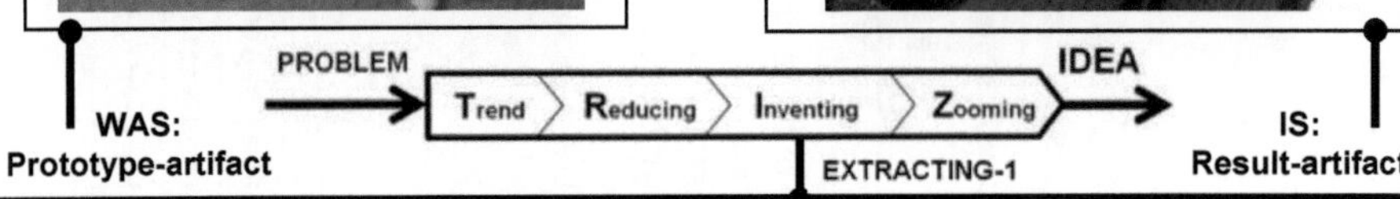

D	No	Navigator	Function / Substantiation
+	12	Local property	Use just a few finger pads to operate with your phone
+	18	Mediator	Use of carbon nanotubes conductor-mediator between the glove and screen.
+	38	Homogeneity	It is necessary to use a conductive fiber, similar in structure to the thread.

REINVENTING

TREND Modern phones have touch screens operated by fingertips. The principle of operation may be different: capacitive, thermal. The use of such phones in the cold is difficult: it is necessary to remove the glove or use gloves without fingers. The result is one: while using the phone, hands are freezing.

REDUCING **Maxi-FIM:** The operating zone by itself ensures the obtaining the following **IFR**:
[*fingers do not freeze when using the phone in the cold*]

SC: Glove's finger ► 10 Ease of use (it's necessary to cut out "holes" in the fingers of the gloves)
VS 13 External damaging factors (fingers are freezing) = 04, 05, 23, 29

RC: Glove's finger ► gapless (not freezing in the cold)
VS with a "hole" (to use your mobile phone)

INVENTING **SC resolution:** *finger glove contains a thread (18 Mediator) conducting current. This type of thread is used locally (12 Local property), only where the glove is in direct contact with the screen of the gadget: on the fingertips. It is important that the gloves do not cause discomfort: tactile, they must be homogeneous (38 Homogeneity).* **RC resolution:** *Separation in material and structure.*

ZOOMING Beautiful compact solution is achieved through the use of new materials – conductive carbon nanotubes.

fig. 10.43. Example from certification work by Anastasiya Grigorjeva (Анастасия Григорьева)

Example 10.23. ***From da Vinci parachute toward the modern parachute.***

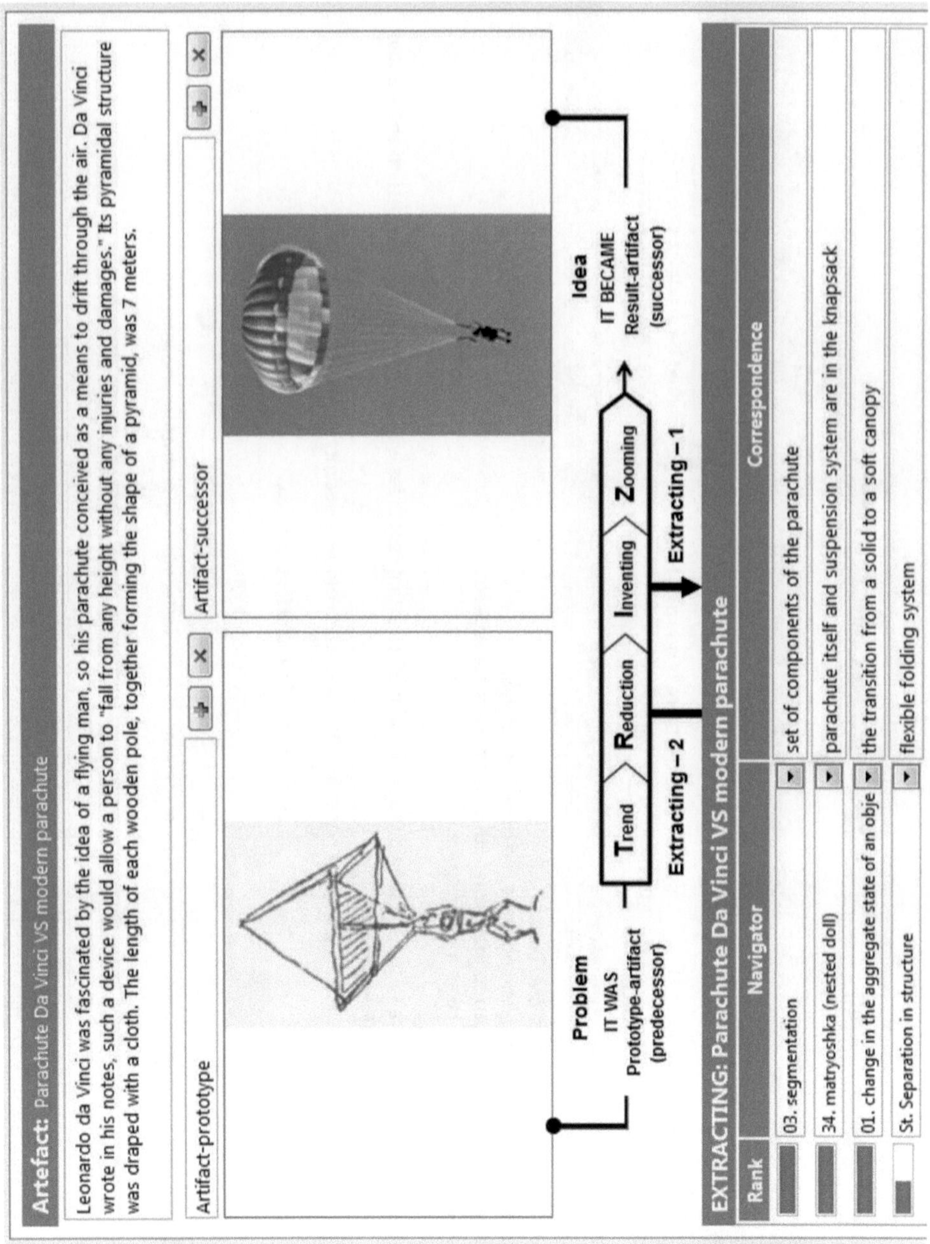

This example was developed by Asel Alkhanova (class 10) from High School "BEST," Almaty, Kazakhstan, absolute winner at UNESCO IITE project "Learning For the Future—Modern TRIZ."

fig. 10.44. Example from certification work by Asel Alkhanova (beg.)

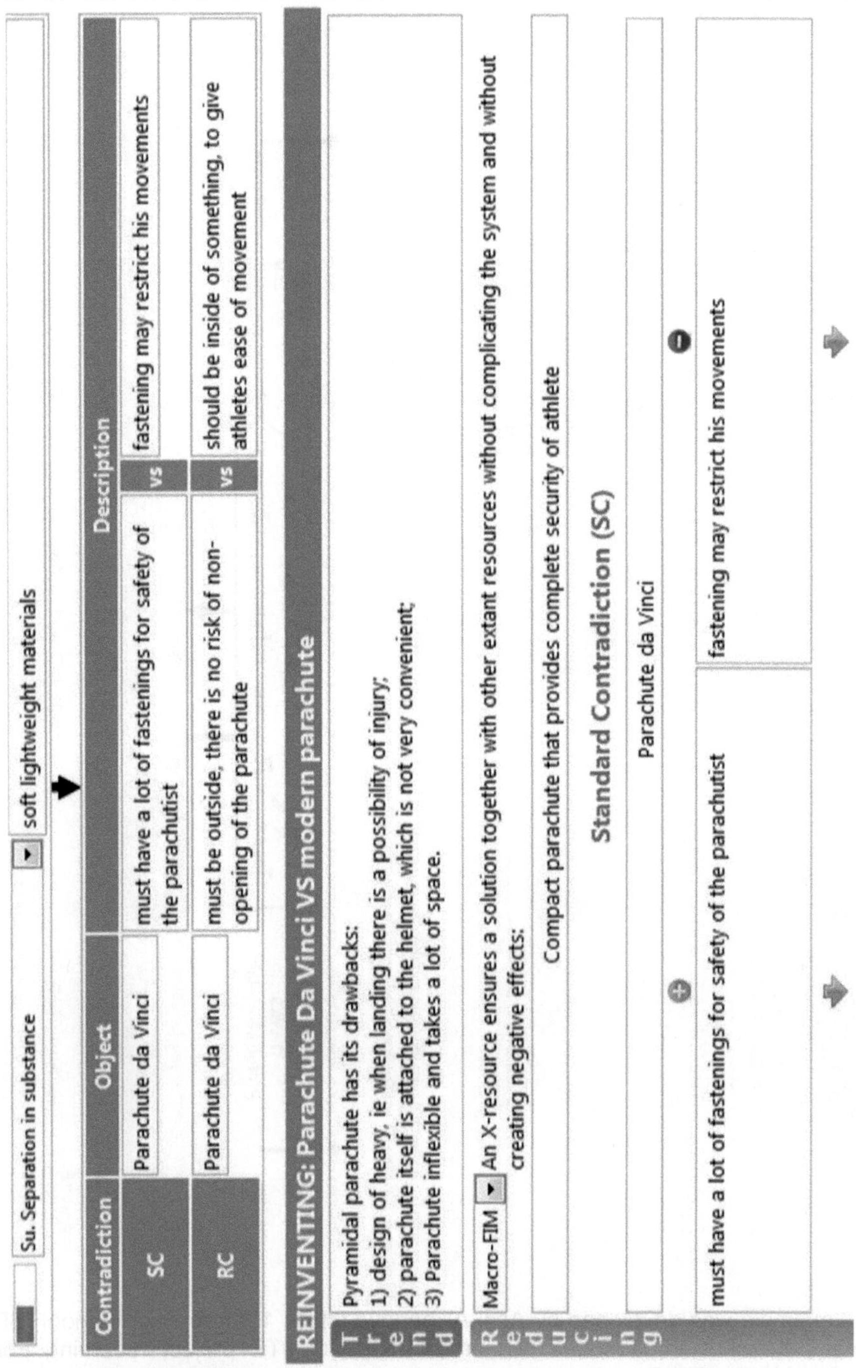

fig. 10.44. Example from the certification work by Asel Alkhanova (cont.)

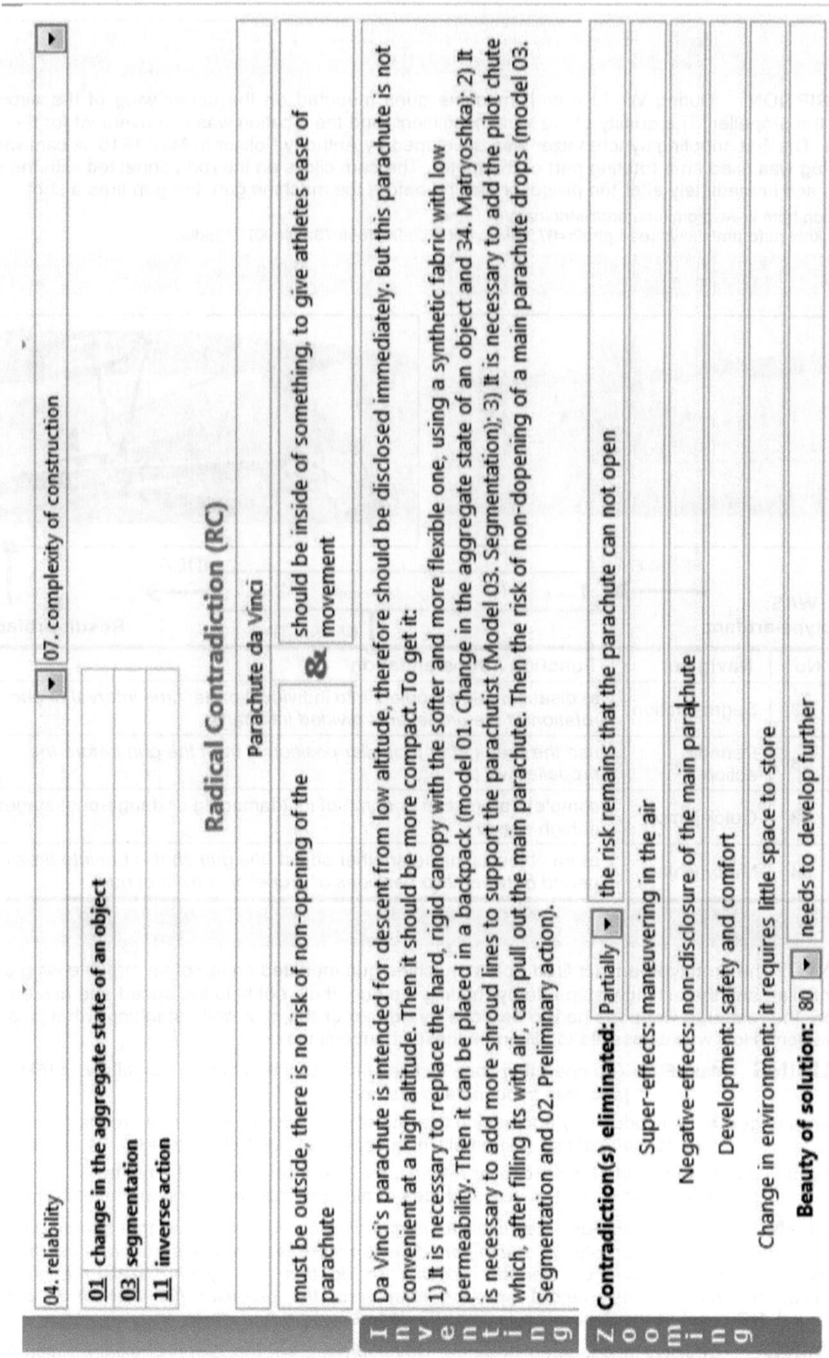

fig. 10.44. Example from certification work by Asel Alkhanova (end).
Screenshots from software EASyTRIZ Junior

Example 10.24. *"Old" TRIZ example: how to shoot through rotating propeller?*

ARTIFACT **SYNC SHOOTING THROUGH THE PROPELLER**

DESCRIPTION. During World War I, machine guns mounted on the upper wing of the airplane above the propeller. The quality of fire was insufficient, and the location was inconvenient for the gun charge. The first shooting synchronizer was developed by Anthony Fokker in May 1915. A cam with a single lug was fixed on a rotating part of the motor. The cam clicks on the rod connected with the gun trigger, and immediately after the propeller placing before the machine gun, the gun fires a shot.

Illustration from www.promzona.org/ru/trip/cam/4/ and www.sukhoi.ru/forum/showthread.php?t=67399&s=635162a9bbf69fc73ad4c0070135d6c1

EXTRACTING

WAS: Prototype-artifact — PROBLEM → Trend > Reducing > Inventing > Zooming → IDEA — **IS:** Result-artifact

EXTRACTING-1

D	No	Navigator	Function / Substantiation
+	03	Segmentation	a) disassemble an object into individual parts: *time interval of one rotation of the propeller is divided into parts*
+	08	Periodic action	*use the break after propeller position in front the gun before the propeller will be back*
+	33	Quick jump	complete a process or some of its (damaging or dangerous) stages at high speed
+	34	Matryoshka	a) an object is inside another object: *the gun shot is put into break period between two positions of propeller in front of gun*

REINVENTING

TREND The prototype-artifact fired from a machine gun mounted on top of an airplane wing upper the propeller, and the aiming was made by rotating the gun. It did not help the speed and accuracy. In addition, the recharge weapons had to be done by gunner or the pilot while standing, what was very inconvenient. How was it possible to eliminate these shortcomings?

REDUCING **Maxi-FIM:** The operating zone by itself ensures the obtaining the following **IFR**: [*accurate shooting and fast recharge*]

SC: Machine gun ► 01 Productivity (increasing the rate of fire, reducing the time to recharge) **VS** 10 Ease of use (extremely inconvenient in the prototype) = 03, 04, 08, 34

RC: Machine gun ► to be out of the propeller rotation (the bullet does not hit the blade) **VS** should be in the area of blades turning (to improve aiming and overcharging)

INVENTING **SC resolution:** *machine gun is mounted on the "nose", so that shooting is done through the plane of propeller rotation, but in periods when there is no blade in front of a machine gun (models 33 Quick jump and 08 Periodic action). To do this, the synchronization engine was introduced, that runs "in resonance" with the rotation of the propeller (06 Use of mechanical oscillations).* **RC resolution:** *Separation in time (03 Segmentation) and structure.*

ZOOMING Effective unexpected decision. Development: set the barrel coaxially within of the propeller shaft (in accordance with the navigator *34 Matryoshka*, recommended by A-matrix).

fig. 10.45. First example from the certification work by Aljona Protsjuk (Алёна Процюк)

Example 10.25. ***Lift by Elisha Otis vs modern lift.***

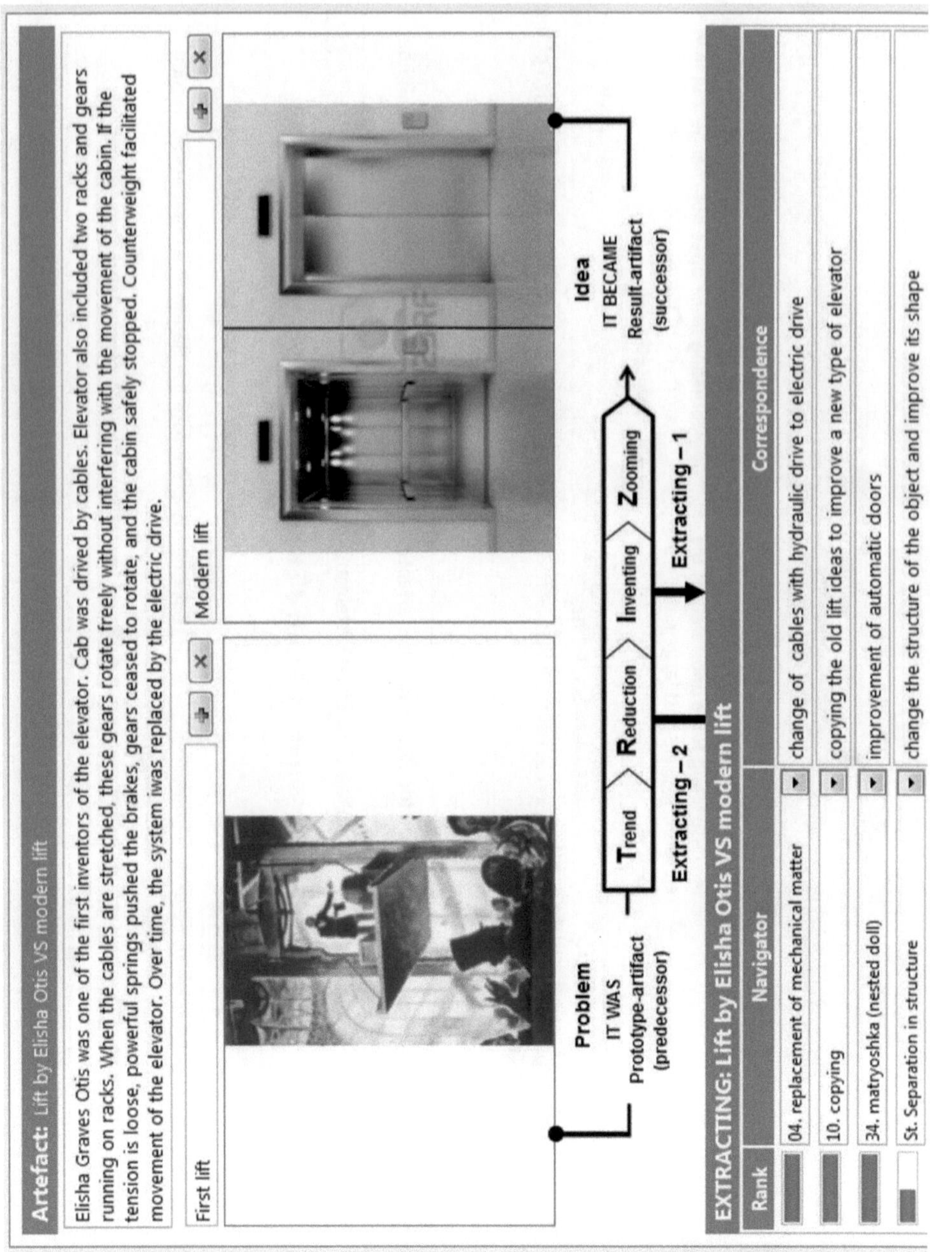

This example was developed by Alexander Kostikov, a teacher of English, from High School "BEST," Almaty, Kazakhstan, absolute winner between the teachers at UNESCO IITE project "Learning For the Future—Modern TRIZ."

fig. 10.46. Example from certification work by Alexander Kostikov (beg.)

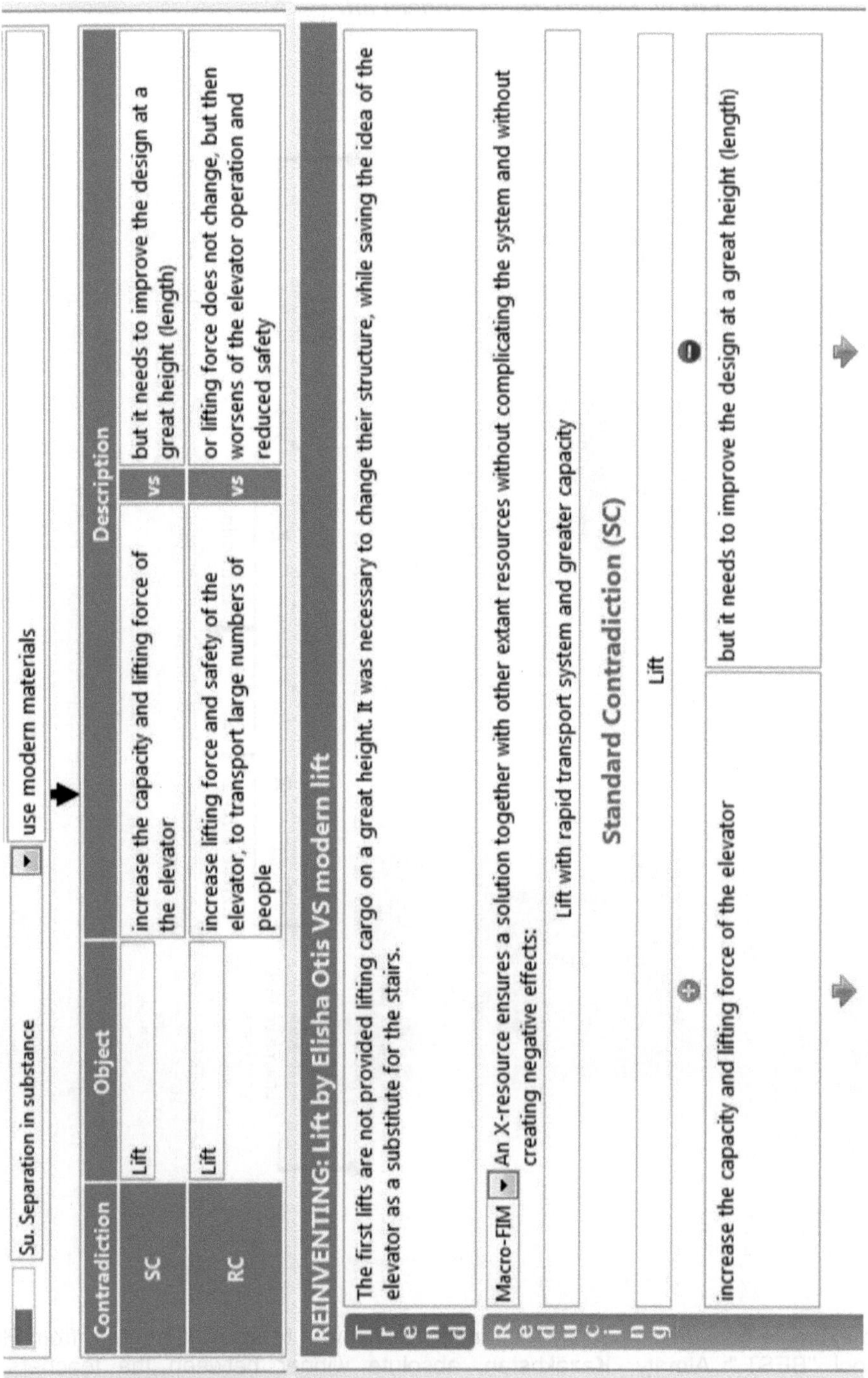

fig. 10.46. Example from certification work by Alexander Kostikov (cont.)

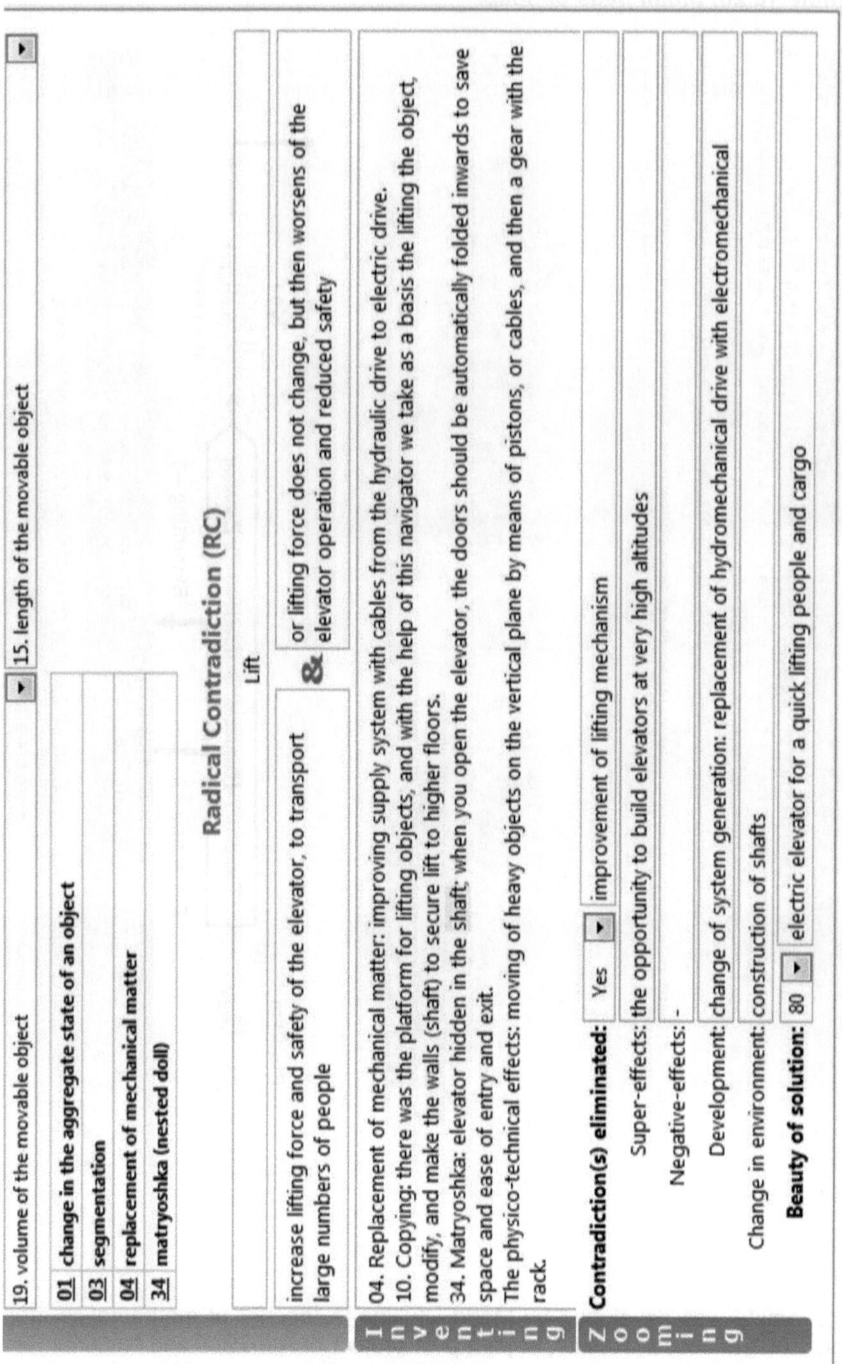

fig. 10.46. Example from certification work by Alexander Kostikov (end).
Screenshots from software EASyTRIZ Junior

Example 10.26. ***Small peas of soap.***

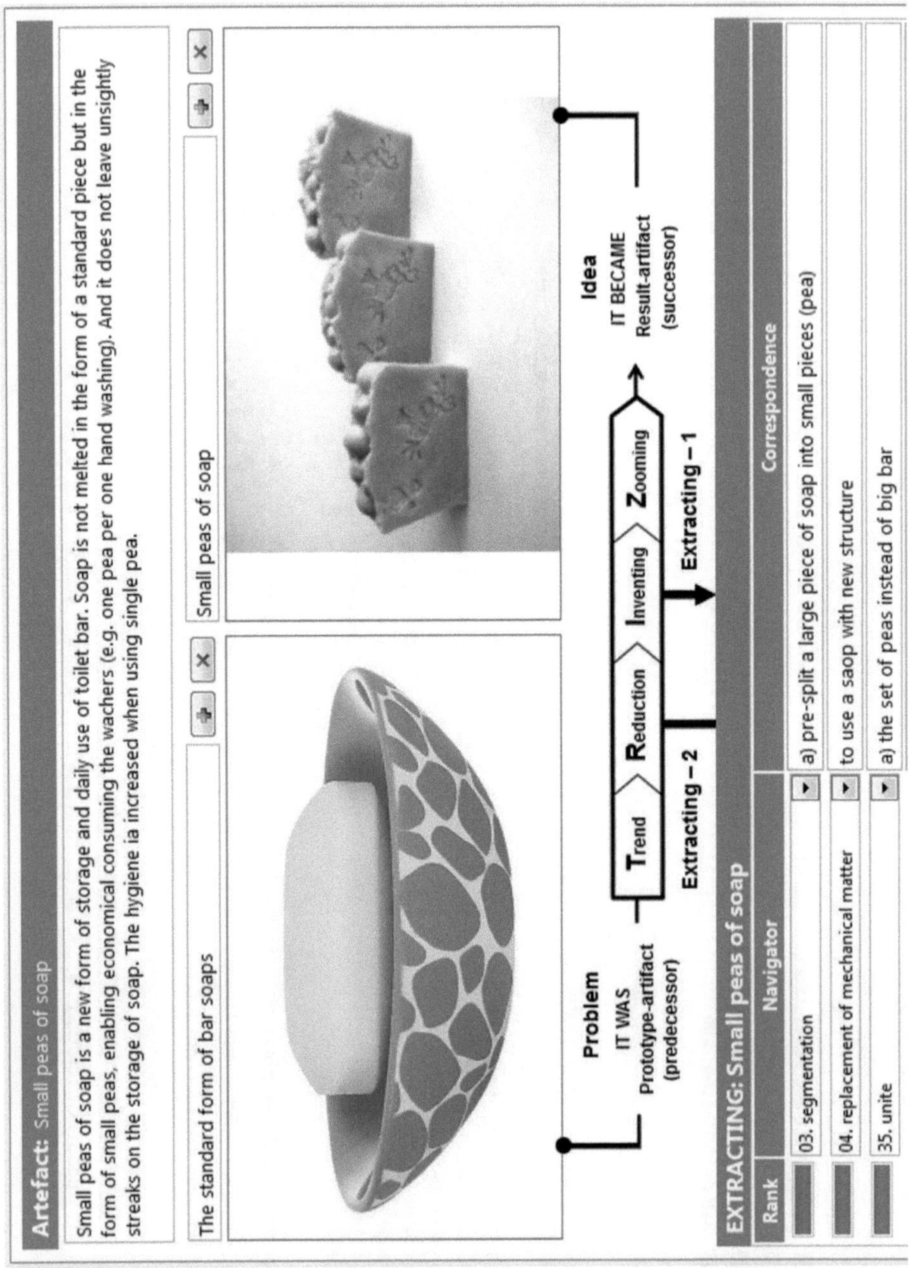

This example was developed by Maria Utukina, a teacher of mathematics, from high school 2091, Moscow, Russia, at joint pilot project of AIMTRIZ and National Research Nuclear University "MEPhI" (Moscow Engineering Physics Institute).

fig. 10.47. Extracting-1 from certification work-1 by Maria Utukina.
Screenshot from software EASyTRIZ Junior

Example 10.27. ***Liquid soap.***

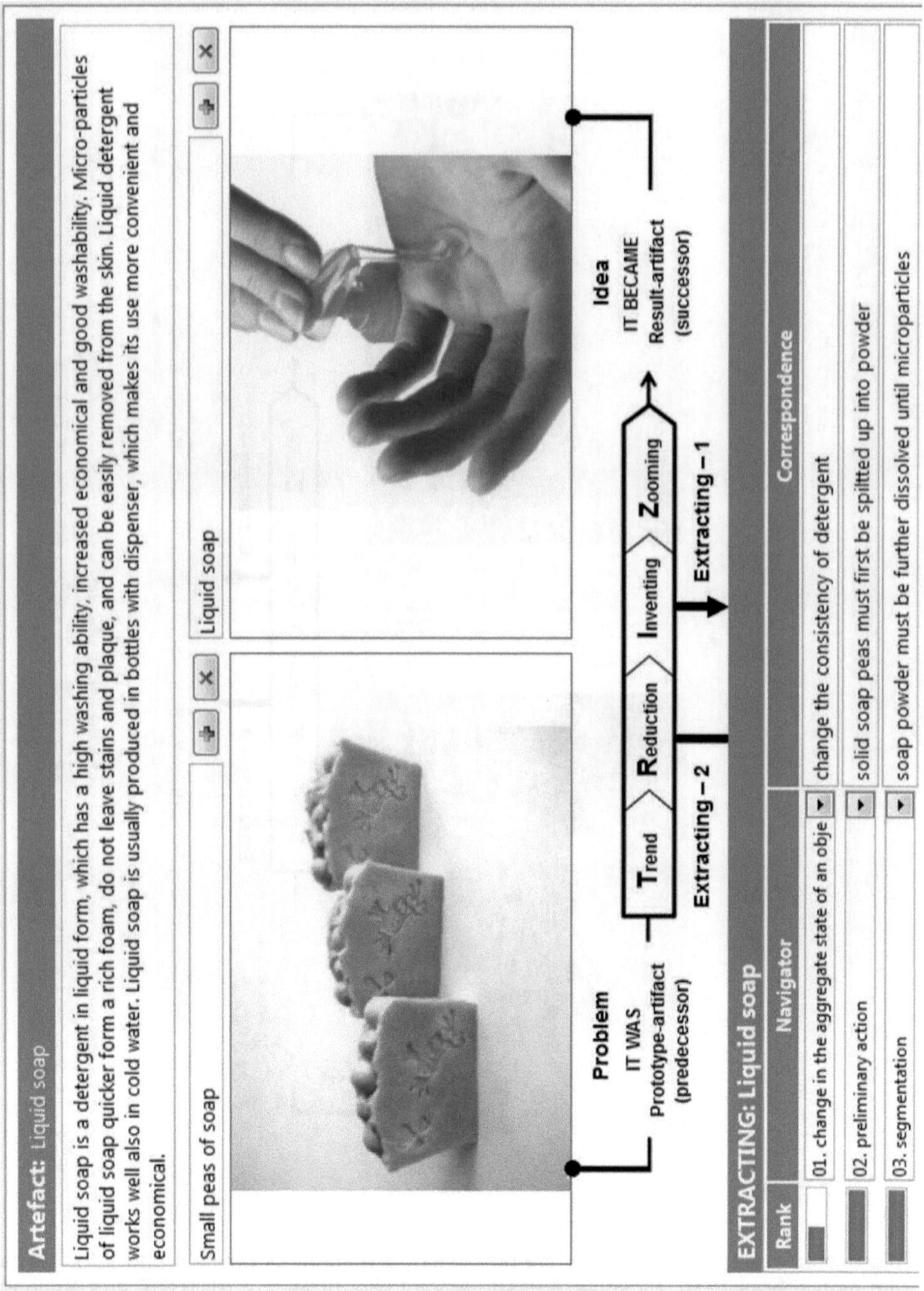

fig. 10.48. Extracting-1 from certification work-2 by Maria Utukina.
Screenshot from software EASyTRIZ Junior

Example 10.28. ***Snow-melting machine.***

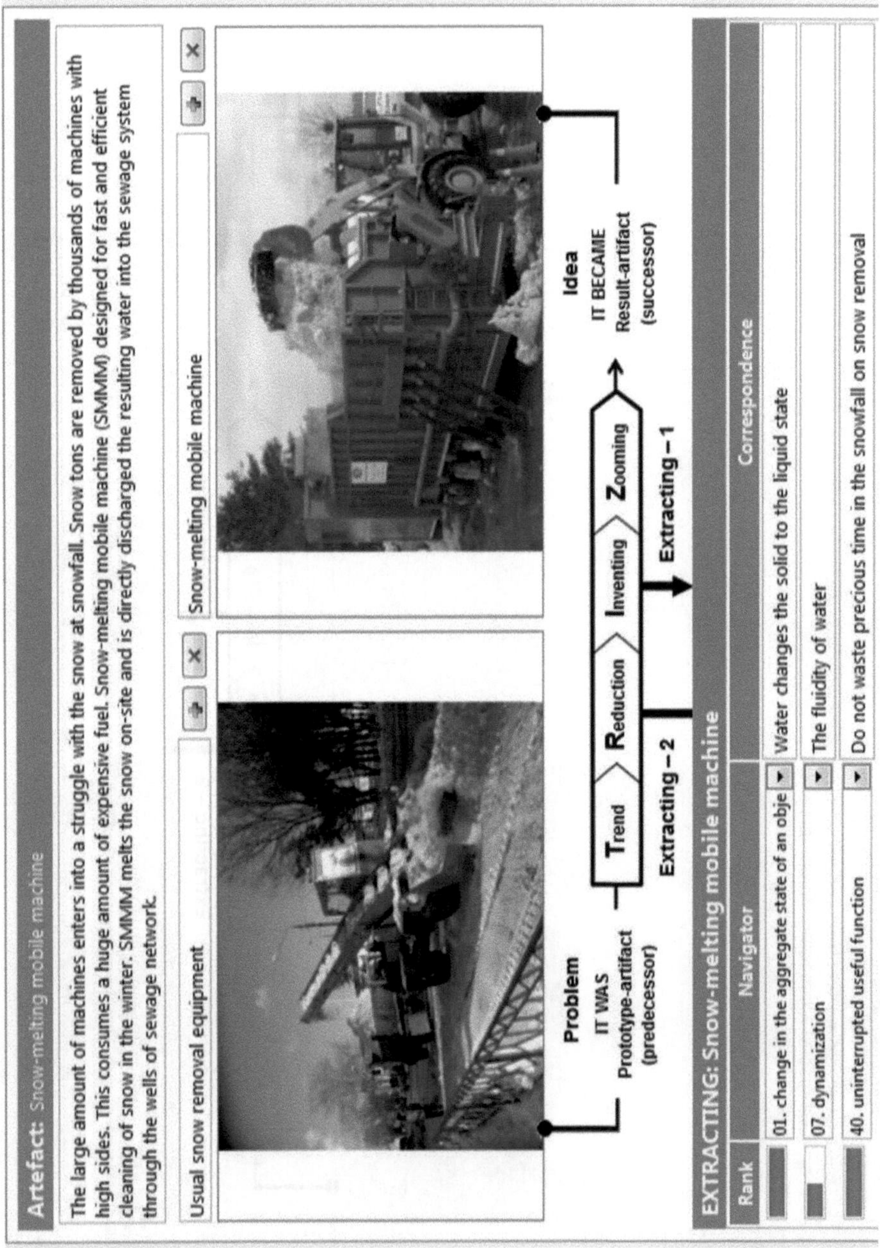

This example was developed by Sergey Holin, a teacher of physics and informatics, from high school 1310, Moscow, Russia, at joint pilot project of AIMTRIZ and National Research Nuclear University "MEPhI" (Moscow Engineering Physics Institute).

fig. 10.49. Extracting from certification work by Sergey Holin.
Screenshot from software EASyTRIZ Junior

10.2.2 Professional Specialization

Professional knowledge is the backbone of innovation and invention. Advanced professional application of TRIZ basics requires certain experience and extended practice, primarily in extracting and reinventing modeling.

Let us consider yet another example of joint work by the Inventor and the Partner.

Example 10.29. It All Boils Down to a Small Pin.

Cycle 1. Pin As... a System.

Chain drives and planetary drives in bicycles have been evolving for a long time, and that evolution is apparently not going to end in the foreseeable future. In the opinion of the author—who does this as a hobby—the possibilities of improving gearing systems and shifting mechanisms are far from exhausted. Incidentally, the author invented a radically different solution as far back as 1996, but for reasons beyond his control it is impossible to publish that solution at this time.

Once the Partner (engineer/developer at a company producing the bicycle components) noted the following problem: When shifting with the help of a cord activated by turning a handle on the handle bar, he sometimes got the wrong gear. The problem was associated with shifting planetary drive gears.

Having analyzed the problem, the Partner established that the error was caused by wrong positioning of a single pin in the shifting mechanism. Adding/removing lubricants and modifying the pin parameters failed to produce the desired effect. And then the Partner went to the Inventor.

When solving the problem, the Inventor applied TRIZ models and a well-known instrument called MAI T-R-I-Z. According to MAI, the solution-finding procedure consists of the following four steps:

Step 1: Trend (Diagnostics). What do we want to get in the end?!

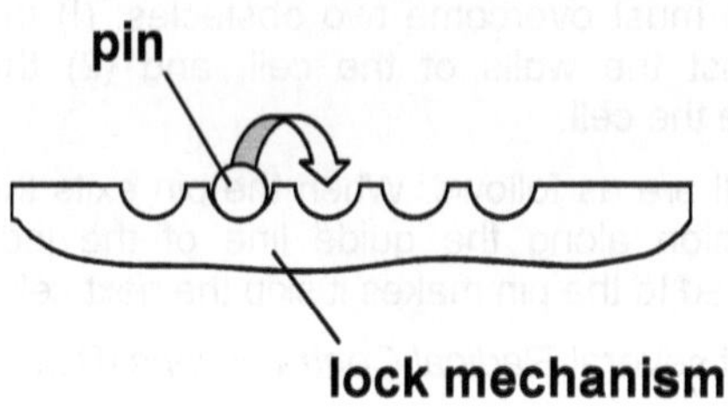

fig. 10.50. Pin lock assembly

A simplified drawing of the shifting assembly is shown in fig. 10.50. When the bicycle rider is changing the gear, the cord is pulling the pin toward the desired new position in the next cell (see the arrow).

Despite the fact that the effort being applied is small (normal), the pin sometimes skips one or two cells, which is unacceptable. If the effort is decreased, it becomes difficult to dislodge the pin from its current cell.

All attempts to change the dimensions of the pin or of the lock cells failed to produce significant improvements.

The pin has a diameter of approximately 3 mm, and the pin is connected to a rod (not shown) that not only moves along the lock mechanism but also rotates, which makes it possible for the pin to exit one cell and enter another cell. The rod is spring-loaded to ensure that the pin is pressed to the lock mechanism at all times. This rod that is put in motion by the cord.

The purpose is to facilitate that motion so that it does not require substantial effort and to improve the shifting so that no "skipping" occurs.

Therefore, the general trend is to improve the "manageability" of the shifting process and reduce the effort applied (energy costs).

Step 2: Reducing (Reforming). What is the cause of the problem?

To better understand the cause of the problem, let us present the initial problem situation through several models featuring different levels and different local conditions.

Thus, at the first level, we must present the problem in the simplest possible way in the form of a contradiction reflecting the conflict of requirements that apply to the general properties of the object under study. Such simplified models are used to provide the initial description of virtually all problems and, therefore, *de facto* constitute the standard for initial presentation of problem situations.

So, the initial *Standard Contradiction* ("direct" SC) may be presented as follows:

A small (normal) pin shifting effort is convenient for the user, but sometimes it leads to the pin skipping the required position.

When the shifting effort is reduced, there emerges a second, "inverse" SC-2:

Insufficient shifting effort eliminates the risk of the pin skipping the required position, but makes it more difficult for the pin to exit the current cell.

It is easy to see that a positive solution to this problem can be presented in the form of the following *ideal result*:

A small (normal) pin shifting effort is convenient for the user because it ensures easy exit of the pin from the current cell ***AND*** *its reliable transition to the required (neighboring) cell.*

However, to "treat" the "disease," we clearly must establish its causes, i.e., find its roots.

So what is the reason for the existence of both model SC-1 and SC-2?

To exit the cell of the lock mechanism, the pin must overcome two obstacles: (!) the friction generated as the pin is sliding against the walls of the cell; and (2) the resistance of the spring that keeps the pin inside the cell.

The causes for the pin skipping the required cell are as follows: When the pin exits the cell, there remains relatively small sliding friction along the guide line of the lock mechanism, and higher rod propelling effort applied to the pin makes it skip the next cell.

Cause modeling can be presented in the form of several *Radical Contradictions* (RCs).

For example, RC-1 with respect to the shifting effort can be formulated as follows:

The pin shifting effort *must be small* to prevent the pin from skipping the next cell, and it *must be sufficiently large* to ensure that the pin moves between the lock mechanism cells.

RC-2 with respect to the position of the pin in the cell can be formulated as follows:

The pin *must be inside the cell* to lock the gear, and it *must not be inside the cell* to facilitate gear shifting.

RC-3 with respect to the negative role played by friction can be formulated as follows:

The pin *must touch cell walls* (to fix its position), and it *must not touch cell walls* (to exclude friction).

Incidentally, lubricants serve to resolve RC-3, but their "intermediary" action proves insufficient.

Step 3: Inventing (Transformation). What must we change to attain the "ideal result"?

Contradictions are the consequence of deficiency or absence of certain resources in the area where problem manifests through a conflict of requirements or properties.

In this case, the *energy resource* is the critical resources throughout. The amount of effort must be optimal subject to applicable ergonomic requirements. Accordingly, a certain set amount of effort apparently represents the "solution" for RC-1. That amount cannot change.

For the purpose of resolving RC-2, placing the pin outside of the cell would be "ideal" because in that case its repositioning would require minimum effort; however, it becomes unclear how to fix the position of the pin. This thinking trend is dominated by the *functional resource (realization of two actions: change and fixation of the position)* and the *spatial resource (position)*. And the actions used cannot be changed either! As for the position of the pin, it is not rigidly set and probably can be varied.

Solutions that eliminate RC-3 might focus on reducing sliding friction. The dominating resources are the *functional resource (change of the operating principle)*, the *structural resource (probable change of component composition)*, and the *spatial resource (shape)*.

Apparently, the pin must be transformed from a simple part into... a complex system?

Theoretically, yes!, because the general trend for the proposed change is the increased manageability of the pin-shifting process. Manageability is impossible without dynamization of critical resources (components) of the object. The "rigid" pin must become... a "dynamized" system to become manageable!

The "composite sketch" of the future solution can be presented as the sum total of properties derived from all contradiction models described above.

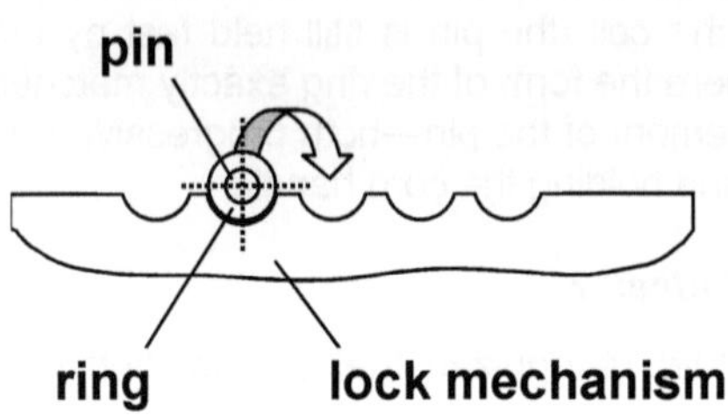

fig. 10.51. First transformation of the pin

The pin must:

1. consist of multiple components
2. be dynamized
3. reside not too deep inside the lock mechanism cell
4. easily move along the guide line of the lock mechanism
5. have reduced sliding friction or use rolling friction
6. permit precise discrete fixation when moving between cells

The first step in designing the future solution is shown in fig. 10.51. Here the pin is made in the form of an axis bearing a free-rolling ring. The influence of sliding friction is reduced due to the fact that the radius of the aperture where the pin touches the inside surface of the ring is smaller than the radius of the cell. The pin inside the ring rolls over the upper edge of the cell, and sliding friction is replaced with rolling friction.

The shifting of the "geometrical center" of the pin (intersection of dotted lines) from the cell to a place above the guide line of the lock mechanism also reduces the effort required for the pin to exit the cell.

This solution meets the first five requirements but not the sixth requirement! On the contrary, the new solution clearly deteriorates the situation with fixing the pin because now it has become even easier for it to skip or, to be more exact, roll from cell to cell.

However, we need the pin to stop in each cell by itself; or, if we look at it from a different perspective, we need each cell to try and stop the pin!

Incidentally, at the level of gear-shifting management it transpired that excessively "easy" repositioning of the cord is not necessarily a good thing! When he started his work, the Inventor was not familiar with that requirement. Now it turns out the bicycle rider must be able to "feel" the moment when the gear has shifted. Accordingly, the new solution had to be made… "worse" or more "crude."

Cycle 2. To Make Worse in Order to Make Better!

Let us skip the modeling stage and present the final solution.

To do that, we can use the spatial resource of the ring and the cells, i.e., change and align their shapes (fig. 10.52 [the elative sizes of the components are approximated without maintaining their actual proportions). The ring is made square!

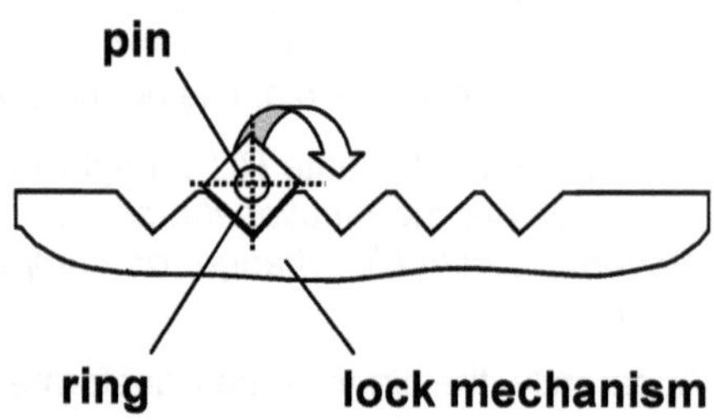

fig. 10.52. Second transformation of the pin

Now, remember: an object that is functionally a ring does not have to have circular shape!

When moved, the "square" ring easily turns and rolls over the crest dividing the cells. Then the ring keeps moving past the crest so that its next edge first lowers itself into the neighboring cell, then props against the wall of the cell (the pin is still held fast by the spring!), and finally comes to rest in a position where the form of the ring exactly matches the form of the cell. When that happens, the movement of the pin—both progressive and rotating (ring) —is arrested, which is felt by the hand holding the cord handle.

Step 4: Zooming (Verification). Is everything "ideal"?

At the level of the pin-lock assembly, the Partner recognized that the solution was efficient.

Naturally, this account gives only a simplified schema and shows just the general direction towards the solution.

Subsequently the Partner's experts streamlined all relevant parameters, improved and tested the design, and made the necessary adjustments.

The final technical solution was patented by the Partner, one of the two world's best known designers of bicycle shifting-management systems.

10.2.3 Nature invents!

What do you think? Do birds or fishes have intelligence? Or instinct only? OK, but if the answer is instinct only, then how could you explain the examples shown below?

Example 10.30. Breakfast of the Clever *Common Blackbird* – modeling the Example 7.15 from the textbook "ABC-TRIZ. Book I: How to learn to invent".

TREND One morning on a hot day the bird flew at the ground near the house to the table of homeowner and began to swirl around the table, as if to say something. And the homeowner knew what the bird wanted to say and what to ask. And he fulfilled the request of the bird. What kind of a fairy tale??

REDUCING **FIM:** The OZ by itself ensures the obtaining of the **IFR**:

[breakfast for the bird will ITSELF come to her at the ground near the house]

SC Ground near the house ► has a food for the birds in the soil
VS but the food is difficult to obtain ►

1) 18 Surface of the fixed object **VS** 10 Ease of use ► [16 and 24 are not appropriate];

2) 26 Quantity of material **VS** 10 Ease of use ► [01, 02, 14, 29 are poorly appropriate]

► the result of the Extracting (!): **02, 10, 12, 18**

RC Ground near the house ► gives the food to the bird
VS there is no food since dry weather

INVENTING **The first dominant models are taken from the sections Time and Structure of Afs-catalogue:** *10 Copying: a) use a simplified and inexpensive copy instead of an inaccessible ... object; 18 Media*tor: a) use another object to transfer or transmit an action; b) temporarily connect an object with another (easily separable) object **– a person helps the bird and copies action of "omnipotent" sky, giving the bird food with the rain!** It is clear that the models *02 Preliminary action* and *12 Local property* are present here: the bird has to ask a person for a help, and the rain will be just on this ground.

ZOOMING Have the contradictions been removed? – Yes. – ~~No~~.

Super-effects: we can now ask ourselves what does this bird understand? And to pay attention to the fact that the deep causal and situational relationships are present in behavior of this bird. Such understanding is not transferred genetically.

BRIEF DESCRIPTION The bird "used" the dominant models *10 Copying* and *18 Mediator* to get their own food through the mediation of the homeowner, who "at the request" of the bird turns on the water to irrigate ground near the house. After that a bird will get the rain-worms from the soil for her breakfast!

fig. 10.53. Extracting and Reinventing for *Breakfast of the Clever Common Blackbird*

Example 10.31. Humpback Whale Invents the "Air Fishing Net" – modeling Example 7.16 from the textbook "ABC-TRIZ. Book I: How to learn to invent".

ARTIFACT **Humpback Whale Invents the "Air Fishing Net"**

Description A humpback whale swims below and around the school of fish. He produces especially small air bubbles that have long do not dissolve in water. The bubbles rise up as loud and white wall! School of fish finds itself in the "air fishing net".

Then the companions of the first whale-beater fly up with open mouths through the school of fish to the ocean surface! Whales feast! Then another whale "works" as a beater and his companions are feasting.

Illustrations: basic pictures are from the popular science film at TV channel N24, Germany

EXTRACTING

A humpback whale and his companions do not want longer to hunt for one fish! What could they do?!

☺

WAS: Artifact-Prototype — PROBLEM → Trend > Reducing > Inventing > Zooming → IDEA — IS: Artifact-Result

EXTRACTING-1

LC	No.	Navigator	Function
++	01	Change in aggr. state	producing especially small air bubbles living long in water
++	02	Preliminary action	The whales had *to arrange an order of co-operation* in advance!
++	09	Change in color	The bubbles rise up as white wall.
+	12	Local property	Whale swam *below and aside* from the school of fish and swims around this school!
++	19	...another dimension	Whale builds *the cylindrical air fishing net* (3D-construction)!
++	34	Matryoshka	School of fish finds itself inside the "air fishing net".
++	35	Unite	The whales have to join together in one hunting team!

REINVENTING

TREND

Usual whale hunting for fish implies that one whale attacks a flock of fish. You can also expect that some whales from different families will compete for prey.

How can the whales arrange hunting for school of fish in order that several whales could not compete for food, but could cooperate in the hunt?

REDUCING **Maxi-FIM:** The OZ by itself ensures the obtaining of the

IFR: [co-operative whales' hunting for fish school]

STANDARD CONTRADICTION (SC) – Extracting-2S

Whales at hunting

(+) *all need to be fed*

(–) *hunting of single whale or few whales, but in competition, is not effective*

From Extracting-1

01 Change in aggregate state
02 Preliminary action
09 Change in color
12 Local property
19 Transition into another dimension
34 Matryoshka
35 Unite

RADICAL CONTRADICTION (RC) – Extracting-2R

Whales at hunting ⇒ *should be in the company that all be fed* **&** *should not be in the company because of competition*

INVENTING

To attain Objective 1 as arranging in advance the order for co-operation, the following models are used: *02 Preliminary action* and *35 Unite.*

To attain Objective 2 as building the 3D "fishing net", the following models are used: *02 Preliminary action, 12 Local Property, 19 Transition into another dimension* and *34 Matryoshka.*

To attain Objective 3 as building the "air wall" of "fishing net", the following models are used: *01 Change in aggregate state* and *09 Change in color.*

Fundamental models: *Temporal* – alternation of stages to bring to bay and to capture fish; *Structural* – arranging the collective hunting; Space – localization and encircle the fish with 3D air net.

Scientific Effects: Change in water density with air bubbles of special small size.

ZOOMING

Contradictions Removed: Yes.

Super-Effects: As a skittish hint: *fantastic enigma for mankind!*

Development Trend: The structure with three (!!!) whales in co-operation as beaters is known!

Beauty of Solution: 100

Highest grade. Substantiation:
1) Such structure was not known before, there are no direct counterparts – only in men hunting, but…;
2) Constructing the 3D air net is fantastic idea of itself required great inventiveness and understanding!

fig. 10.54. Extracting and reinventing for *Humpback Whale Invents the "Air Fishing Net"*

Example 10.32. Humpback Whale Invents "Fishing in Diagonal"

Humpback whales have invented yet another incredible way to hunt. Several whales swim in a diagonal (fig. 10.55, a) line. What for? It turns out that the whales with an open mouth are driving a school of fish (fig. 10.55, b)!

At first they line up for hunting, then the "line" moves forward. The fish become scared and move "downward" (fig. 10.56, a). Therefore, each next whale, including the last, receives food directly to the mouth (fig. 10.56, b)! Here is a form of 2D-hunting!

fig. 10.55. Diagonal 2D-hunting of the humpback whales:
a) At first several humpback whales line up and b) then they move forward with open mouths
(Basic pictures are from the popular science film at TV channel N24, Germany)

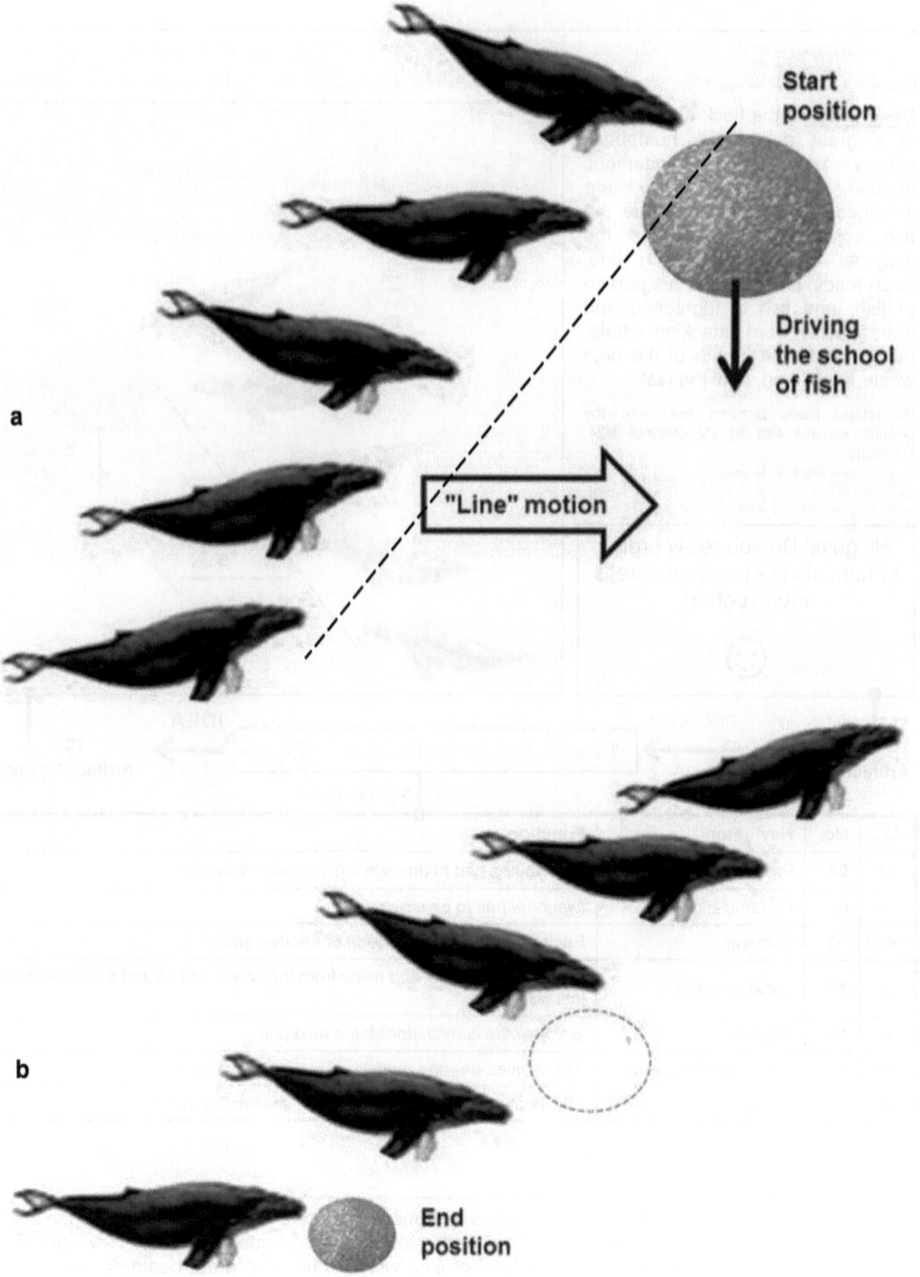

fig. 10.56. Diagonal scheme of humpback whales' 2D-hunting:
a) Motion direction of the whales and school of fish;
b) one of the intermediate (dash line) and end positions of the fish school

ARTIFACT Humpback Whale Invents the "Diagonal Fishing"

Description If the flock of fish is not at a great depth, the humpback whales use another ingenious method of hunting. They are arranged in diagonal and drive the fish from the front end of the diagonal along the moving line. Each leading whale eats his portion of fish, and fish is frightened and swims away from attacking whale, but gets into the mouth of the next whale, and so on, until the last.

Illustrations: basic pictures are from the popular science film at TV channel N24, Germany

EXTRACTING

Hi, guys! Do you really prefer to hunt alone? Lunch in core is much better!

☺

WAS: Artifact-Prototype → PROBLEM → Trend > Reducing > Inventing > Zooming → IDEA → IS: Artifact-Result

EXTRACTING-1

LC	No.	Navigator	Function
++	02	Preliminary action	The whales had *to arrange a diagonal* in advance!
++	07	Dynamization	*Diagonal has to be movable!*
++	10	Copying	Each whale *copies the action* of another one.
+	12	Local property	Whale swam *below and aside* from the school of fish and swims around this school!
++	18	Mediator	*Each whale is mediator for the next one.*
++	19	...another dimension	The whales *arrange a diagonal line* (2D-construction)!
++	35	Unite	The whales have to join together in one hunting team!

REINVENTING

TREND

Usual whale hunting for fish implies that one whale attacks a flock of fish. You can also expect that some whales from different families will compete for prey. How can the whales arrange hunting for school of fish, which is not at a great depth, in order that several whales could not compete for food, but could cooperate in the hunt?

REDUCING **Maxi-FIM:** The OZ by itself ensures the obtaining of the

IFR: [co-operative whales' hunting for fish school at not a great depth]

STANDARD CONTRADICTION (SC) – Extracting-2S

Whales at hunting

(+) *sometimes the flock of fish is not at a great depth*

(−) *hunting of single whale or few whales, but in competition, is not effective*

From Extracting-1

02 Preliminary action
07 Dynamization
10 Copying
12 Local property
18 Mediator
19 Transition into another dimension
35 Unite

RADICAL CONTRADICTION (RC) – Extracting-2R

Whales at hunting ⇒ *should be in the company that all be fed* & *should not be in the company because of competition which is especially ineffective if the fish is not at a great depth*

INVENTING

To attain Objective 1 as arranging in advance the order for co-operation, the following models are used: *02 Preliminary action* and *35 Unite*.

To attain Objective 2 as building the 2D "fishing diagonal", the following models are used: *12 Local Property* and *19 Transition into another dimension.*

To attain Objective 3 as building the "fishing process", the following models are used: *07 Dynamization, 10 Copying* and *18 Mediator.*

Fundamental models: *Temporal* – alternation of stages to arrange the diagonal and to capture fish; *Structural* – arranging the collective hunting; *Space* – localization and driving in the fish with 2D "fishing diagonal".

Scientific Effects: Maybe, effective deviation angle of diagonal and proper velocity of line motion ☺

ZOOMING

Contradictions Removed: Yes.

Super-Effects: As a skittish hint: *fantastic enigma for mankind!*

Development Trend: What will they invent else? Who knows!

Beauty of Solution: 100

Highest grade. Substantiation:
1) Such structure was not known before, there are no direct counterparts – only in men hunting, but…;
2) Constructing the 2D movable fishing line is fantastic idea of itself required great inventiveness and understanding!

fig. 10.57. Extracting and reinventing for *Humpback Whale Invents the "Fishing in Diagonal"*

10.3 Advanced Reinventing Cases

10.3.1 Originated from Siemens

> Only when it is closely associated with solutions to problems of the day does innovative research become useful and assured of success.[94]
>
> *Werner von Siemens*

These words were spoken by an outstanding inventor, a stalwart manager, and a wonderful human being who gave his name to SIEMENS, one of the world's largest concerns. Together with his brothers, he launched it on a glorious path of achievement more than 150 years long.

I will tell you only about three of the concern's numerous inventions. They span two centuries and lie a hundred years apart. They are neither momentous nor exceedingly complex. But they embody trailblazing ideas worthy of industry leaders!

There were numerous attempts to contest their validity. These attempts, however, do not diminish their significance, even if similar inventions were concurrently made in America, Russia, or England. Moreover, all disputes regarding their validity (and they were very tense) only add weight to the words uttered by their author that I used as the epigraph: There is no doubt that all inventions and innovations by Werner von Siemens contributed to solving burning issues that affected all humanity.

By the same token, there is no doubt that research performed by innovative SIEMENS scientists today are equally and eminently relevant.

Example 10.33. Invention of Tramway and Trolleybus

By the second half of the 19th century, railroad trains (fig. 10.58, a), drawn by powerful steam engines, had already earned recognition and respect from travelers and businessmen in all industrial countries.

Railroads came into cities as well. They were used by carriages and wagons with iron wheels with horses rather than steam engines being the source of traction power (fig. 10.58, b).

The first horse-drawn tram appeared in Berlin in 1865. It ran from the Brandenburg Gate (exit from Berlin for westbound travelers) to the Charlottenburg suburb along a road, which today has been transformed into a wide highway (today *Strasse des 17. Juni*).

The thing is, city streets are no place for a steam engine with its smoke-spewing stack and hot steam jets rhythmically rushing out of engine cylinders. People peacefully walking by would be very unhappy!

---

[94] *Ernst Werner von Siemens* (13 December 1816, Lenthe, near Hannover, on 6 December 1892. Charlottenburg, now part of Berlin. Outstanding German engineer, inventor, researcher, industrialist, cofounder the company Siemens, public and political figure (см. www.wikipedia.com). The citation in the epigraph was replicated from the article "160 years of Siemens," October 2007 (www.siemens.com/history/pool/en/history/1847-1865_beginnings_and_initial_expansion/160j_e.pdf)

fig. 10.58. Train (a) and horse-drawn tram (b) in second half of the XIX century.
Illustration: http://prod-pub.e4.ratry.ru/blog/varlamov_i/802639-echo.phtml; en.wikipedia.org/wiki/Berlin_tram

And so the world's first solution of this problem (fig. 10.59) was shown by Werner von Siemens and his Siemens and Halske partners in 1879 at the Berlin Industrial Exhibition. The world's first tramway made its maiden voyage on May 16, 1881, to the Lichterfelde suburb.

fig. 10.59. The world's first electrical "train." (a) и Tramway (b)

Illustration: http://w3.siemens.ru/about_us/history/historic_photos/; en.wikipedia.org/wiki/Gross-Lichterfelde_Tramway; de.wikipedia.org/wiki/Elektromote

The first horse tram and the tramway shared a dangerous drawback: They both used contact rails with live direct current. An unwary pedestrian could be electrocuted. In addition, some rash youngsters started to have fun short-circuiting the two rails with a piece of wire to "make sparkles."

Then it was decided to lift the source of power using wires. In 1882, a 500-meter track for the world's first trolleybus (fig. 10.60)—or as its author called it, *"Elektromote"*—was laid toward the Halensee suburb.

Power was fed into the electric engine from wires strung between poles. A four-wheeled cart rolled along each wire. Two joined carts made up a contact car. In German, that contact car was called *"Kontaktwagen",* while its English name—*"trolley"*—gave rise to the appellation that we now commonly apply to this type of vehicle, *"trolleybus."*

Competing ideas emerged on both sides of the Atlantic almost immediately.

fig. 10.60. The world's first trolleybus
Illustration: http://W3.Siemens.Ru/About_Us/History/Historic_Photos/; en.wikipedia.org/wiki/Gross-Lichterfelde_Tramway; de.wikipedia.org/wiki/Elektromote)

And now let us consider the reinventing of the solutions (fig. 10.61 and 10.62) that eventually assured global expansion of trams, trolleybuses, and, later, electric trains.

Stunningly, some 10 years later a tram moving along a similar railroad track accelerated to a record-breaking speed of more than 200 km/h! Unfortunately, the development of electric transport was obstructed for a century because of the barriers erected by lobbyists who represented oil companies and manufacturers of vehicles running on internal combustion engines.

And now that you have reviewed the reinventings, I would like to remind you of a well-known psychological effect: All explicated solutions lose their mysteriousness and unpredictability and henceforth seem unsophisticated and mundane. A TRIZ expert, however, is well aware of the difference between a powerful solution whose beauty invariably serves as a source of astonishment, or even exalted admiration, and a truly simple solution of a simple problem explicated—always bearing in mind that BOTH solutions are useful.

The solutions proposed by Werner von Siemens in the two examples above are not very complex. But they both were the first in the world! They both solved problems that had not even existed before! And they both served to assure the attainment of engineering and commercial purposes that could have been foreseen only by Werner von Siemens, who clearly discerned their relevance and prospects, and not by any other man! This is what makes it all wonderful: the purpose, the idea, the constructive realization, the method and place of demonstration, the organization of production and global commercial distribution!

REINVENTING 1: TRANSITION TO THE SYSTEM OF STRETCHED WIRES WITH MOVABLE CURRENT COLLECTOR (ELECTROTROLLEY)

TREND Power transmission to railway locomotives on the rails associated with a number of disadvantages: 1) the need for insulation of the rails from the ground; 2) the need for measures against short circuit between the rails; 3) the risk of electric shock when people touching the rails which are at high operating voltage. What can be offered?

REDUCING **Maxi-FIM:** The X-resource, without making the system excessively complex and without causing inadmissible negative effects, ensures, together with other available resources, the obtaining of the IFR: [*safe power supply to electric locomotive*]

STANDARD CONTRADICTION (SC) – Extracting-2S

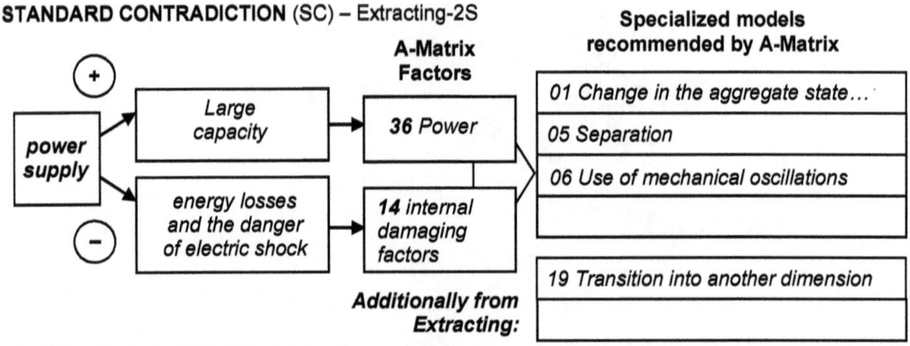

RADICAL CONTRADICTION (RC) – Extracting-2R

power supply	⇨	*must be continuous and safe all over the road*	VS	*should not be a continuous and safe as possible short circuits and electric shock when the unwary pedestrians touching of conductive parts (e.g. rails)*

INVENTING In order to radically reduce the risk of electric shock, the power supply system is designed as a taut contact wires, raised high on poles, and electric power transmission to locomotive vehicle is realized through moving current collector (trolley), rolling on the mounted wires and connecting with engine through flexible insulated wires – the dominant principles correspond to *05 Separation: select only the desired property*; and *19 Transition into another dimension: a) the transition from a surface to a three-dimensional space; b) do construction on several floors.*

ZOOMING **Contradictions removed:** Yes.

Super-effects: The possibility of low-cost arrangement of urban, regional and continental – passenger and freight – electrified roads.

Negative effects: Requires the construction of track structure (for rail), because old vehicles and apparatus roadway unacceptable.

Development trend: The possibility of organizing the network of roads, at least for "trolley" transport!

Changes in ambient systems: The construction of transport terminals and portals.

Extended uses: Vehicles in mines and buildings, mountains and open pits, park amusements and rides in a rough terrain with a difficult relief.

Beauty of solution: **100** The highest score, because the idea has great backbone, functional and economic potential.

fig. 10.61. Reinventing the idea of electric wires stretched on the poles and used for movement on these wires the contact current collector of *"Elektromote."*

REINVENTING 2: TRANSITION TO BOW COLLECTOR

TREND Outstanding American inventor Frank Julian Sprague (1857-1934) patented in 1880, the spring-loaded trolley rod, known today throughout, with a roller pressed to the contact wire upwards. Problem: The roller is sometimes jumps from the wire that leads to the loss of contact.

REDUCING **Macro-FIM:** The X-resource, without making the system excessively complex and without causing inadmissible negative effects, ensures, together with other available resources, the obtaining of the **IFR**: [*exclusion of loss of contact with the wire*]

STANDARD CONTRADICTION (SC) – Extracting-2S

Trolley rod ► to be pressed against the wire **VS** loss of contact = 30 Force **VS** 04 Reliability = 01, 11, 12, 33

RADICAL CONTRADICTION (RC) – Extracting-2R

Contact element ► must be strongly pressed against the wire for good contact **VS** should not be strongly pressed to the wire because of the increased danger of roll jumping off from wire and loss of contact

Sprague' contact

Sliding contact
(as a proto-prototype)

INVENTING **Dominant models to resolve SC:** in order to radically resolve the problem of roll jumping off from wire and loss of contact, the principle *11 Inverse action* is applied: *a) instead of an action prescribed by the conditions of an assignment, complete a reverse action*: instead of the roller with a groove along the entire circumference for trapping wire, it's offered a crossbeam in the form of half-arches, mounted on a frame in the form of a spring-loaded bow; here the wire slides and is in constant contact with the crossbeam along its entire breadth in any turn and tilt of the locomotive.

ZOOMING

Super-effects:	The growth speed. The lack of direction switch for the prototype.	
Beauty of solution:	100	Surprisingly simple and effective principle and the key design, being in use today at the majority of current collectors in the world, including pantograph!

With this solution, Siemens also overcame the limitations that patent of Sprague imposes on the commercial application of rod current collectors.

fig. 10.62. Reinventing the idea of bow (hoop) current collector.

Example 10.34. Veterans' Experience—to Young!

And now I will tell you a story that casts Modern TRIZ in a somewhat unusual light, presenting it as a way to accumulate creative knowledge residing in experienced veteran workers and pass it on to young workers.

Let us recall that each reinventing is basically structured knowledge of the methods and procedures used to create an efficient idea and solve a complex problem. In addition, each reinventing performed on the basis of MAI T-R-I-Z contains a standard description of those procedures that is distinct, clear, and easy to understand for anyone who wishes to study that reinventing. And this means that reinventings become a common model language and a common tool for all think-tank team members including people excelling in various areas of expertise.

This guarantees accurate and psychologically comfortable mutual understanding within the team, streamlined formulation and systemically managed solution of problems, and standardized record-keeping, thus assuring the efficient exchange of outcomes among experts and units within the company.

This is particularly important for international companies that have partners and subsidiaries in different countries. In addition, standardized modeling helps to overcome language and education differences to create a uniform environment for designing and modeling thereby accelerating the development of solutions and enhancing their efficiency.

The standardization underpinning Modern TRIZ is necessary to organize large-scale instruction in TRIZ basics at any educational establishment. Evidently, it is possible to exchange examples between schools of different types and, even more importantly, between schools and the industry.

One evening in 1995, when I came to Berlin from Essen where the German-Belarus design company that I started together with my German partner had its head office, I opened a Berlin newspaper and read a short article about a new SIEMENS gas turbine with record-breaking performance characteristics.

The performance boost had become possible on the application of a new idea: The engineers had found a special arrangement for gas burners and increased their number. Naturally, the solution itself was not disclosed, but there was a photograph of the turbine stripped of its outer casing. Meanwhile, the secret was in the new construction of that casing. The article just mentioned that hot gas generated by the burners exerted more pressure on turbine blades, which boosted the productivity and reliability of the turbine to record levels!

I opened a machine engineering encyclopedia and read the article on gas turbine operating principles and designs. I also learned about issues related to the durability of turbine blades and improvement of their maintainability and overall productivity.

I was so excited with the problem that I continued the modeling exercise and eventually formulated the contradictions subject to hypothetical requirements and limitations (which were not known to me at the moment). I also defined the ideal final result. After that I quite logically arrived at an apparently reasonable solution.

I drew several simple sketches, and the next day called the Berlin-based SIEMENS branch that had developed the new turbine. I wanted to tell the authors how TRIZ makes it possible to model competitor solutions without talking to the authors—just by studying patents, articles, and generally any available materials—and how those solutions can be improved.

Soon a meeting was set up with the members of the design team, and we conducted our first improvised workshop in a negotiating room. I will tell you about what happened after you have reviewed this severely abridged reinventing (fig. 10.63).

At one of the subsequent meetings, I spoke with one of the turbine designers with other participants intently following the conversation:

– Tell me, professor, are you an expert in turbines?

– Oh no! I specialize in computer science, systems engineering, optics, mechanics, electronics, etc., and I am also a TRIZ proponent and an advisor on the basics of that theory.

– Then you must have advised turbine producers in the past.

– No, I never did that. What is this all about?

– You see, nobody here believes that you are not a turbine expert.

– But I am not a turbine expert. I told you that!

ARTIFACT **REINVENTING: SIEMENS NEW GAS TURBINE**

TREND Gas turbine blades are suffered from impacts of thermal and mechanical loads. With increasing temperature, the efficiency of the turbine increases, but blades wear out faster. Failure of multiple burners requires repair and stop the turbine sometimes. The destruction of several of the blades also results in stopping the turbine, which is highly undesirable. **What could be done?**

REDUCING **Maxi-FIM:** The X-resource, without making the system excessively complex and without causing inadmissible negative effects, ensures, together with other available resources, the obtaining of the IFR:

[*reducing the load on the turbine blades*]

SC: Burners ► 01 Productivity **VS** 04 Reliability = 01, 02, 03, 30

RC: Blade ►***should be*** under high temperature and pressure for the growth of productivity **VS** ***should not be*** under high temperature and pressure to serve longer.

Prototype-turbine:
2 chambers by 8 burners

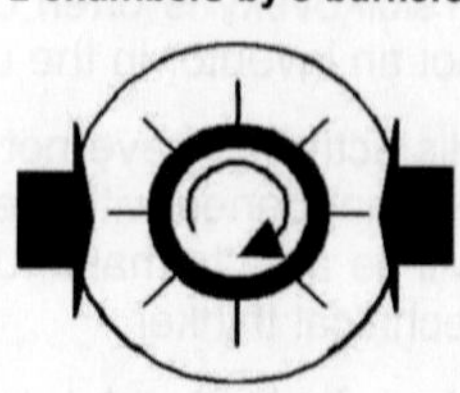

INVENTING **SC resolution:** to significantly reduce the problem, the principle ***03 Segmentation*** is applied: *a) to divide into independent parts:* the concentrated sources of hot gas (burner) are divided into several groups (chambers) and these new chambers are installed uniformly around the circumference of the turbine housing.

RC resolution: there is "empty" space in the housing of the prototype, where you can install additional burners (*spatial resource is used*); by this, the blades will not have time to cool down before the next impact (*time resource is used*), and thus will not get a big temperature drop (thermal impact wears off!); the same about the mechanical impact because of smaller but uniform impacts are enough for the desired rotating speed!

Result-turbine:
8 chambers by 3 burners

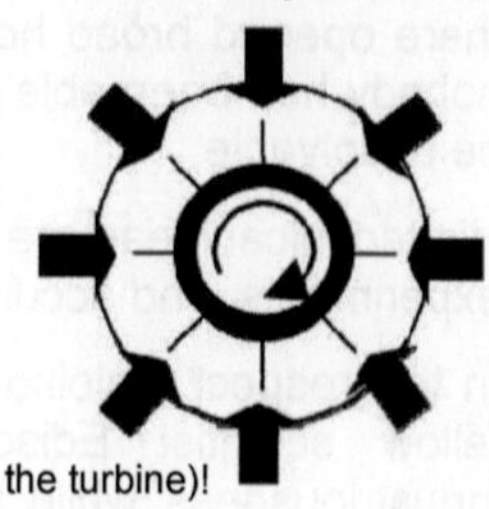

ZOOMING **Effect:** record growth of productivity!

Super-effect: Increase of durability and maintainability (and without stopping the turbine)!

fig. 10.63. Reinventing the innovative gas turbine

– The thing is, after your presentation, we were, well, shocked, and everybody said more or less the same thing about you: It is like he have been inside our heads for several months while we were looking for the solution because you described all of the key contradictions that we identified as well as all of the key arguments to support the new solution that we came up with after tedious discussion and endless research. It is just that we structured it all in a different way.

– Thank you; I see now.

– We simply could not believe that somebody who is not an expert in this field could model our entire long-drawn-out work during the course of one evening and then tell us not just about our results, but also—and this is the most important thing!—about how we arrived at those results. And that is very close to what we had lived with for several months before we found the solution.

I think it is quite clear from this story what "veteran rescue" I am talking about: Modern TRIZ modeling can be used to restore and record creative experience, share it with other company experts, make it available to university students majoring in relevant disciplines, convert it into an efficient warehouse of empirical knowledge that can be used to train young specialists, and competently solve new problems!

10.3.2 *Per Aspera Ad Astra!*

> The main motif of my life is this: I must do something useful for the people, make my life meaningful, and help humanity to move ahead, even if a little. That is why I went in for things that yielded neither bread nor power. But I hope that my works—be it very soon or in the more distant future—will give society mountains of bread and endless power.[95]
>
> *K. E. Tsiolkovsky*

I invite you to share some of the thoughts and feelings about Tsiolkovsky that were expressed, in profoundly heartfelt words, by a person whose deeds and destiny are also worthy of abiding memory and deepest respect.

"Tsiolkovsky is often called—for lack of a better word—an *inventor*. However, he was not an inventor in the usual meaning of this word.

His activities have not yet brought tangible results to become part of our daily lives as has happened with steam engines, telegraph, or electric lights. Only future generations will be able to make full use of his work. He is a creator of brave ideas, an outstanding technical thinker.

Moving along unbeaten paths, Tsiolkovsky always applied his inventiveness to large-scale problems. He was attracted to extreme heights of technical thought from whence there opened broad horizons in space and time. He fearlessly tackled problems that nobody had been able to solve before him, problems that sometimes were believed to be unsolvable.

His technical ideas are the fruit of systematical thinking, elaborate research, numerous experiments, and accurate mathematical calculations.

In this respect, Tsiolkovsky can serve as a better model for inventors than his luckier fellow scientist Edison. The American inventor also showed unprecedented industriousness while working on his inventions. His every success is, in his own words, ***1 percent inspiration and 99 percent perspiration***.[96]

However, Edison groped his way ahead, seeking to obtain results in a purely experimental way, whereas Tsiolkovsky used test results to develop a general theory that enabled him to foresee future outcomes.

Naturally, the original creative impetus was engendered by imagination." [97]

And in his book, Perelman goes on to cite from Tsiolkovsky himself:

"First, inevitably, there come thought and fantasy. They are followed by scientific calculation. And only in the end the thought is crowned with performance."

---

[95] K. E.Tsiolkovsky, *First Model of an All-Metal Aeronaut of Corrugated Iron*. – Kaluga, 1913 (from the archives of the State Public Scientific and Technical Library of Russia, Moscow); quoted also in J. I. Perelman, 1937, and in: S. P. Korolyov, *Life and Activities of Tsiolkovsky*. – Report delivered on September 17, 1947, at a meeting at the Central House of the Soviet Army conducted by the Academy of Artillery Sciences and dedicated to the 90th birth anniversary of Tsiolkovsky.

[96] Thomas Alva Edison (1847 to 1931) – outstanding American inventor and entrepreneur who received 1,093 patents in the USA and approximately 3,000 patents in other countries.

[97] Compiled by me (M.O.) from the following book: J. I. Perelman: *Tsiolkovsky. Life and Technical Ideas*. – Moscow – Leningrad: ONTI, 1937 – (Перельман Я.И. *Циолковский. Жизнь и технические идеи*. – Москва – Ленинград: ОНТИ, 1937)

Perelman continues: "Tsiolkovsky's inventing style, where each step, each conclusion is backed up by a solid theoretical and experimental foundation, can serve – I repeat – a model for all inventors: this is how you must work and invent!

The scientific research field would be barren as the Sahara desert, if scientists did not resort to imagination, if they did not know how to take their minds off the visible world to create intangible mental images. Science cannot make a single step without imagination; it keeps feeding on the fruits produced by fantasy, but it is science-based fantasy capable of drawing ideal images with utmost clarity."[98]

Tsiolkovsky's major contributions to the development of aeronautics and astronautics are summarized in books by Perelman, a talented promoter of his scientific pursuits, a promoter of knowledge in general, particularly school knowledge, who created incredibly engrossing and useful handbooks for schoolchildren (and for teachers!) under headings that always included the word "entertaining," i.e., *capable of entertaining a person's attention and imagination.*

Here are those contributions[99]:

Tsiolkovsky's Publications in Russia		**International Publications**	
		Dirigible	
1892	Publication *Metallic Dirigible Aerostatic Balloon*	**1895**	First Zeppelin project (Germany)
		Airplane	
1894	Article *Aeroplane, or Birdlike Flying Machine*	**1896**	Model airplane flight, Samuel Langley (USA)
		1903	First airplane flight, Wright brothers (USA)
		Rocket	
1896	Development of reactive device theory	**1919**	Extreme altitude rocket, Professor Goddard (USA)
1903	First published work on reactive locomotion theory and interplanetary rockets	**1923**	Book on interplanetary rocket, Professor Obert (Germany)

fig. 10.64. List of Tsiolkovsky's publications.

Modern TRIZ also teaches you to "resort to imagination," "take your minds off the visible world" so that you can "create intangible mental images," i.e., creative-thinking models inherent in artifacts, which *remain invisible without proper MTRIZ-training.*

In the following two sections, we will take a look at some of Tsiolkovsky's ideas and extend them into today to merge with modern technical ideas and… fantasies.

[98] J. I. Perelman, *Interplanetary Travel.* – Moscow – Leningrad, ONTI, 1935. – (Перельман Я.И. *Межпланетные путешествия.* – Москва – Ленинград, ОНТИ, 1935)

[99] Compiled by me (M.O.) from the following books: J. I. Perelman, *Tsiolkovsky. His Life, Inventions and Scientific Works.* – Moscow – Leningrad: ONTI (United Scientific and Technical Publishers), 1932; J. I. Perelman, *Tsiolkovsky. Life and Technical Ideas.* – Moscow – Leningrad: ONTI, 1937. – (Перельман Я.И. *Циолковский. Его жизнь, изобретения и научные труды.* – Москва – Ленинград: ОНТИ, 1932; Перельман Я.И. *Циолковский. Жизнь и технические идеи.* – Москва – Ленинград: ОНТИ, 1937)

AIR OCEAN

Example 10.35. AERONAUT (DIRIGIBLE)

Tsiolkovsky wrote in his autobiography: "When I was 15 or 16 years old, I ... became interested in aerostatic balloons... Since then, the idea of making a metal aerostatic balloon has stuck in my brain." [100] When he reached the age of 28, he revisited his childhood dream. and over the course of two years developed a metal dirigible.

ARTIFACT **Tsiolkovsky's Aeronaut (Dirigible)**

Description Back in 1890, K.E.Tsiolkovsky came up with an idea of a full-metal aeronaut (dirigible) with variable size and moveable shell to replace prototypes with soft inflatable shell. Soon other initiators began to build full-metal dirigibles, too. Tsiolkovsky proposed a dirigible with elastic corrugated (and hence more durable) metal "sides" compressed until they joined (pos. 5 in the figure), or "inflated" until they became spindle-shaped (pos. 4). He also suggested that the gas inside the dirigible be heated to create additional lifting force.

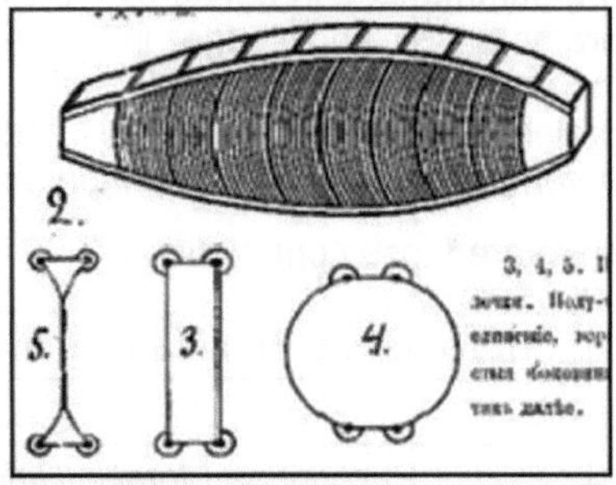

EXTRACTING

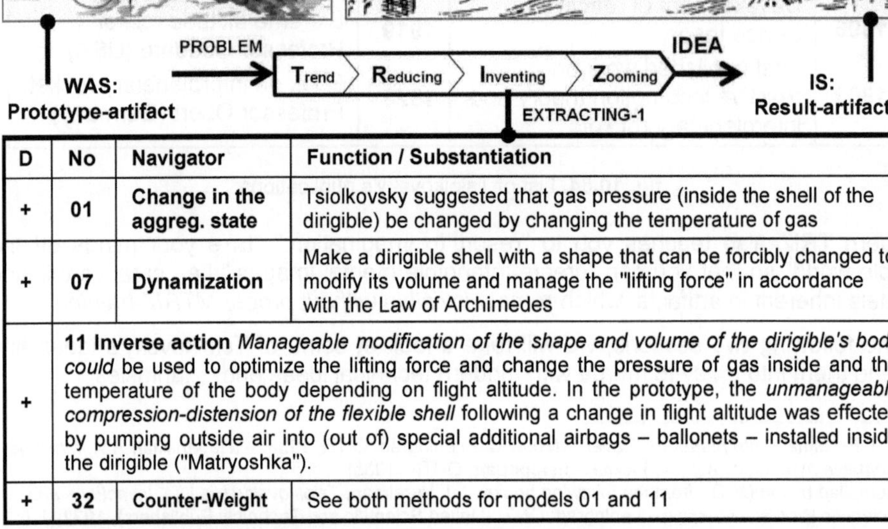

D	No	Navigator	Function / Substantiation
+	01	**Change in the aggreg. state**	Tsiolkovsky suggested that gas pressure (inside the shell of the dirigible) be changed by changing the temperature of gas
+	07	**Dynamization**	Make a dirigible shell with a shape that can be forcibly changed to modify its volume and manage the "lifting force" in accordance with the Law of Archimedes
+	**11 Inverse action** *Manageable modification of the shape and volume of the dirigible's body could* be used to optimize the lifting force and change the pressure of gas inside and the temperature of the body depending on flight altitude. In the prototype, the *unmanageable compression-distension of the flexible shell* following a change in flight altitude was effected by pumping outside air into (out of) special additional airbags – ballonets – installed inside the dirigible ("Matryoshka").		
+	32	**Counter-Weight**	See both methods for models 01 and 11

[100] J. I. Perelman, *Tsiolkovsky. His Life, Inventions and Scientific Works...* 1932

REINVENTING

TREND Known dirigibles with soft flexible shells had the following major drawbacks: 1) risk of gas venting or risk of fire (especially if hydrogen was used) in the event of violation of the shell's integrity; 2) changes in the lifting force due to changes in the shell's volume when the body heated up in the sun or, conversely, when it cooled down; 3) increased weight due to the use of reinforcing internal frames. What can be done to remove these critical deficiencies?

REDUCING **Maxi-FIM:** The operating zone by itself ensures the obtaining the following **IFR**: [*managing the dirigible's lifting force and high reliability*]

STANDARD CONTRADICTION (SC) – Extracting-2S

Dirigible body ► 04 Reliability **VS** 02 Universality, Adaptability = 01, 11, 18, 32

Extra-navigator from Extracting: *07 Dynamization*

RADICAL CONTRADICTION (RC) – Extracting-2R

Dirigible body ► must have constant volume, as determined by its dimensions and construction **VS** must have variable volume, as its shape changes as it is heated up in the sun or cools down with the change of flight altitude

INVENTING **Dominant models used to resolve the SC:** To attain Objective 1 (manageable modification of gas pressure in the dirigible's body), the body is made in the form of a full-metal moveable shell – dominant models *07 Dynamization* and *11 Inverse Action*. To attain Objective 2 (creation of additional lifting force), the gas inside the shell is heated – dominant models *01 Change in the Aggregate State* and *32 Counter-Weight*.

Dominant resources used to resolve the RC: *01 Space* – introduction of body shape and volume management; *03 Structure* – the body is made of elements assuring manageable modification of its shape; *04 Material (Energy)* – 1) the body is made of thin corrugated metal sheet; invention made by Professor Dr. F. Mirtsch (Berlin, Germany) – advantages of such structures are described in Example *13.8. Self-Organization: By the Law of Nature* in the book by M. Orloff *Modern TRIZ*. – Springer; 2) it was suggested that the gas inside the body could be heated to create additional lifting force.

Scientific effects: Managed modification of body volume and shape to control internal gas pressure; change of gas temperature to modify lifting energy.

ZOOMING

Super-effects: High durability and fire resistance of the full-metal structure (with use of corrugated metal sheets). Possibility to make huge dirigibles with large load-lifting capacity.

Negative effects: Relative complexity of construction in the second proposed option.

Development trend: Exciting prospects that largely remain unrealized – see examples below.

Changes in ambient systems: Radical improvement of many transportation operations within a wide range of cargo weights. Possibility to deliver cargoes and passengers to inaccessible regions due to the dirigible's vertical takeoff/landing feature.

Beauty of solution: **100** Highest grade. Substantiation:
1) pioneering solution, interesting, unexpected idea;
2) development of the "full-metal body" idea resulted in subsequent practical implementation of such constructions.

Illustrations: prototype artifact – Giffard dirigible, 1852, www.spiraxsarco.com/ru/steam-academy; ideal artifact – K. E. Tsiolkovsky, Aerial Transport. – 1916; J. I. Perelman, Tsiolkovsky. – Moscow: ONTI (United Scientific and Technical Publishers). 1937.

fig. 10.65. Reinventing of Tsiolkovsky's idea of a full-metal dirigible with variable body shape (and volume).

Now, 100 years later, Tsiolkovsky's ideas are researched, with redoubled energy, by dozens of companies all over the world. Some of those ideas (normally with indications as to their purpose and certain parameters) are listed below. Naturally, modern materials, construction, knowledge, modeling, and manufacturing opportunities are much more advanced than they were in Tsiolkovsky's time. This makes his laboriousness and sagacity even more surprising and worthy of historical recognition and deep human respect. It does not matter that modern authors do not have to provide references to the sources of their solutions because they have already walked their own paths of discoveries, victories, and defeats. What is really interesting and important for us is Tsiolkovsky's ideas in the context of their historical and personal dimensions, in their scale, which combines in one and the same person, a school teacher, a titanic laborer, and an ingenuous dreamer.

EXAMPLES OF MODERN PROJECTS (2011)

Example 10.36. Cargo-Passenger Dirigibles

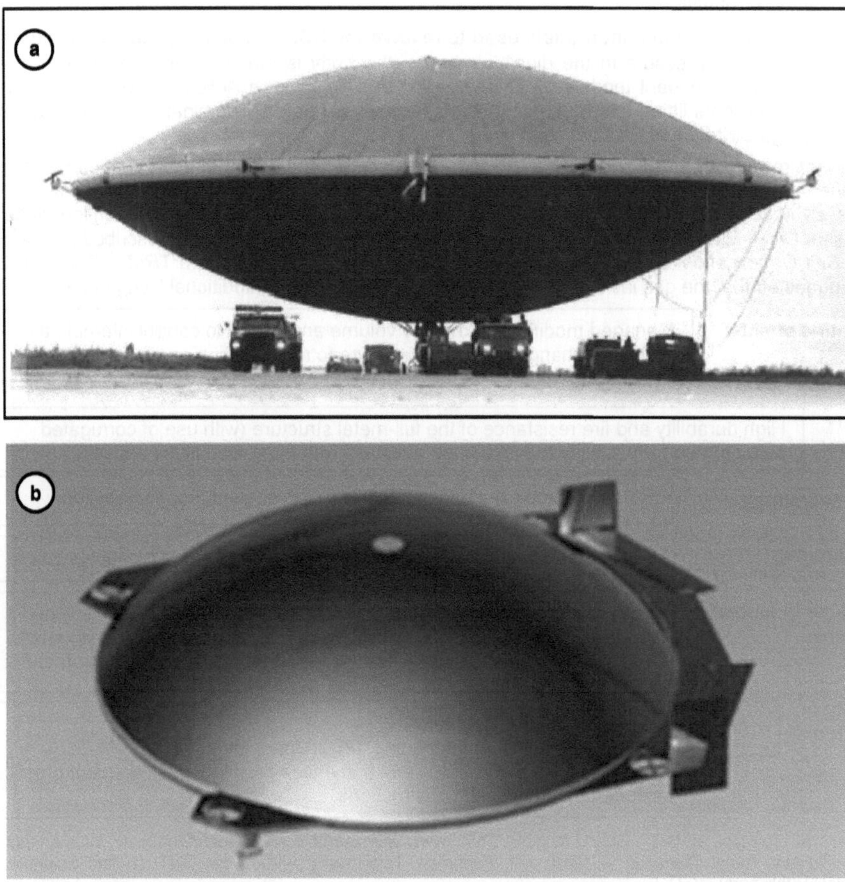

fig. 10.66. Hybrid gas-and-heat (according to Tsiolkovsky) dirigible "Thermoplan"

a) Joint design by the Moscow Aviation Institute, Design Bureau Thermoplan, and Ulyanovsk Aviation Plant; late 1980-es; b) today this idea is promoted by the LOCOMOSKY company; it is expected that the craft will be able to transport up to 600 tons of cargoes for distances of up to 5,000 km! www.locomosky.ru

fig. 10.67. "Flying saucers": a) Today (model); b) in the future with a diameter of 200 m!
Dolgoprudnenskoye Automation Design Bureau (DKBA)); www.dkba.ru; http://aerocrat.livejournal.com

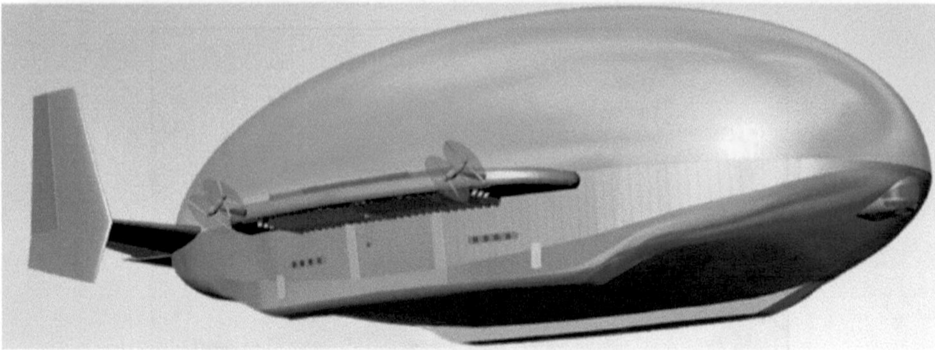

fig. 10.68. Multipurpose craft "Atlant-100"
Designer: Augur (Авгуръ) RosAeroSystems; payload up to 60 tons; flight range – 6,000 km; http://rosaerosystems.ru/files/newss/doks/ATLANT100.pdf

Example 10.37. Stratospheric Geostationary Dirigible Satellites

Objective: The creation of high-altitude (20 to 22 km) stationary unmanned platforms to support telecommunications and internet communications as well as meteorological and environmental observations, etc. (there is a long list of special applications, including police and military applications, but they are not disclosed here).

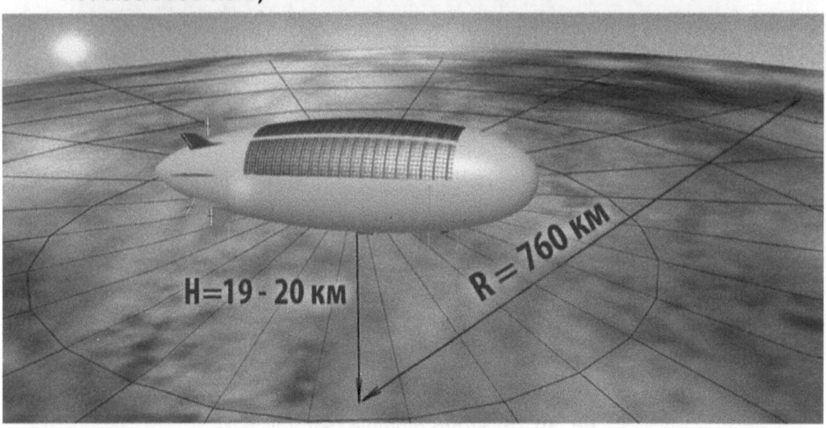

fig. 10.69. High-altitude aerostatic platform "BERKUT" ("GOLDEN EAGLE")
Designer: Augur (Авгуръ) RosAeroSystems; payload up to 1,200 tons; fast internet within a 1 mln km2 area (~ territory of France or England); http://rosaerosystems.ru/projects/obj687

fig. 10.70. High-altitude aerostatic multipurpose platform "VAMP"
Designer: DKBA; payload up to 2,500 tons; http://lj.rossia.org/users/aerocrat/115606.html

fig. 10.71. Strato-dirigible "Stratellite 1A"
Design: Sunware; http://en.wikipedia.org/wiki/Stratellite

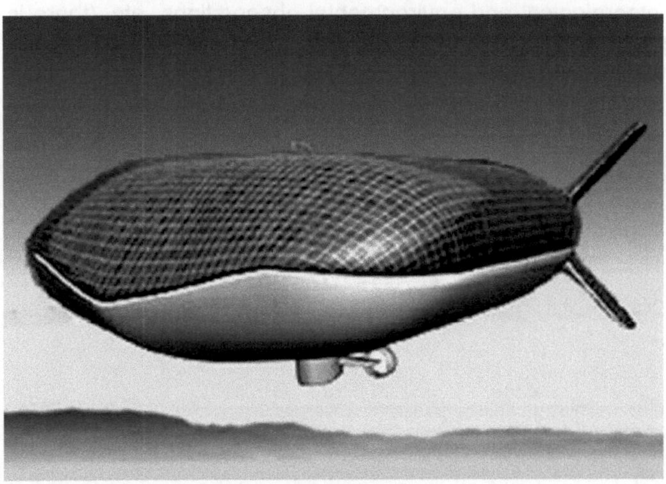

fig. 10.72. Strato-dirigible "Stratellite 2A"
Design: Sunware; www.theregister.co.uk/2005/04/13/broadband_airship

Example 10.38. Exotic Hotels-Offices

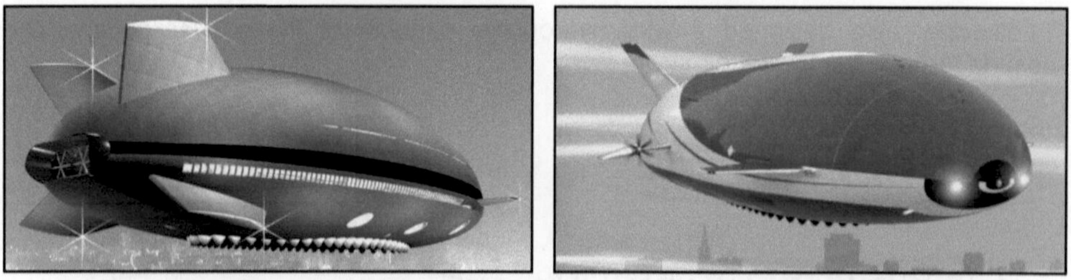

fig. 10.73. One of the most realistic cargo-luxury projects "Aeroscraft ML866"
Designer: AEROS, USA; founder (1994): Igor Pasternak (b. 1964, Lvov); www.aerosml.com

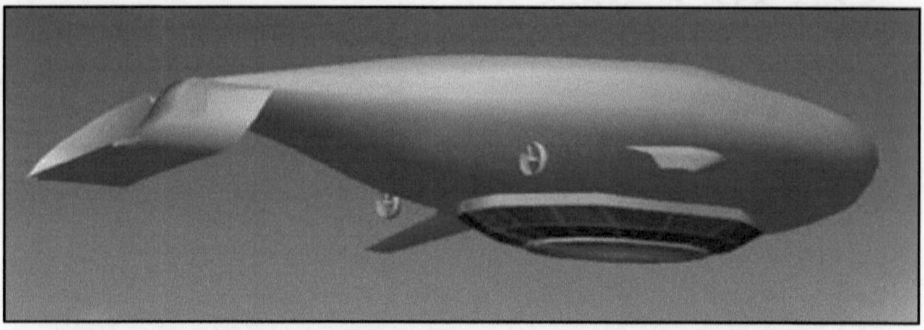

fig. 10.74. Conceptual design of a luxury hotel "Manned Cloud", 2005
By 2020, ONERA expects to give 5,000-km aerial rides to 40 passengers using this *Manned Cloud* from a French designer *Jean-Marie Massaud*; www.dezeen.com

fig. 10.75. Conceptual design of a luxury hotel "Aircruise", 2007
This project, conceived by the English company *Seymourpowell*, drew the attention of *Samsung Construction and Trading*; height of the hotel (minus mooring berth): 256 m; www.seymourpowell.com

fig. 10.76. Conceptual design "Strato Cruiser" (up to 400 km/h), 2007
Luxury hotel-office designed by *Tino Schaedler* and *Michael Brown* inspired by Richard Branson's idea that the skies and the space will be conquered by universal liners; www.dezeen.com

Example 10.39. INTEGRATION OF ALTERNATIVE SYSTEMS

Tsiolkovsky also designed a wing-hydroplane capable of flying over oceans and deserts through the lowest air strata.[101]

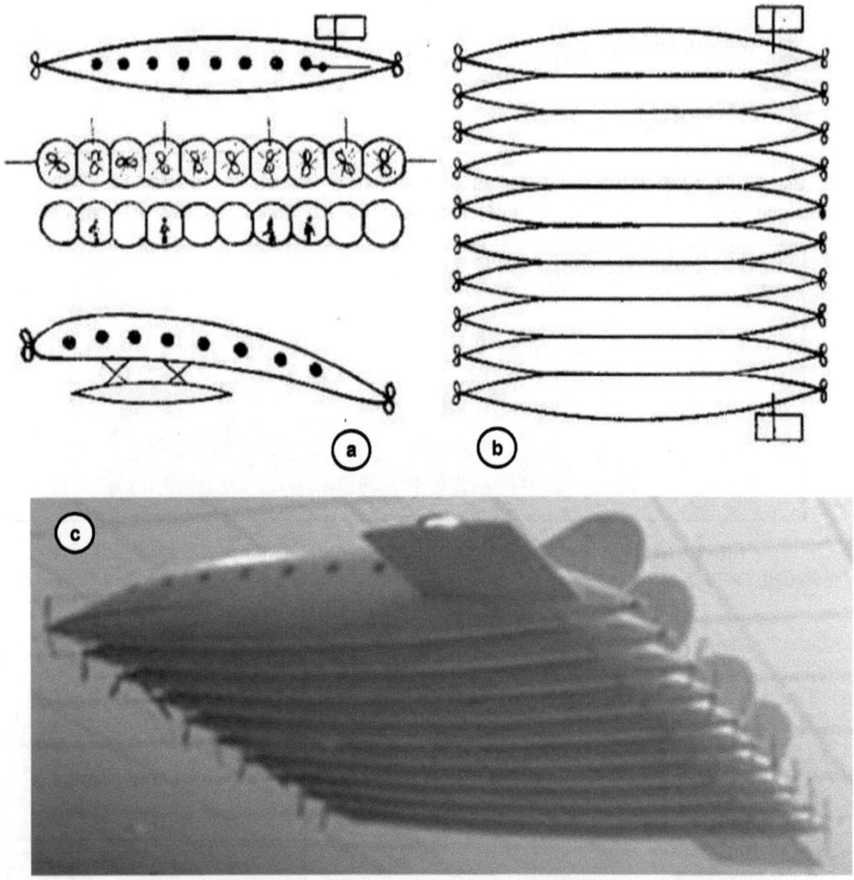

fig. 10.77. Dirigible airplane "Flying Wing."
a) Front and side views; b) plane view (top/bottom view); c) photograph from Tsiolkovsky Museum in Kaluga.

Tsiolkovsky gives the following description (fig. 10.77):

"The gigantic wing-hydroplane that I am proposing is a flying and floating wing which holds the crew, passengers, fuel, etc.

The wing has neither the towers that usually house engines, nor floats... This makes it more streamlined.

The entire body of the wing is divided into straight and oblique cells which can be used as passenger cabins, fuel storage units, etc. The wing-hydroplane can move huge payloads, and is capable of reaching high speeds... ranging from 324 km/h to 592 km/h.

[101] See above: J. I. Perelman, 1937.

It can fly across the Atlantic Ocean in 24 hours. It can take off from, and land on, water and snow surfacescнer."[102]

REINVENTING

TREND Dirigibles can be effectively used for long-distance transportation of heavy cargoes. However, construction of large-size dirigibles presents a complex technical problem; besides, they will be idling.

What can you do to increase the lifting capacity and reduce the complexity of the airborne device (AD)?

REDUCTION

FIM: the operating zone BY ITSELF ensures the obtaining of the following **IFR**:

[*aircraft with required lifting capacity and reasonable complexity*].

Standard Contradiction

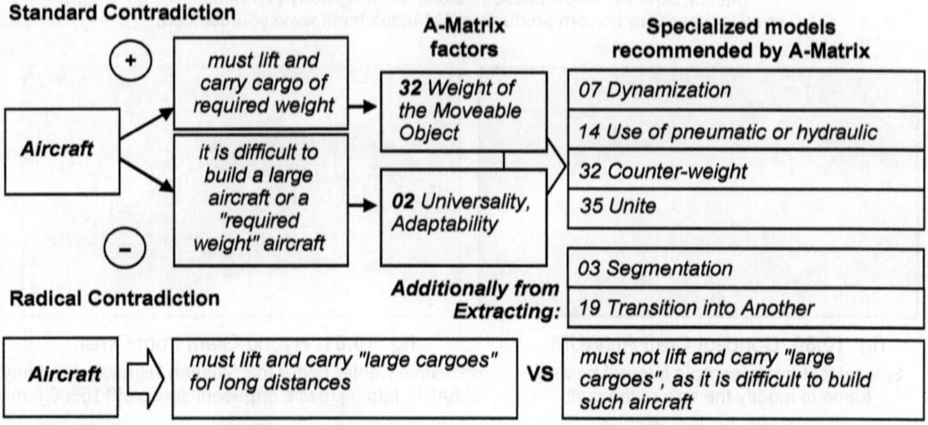

INVENTING When solving the problem, four dominant navigators make up the "portrait" of the future solution, and then are incorporated into that solution: *03 – divide a "large" aircraft into parts (for example, several separate aircraft); 35 – unite such separate aircraft into one "large" aircraft to increase its static lifting force; 19 – perform such join "broadwise" so as to obtain a "flying wing" and additional (dynamic) lifting force using 32.*

ZOOMING Have the contradictions been removed? – Yes.

Super-Effects: joining several identical dirigibles into a "wing" makes it possible to use the dynamic lifting force generated by the "wing" when the aircraft are moving.

Negative Effects: *the construction may become more complex when variable-size aircraft are built.*

BRIEF DESCRIPTION

To assure required lifting capacity, dirigibles can be joined into a "flying wing" of the required size (development of the original idea expressed by K.E.Tsiolkovsky). Dominant Models: 03 Segmentation, 35 Unite, 19 Transition into another dimension and 32 Counter-weight.

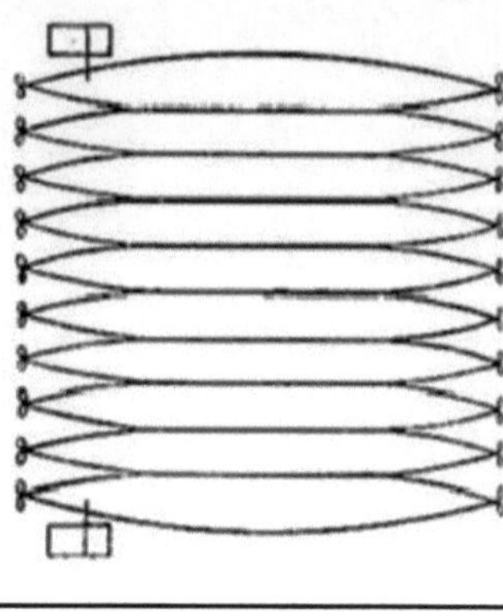

Illustration: K. E. Tsiolkovsky – New Aeroplane. – Kaluga,

fig. 10.78. Reinventing of the "flying wing" dirigible airplane (slightly modified)

[102] Quote and drawings: Tsiolkovsky, New Airplane. – Kaluga, 1929 (from the archives of the State Public Scientific and Technical Library of Russia, Moscow). – Циолковский К.Э. *Новый аэроплан.* – Калуга, 1929 (из архива ГПНТБ, Москва)

Example 10.40. Hybrids: Powerful speedsters and... taxi cabs, if necessary

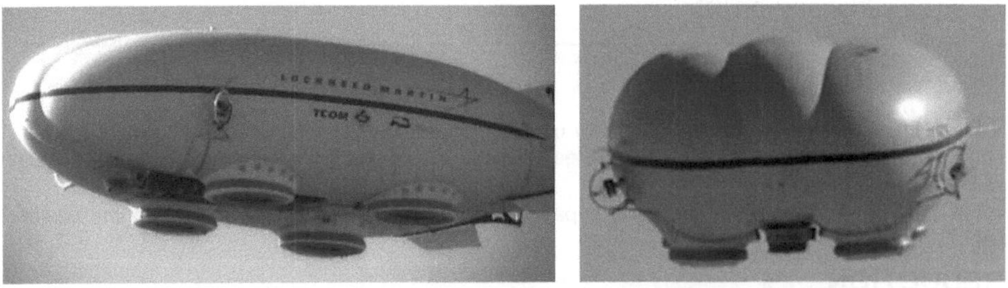

fig. 10.79. P791 Hybrid Craft Designed by Lockheed Martin (2006)

The aircraft is built using the "heavier-than-air" principle: 80% of total weight is represented by helium (static), 20% by airflow passage along the wing-body (dynamic); www.lockheedmartin.com/products/p-791/index.html; www.youtube.com

fig. 10.80. Concept Craft ANVIUM

Special feature: telescopic internal-bearing frame to modify the size of the craft

fig. 10.81. Hybrid Giant from Omsk

Innovative Center under the Siberian Highway Academy (SibADI); http://lj.rossia.org/users/aerocrat/115606.html

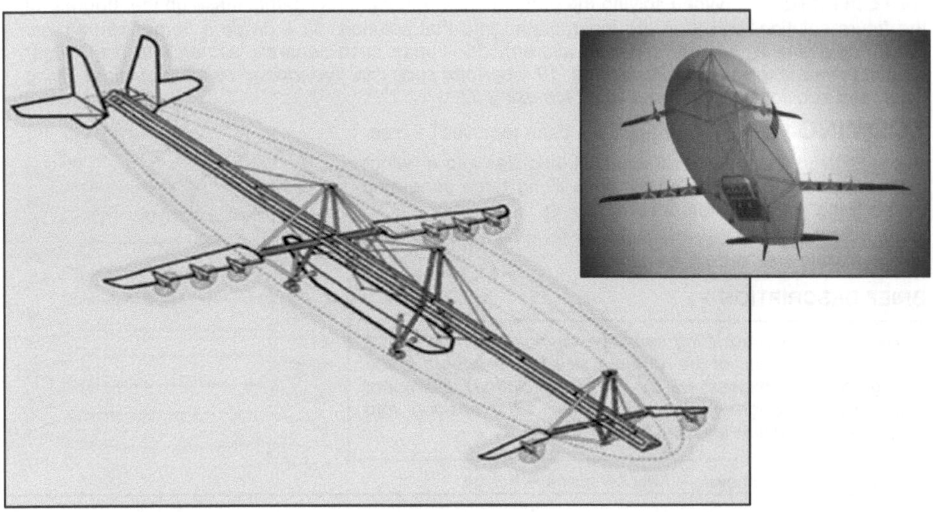

fig. 10.82. Hybrid Craft Dynalifter from Ohio Airship Company, early 2000s

Outstanding invention – "hanging bridge" as the skeleton for the entire construction. Buoyancy: 50% helium, 50% wings. Large payload and high speed; www.dynalifter.com

Example 10.41 (a). Experimental ideas

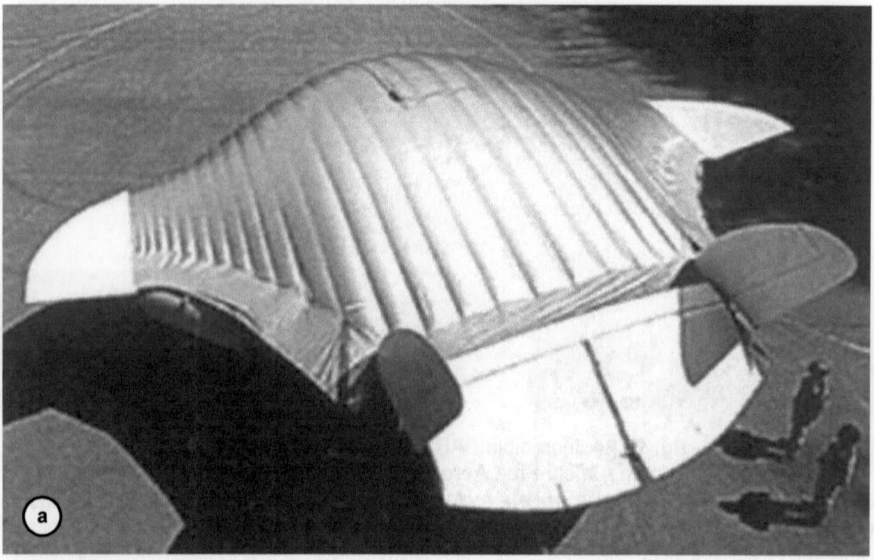

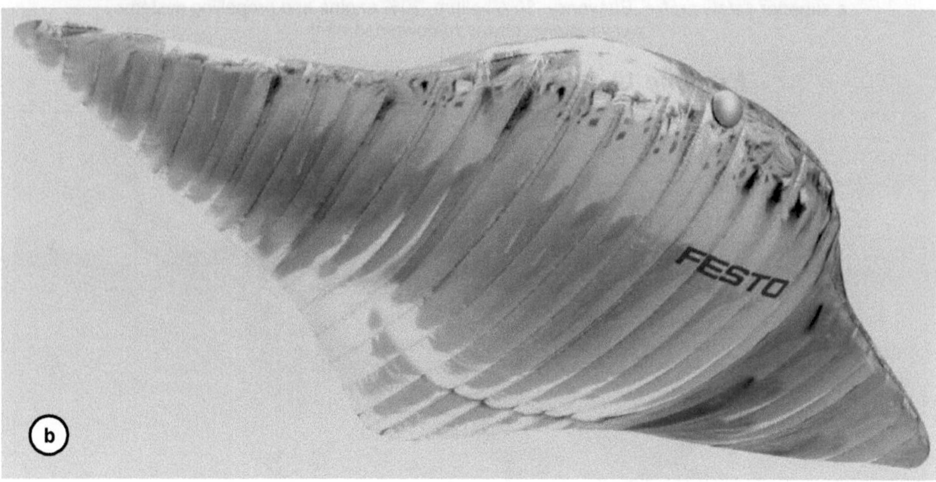

fig. 10.83. Bionic Craft STINGRAY from FESTO Company (Germany)
a) www.prospective-concepts.ch: "stingray" designed by the Swiss company Prospective Concepts to the Order of FESTO;
b) FESTO design – stingray floating through the air by beating its wings; see wonderful ideas at www.festo.com/cms/en_corp/9647.htm; nice video clip: "Festo Air_ray" at www.youtube.com

Example 10.41 (b). Experimental ideas

fig. 10.84. Incredible BARS (Leopard) Project from
Tyumen Airfield-Free Aerostatically Unloaded Aircraft

Inventor Alexander Filimonov: **FOUR-IN-ONE SYSTEM: aircraft, helicopter, dirigible, and air-cushion vessel.** The craft is a mixed-type "flying wing" with a discoid center wing as its core element. It is basically a container for lifting gas (helium) with a central canal housing a hoisting system and a cargo cabin (section) and marginally mounted pilot/passenger cabin, wing consoles, and tail assembly. Outstanding performance characteristics: payload – up to 500 to 600 tons (in the future) combined with high speed and a superior safety profile. Buoyancy: 80% helium, 20% engine and propelling systems; www.tumenecotrans.ru/download.html

Example 10.42. Atomic[103] and vacuum dirigibles

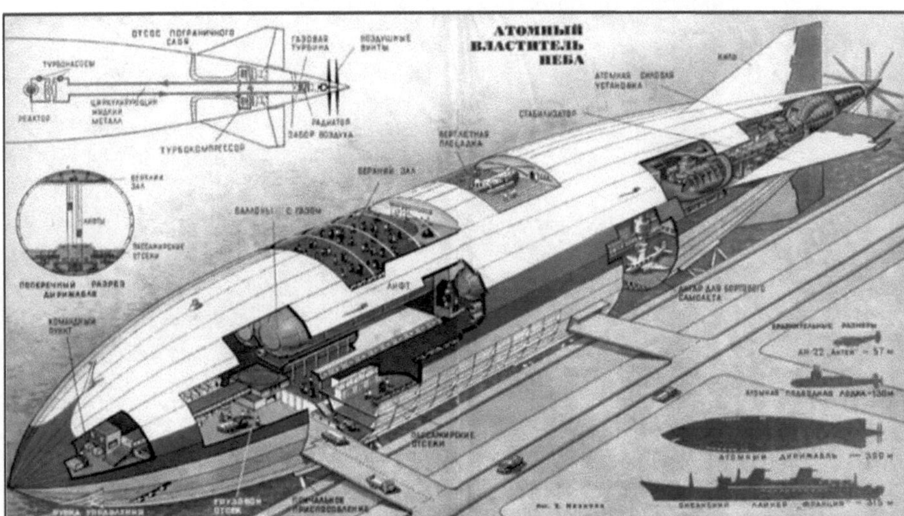

fig. 10.85. "ATOMIC RULER OF THE SKY" – the project of the early 1950s (the USSR)

---

[103] http://longstreet.typepad.com/thesciencebookstore/2011/05/differentially-powered-dirigibles-with-built-in-airports-atomic-solar-vacuum.html; article by Frank Tinsley *"Why Don't We Build An Atoms-For-Peace Dirigible," Mechanix Illustrated*, March, 1956

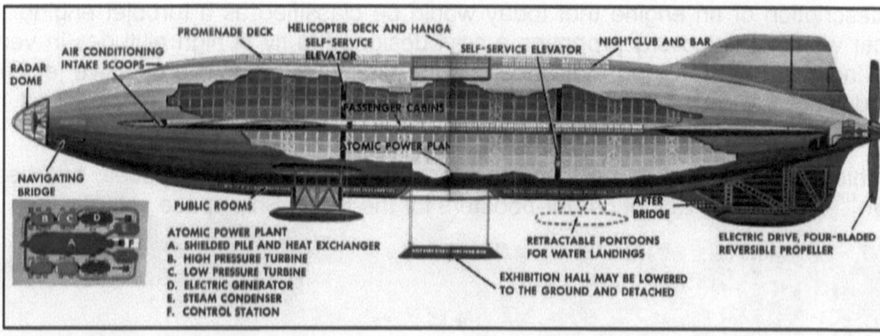

fig. 10.86. Dirigible "Atoms-For-Peace" – the project of the mid-1950s (the USA)

Some words about vacuum dirigibles: The idea is that it is possible to increase buoyancy by another 10% to 12% by discarding helium or hydrogen and creating a deep vacuum inside the craft. What do you think will happen to the dirigible's body?

Example 10.43. Triple-hull stratoplane

> Breakthroughs are what define our species.
>
> Our best performance comes from structuring a specific blend of creative talent with others whose passion is to apply an innovative breakthrough to the building and testing of a real product...
>
> *Burt Rutan*

Tsiolkovsky proposed to use gyroscopic devices to equip aircraft with local coordinate systems. He described and calculated the main parameters of an air-breathing jet engine to propel aircraft through the upper air strata, which was later patented by many engineers in various countries.

Last but not least, Tsiolkovsky produced a very realistic sketch of an airplane (1895) resembling a bird.[104] In line with the modern views, such an airplane could actually fly, especially taking into account the well-drawn swelling of the leading edge of the wing and the curved wing profile like that of a soaring bird (fig. 10.87).

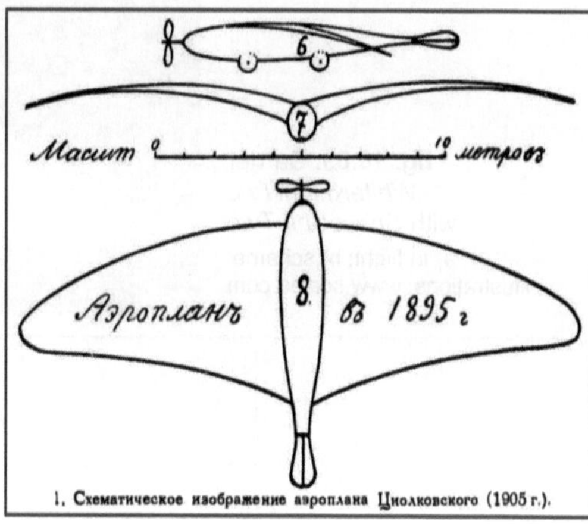

fig. 10.87. Airplane by K. E. Tsiolkovsky

This is what Tsiolkovsky had to say about the general construction of a stratoplane[105]: "My stratoplane consists of three streamlined hulls. They are joined by one common wing. The system has a direction rudder, an elevation rudder, and a lateral stability rudder. The two outer hulls are impermeable to gas, and are used primarily to carry people and fuel. The middle hull is opened at both ends. It contains..." What follows is

[104] See above: J. I. Perelman, 1932
[105] See above: J. I. Perelman, 1937

a description of an engine that today would be classified as a turbojet engine. In his other works, Tsiolkovsky proposes a craft designed to fly at high altitudes in very thin air and describes an idea that underlies the design of modern propulsive jet engines occupying a wide stretch of the continuum between turboprop engines and rocket engines.

At this juncture, it would be fitting to mention, for illustrative purposes, two beautiful craft[106]—the White Knight carrier-boosters for the SpaceShip space vehicle.

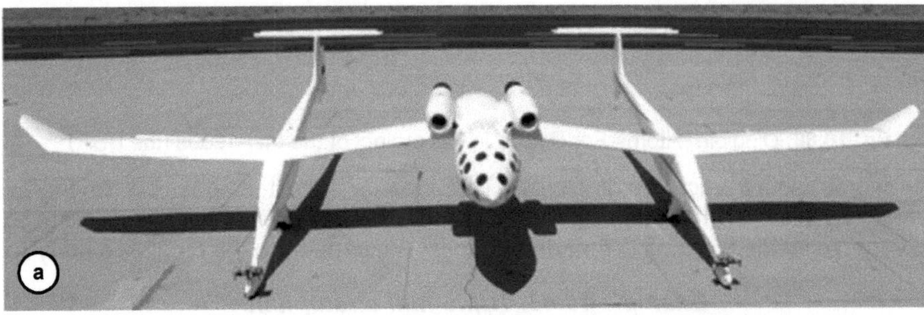

fig. 10.88. Carrier *WhiteKnightOne*
a) carrier;
b) carrier with *SpaceShipOne*;
illustrations: www.scaled.com

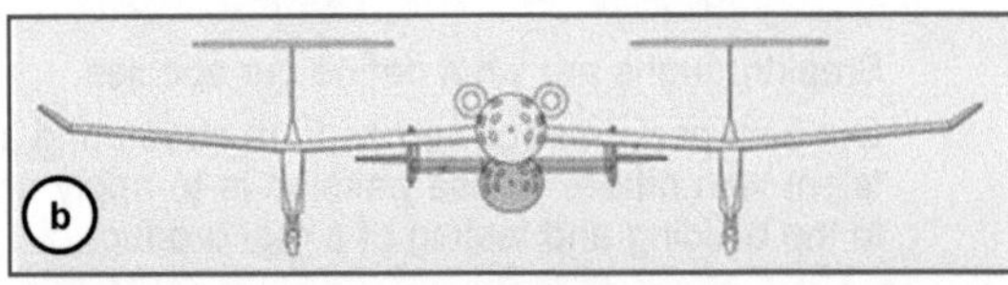

fig. 10.89. Carrier *WhiteKnightTwo* with *SpaceShipTwo*
a) in flight; b) scheme;
illustrations: www.scaled.com

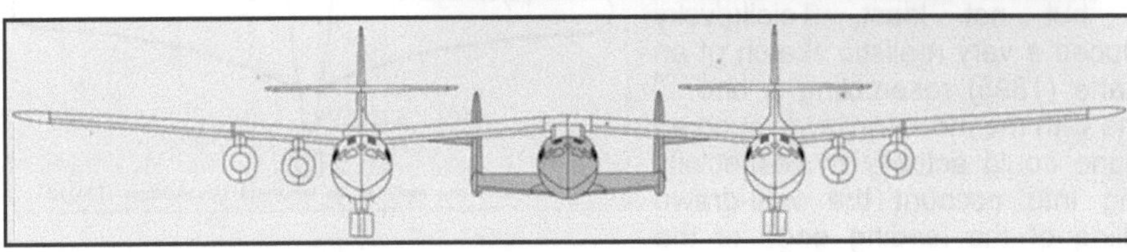

In my opinion, these wonderful modern machines are quite concordant with the dream voiced by Tsiolkovsky more than 80 years ago…

And now retrace your steps and review the reinventing of the composite "flying wing" dirigible in fig. 10.77-10.78. Would you agree that going this way you could have invented something similar to the *White Knight*? Why or why not?

---

[106] Elbert Leander "Burt" Rutan, b. 1943) – American aerospace designer, developer of several outstanding aircraft, including Model 76 Voyager (1986 world record: 9-days' non-stop flight around the world), Virgin Atlantic Global Flyer (2005 world record: 3-days' non-stop flight around the world), etc.

OUTER SPACE

> Two ideas above all others held sway over Tsiolkovsky's inventive mind: the idea of a dirigible metal aerostatic balloon, and the idea of flying beyond the boundaries of the Earth's atmosphere, into the outer space.[107]
>
> *J. I. Perelman*

Example 10.44. Rockets

"The most wonderful, audacious and original product of Tsiolkovsky's creative mind is his ideas and works dedicated to rocket technologies. He had no predecessors in that field and was far ahead of the global scientific community and, indeed, of his time.

"In 1903 he published his first work called *Exploration of Space with Reactive Devices*, followed by an unbroken chain of related publications.

"In those works Tsiolkovsky showed that a rocket engine is usable and beneficial at high flight velocities, and that only such engine is capable of propelling a body in an airless or extremely rarefied environment.

"To assure sufficient acceleration, he proposed so-called "space rocket trains", where a whole family of related rockets is launched, and then redundant parts are jettisoned as fuel is burned off, while the moving system as a whole retains the necessary mass ratio and reaches the required velocity.

"The idea that man can fly into space permeates all Tsiolkovsky's writings. He examined issues related to that problem and, in particular, he wondered how human organism would behave at high altitudes, above the Earth's atmosphere, at high accelerations, etc. Tsiolkovsky believed that ultimately man would break the chains of the Earth's gravitation, and he can justly be called the greatest ... forerunner of all future astronauts.

"He undertook a detailed study of routine matters that future interplanetary travelers would have to deal with in their daily life, he considered the possibility of creating an artificial satellite of the Earth to serve as an intermediate interplanetary island or station to be built along future space routes.

"This was fantastic and absolutely mind-blowing. It sounds brilliant even today, in our 'age of miracles', except that we have to admit that this is a scientific truth – and a scientifically sound projection of a not-so-distant future."[108]

Sergey Pavlovich Korolyov, who wrote these words, working together with dozens of outstanding rocket engine builders and rocket designers launched into space the *Vostok-1* rocket carrying the first *Sputnik* (October 4, 1957) and the first cosmonaut (April 12, 1961).

For now, we will consider just two of the many ideas advanced by Tsiolkovsky: (1) the creation of a multistage rocket; and (2) the use of gas rudders (jet vane) to change direction.

---

[107] See above: J. I. Perelman, 1937

[108] Compiled by me (M.O.) from the following book: S. P. Korolyov, *Life and Activities of Tsiolkovsky*. – Report delivered on September 17, 1947, at a meeting dedicated to the 90th birth anniversary of Tsiolkovsky. – (Королев С.П. *Жизнь и деятельность К. Э. Циолковского*. – Доклад 17 сентября 1947 года к 90-летию К.Э.Циолковского).

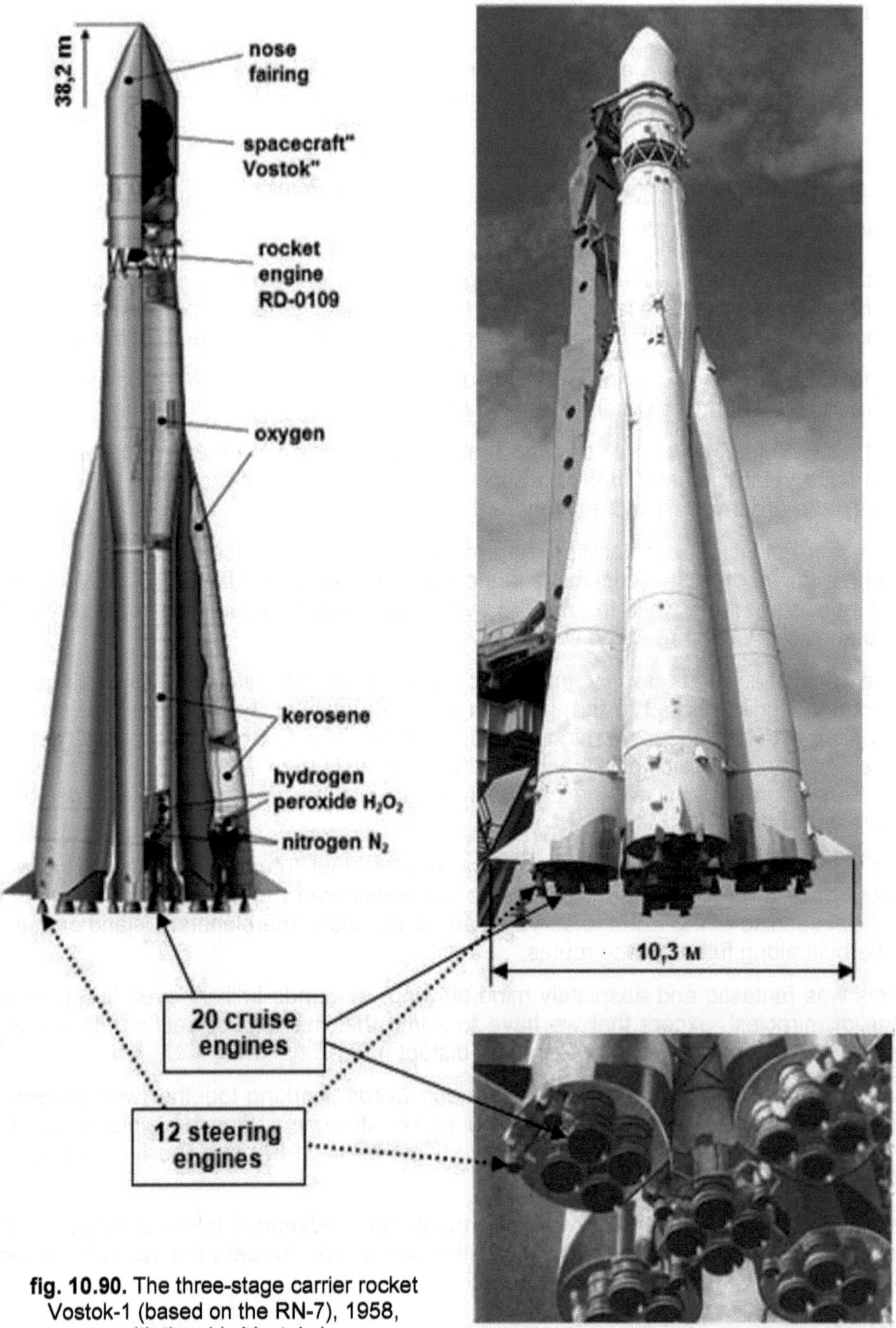

fig. 10.90. The three-stage carrier rocket Vostok-1 (based on the RN-7), 1958, with the ship Vostok 1

Illustration: http://galspace.spb.ru/start-4.htm, http://en.wikipedia.org/wiki/Vostok_(rocket_family)

TREND To change the travel direction of a dirigible or airplane, they can be fitted with elevation rudders (horizontal "plates", usually placed in the aft portion of the aircraft's body) and direction rudders (vertical "plates"). But how do you change the travel direction of a rocket in outer space?

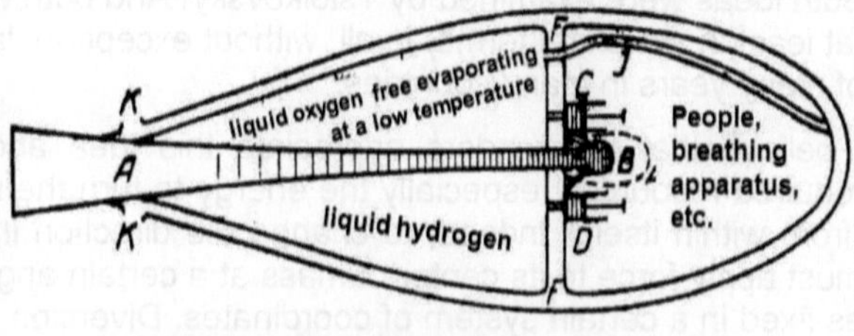

Diagram of the Tsiolkovsky rocket ship

REDUCTION

FIM: the operating zone BY ITSELF ensures the obtaining of the following **IFR**:

[*required change of rocket travel direction*].

Standard Contradiction

Rocket ► travel direction must be changed **VS** airplane-like rudders do not work in outer space = *10 Ease of use* **VS** *30 Force = 01, 04, 11*, as well as *18, 19, 29 и 34*

Radical Contradiction

Rocket ► must change direction (to be steerable)
VS must not change direction, and must continue to move by inertia or as a result of operation of the engine

INVENTING When solving the Standard Contradiction, several dominant navigators make up the "portrait" of the future solution, and then are incorporated into that solution: *04,c) give a certain structure to the jet issuing from the engine, for example, divide the jet; 11,b) make the jet capable of changing direction, for example, divert the jet by turning the engine's nozzle; 18,a) install diverting plates inside the jet; 19,a) assure that the jet can not only move along a straight line relative to the body of the rocket, but also be diverted within a certain spherical sector; 29) the rocket must BY ITSELF change the travel direction without resorting to its environment; 34) install controlled diverting plates in the way of the jet inside the engine's nozzle.*

When solving the Radical Contradiction, the key fundamental model is *03 Segmentation in structure: the jet is divided into parts, and each such part can be diverted when exiting the nozzle.*

ZOOMING Have the contradictions been removed? – Yes.

Negative Effects: *the rudders can be destroyed by the jet's powerful mechanical and thermal impact.*

gas rudders

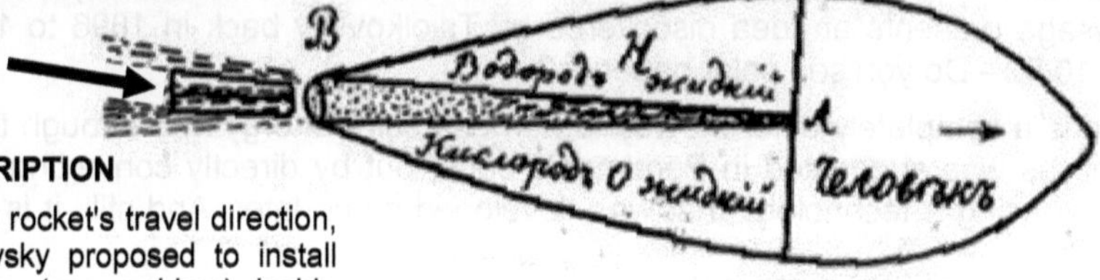

BRIEF DESCRIPTION

To change the rocket's travel direction, K. E. Tsiolkovsky proposed to install diverting plates (gas rudders) inside the jet. Dominant Models: 04,c; 11,b; 18,a; 19,a; 29 и 34.

Illustrations: from works by K. E. Tsiolkovsky and J.I.Perelman. Note: Werner von Braun used such rudders in his V-2 rockets (late 1930-es - mid 1940-es)

fig. 10.91. Reinventing the idea of Tsiolkovsky "jet vane of missile"

As concerns the first idea, we… well, we already did it! Look again at the "flying wing" reinventing: There is a lot to learn there again! Try transformation model *19 Transition into Another Dimension* to create a multistage rocket design. You can easily wind up with a "rocket pack" idea ("width" modification) or a multistage rocket idea ("height" modification).

Both ideas were examined by Tsiolkovsky! And both were developed and implemented (at least in structural terms) in all, without exception, large rockets built over the course of many years in many countries.

I believe that the readers appreciate this idea and judge it beautiful because all required resources, especially the energy to turn the rocket, is extracted by the device "from within itself." Indeed, to change the direction in which the rocket is moving, we must apply force to its center of mass at a certain angle relative to the current direction as fixed in a certain system of coordinates. Diversion of a portion of the jet stream with "gas rudders" results in the rocket turning relative to its center of mass. Subsequent operation of the jet stream changes the rocket's direction.

We all remember that the first cosmonaut was Yuri Alekseyevich Gagarin. This is what he said about Tsiolkovsky: "I was a technical school student when I was told to make a report on *K. E. Tsiolkovsky and His Theory of Rocket Engines and Interplanetary Travel.* I read several works written by this founder of space science. Tsiolkovsky turned me inside out. It was way beyond Jules Verne, Herbert Wells and other science fiction writers. I was amazed by the vigor with which the scientist's ideas pierced space in an unyielding, boss-like manner..." [109]

Example 10.45. Satellites and space stations

"Interplanetary travel could be greatly facilitated, if we had realized *one idea proposed by Tsiolkovsky, an idea unparalleled in its audaciousness and originality in the entire history of technical thought.*

"We refer to the idea of creating an *artificial Earth satellite*, a small new Moon. This is not as far-fetched as one might think at first sight; foreign proponents of the stellar navigation theory, working after Tsiolkovsky and apparently quite independently from him, came up with the same idea and developed a detailed project envisaging creation of an *extraterrestrial station*, a name often applied to the artificial satellite of the Earth...

"The living conditions on or, rather, inside such planetary island will be quite unusual and, to some extent, evocative of those on board a submarine. But, unlike on board a submarine, here one will have ample opportunities to use the energy of solar rays." [110]

And here is another question for you: Are you not astonished by the fact that this passage presents an idea discovered by Tsiolkovsky back in 1896 to 1898? Look at fig. 10.92—Do you see solar batteries?

This is a completely different way to harness solar energy: not through the medium of heat, as was suggested in Perelman's book, but by directly converting it into electric power using a technology that was developed much later. And still, it is the energy of the Sun!

Now look at this wonderful merger of space station ideas and dirigible ideas! This is an integration of alternative systems in a new environment!

The Bigelow Expandable Activity Module (BEAM) is an experimental expandable space station module developed by Bigelow Aerospace, which was berthed to the International Space Station (ISS) on April 16, 2016 and was expanded and pressurized on May 28, 2016. Great event!

---

[109] Y. A. Gagarin: Preface to the book by M. S. Arlazorov, *Tsiolkovsky*. – Moscow: Young Guard Publishers, Life of Outstanding Individuals Series, 1963. – (Гагарин Ю.А.: Предисловие к книге Арлазорова М.С. *Циолковский*. – М.: Молодая Гвардия, Серия "Жизнь замечательных людей", 1963)

[110] See above: J. I. Perelman, 1937

fig. 10.92. Laboratory – Plant – Hotel … in Space!
Robert Bigelow, a billionaire entrepreneur from *Las Vegas, USA*, has been engaged, since 1999, in a project envisioning creation, in space, of manned stations using inflatable modules to conduct scientific research, manufacture nano- and biotechnological products in an environment of micro-gravitation, and provide space access to the general public; www.bigelowaerospace.com; http://aerocrat.livejournal.com/130782.html

Now let us consider Tsiolkovsky's ideas as to how rockets can be launched into outer space.

Example 10.46. Launch from Earth's Surface

"Departure of the interplanetary rocket from the Earth will take place from a mountainous area. A straight flat runway must be prepared going up at an angle of 10-12 degrees… Tsiolkovsky suggests that another rocket be used for the takeoff run.

"He calls this auxiliary rocket an *earth* rocket to distinguish it from the *space* rocket designed to make the interplanetary flight. The space rocket must be temporarily placed inside the earth rocket, and then the latter, without ever leaving the surface of the earth, will accelerate the space rocket to the necessary velocity, and at a certain moment release it for independent flight into outer space. The earth rocket … will be … sliding along special … rails.

"The best deceleration method is to have special brake planes to extend from the earth rocket at right angles to its body: at high velocity wind resistance will be overwhelming, and the rocket will soon halt. Having thus completed the takeoff run, the rocket [space rocket – M.O.] will commence its independent flight propelled by flammable substances exploding inside its body." [111]

---

[111] J. I. Perelman, *Interplanetary Travel.* – Leningrad, ONTI, 1915. – (Перельман Я.И. *Межпланетные путешествия.* – Москва - Ленинград, ОНТИ, 1915) – see the cover in section *3.4. Jacob Perelman*

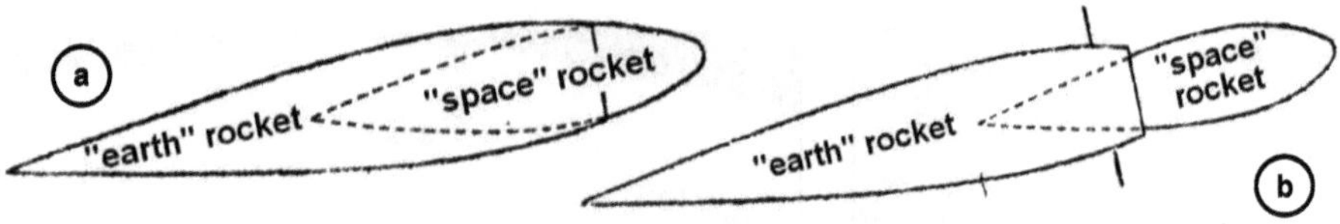

fig. 10.93. Reinventing the idea of Tsiolkovsky for two-stage rocket start:
a) "matryoshka" of two rockets during acceleration;;
b) moment of braking the "earth" rocket and separation from it the "space" rocket.

Such a launch was later researched by many leading designers in various countries. One such project was developed by Professor Eugen Sänger in Germany in 1935 to 1941 and after 1945 (see the author's other book, *Modern TRIZ*).

Example 10.47. Launch from Adjacent Space

"According to Tsiolkovsky, a settlement near Earth must be established outside the boundaries of the planet's atmosphere, at a distance of 1,000-2,000 km from its surface... The existence of such extraterrestrial station will make it easier for the astronauts of the future to embark on their interplanetary travels. The station will be made up of multiple rockets sequentially launched into orbit around the Earth and then joined into one unit. The most difficult part will be the creation of that celestial island: it will be relatively easy to leave it and go on further out into space, as the mass of the artificial moon is negligible, and minute amounts of energy will be quite sufficient to overcome its gravitation." [112]

Such stations have not yet been built. A modern approximation of such a launch could be described in terms of launching the fourth (!) stage of a rocket that had been put into orbit by the first three stages for a flight, say, to Venus, after several revolutions around the Earth moving along an artificial satellite orbit. This considerably improves launch accuracy because deviation from the required velocity of, say, 11.2 km/s by one hundredth of a percent (that is, by 1 m/s) (!) may prevent the rocket from "locking into" the design flight trajectory!

The ***world's first*** Venus-1 station was launched from a satellite orbit on February 12, 1961.

It was before Yuri Alekseyevich Gagarin's flight... Everything was just beginning! And still, the first spacecraft had already started on their journeys to the Moon, Venus, and Mars!

Example 10.48. Launch from the Stratosphere

We already described the launch of a *SpaceShip* from the *White Knight* mothership, which first "lifts" the spacecraft to an altitude of 15 to 20 km.

However, there are two more strategies for pulling off a so-called "altitude launch," which itself is an intermediate option wedged in between the earth-surface launch and the space-station launch:

1) Launch from a (moving) airplane or a (stationary) dirigible by simply "dropping" the rocket or extending it several meters to the side on a rod (the "rod option").

---

[112] Compiled by me (M.O.) from the book by J. I. Perelman, 1937.

2) Launch a heavy rocket from an aerostatic platform made up from a group of dirigibles joined into a rigid carrier/launcher construction (the "platform option").

Strategy 1. The Rod Option.

A space aerostat complex is being designed within the framework of the Russian High Launch program to perform dirigible-based orbit launches of micro- and nano-satellites.

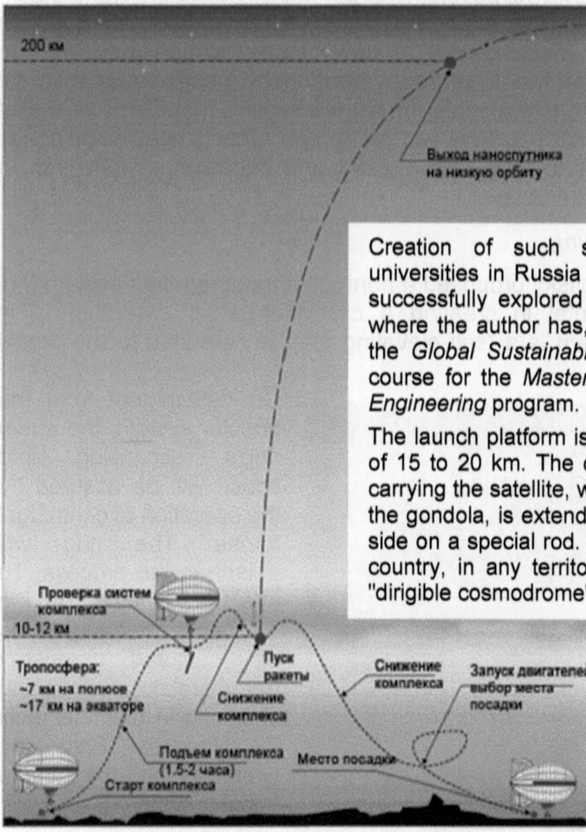

Creation of such satellites is underway in many universities in Russia and other countries. This option is successfully explored at the Berlin Technical University where the author has, for the last 5 years, been reading the *Global Sustainable Innoventions with Modern TRIZ* course for the *Masters of Science in Global Production Engineering* program.

The launch platform is a dirigible that reaches an altitude of 15 to 20 km. The dirigible then stops, and the rocket carrying the satellite, which is attached to the lower part of the gondola, is extended for launch several meters to the side on a special rod. Such a launch can be made in any country, in any territory—All one has to do is take the "dirigible cosmodrome" to the required point in space.

fig. 10.94. Dirigible-based satellite launch
http://dlib.eastview.com/browse/doc/12195139

Strategy 2. The Platform Option.

A sketch picturing such a launch was presented in 2001 at the All-Russian Space Competition by Mikhail Sinitsyn, a student of the Kirov Aviation Technical School, and Aleksey Melchakov, a student of the Forest Industry Technical School Such a "cosmodrome" will pick up the rocket by itself and carry it to the optimal launch location, for example, to the equator.

fig. 10.95. Aerostatic platform-based launch
http://www.pereplet.ru/pops/sojuz/sojuz1.html

10.3.3 *Per Aspera Ad Orbis*: Into Space... on a Wheel![113]

After more than a century of space systems evolution, there emerged a rather intricate problem: Rockets proved to be powerful pollutants of air, land, and water. The creation of a circumterrestrial civilization (for starters consider the transfer of certain production processes into near space; the development of new communication systems; the accumulation of solar energy for the Earth; the creation of weather-management Systems—This can happen in the foreseeable future) will require thousands of rocket launches per year, and those future rockets will be carrying much heavier payloads than today. This option appears to lead into a dead end because, at minimum, it threatens to destroy the Earth's ozone layer.

Example 10.49. String Space Ring.

Back in the late 1970s, A. E. Yunitsky proposed a nonrocket transportation system (fig. 10.96) that could be instrumental in creating a circumterrestrial civilization. The propulsion unit for the new system is a "self-elevating ring" accelerated to the orbital velocity along the equator.

On detachment from the booster system, the rotary ring's ascension into space will be assured by the operation of centrifugal forces. The ring will stretch in the process. To effect the landing, the ring will be decelerated giving off stored energy (recuperation) and shrinking in diameter until it descends onto the platform of the same booster system.

fig. 10.96. String Space Ring by Yunitsky

Illustrations in fig. 10.97: Artifact results – Cover of open monograph Yunitsky A.E., *String Transport Systems on Earth and in Space*. – Gomel, Infotribo. – 1995. – 337 pp. (Юницкий А.Э., *Струнные транспортные системы на Земле и в космосе*. – Гомель, Инфотрибо, – 1995); www.yunitskiy.com/author/press_monograph.htm

[113] A. E. Yunitsky, *Into Space… on a Wheel.* – Technical Knowledge for Youth Magazine, Moscow, 1982, Issue No. 6, pp. 34-36 and back cover (drawing). – (А.Э. Юницкий, *В космос… на колесе.* – Журнал "Техника – молодежи", Москва, 1982, 6)

ARTIFACT **YUNITSKY SPACE TRANSPORT –** ***Yunitsky Ring***

Description In the late 1970s A.E.Yunitsky proposed the idea of ***a string space vehicle*** – "pan-planet vehicle". The special ***platform-ring – a vehicle!*** – on the basis of the string structure by A.Yunitsky, is set around the Earth following the Equator (there are other options). Ring is accelerated for a few weeks to orbital (first cosmic) velocity, is released from the holding fields and takes off into space by centrifugal force. When taking off, the ring expands. The landing and maneuvering the ring(s) are possible. Potentially, the ring can transport millions of tons of cargo in both directions.

D	No	Navigator	Function / Substantiation
+	07	Dynamiza-tion	During the launch and the landing, the rotor changes in size by up to 5-10%
+	11	Inverse act.	Increase – not decrease! – the payload by millions of times
+	19	Transition into another dimension	The ring moves horizontally, rotating in a circle, and vertically, in radial direction (vertical movement raises it above the Earth as an expanding/collapsing rotor)
+	29	Self-servic.	Yunitsky ring is the only self-transporting vehicle!
+	37	Equipoten-tiality	The ring is imparted equipotential acceleration above the equator (unlike a rocket which is launched vertically)
+	F	01 In Space	Each component of the ring is rigid, while the ring as a whole is dynamic in terms of its dimensions
+	F	03 In Structure	The ring is segmented. The "Matryoshka" principle is used to create a telescopic structure
+	F	04 In Material	The system uses the string concept to form a rotor which self-organizes into a regular ring when in motion

Contradiction / Description	
SC	The general planetary conveyance system must assure ***gigantic lifting capacity and attainment of the circular velocity to enter the Earth's orbit***, but none of the existing rockets ***have sufficient energy*** to support the required cargo traffic.
RC	The general planetary conveyance system ***must*** transport to space and back hundreds of thousands and millions of tons of cargoes, and it ***must not*** do it, as the rocket method of transportation threatens to inflict irreparable damage on the planet Earth.

fig. 10.97. Extracting for Yunitsky's "Pan-Planetary Transportation System"

10.3.4 *Let's do it!*

Example 10.50. Virgin SpaceShipOne

Have you ever ridden a Ferris wheel? Maybe even the famous *London Eye* (fig. 5.38) standing 135 meters tall?

Well, *Virgin Galactic*, a US company, suggests that anyone can look[114] at the Earth from an altitude of ... 120 km, which is approximately 1,000 times higher! There is a snag, though: To get a ticket, you will have to shell out 10,000 times more (US$ 200,000). I hope they will launch one within the next 10 years by, say, 2020. Shall we wait?

fig. 10.98. *London Eye*
– photo by the author, 2008

fig. 10.99. The world's first *Spaceport America* in New Mexico is waiting for you!

For now, let us take a closer look at the reusable space vehicle *SpaceShipOne* [115] (рис. 10.100, a), which was used to test this system.

---

114 Illustrations: *www.virgingalactic.com*; *www.spaceportamerica.com.*
115 See M. Orloff, *Modern TRIZ* (2012)

First the vehicle was spiraled up some 15 km above the space port (or, to be more exact, above an ordinary test airport) on board the *White Knight One* mothership. Then the vehicle's rocket engine kicked in to take it to the 50-km mark over the course of 10 seconds. Then the vehicle used the momentum generated by the rocket engine burn to glide further up to an altitude of more than 100 km! The vehicle spent some 3 minutes in space moving along a parabolic trajectory.

Then it commenced its wondrous descent: You see, *there was no more fuel on board!* Its relatively small weight permitted the momentum-driven jump to a >100-km altitude!

Well, to use its body to slow down during the descent, the vehicle... raised its wings (in advance, during the inertial flight; fig. 10.100, b) and went down "flatwise"!

Only at the altitude of 10 to 15 km did the wings return to their normal position, thus enabling the vehicle to glide down and land at the same space port.

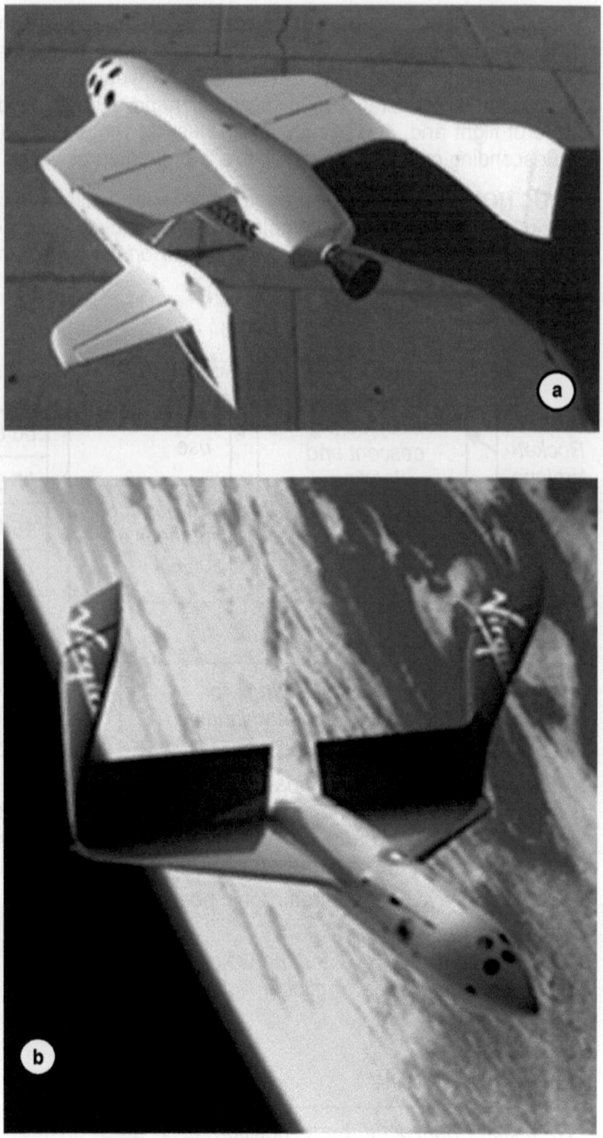

fig. 10.100. *SpaceShipOne*

a) wings in flight position;
b) wings raised – descent!

Example 10.51. Virgin Galactic SpaceShipTwo

In the opinion of *Sir Richard Branson*, the founder and leader of a group of companies operating under the umbrella brand of *Virgin*, including *Virgin Galactic*, tourists yearning for weightlessness will make their first suborbital flight on board the *SpaceShipTwo* vehicle, which will be launched from *Spaceport America* soon.

TREND To descend upon completion of a sub-orbital flight, a rocket airplane could glide down using its wings. However, at the initial stage of the descent it would be better for the craft to move not "nose first", but "flatwise". It would make is possible to faster reduce speed using higher air resistance. On the other hand, smooth landing is predicated on the gliding nature of flight and, accordingly, on the use of wings. PROBLEM: what can you do to orient the descending craft horizontally, if this is prevented by wings forcing it into a nosedive?

REDUCTION **FIM**: X-resource, together with available or modified resources, and without making the object more complex or introducing any negative properties, guarantees attainment of the following **IFR**: [*horizontal orientation of the rocket airplane's body during descent and landing*].

Standard Contradiction

		A-Matrix factors	**Specialized models recommended by A-Matrix**
Rocket Airplane	(+) *Horizontal orientation of the body during descent and landing*	***10*** *Ease of use*	*06 Use of mechanical oscillations*
	(-) *Wings force the craft into vertical position*	***18*** *Surface of the fixed object*	*07 Dynamization*
			16 Partial or excess effect

Radical Contradiction

Rocket Airplane ⇨	***must have wings*** *for gliding flight and landing*	**VS**	***must not have wings*** *to avoid a nosedive at the initial stage of descent*

INVENTING The dominant navigator is *07 Dynamization: a) the characteristics of an object or an environment are changed to optimize every work procedure; b) disassemble an object into parts that are moveable among each other; c) make an object moveable that is otherwise fixed*: **make the wings capable of turning!**

ZOOMING Have the contradictions been removed? – Yes.

Super-Effects: *fuel saved, and the weight of the craft is reduced due to gliding flight and landing.*

Negative Effects: -

BRIEF DESCRIPTION

To assure horizontal body position of sub-orbital rocket airplanes Virgin Galactic SpaceShipOne and SpaceShipTwo for deceleration at the initial stage of descent and during the gliding landing, the wings of the craft are designed so that they are capable of turning (07 Dynamization).

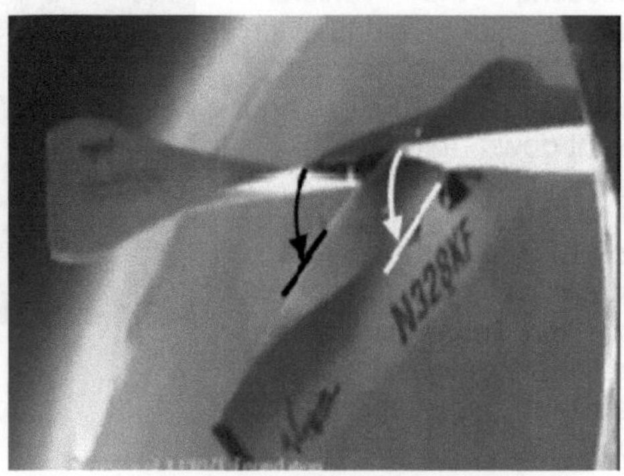

Illustration: www.virgingalactic.com and Virgin Galactic SpaceShipTwo Animation at www.youtube.com

fig. 10.101. Reinventing the transformation idea of the *SpaceShipOne* and *SpaceShipTwo* by private company *Virgin Galactic, USA*

To make it possible for the vehicle to be used by tourist passengers, rather than by well-trained test pilots, designers had to make a number of unusual transformations to the cabin (fig. 10.102).

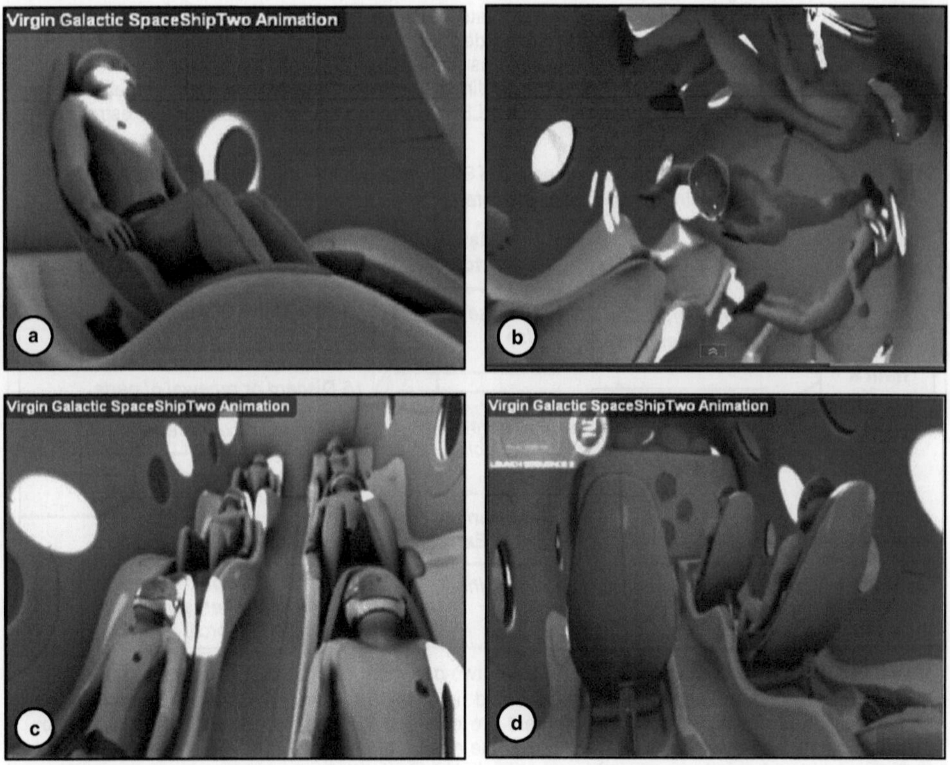

fig. 10.102. Chair transformation in *SpaceShipTwo* space vehicle

Because during the launch astronauts (in the US this term is used to denote people who enter outer space, not just those who make orbital flights) are exposed to enormous G-loads, it is necessary to assure that their bodies are positioned safely so that the load is applied flatwise (i.e., "at right angles" to the backbone). The astronaut must go up in a "crouch" (fig. 10.102,a) because the angle of attack may range from 60° to 80°.

On the other hand, after the rocket engine is cut off, it would be nice for the chairs to "disappear" to make more space inside the cabin so that space tourists can freely enjoy weightlessness (fig. 10.102,b). So the chairs are installed into special cavities (boxes).

During the fall/descent (before the wings are lowered), the astronauts are strapped in "submerged" chairs (fig. 10.102,c).

When the vehicle begins its gliding flight, the chairs lift the astronauts into the sitting position (fig. 10.102,d).

REINVENTING

TREND During the steep vertical ascent of the rocket airplane after launch, the astronauts' bodies must be in a position which is intermediate between the "lying" position and the "sitting" position relative to the "floor" of the craft. During deceleration in the dense atmospheric layers, the astronauts' bodies must be in a "lying" position, as the body of the craft during descent is oriented horizontally. Both positions make it possible to direct G-forces "flatwise", at "right angles" to the spinal column, rather than along its length. PROBLEM: what can you do to maintain the astronauts' bodies in a safe position?

REDUCTION **FIM:** the operating zone BY ITSELF ensures the obtaining of the following **IFR**: [*orientation of the astronaut's chair to allocate G-forces throughout the entire back surface*].

Standard Contradiction		A-Matrix factors	Specialized models recommended by A-Matrix
Astronaut's chair	(+) *Neutralize negative G-Force effects*	***10** Ease of Use*	*03 Segmentation*
			07 Dynamization
			15 Discard or renewal of parts
	(–) *Ordinary chairs have limited mobility*	***02** Universality, Adaptability*	*16 Partial or excess effect*
		Additionally from Extracting:	*34 Matryoshka (nested doll)*

Radical Contradiction

Astronaut's chair	⇨	*must be "mobile" to optimize the astronaut's body position*	**VS**	*must be "immobile" to be sturdy and reliable when exposed to excessive G-Force*

INVENTING When solving the problem, two dominant navigators make up the "portrait" of the future solution, and then are incorporated into that solution: 07 Dynamization: a) the characteristics of an object or an environment are changed to optimize every work procedure; b) disassemble an object into parts that are moveable among each other; and 34 Matryoshka (Nested Doll): a) an object is inside another object that is also inside another, etc.

Solution: the chairs have a large range of rotation; during descent, they are embedded into the "floor" (matryoshka principle) to increase durability and safety of the entire construction.

ZOOMING Have the contradictions been removed? – Yes.

Super-Effects: it becomes possible to place the chairs in an optimal position both for the launch/descent, and for looking through portholes (during the weightlessness phase, the chairs are also embedded into the "floor" to free more space where the astronauts can "swim around".

BRIEF DESCRIPTION

To place the astronaut's body into an optimal position to neutralize the negative effect of excessive G-Force during ascent and descent (deceleration in the atmosphere), the chair is made capable of rotating, and can be embedded into, and fixed in, the "floor". Dominant Models: 07 Dynamization and 34 Matryoshka (Nested Doll).

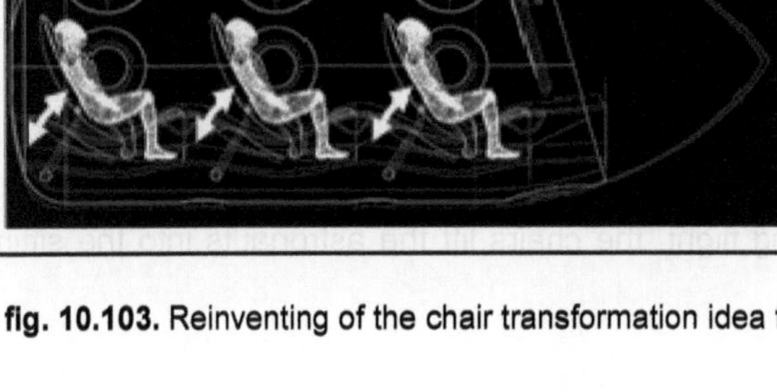

Illustration: www.virgingalactic.com and Virgin Galactic SpaceShipTwo Animation at www.youtube.com

fig. 10.103. Reinventing of the chair transformation idea for the *SpaceShipTwo* space vehicle

11. The Origins of Creative Personality

How to Become a Genius: Ways of Talented Persons

To keep this book compact, I had to set limits on the volume of each section, including this section containing essays about outstanding people, which could run on and on for hundreds of pages. I have selected those moments that I believe will affect you in the same way they have been affecting me for years... Thus, I asked myself: Where is the starting point of creation? Is it interest in creation? Commitment to creation? I hope that these sketches will provide you with some answers.

I will tell you about two outstanding practical researchers and entrepreneurs from 19th-century Germany who not only created certain inventions but also—even more importantly—laid at their originally minute enterprises the foundations of future global industries.

Then I will tell you about two people from prerevolutionary and postrevolutionary Russia who had one thing in common: a fantastic idea that people could fly to the stars. Space travel has become reality even though those people did not live to see the accomplishment of the idea they had pursued all their lives.

I will introduce you to two brilliant entrepreneurs of US and British origin who changed the modern world. Incidentally, one of them will probably be taking many would-be astronauts up to the stars in the literal meaning of the word by offering tourist trips to near outer space.

I will also introduce you to a modern designer *cum* inventor *cum* entrepreneur from Belarus who spanned both the globe and outer space with his breakthrough transportation ideas.

I will tell you about a man whose ideas I have been admiring for dozens of years. I am working to bring ideas that he uncovered closer to all my contemporaries. I believe that TRIZ star ideas will be available to every expert, student, and schoolchild and that it will happen in the foreseeable future.

11.1 Josef von Fraunhofer[116]

I would like, for several minutes, to take you ... two centuries back in time!

1. Josef was the eleventh and last child in the family of Franz Xavier Fraunhofer. Seven of his brothers and sisters died at a very young age.

2. His father and paternal grandfather were glass smelters and polishers, and along the maternal line the glass-making tradition went almost two hundred years back to the early 1600s.

Josef von Fraunhofer in 1825
– a 38-year-old professor of the Munich University and a member of the Bavarian Academy of Sciences

[116] Josef von Fraunhofer (March 6, 1787, Straubing – June 7, 1826, Munich) – famous German physicist, optician, experimenter, researcher, and entrepreneur

M.A. Orloff, *ABC-TRIZ*, DOI 10.1007/978-3-319-29436-0_11

3. In 1796, his mother died at 54, and a year later so did his father at 56. The 10-year-old Josef became a turner's apprentice, but he was not strong enough, and he switched to training him in his family trade.

After his father had died, the guardian sent him, under a 6-year apprenticeship contract, to a Munich-based glass-polisher named Weichelsberger.

4. Three years later, in 1801, tragedy struck: The workshop in which he worked collapsed, killing Weichelsberger's wife, while Josef was buried under the rubble and then miraculously rescued.[117] At the site of the tragedy, the saved boy was noticed by Prince-Elector Maximilian IV, the future King of Bavaria, and his companion Josef von Utzschneider, entrepreneur and politician, who had both played an important role in Fraunhofer's life. Prince-Elector gave the boy 18 ducats, enough to buy a metal cutter and a glass polisher to work with glass.

5. Weichelsberger would not permit Josef to attend Sunday school and read books, and in 1804, the 16-year-old boy left him and tried to earn his living by drawing, printing, and engraving visiting cards. He bought the required equipment with the rest of the money he got from the Prince-Elector. Utzschneider helped Josef by lending him books on optics and physics. Unfortunately, his salary—even with the allowance he received from his elder sister—was not enough to live on, and Josef had to return to Weichelsberger to become his apprentice one more time.

6. In May 1806, Ulrich Schiegg, a Benedictine monk, provided him with references that enabled him to go to work at the newly opened Optical Institute started by Utzschneider and his partner Georg von Reichenbach. The institute became a part of the Mechanical Institute, which the partners, together with the well-known watchmaker Liebherr, had founded in Munich in 1804. Before being admitted, Josef passed a harrowing entrance examination that lasted several days!

The institute and the workshop (manufacture) were involved in the production of geodetic instruments that were used to make new maps, particularly for the Army. Reichenbach had studied at the Military Academy at Manheim and visited England where he explored the latest technical novelties such as James Watt's steam engine and Ramsden and Dollond's optical instruments. He created a dividing engine that could grave precise circles; however, to improve angle measurement, the engine needed better optics. Josef Fraunhofer made the lenses that did the job.

7. Fraunhofer also improved the polishing machine, which made it possible to control the quality of lenses and make the process less dependent on the workers' skills. He also developed new polishing pastes and glues to manufacture composite lenses.

This boosted both quality and productivity, which was very important for the growing production of optical instruments.

8. In late 1807, Utzschneider moved the production facility to a monastery at a place called Benediktbeuern. Glass production was housed in a special guarded building (Glashütte) on the monastery grounds. Fraunhofer lived and conducted his famous experiments on the top floor. The ground floor was occupied by the workshop.

---

[117] Dr. Wolfgang Jahn *Er hat uns die Sterne näher gebracht. Zum Leben des Joseph von Fraunhofer.* – In: "Fraunhofer in Benediktbeuern Glashütte und Werkstatt", Fraunhofer-Gesellschaft, München, 2008; illustrations dedicated to Fraunhofer are also taken from that work

With cutthroat competition in the European optics market, workshop locations had to be kept secret to protect production methods and recipes.

On the other hand, it was necessary to advertise the instruments for sale. Accordingly, rapidly growing product and price lists were published in numerous industry papers. Legends about the superior quality of new optical components and instruments quickly spread throughout Europe.[118]

9. In 1805, the workshop was headed by a Swiss master Pierre Guinand. His secret method of making pure glass (without heterogeneity lines, admixtures, or bubbles) was improved by Josef. For example, he suggested that a special moving rod plastered with fire-resistant clay be immersed into molten glass mass to stir it while it cooled down.

10. Fraunhofer proposed a completely new method of lens processing, the so-called "radial polishing method," and improved the "trial glass method" used to measure lens-surface curvature.

Using the trial glass method, Fraunhofer achieved much higher lens-processing precision. He did this first by observing numerous refracted lines that made up the color spectrum with the help of prisms. Then—and this was a truly revolutionary development[119] and a trade secret—he improved upon his first work by observing interference lines that greatly exceeded all previous observations in terms of both quantity and quality.

11. In 1809, the 22-year-old Fraunhofer was appointed the general manager of the workshop. Master Guinand refused to accept that appointment, and he left the institute in 1813. In 1814, Fraunhofer became a co-owner of the Optical Institute together with Utzschneider. It should be noted that later in the 19th century, Guinand and his descendants also scored several major successes in a glass instrument–making business in Europe and elsewhere.

12. By observing sunlight, Josef Fraunhofer discovered and described 574 dark interference bands of various intensity (see his hand-made drawing). In 1815, he

---

[118] Jackson, Myles W. *Spectrum of Belief: Joseph von Fraunhofer and the Craft of Precision Optics* (Cambridge, MA and London: The MIT Press, 2000. – 284 pp.); German translation: *Fraunhofers Spektren: Die Präzionsoptik als Handwerkkunst* (Göttingen: Wallstein Verlag, 2009)

[119] Carl R Preyß, Gründungsmitglied der Fraunhofer-Gesellschaft *Fraunhofers Bedeutung*. – in "Fraunhofer in Benediktbeuern Glashütte und Werkstatt", Fraunhofer-Gesellschaft, München, 2008

submitted that description to the Academy of Sciences in Munich... thus immortalizing his name in the annals of science (later the phenomenon was called *Fraunhofer lines*).

In his another work published in 1822, Fraunhofer described a method that could be used to determine the length of light waves based on diffraction grating. He learnt to determine glass composition using the observable spectrum. He discovered differences in spectrums of various planets and stars.

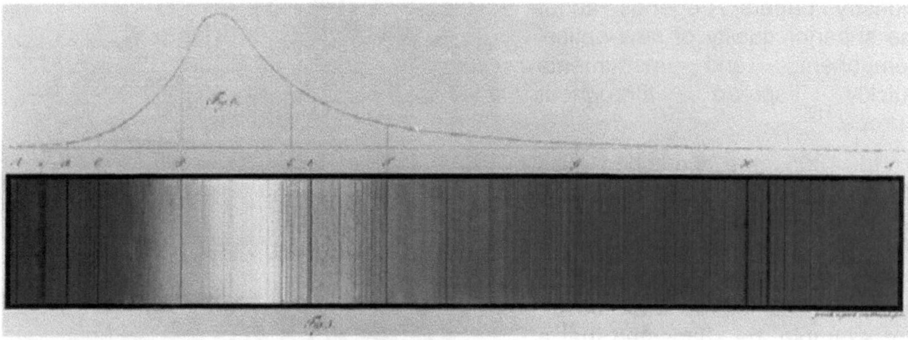

Today spectral analysis is one of the most widespread critical instruments employed both in scientific research and production and spans a broad range of applications from the nano-world to the farther reaches of outer space.

13. In 1816, the great mathematician and physicist Karl Friedrich Gauss (1777 to 1855), who at the time was the head of an observatory (from 1806 and until his death) and a professor at the University of Göttingen, visited Benediktbeuern and personally inspected the institute's achievements. He wrote *The Optical Institute operates under the direct management of a talented and energetic man – Fraunhofer.*

14. Still, it took time for Fraunhofer to be accepted as a full member of the Academy. Historically, craftsman's skill was regarded as the opposite of "inspired genius."[120] It was believed that inspiration, rather than "shameful" adherence to rules and methods required of a craftsman, was the only superior creative form of knowledge. Disputes about the value of "practical knowledge" and about whether it belonged to science were extremely heated, and that dilemma fully applied to Fraunhofer, the more so because Fraunhofer had never gone to school, let alone a university. In other words, he was a fully "self-taught man" in the direct meaning of that expression

Thinking about Fraunhofer, Michael Faraday (1791 to 1867), a great physicist, the founder of the electromagnetic field theory and another self-taught man, wrote that mastery and knowledge can only come from practice or, we might say, from trial and error.

Fraunhofer believed that high craftsmanship can be achieved first and foremost by extracting knowledge that lay deep within the culture and skill of great masters.

He built theory from practice. TRIZ started the same way. We are doing the same thing in Modern TRIZ. The great Faraday, basically, did the same thing, too.

[120] See Myles W. Jackson

15. In 1819, a 9-inch telescope (see picture with an analogue at The Deutsches Museum in Munich, Germany), the world's largest at the time, was put into operation in the Russian Czar's observatory in Tartu (Dorpat).

During a period of 3 years, an outstanding Russian (again self-taught!) astronomer of German origin Friedrich Georg Wilhelm von Struve (1793 to 1864) or, as the 27-year-old director of the observatory was called by his colleagues, Vasily Yakovlevich, used that telescope to create the first binary star catalogues.

After Fraunhofer built the telescope, he was awarded the title of Academician (1821) and an honorary doctorate from the University of Erlangen (1822), and invited to work as a professor and the curator of the physical laboratory at the Bavarian Academy of Sciences (1823). Maximilian I, the King of Bavaria, made him a noble and gave him the Civil Merit Order (1824).

Josef von Fraunhofer demonstrating the spectroscope to Josef von Utzschneider, Georg von Reichenbach, and Georg Mertz, the future owner (since 1839) of Utzschneider and Fraunhofer's workshop who learnt their secrets and multiplied their achievements

16. Josef von Fraunhofer died on June 7, 1826, at the age of 29, from tuberculosis, which at the time was the professional glassmaker disease.

17. Today Fraunhofer's name lives on in the title of Germany's largest applied research association whose operations have already acquired a global dimension—*Fraunhofer-Gesellschaft*—a conglomerate of 80 research teams of more than 18,000 scientists working worldwide with an annual budget of approximately € 1.6 billion.

18. I believe that the best tribute to Josef von Fraunhofer is this expressive and fittingly appreciative two-word epitaph inscribed on the monument that was erected to commemorate this Honorary Citizen of Munich, Bavarian capital, in the city's Southern Cemetery[121]:

Aproximavit sidera* ► *He brought the stars closer

[121] *Aproximavit sidera* (Latin) → *Er brachte die Gestirne näher* (German) → *He brought the stars closer.* I would like to share with my readers another – stunning! – interpretation of this epitaph in the title of an article written by Dr. Wolfgang Jan (see the footnote a couple of pages back): ***He brought us closer to the stars.***

11.2 Werner von Siemens

I am not selling the future for the sake
of gaining short-living benefits.[122]

Werner von Siemens

1. ""I had just turned five.[123] Once, when I was playing in Father's room, Mother came in with my 7-year-old sister Matilda in tow. Matilda was crying her heart out. She was due for a knitting lesson at the pastor's house but, she complained, a dangerously looking gander denied her entrance into the yard and repeatedly pinched her.

Werner von Siemens,
26-year-old artillery lieutenant, 1842

"Despite all Mother's reassurances, Matilda refused to go to her lesson without an escort... And then Father gave me his walking stick which was longer than I was tall, and said: *Then Werner will take you there. I hope he is braver than you.* I knew at once it was quite dangerous, because Father gave me this warning before I took off to do that errand: *If the gander approaches you, bravely walk up towards him and beat him up well, then he will run away.*

"No sooner had we passed through the gate than the gander ran in our direction stretching his neck and hissing in a most horrid manner. My sister uttered a terrified cry and turned back, and I would have gladly followed her, but I believed what Father had told me, so I moved towards the monster – with my eyes closed – desperately beating the ground around me with the stick. And lo and behold! – the gander got scared and, with loud honking, rushed back towards a flock of fleeing geese.

"That first victory made a surprisingly deep and lasting impression on my mind... Later in life, when I found myself in innumerable tight spots, that victory over the gander subconsciously encouraged me not to yield to looming dangers, but vanquish them bravely advancing towards them."

2. Werner and his brother Hans often hunted ravens and birds of prey with their self-made bows, and eventually their archery skills became quite good... Once the brothers had a falling out, and Hans suggested that the winner be determined in a duel rather than according to the law of force because Werner, being the elder brother (at the time

---

[122] My (M.O.) translation from German of one of the best known maxims by Werner von Siemens (source: corporate web site www.siemens.com): *"Für augenblicklichen Gewinn verkaufe ich die Zukunft nicht"*

[123] All fragments of this sketch, including the illustrations, are reproduced from the autobiographical book by Werner von Siemens *Lebenserinnerungen*. – Herausgeber: Prof. Dr. W. Feldenkirchen; Redaktion: Siemens Corporate Archives, München; Piper Verlag GmbH, München, 2008; the memoir was started by the 72-year-old author in June 1889, and completed in late autumn of 1892, not long before his demise; translation from German (with abridgements and redaction) is mine (M.O.); exact quotes are given in quotation marks to differentiate them from freeform rendition of the original; most omissions are not marked out.

Werner was 9 or 10, while Hans was 2 years younger – M.O.), was clearly stronger. Werner agreed it was fair, and the brothers proceeded with the duel, going by the rules they remembered from their father's student stories.

"They each counted off 10 steps, and at the signal released their arrows tipped with needles secured to arrowheads with sewing thread. Hans' arrow hit Werner right on the tip of his nose, piercing the skin and wedging itself deep in the flesh. The boys' shouts summoned Father who extracted the arrow, and then unsheathed his pipe stem to mete out requisite punishment. This, however, ran counter to Werner's idea of fairness, so he resolutely stood between Father and Hans and said: *Father, Hans had nothing to do with it, it was a duel.* Father looked perplexed – he used to do the same thing when his honor was at stake. He never administered the well-deserved beating. Instead put the pipe stem back into the sheath and said: *I do not want to hear about that foolishness again.*

3. "My life took a drastic turn on Easter Day of 1829, when Father hired a private tutor. His choice proved to be exceptionally fortunate. Sponholz was still a young man.

"It took him only a couple of weeks to gain unlimited power over us, semi-wild youngsters, and even today I have no idea how he pulled that off. He never punished us, never uttered a word of rebuke, almost always took part in our games and somehow, in a playful manner, managed to nurture our strengths while mitigating our weaknesses. He knew how to set reachable objectives and feed our energy and ambition with the joy of achievement. And he sincerely shared that joy.

"Thus, in a few weeks, he succeeded in turning youngsters who were running wild and dodged work as best they could into assiduous and diligent pupils who, instead of being forced to labor, had to be restrained from laboring too much. Especially in me, he awakened an unquenchable feeling of satisfaction with useful work, and an ambitious desire to do it well.

"One of his most efficient tools was his stories. Late in the evening, when our eyelids started to grow heavy, he pointed, with a nod of his head, to the corners of an old leather sofa on which he used to sit behind our worktable, and we cuddled in those corners while he was conjuring up visions of our future. We were presented with the key points of our social careers, both those to which we had to ascend through our industriousness and wise application of our moral qualities, and those which we were to avoid at all costs. Thus, we were to abate the hardships faced by our parents – which was particularly relevant for those who were engaged in agriculture in those difficult times. Or we were given a taste of the ill fate that would befall us if we weakened our efforts and failed to resist the enticements of evil.

"Unfortunately, that happiest period of my boyhood lasted less than a year… The pain we suffered upon the loss of our beloved friend and tutor was unrestrained.[124]

"But I have preserved my love and gratitude to him until the present day."

4. "The second invited tutor was the exact opposite of his predecessor. His education system was formalistic. That ageing gentleman insisted mostly on obedience and refined manners… and demanded that we do not disturb him during his off hours. The poor man was ill, and two years later he died in our home from tuberculosis.

[124] The young teacher suffered from depression; on a winter night, he left home, went into the forest, and killed himself with a hunting gun (M.O.)

"...However, those two years were not lost in vain thanks to what I had learned from Sponholz. He had been so successful in shaping my aspiration to do my duty in a conscientious manner and my predilection for sound learning that instead of losing my fervor I kept spurring my new tutor into action.[125]

"In later years I was often plagued by memories of robbing that poor sick man of deserved rest by making him stay at his work place for hours on end, and ignoring the pathetic little ploys he used to get rid of me."

5. On graduation from the Lübeck gymnasium, Werner decided to enter the artillery school in Berlin. His father said: *Things in Germany cannot stay the way they are for long... The only solid thing is the state founded by Frederick the Great and his Prussian Army, and in these times it is better to be the hammer than the anvil.*

Seventeen-year-old Werner embarked on a trip to Berlin, on foot and with a handful of coins in his pocket, to find his place among ... *"future hammers."* The distance from Lübeck to Berlin is approximately 145 miles (234 km)... Werner's application to the country's best known artillery school was turned down. Later he entered another artillery school in Magdeburg. However, in view of his special technical talents, he was assigned for his tour of practice to... the same Berlin Engineering and Artillery School! He thought those years— from the autumn of 1835 to the summer of 1838—were the happiest years of his life.

6. His mother died on July 8, 1839, and 6 months later, on January 16, 1840, the father also died. By that time the Siemens family had 12 living children (2 had died soon after birth). Werner was preceded only by Lüdwig (b. 1812) and Matilda (b. 1814). Werner's officer pay was the only permanent source of income... With some help from his relatives, he assumed the burden of caring for his younger brothers and sisters.

7. "I was grateful to young Berlin physicists for letting me take part in the establishment of the Physical Society. Acquaintance and joint work with those young people, blessed with outstanding talents and firm aspirations, reinforced my passion for scientific research and development, and encouraged me to dedicate my future to serious science.

"However, circumstances prevailed over my wishes. My innate urge to put acquired knowledge to good use rather than let it stay dormant forced me back to technology. And so it was all my life. My love belongs to science, while my work and my achievements lie mostly in the technical realm.

"My scientific pursuits convinced me that technical progress is possible only through dissemination of scientific knowledge among technical experts.

"At that time[126] science and technology were still divided by an impassable chasm."

---

[125] Werner was 14 or 15 years old (M.O.)

[126] I cannot refrain from making this comment: 120 years ago (the lines quoted here were written approximately in 1890), Werner von Siemens defined what today is perceived as an absolutely organic notion and an objectively existing phenomenon—I am referring to the fact that science and technology are deemed to be inseparable—as something that would only evolve in the future, and it is simply amazing that it was he, of all people, who sensed that future, foresaw it, and attempted to merge the one and the other in his practical pursuits, which led him to astonishing solutions and wonderful victories. He also recalls the divide as something that existed approximately 50 years ago, in the first half of the 19th century, informing his conclusion with his own experience and with modern global drive to integrate science and technology in the second half of the 19th century (M.O.)

8. "In Berlin my attempts to earn money with my inventions met with success, even though as a military officer I was restrained in selecting my business enterprises.

"...Soon we realized that speculative inventions not supported with solid knowledge and sufficient funding were a very unreliable thing which seldom, if ever, yielded tangible benefits.

"...Manufacturers often came to me for advice, and thus I came to know about the devices they used and their operating principles. It became clear that technology cannot advance in leaps and bounds, as it often happens in science under the influence of fruitful ideas generated by few outstanding individuals. A technological invention acquires value and significance only when the technology itself is so far advanced that the invention is workable and has become a necessity."

9. "Meanwhile, originally my theory of electrostatic interaction between closed and opened conductors did not earn credibility in the scientific community as it contradicted the dominant ideas of the time.

"Generally, today it is difficult to imagine how a civilized person can live without railroads and telegraph, and it is not easy to mentally revisit the old point of view and understand the difficulties that we encountered then dealing with things which today are taken for granted. Concepts and means now familiar to any schoolchild had to be mastered with effort and hard work."

10. "While building that line[127], I met Mr. Reuter, an entrepreneur who had set up a pigeon mail service between Cologne and Brussels – a useful and profitable business which had to be mercilessly destroyed by electric telegraph.

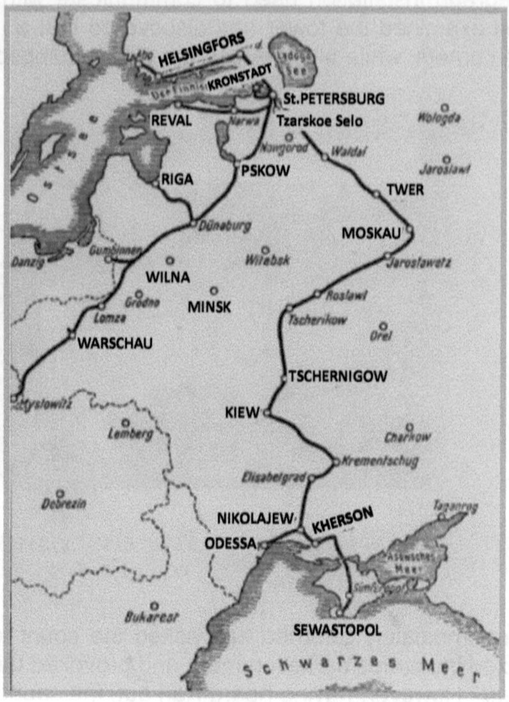

Russian telegraph lines built by SIEMENS in the early 1850-es

"When Mrs. Reuter, who always accompanied her husband on his trips, complained to me about the imminent demise of their business, I told the spouses that they should go to London and start up a telegraph bureau like the one that my cousin, Justice Counselor Siemens, and Mr. Wolff founded in Berlin.

[127] Leaving out the details, this is about a new telegraph line between Cologne and Brussels, one of the first projects (1848) implemented by a newly-created company, Siemens & Halske, that was established in Berlin in 1847 and evolved into the global concern SIEMENS (M.O.)

"The Reuters followed my advice with eminent success. The Reuters Telegraph Bureau[128] and its founder, Baron Reuter, are now world-renowned."

11. "The immense practical significance of electric telegraph was recognized by railroad managers who proceeded to install telegraph lines to transmit messages and signals, which increased productivity of railroads and improved their operating safety record."

12. Count Kleinmichel [129] was rather vexed upon seeing that Karl, Werner's brother, was a very young man. Karl, who then headed the company's Saint-Petersburg office, was a little over 20 years old. However, the count did not revoke his decision, as agreed during an earlier meeting with Werner, and Karl was granted permission to design and lay a telegraph line to the Czar's study in the Winter Palace provided that no alterations were made to the tower, which hosted both the study and the optical telegraph installation used to communicate with Warsaw. Leaving the count's office, Karl examined the tower and discovered that a rainwater pipe was missing on one of the corners, while all other corners were equipped with such pipes.

He immediately returned and asked the count to grant him another audience, which vexed the statesman even more and provoked this caustic question: *NOW what do you want?* However, having heard Karl out, the count summoned the officer responsible for the line and heavily reprimanded him because, according to an earlier proposal, a duct had to be cut out in the tower wall.

Karl proposed to install the missing rainwater pipe and lay insulated telegraph cables inside that pipe (is this not an ideally simple TRIZ solution based on models *Separation*, *Local Property*, *Copying* and, naturally, *Matryoshka (Nested Doll)*?! – M.O.). From then on both Karl and the Siemens company invariably enjoyed the count's favor and support.

---

[128] Even today, in the 21st century, *Reuters Group* (since 2008: *Thomson Reuters*) is one the world's largest international news and financial information agencies (see Wikipedia)

[129] Count Pyotr Andreyevich Kleinmichel (1793-1869) – Russian statesman who carried out orders of the Russian Emperor Nicholas I to build the Nikolayevskaya Railroad (since 1923: Oktyabrskaya Railroad) from Saint-Petersburg to Moscow (1851) and Russia's first telegraph lines (early 1850-es)

13. "Even though I was heavily biased in favor of scientific research, I saw that I must first direct all my energy towards technical work, as its outcomes could – and did – provide me with the means and opportunities to engage in my scientific pursuits. Research and inventing activities in those labor-laden times were almost exclusively dictated by technical requirements."

Photo made in 1864

14. Newspaper[130] *"Illustrierte Zeitung,"* December 8, 1866:

"...In the land where the Goddess of Invention was born (according to the newspaper, it meant England. – M.O.), where practical application follows closely on the heels of scientific discoveries, one of the most magnificent telegraph cable works in the world has been set up by a German.

"The Siemens works at Woolwich (at the end of the XIX century, Woolwich was a suburb, and now it is the area on the Thames in east London. – M.O.) not only supplies submarine and land cables and telegraphic apparatus of all kinds; the establishment is at the same time a scientific laboratory, a place of continuous experiments and clever inventions. Cables from this factory loaded onto Thames steamers and shipped to all parts of the world connect St. Petersburg with Kronstadt and France with Corsica and Algeria; they run for the Pasha of Egypt over the bed of the Nile; they are working in India as well as in Brazil and La Plata[131], at the Cape[132], in Turkey and in Spain.

"Their total length is 6,000 nautical miles, a good part of the girdle which Puck promised to put around the Earth. And this piece of the girdle has been woven in the short space of time since the establishment of the factory, namely in 1859; a proof of the high reputation which the name of Siemens enjoys all over the world...

"That the name of the founder of the works commands the highest respect in scientific circles in England is common knowledge. If educated Englishmen are beginning to renounce the old belief that the German is an essentially unpractical philosopher, the credit is largely due to men like Siemens."

15. Completion [133] of the Indo-European telegraph line could easily be compared with such global accomplishments as construction of the first trans-American railroad from New York to San Francisco; the construction of the Suez Canal, which considerably

---

[130] Direct quotation from Werner von Siemens, *Recollections.* – Piper Verlag GmbH, Munich, 2008 / Transl. from the German by W. C. Coupland, Edited by W. Feldenkirchen.

[131] The capital of the province of Buenos Aires in Argentina.

[132] Abbreviation for Peninsula Cape Cod in eastern Massachusetts, USA; the final part of a telegraph lines laid by Siemens from Europe across the Atlantic Ocean have been placed here later (in 1879); total of eight cable lines were laid on the bottom of the Atlantic Ocean.

[133] Weiher, Siegfried von / Herbert Goetzeler (1984) *The Siemens Company, Its Historical Role in the Progress of Electrical Engineering 1847–1980*, 2nd ed. Berlin and Munich; quoted from: S. Weiher, E. Schroeder, *Great Industrialists. Werner von Siemens. Krupp.* – Series: *Historical Silhouettes.* Publishers: PHOENIX, 1998.

shortened the sea route to India; and, in terms of information transfer, the permanent transatlantic cable line, which was finally completed in 1866.

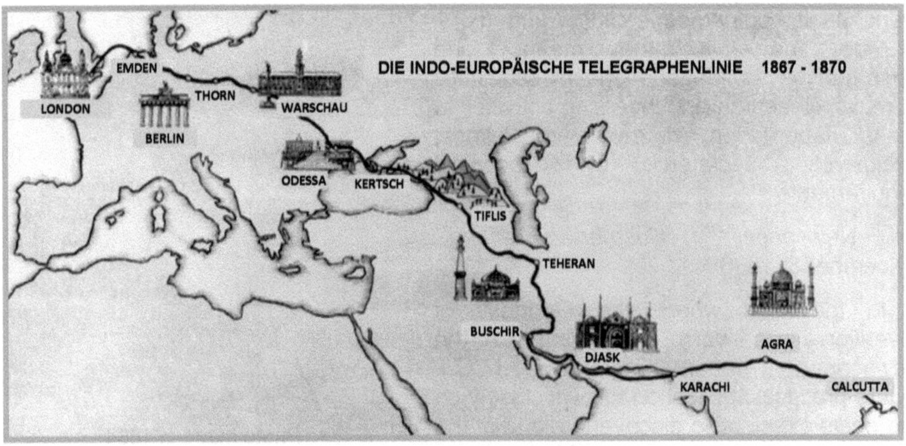

London-Calcutta Telegraph Line (11,000 km)
built by SIEMENS in late 1860-es – early 1870-es

16. After several attempts,[134] Siemens engineers finally managed to make an impeccable cable for the direct line connecting Europe and the USA. Everything they had learnt in expeditions confirmed the cable theory developed by Werner back in 1857 and proved useful in making an even more perfect cable.

Meanwhile, competitors stopped at nothing trying to regain lost ground. They even fished the cable laid by the Siemens brothers out of the sea to cut it into pieces and see what it was made of. However, with time it became clear that such primitive things were not enough to remove a formidable opponent that the Siemens brothers had become.

Within 10 years after the first direct line to the USA, the company had laid five more transatlantic cables.

17. From the letter written by Werner to his brother Karl on December 25, 1887: "Since my youth, I aspired to build a world-class enterprise, like that created by the *Fuggers*[135], that would give not only to me, but also to my heirs, power and authority throughout the world, assuring better lives for my brothers, sisters and close relatives.

"This feeling can be traced back to the stories we heard from Sponholz, our tutor, who used his *life tales* – each time giving us an opportunity to eliminate, in one fell swoop, all hardships suffered by our parents – to encourage us, lazy young boys, to persist in our efforts. That idea was firmly lodged inside me, gaining strength as I moved along my life path which made me responsible for my brothers and sisters."

18. "...I am greatly indebted to lucky happenstance. It was a happy coincidence that my young years concurred with a period of rapid development of natural sciences, and that I devoted myself to electrical technologies – as they were still underdeveloped, which

---

[134] Ibid.
[135] A major group of German industrial, trading and financial enterprises of the late Middle Ages owned by the Fugger family whose descendants are still around today.

created an environment conducive to inventions and improvements. On the other hand, I often had to counter exceptional woes. It was an incessant struggle against completely unexpected difficulties and accidents which, as a rule, impeded my undertakings from the very start – but which, owing mostly to good luck, I always managed to overcome...

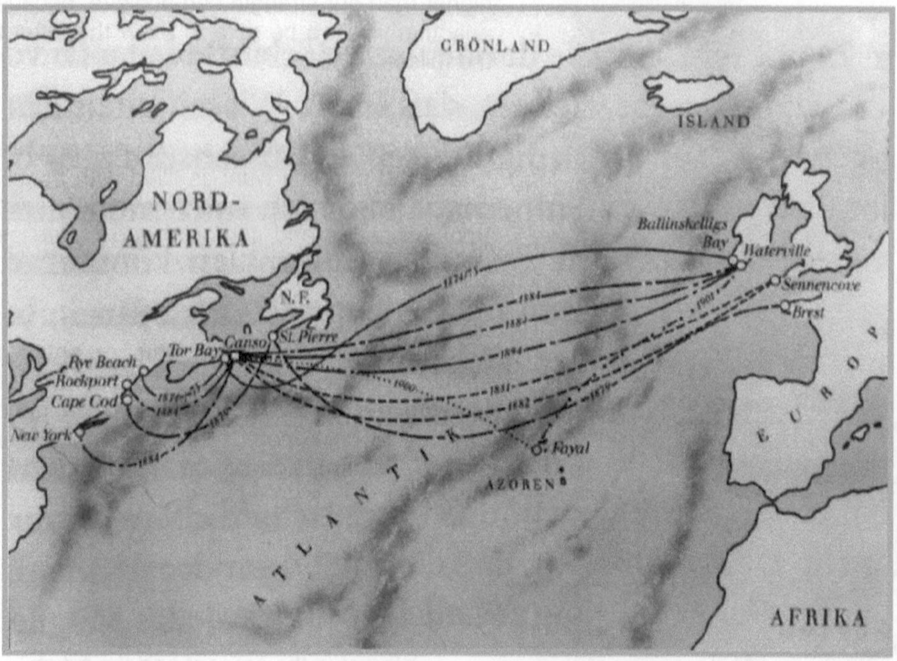

Transatlantic telegraph lines
laid by SIEMENS in 1874 to 1902

"In human life, success and failure, victory and defeat often depend entirely on the ability to promptly and correctly use a good opportunity as it arises. My ability to quickly decide what must be done at the critical moment, and to do the right things without thinking too long, has stayed with me throughout my life, despite my somewhat dreamy mental disposition to which I succumbed on numerous occasions or, I should say, almost habitually. Time after time, that ability protected me from harm and guided me in difficult situations...

"I lacked good memory, and I had no feeling of order, of consistent and inexorable austerity. Nevertheless, I founded large business enterprises and managed them with exceptional success, which proves that diligence combined with energy often vanquish or disarm our deficiencies."

19. "I believe that prompt prosperity of our factories is predicated on the fact that our production is mostly based on our inventions. And even though in most cases they were not protected by patents, they still gave us an edge over our competitors who kept running late, while we kept gaining through the use of our latest improvements...

"My life was wonderful, full of successful efforts and useful labors, and ultimately the only thing that saddens me is that... I will be denied the chance to continue fruitful work to fully realize the *Era of Natural Sciences*."

Monument in the Territory of the Berlin Technical University

Photo by the Author, 2012

20. "Our aim and objective must be always higher than our strength can achieve, because only then will we be able to exert the latter to the fullest." – 1852 [136]

21. "For someone who is serious and prepared to act, the phase 'I want' is filled with magical power! Indeed, one should never shy away from obstacles and setbacks and never lose sight of one's goals!" – 1854

22. "Personal accomplishments of any kind should only be judged according to the benefits they create for others. Such deeds deserve to be honored only when they have made a contribution to the public good." – 1872

"I have always had more express interest in the present and the future, than in the past. I also think that it may prove more useful and stimulating for the coming generation if I can convincingly demonstrate that a young man without inherited resources or influential guardians, moreover, without proper prior training, relying exclusively on himself, can rise and do something useful in the world."

Werner von Siemens

Harzburg, June 1889

---

[136] Paragraphs 20-22: excerpts from Special Issue *160 Years of Siemens*, October 2007, SIEMENS Internet Archive (www.siemens.com/history/pool/en/history/1847-1865_beginnings_and_initial_expansion/160j_e.pdf).

11.3. *Konstantin Tsiolkovsky*

K. E. Tsiolkovsky: ***I learnt by creating...***

Konstantin Eduardovich Tsiolkovsky (1857-1935)

1. For Tsiolkovsky, to "live" meant to "invent"...[137] Until his extreme old age, the scientist treasured the memory of his first childhood experience of aerial navigation. He was 8 or 9 years old when his mother gave the children a toy air balloon made of blown-out collodion (a solution of nitrocellulose in ether or alcohol, which produces a thin film after the solvent evaporates – M.O.) filled with hydrogen. The future champion of dirigibles solemnly paced the rooms of the house and then walked through the garden, pulling around a minute aerostat tied to a thread...

2. At age 10 years, Tsiolkovsky contracted scarlet fever, which left him deaf for the rest of his life.

Wooden house in Kaluga, on the banks of the Oka, where K. E. Tsiolkovsky lived and created many his works.

3. The enterprising adolescent was extremely enthusiastic about self-made 3. The enterprising adolescent was extremely enthusiastic about self-made mechanisms. He always kept working at one thing or another. He put together wind motors, weight clocks, a workable lathe; he made a self-propelled cart fitted with a wind engine that could move against the wind. Another successful do-it-yourself contraption was a primitive automobile (just a reminder: that was in the early 1870s!) that was propelled by a small steam turbine.

4. In Moscow, the 17-year-old young man found himself living in abject poverty. The 10 or 15 rubles allocated by his father out of the meager family budget could only support an existence bordering on starvation, especially taking into consideration the fact that

[137] Paragraphs 1 through 5 are based on: Perelman, 1937.

Tsiolkovsky spent only a small part of that money on food; most of it went to buy books and materials for his experiments. *He ate only brown bread, denying himself even tea and potatoes. "I remember vividly that there was nothing but water and brown bread. Every three days I went to the bakery* and bought 9 kopecks worth of bread. Thus I subsisted on 90 kopecks per month… Still, I was happy with my ideas, and my brown bread diet was never cause for distress" (from Konstantin Eduardovich's autobiographical notes).

5. Many hold the erroneous notion that Tsiolkovsky invented the rocket; in reality, he just substantiated the possibility of its practical utilization. Moreover, Tsiolkovsky predicted the possible evolution of the rocket; for decades to come, he blueprinted the changes that the rocket would go through to increase its power and become capable of performing exciting tasks that could not be performed by any other technical means.

6. On November 17, 1919, the Tsiolkovskys had five unexpected and unwelcome visitors. They searched the house, arrested the head of the family, and took him to the Lubyanka prison in Moscow. There he was interrogated for several weeks.

October 4, 1957

7. "It is necessary for several generations to complete a stupendous amount of work in several branches of science and technology to turn Tsiolkovsky's ideas from fantasy into reality. In 1926 Tsiolkovsky wrote: ***The first great step of mankind is to leave the atmosphere and become an Earth's satellite.*** We now stand on the verge of that step." [138]

8. "Jacob Isidorovich (Perelman) recalled that the article in the *Air Navigation Messenger* literally overwhelmed him. A man named Tsiolkovsky, of whom he had never heard before, had made a great discovery. The road to space was marked. Perelman realized that he would do a lot of good by disseminating Tsiolkovsky's ideas and presenting them in a way that would convince even people whose daily pursuits are far removed from science that space flights are not just castles in the air. That sooner or later people would overcome gravitation and visit the Moon and remote planets.[139]

9. In 1923, a *15-year-old boy Valentin Glushko* wrote to Konstantin Eduardovich:

"Esteemed K. E. Tsiolkovsky! I have a request for you, and I will be very grateful if you do what I ask. This request concerns the possibility of interplanetary and interstellar travel. I have been interested in this matter for more than two years. I have done a lot of reading on the topic. Everything became clearer when I read a wonderful book by Perelman called *Interplanetary Travel.* But then I felt a need to do certain calculations. Without any manuals, just by myself, I began to make those calculations. And then I was lucky to get your article in the *Science Review* magazine (May 1903) entitled *Exploration of Space by Reactive Devices.* Unfortunately, the article was very brief. I know that there is another article under the same title – it was published separately

[138] Paragraph 7 is based on a report made by Valentin Petrovich Glushko (1908 to 1989), an outstanding rocket engine designer, at a meeting dedicated to the 100th birth anniversary of K. E. Tsiolkovsky in 1957, *2 weeks before the launch of the first artificial satellite of the Earth* (!!!).

[139] From the book by G. T. Chernenko *Invitation to Go on an Interplanetary Trip.* – (Г.Т.Черненко *Приглашение в межпланетное путешествие*); see www.fandom.ru.

and was much more detailed. This is what I have been looking for, and this is what my request to you is about. The separate article *Exploration of Space by Reactive Devices* and your other work called *Beyond Earth* are not the only things that prompted me to write this letter to you, there are very many other very important questions that I would to ask you ..."

Tsiolkovsky wrote back to that Odessa schoolboy, sent his books to him, and enquired how serious he was in his passion for space travel.

A happy Valentin replied right away: "As to the question about how interested I am in interplanetary travel, I can only tell you that *this is my ideal and the purpose of my entire life which I wish to dedicate to this great work.*[140]

Presidium of the Festive Meeting Dedicated to the 100th Birth Anniversary of K. E. Tsiolkovsky: first row at the table: V. P. Glushko, 2nd from the left; S. P. Korolyov, 3rd from the left

10. On August 2, 2007, a *Progress* cargo spacecraft was launched from the Baikonur Cosmodrome in Kazakhstan to deliver fuel, food, and materials to the International Space Station (ISS). The launch was dedicated to the 150th birth anniversary of K. E. Tsiolkovsky. On that day, a portrait of the great explorer and dreamer went up into space; it was stenciled on the hull of the craft.

11. I wish to remind you the meaning of the Latin words that I used in the heading:

Per Aspera Ad Astra ► Through hardships to the stars

[140] In March 1938, V. P. Glushko was arrested, and for the duration of the investigation (which ended in August 1939), he was kept in the internal NKVD prison at Lubyanka and in Butyrka prison. On August 15, 1939, he was sentenced to 8 years by the Special Council under the NKVD of the USSR and was later reassigned to the Technical Bureau. On August 27, 1944, the Presidium of the Supreme Council of the USSR ordered his early release, and the record of his conviction was expunged. He was fully exonerated in 1956.

11.4. *Jacob Perelman*

> ***To be able to find the new in the old***
> ***is the sign of a genius.***
> *J. I. Perelman*

1. Białystok. In this city, a second son named "Jacob" was born on November 22, 1882, into the family of Isidor Perelman, a bookkeeper at one of the local cloth factories. The family leased a modest apartment, and the father's meager salary was barely enough to make ends meet. Real hardships began in September 1883 when the father died.

Jacob Isidorovich Perelman
(1982 to 1942).

There was not enough money to live on, and the mother, an elementary school teacher, was forced to give private lessons. The heavy burden of nurturing and rearing the sons fell on her fragile shoulders. Despite the family's financial plight, she decided she would give her children good education. Since their tender years, she inured them to household chores, directed their reading, helped them do their homework and, knowing French and German, taught them those languages.[141]

2. On September 23, 1889, the *Grodno Province News* carried a sketch entitled *Regarding the Anticipated Rain of Fire* that was signed with the author's initials – J. P. Those two letters concealed a student of nonclassical secondary school, then an unknown 16-year-old boy, now a recognized classic, and, furthermore, the founder of the "entertaining science" genre, and the author of more than a thousand popular science articles, more than a hundred books and brochures, and almost two dozen textbooks and manuals—Jacob Isidorovich Perelman. But why "J. P."? Why did the beginning author need a pseudonym? Why did he have to be so modest? The real reason was not modesty, though. According to the rules that existed at the time, schoolchildren and students were prohibited from publishing their works in newspapers and magazines under pain of expulsion.[142]

3. Jacob did not just merely read a lot; he had a reading system: His mother had compiled a long list of books, which he had to read exactly in the prescribed order. In autumn 1901, Jacob donned the uniform of a Foreign Institute student. His life was not easy. He had to pay for his tuition, buy his food and uniform, and pay the rent. His mother could no longer help; lately she had been prone to disease. In May 1903, trouble knocked on Jacob's door: his mother had suddenly died, and the brothers[143] petitioned for urgent leaves of absence and went to Białystok to bury the person dearest to them…

---

[141] Here and in several paragraphs below, I have used excerpts from the following book: G. I. Mishkevich, Doctor of Entertaining Sciences. – Moscow: Knowledge Publishers, 1986. – (Мишкевич Г.И. *Доктор занимательных наук*. – М.: Знание, 1986)

[142] N. M. Karpushina, *Jacob Perelman: Draft Sketch*. – Moscow: *School Mathematics Magazine*, Issue No. 5, 2007. – (Карпушина Н.М. *Яков Перельман: штрихи к портрету*. – М.: журн. "Математика в школе", 5, 2007).

[143] In 1913, his elder brother Joseph (1878 to 1959) went to the USA to become a well-known playwright.

4. In 1912, the Saint-Petersburg magazine, *Air Navigation Messenger*, carried an article by K. E. Tsiolkovsky called *Exploration of Space by Reactive Devices*. Since then J. I. Perelman turned into an outstanding promoter of Tsiolkovsky's ideas of space exploration. He published several books to support and clarify those ideas. The grateful Tsiolkovsky wrote back to Perelman: *Special works are read by few, those that are generally available works can be accessed by millions.*

5. On November 20, 1913, Perelman delivered the world's first report[144] ***On Possibility of Interplanetary Travel*** (s. cover) at a session of the *Russian Society of Globology Amateurs.* The report contained one extremely interesting piece of information: During the launch, it said, people in a spaceship must assume a horizontal position to minimize the effect of multiple G-force load on the human organism. And this is the launch position currently assumed by all modern cosmonauts and astronauts!

6. Perelman's book Interplanetary Travel. Flights into Outer Space and Reaching Celestial Bodies, which had numerous subsequent editions, was published for the first time by the Saint-Petersburg P. P. Soykin Publishing House in 1915. The 10th edition (1935) featured a list of people who had designed Soviet rocket equipment and sent up the first spaceships: S. P. Korolyov, V. P. Glushko, M. K. Tikhonravov, and Yu. A. Pobedonostsev.

7. S. P. Korolyov (who at the time was 25 years old) wrote to J. I. Perelman on July 31, 1932[145]: "Despite our huge experimental load, we all are extremely preoccupied with what we do for the general public. Clearly, it would be completely wrong ... to focus exclusively on the military ... aspects of our work ... while there is absolutely nothing for the members of the GIRD (Russian acronym for "Rocket Propulsion Study Group") anxious to learn, to do some useful work... I would love to see your wonderful books among works that promote rocket science, fight to assure that it is thriving, and educate others. *And if that happens, there will come a time when the first terrestrial vessel will leave the Earth for the first time."* Perelman replied to that request by publishing a brochure called *By Rocket to the Moon* and a new edition of *Interplanetary Travel.*

8. In the early 1930s, J. I. Perelman was one of the first members of the GIRD, a voluntary association of initiators and enthusiasts of rocket science development with chapters in Moscow and Saint-Petersburg and later in other cities. In Moscow, the group's founders and promoters included outstanding researchers and designers S. P. Korolyov, F. A. Tsander, Yu. A. Pobedonostsev, M. K. Tikhonravov, and many others; in Leningrad the group was represented by V. V. Razumov, N. A. Rynin, J. I. Perelman, N. I. Tikhomirov (died in 1930), B. S. Petropavlovsky, V. A. Artemyev, and many others.

---

[144] G. T. Chernenko, *Invitation to Go on an Interplanetary Trip.* – In Anthology: *The Globe*, Leningrad: Children's Literature Publishers, 1988. – (Г.Т.Черненко *Приглашение в межпланетное путешествие*. – в сб. "Глобус", Ленинград: Дет. лит., 1988); quoted from http://www.fandom.ru/about_fan/chernenko_1.htm.

[145] Sergey Pavlovich Korolyov (1907 to 1966), outstanding rocket designer, graduated from Moscow Higher Technical School named after Bauman in 1929; in 1938 he was arrested, convicted, and exiled to Kolyma where he worked as a wheelbarrow operator carrying gold ore to the washing plant. His survival was a true miracle. He was returned to Moscow in 1939, and until 1944 worked in a "*sharaga,"* a closed special design bureau together with many other convicted aviation and rocket designers.

GIRD, the Russian acronym of the group's name, was often jokingly deciphered as *Group of Engineers Working for Free.* And at first it was true: GIRD members often relied on their own funds to pay for welding or soldering works or to buy instruments. S. P. Korolyov, who headed GIRD's Moscow chapter, used to say to his mother: *You see, I did get my salary, but I had to pay for the drawings…* On another occasion he, with her permission, took several old silver spoons to be used as soldering alloy.

9. **THE ART OF BEING AMAZED**[146]

We soon cease to be amazed, and we lose the ability to take interest in things that do not directly affect our existence. Habit kills interest; we find it hard to even focus on things that are happening around us every minute. **To be able to find the new in the old is the sign of a genius.**

We have even gotten used to parachutes, and are not surprised by a living person descending from the sky to the earth.[147] **To overcome the indolence of routine thinking, to draw attention to overly familiar objects, such objects must be shown in a new light, and their unknown sides must be exposed.**

Where there is interest, the door is wide open for new perceptions and new knowledge.

Does this mean that learning must be turned into some kind of entertainment? No, and entertaining science is not guilty of that sin. *The role of the entertaining element is exactly the opposite: Instead of turning science into entertainment, the element of entertaining seeks to use entertainment as a learning tool.* In addition, by exposing unexpected facets of familiar objects, *entertaining science helps deepen understanding and improve observation powers.*

10. In 1935, Perelman initiated the opening in Leningrad of an *Entertaining Science Museum* (it was ruined during the blockade in WWII). One of those who visited the museum was a future cosmonaut, then a Leningrad schoolboy, Georgi Mikhailovich Grechko (b. 1931), later (1955) a graduate of the Baltic State Technical University.

11. Eventually literary critics will probably accord Perelman the title of a genre-creator. A new genre is, in and of itself, an outstanding literary phenomenon that is sufficiently momentous to glorify its creator.

But the true significance of Perelman's discovery goes far beyond the boundaries of purely literary notions. ***Most importantly, he essentially developed and validated a new kind of entertaining education!*** Even the most rigorous critics did not find in his books anything that would even remotely resemble profanation or deformation of science.[148] ***On the contrary, everybody unanimously agreed that those books represented a new type of textbook—one that is available to millions of people and is ingenious, demonstrative, jovial, and yet instructional.***

---

[146] Compiled from the book by J. I. Perelman, *What Is Entertaining Science* (1939), published in the e-version of the *School Libraries Newspaper*, Issue No. 7, 2008, on a wonderful web site lib.1september.ru run by September the First Publishers (www.1september.ru). – (Перельман Я.И. *Что такое занимательная наука*, 1939).

[147] Note that today most people take for granted such devices as notebooks, GPS navigators, or mobile phones, even though none of these three artifacts, at least in its latest implementation (2011), is "older" than 30 years!

[148] The list of outstanding books written by J. I. Perelman includes *Entertaining Physics*, *Entertaining Geometry*, *Entertaining Arithmetic*, *Entertaining Astronomy*, *Entertaining Mechanics*, *Entertaining Algebra*, and others. – (Я.И.Перельман: "Занимательная физика," "Занимательная геометрия." "Занимательная арифметика," "Занимательная астрономия," "Занимательная механика," "Занимательная алгебра" и другие).

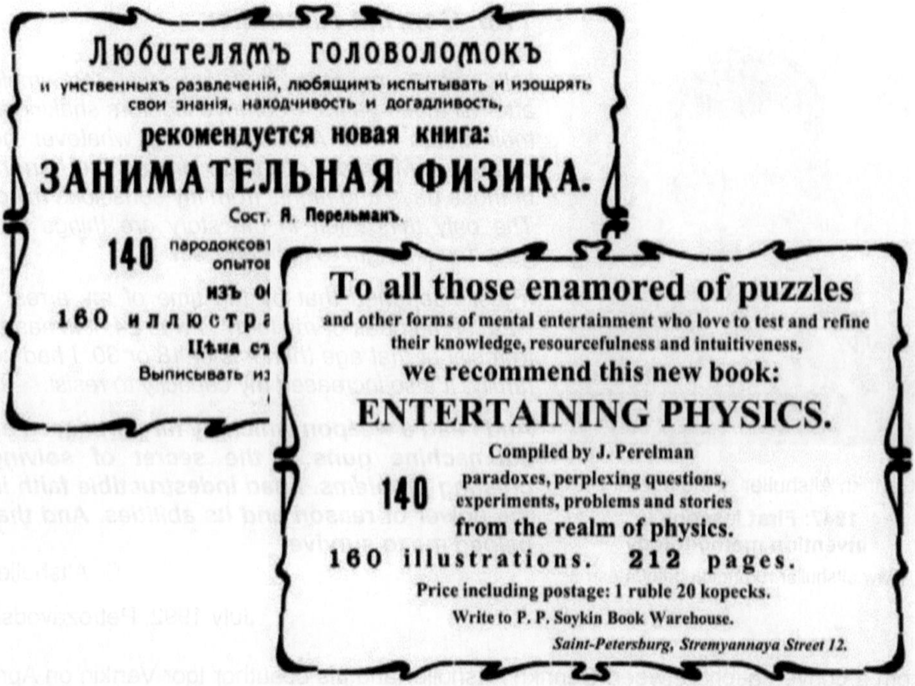

The meaning and scale of that discovery will gain additional weight and significance if we remember that it was published in a country where many millions of people were illiterate.

When asked about his predecessors at one of reader conferences, Jacob Isidorovich replied: *Russia is rich in marvelous science promoters. I learnt many things from them, but I cannot write as they did…*

12. CHRONICLES OF THE LENINGRAD BLOCKADE[149]

July 1, 1941, to February 1942: Read lecturers on ways to find one's location without instruments to military intelligence officers of Leningrad Front and Baltic Fleet and to local partisan fighters.

January 18, 1942 – Anna Davidovna died of starvation while on duty at the hospital.

March 16, 1942 – Jacob Isidorovich died of extreme emaciation caused by starvation experienced in Leningrad during its siege by German Nazi troops.

Jacob Isidorovich Perelman with his wife Anna Davidovna Kaminskaja-Perelman. **Leningrad, May 1941.**

[149] ru.wikipedia.org/wiki/Перельман,_Яков_Исидорович

11.5. *Genrikh Altshuller*

Genrikh Altshuller at the age of 21.

1947: First lessons in invention methodology

www.altshuller.ru/photo/photo04.asp

I did not like my story[150]. It looks way too simple after all these years. Poor investigators shaking in their boots, with Altshuller doing whatever he pleases… My brain obligingly erased the horrors of those days and nights from my conscious mind. The only things left in the story are things that gave me strength to fight and win.

It just happened that by the time of my arrest I was "at the peak of my form". I was 24: it is easier to resist at that age than it is at 18 or 30. I had no family. It also increased my capacity to resist.

And I had a weapon which by far surpassed all submachine guns – the secret of solving creating problems. I had indestructible faith in the power of reason and its abilities. And that helped me to survive.

G. Altshuller

July 1992, Petrozavodsk

From a conversation between Genrikh Altshuller and his coauthor Igor Vertkin on April 28, 1986:[151]:

1. I remember we were going from Krasnovodsk to Baku by steamer. In the canteen, I dropped a fork (or it may have been a spoon). Terribly afraid, I asked: "Did the captain hear the noise?" They assured me that he did. I was flabbergasted… This must have been the first time when I felt reverence for all Captains…

2. It was approximately in 1928. Baku. We lived on the second floor, in a glazed verandah. Each morning the sun rose from behind the wall of the house across the street, and the windowpanes flashed "gold"…

3. We did not have an apartment in Baku, and we had very many books. Moving from one apartment to another, we dragged them behind. I remember those books very well.

4. The yard was a good one. I find it hard to say what exactly shaped my character. Was it books that I was reading, little by little beginning to understand them, or was it the yard community? The thing is, that community was special, it was competitive. If someone got something, the others tried to catch up and take over. Say, someone started to play chess. There was this guy who carried out a chessboard and, with a heroic air, offered us to have a go at him. He kept winning for about two weeks, and then the others, just to spite him, rose to his level. The chess was forgotten and

---

[150] Compiled from G. S. Altshuller, I. M. Vertkin, *How to Become a Genius. Life Strategy of a Creative Personality.* – Minsk: Belarus, 1994. – (Альтшуллер Г.С., Верткин И.М. *Как стать гением. Жизненная стратегия творческой личности*. – Минск: Беларусь, 1994). In 1950, G. S. Altshuller was arrested by state security bodies together with his colleague Raphael Shapiro, and both were sentenced to **25 years** (!) in Vorkuta camps for a letter to Stalin and several ministries where they described the drawbacks of the USSR patent system and proposed a new inventing methodology (the future TRIZ) designed to improve the quality of inventions and reduce inventing time frames.

[151] http://www.altshuller.ru/interview/interview5.asp

replaced with the bicycle. Everybody got their bicycles, and we started a race. Then someone got an "A" in drawing and started a drawing society. Then there was photography...

It was not important what you did, what mattered was that you were the best at what you were doing. I was the youngest boy in that company. I was one or two years younger than the rest, and it was hard for me to get through. It was probably then that I first had an idea that later held up me throughout my entire life. You see, someone was the best cyclist, someone else was the best chess player, etc. Unfortunately, I realized that I had no outstanding talents that I could boast or show off. For starters, at least in my yard. Then I made a second discovery – when your read to someone, all desires passed on into another plane, and there the outcome did depend on preliminary preparation. Not on physical abilities of the person, but on how well he was prepared. I will not say that this idea was clearly formulated in my mind, but it did twist and turn somewhere deep inside.

I needed a method. And books became my method.

3. I believe that it was then that I came up with my first innovation proposal. It was a time of beautiful revolutionary holidays. Everyone wanted to attend the celebrations. May Day, anniversary of the October revolution... But there were not enough toilets... People would do it in houses, behind front entrance doors. Naturally, those who lived in those houses set up their defenses – caltrops, barbed wire, locks – you name it... And then I had this idea of a movable toilet mounted on a car trailer. I decided to make a small-scale model. I had a children's tool set (Father had bought it for me). So I made all parts from plasticine and solemnly presented the whole contraption to Father. He said it was too good to be new. I resented the implication. After all, I had thought it all up myself! I had been out in the street, and then I had thought it all up... Father went to the bookshelf and produced an architecture manual with a picture of a similar street toilet in Paris. I was impressed. I realized it was not all that simple to make an invention.

4. ...I did not even suspect that I was acquiring an explosive character. No, not just "explosive", but "insolent-lingering-explosive"...

5. Not only did I have no musical talent, I was also phenomenally stupid when it came to learning foreign languages. It was terrible! I could not make myself learn four rhymed lines...

If there was an algorithm for understanding, I did it easily. If something just had to be mechanically learnt by heart, it was an almost insurmountable obstacle for me.

6. Mather and Father often went away. Somebody had to care for me while they were gone. Sometimes the landlady did it, sometimes it was the neighbors. I should say it was difficult to starve to death in those times. There were grapes growing in our yard, and they were good grapes, too. Bunches of roach strung on ropes were hung out in the yard to dry, and you could always get a plate of soup if you came to someone's home. But Mother thought it was not enough, and so she decided that she would entrust my care to Lavrov. Lavrov was an alcoholic professional artist. He had graduated from the Academy of Arts. What did he look like? He looked like a character from Gorky's play *At the Bottom*: barefoot, disheveled, unkempt hair and beard, plain cloth shirt.

He made portraits to order from photographs. Typical hack! But people said he used to be a good artist. I was irritated by that duality. So once I told him: "You are no artist at all!" And I said it in a cruel, pitiless manner. Now it all has passed, but then I was hungry... "You do not even remember any longer how to hold a brush." I said many beastly things, keeping close to the door all the time so that I would have a retreat route if he started a fight. He said: "Do you want me to paint something for you?" "I do", I replied. "What? You choose, and I will paint it."

Then I said I wanted him to paint the night sky. I realized that it was one of the hardest things to paint. He started his preparations which lasted well into the next evening. He even temporarily stayed off alcohol – well, maybe he took a little beer... He was pensive. He kept brushing his beard with a piece of broken comb. He put together his paints. And then the evening started with a painting séance. To my surprise, he did not go out to look at the sky, but instead started to paint right away from memory. He worked for a long time. I fell asleep, then woke up – he was still painting – then dozed off again. It lasted for five or six hours. He painted a cloudy sky. The clouds in the middle were parted, and one could see the moon in the opening. There was also a small star next to the moon. And that was all. Lavrov complained he did not have the paints that he needed to do it right. But the painting was breathtakingly beautiful.

When I woke up, I saw the finished piece... it was like a postcard, maybe one and a half post cards. And he had lost all his strength, he was just sitting there with a blank face staring at that fragment, and he could not do anything else. What could I say?... I realized the creation was akin to hard labor. For several days I lived next to that artist. I clearly saw that his undeniable talent had been ruined by alcohol. I could not explain it. But for me even a glass of beer, ordinary beer, became poison.

Genrikh Altshuller's workshop.
Documentary Film Invention Algorithm.
Central Scientific Film Studio, 1974
www.altshuller.ru/img/altshuller/gsa019.jpg

7. If I were to range those factors, the parents'[152] opinion would come out on top. Another factor that may affect the shaping of a creating personality is age. I believe that childhood is the most favorable time for the emergence of creative properties. The child must grow up in an environment where there is plenty of spare time, and be exposed to powerful creative stimuli. Competitive environment and good school teachers – these are important things, too. The home. Next... they kept telling me that life has any meaning only if you do something for the people. Something big, something significant, like discovering the North Pole... And this takes a lot of preparation, this takes knowledge, and you must prepare, prepare, prepare... All those lines crossed, converged at that point.

Moreover, I saw my father work. He could work on a small newspaper article or on a large magazine article for days, even weeks on end. In other words, I grew up in a "non-hack" environment. And even it was clear that the "non-hack" way is a difficult way, it was only the only worthy way. There was no choice there.

---

[152] Saul Efimovich, Genrikh Altshuller's father, journalist, and writer, died in late 1942 after a prolonged disease. Rebecca Yulyevna, his mother, also a journalist, committed suicide in 1954 succumbing to lawlessness, hopelessness, and despair believing that she would never see her son again. She died 6 months before Genrikh—convict no. 1-Ch-502—was released from the Vorkuta camps.

8. It should be noted that writing books and, generally speaking, other literary pursuits were far from my mind. Thus I finished the sixth grade, and it seemed things were beginning to work out. Then there came a time[153] when we all had to forget about writing books. The best thing was to refrain from sticking your neck out. And my father did just that. He was engaged in run-of-the-mill publishing work.

It was a perilous and portentous time. They would tell us: open your textbooks at page such-and-such and cross out Tukhachevsky's (Marshal of the Soviet Union, repressed and executed in 1937. – M.O.) portrait, he has proven to be an enemy of the people, and our valiant national security agents have exposed him for what he really is.

Parents commented on those events in the most negative terms. They saw a wave of arrests rising. Being newspaper people, they knew more. It was the time. Ominous, unforgiving.

9. Then I became interested in space rockets. Naturally, it all started with sci-fi novels by Jules Verne. Probably, with *Nautilus*... Anyway, I scrutinized several dozen books... to the extent of my understanding. I did not understand mathematical books at all, but I knew Max Valier's *Der Vorstoß in den Weltenraum* (*The Advance into Space*) almost by heart... So I thought why not... continue what Valier had started. Build a boat equipped with a rocket engine. We were going to accelerate to about 400-500 km/h. The world record at the time was 600 km/h. What was the sense of building a boat if it was not going to break the world record?

...The pipe (for pumping outboard water into the "carbide" engine while the boat was in motion – M.O.) was bent so that its mouth was facing the direction in which the boat was moving. We borrowed that from an American invention (which enabled high-speed trains to have water pumped into their tanks without stopping). That invention was described in Perelman's books. We had made ample use of his *Entertaining Physics* and other books. They had been one of the main sources of inspiration for our boat. And for a vessel designed on the basis of Perelman's entertaining science books, our boat looked quite presentable.

...Our chemistry teacher had a laughing fit. He said, a stupid head does not let the feet rest... I remembered that proverb well. We should have spent more time thinking that thing through...

10. Each day there was a zillion of questions. We grasped the first solution that came to mind. I did not know then what a technical contradiction was. I had no idea that manifestation of technical contradictions is an objective law. That you cannot expect them to just disappear all by themselves.

11. I already spoke about how it was in my childhood – competitive environment, etc. A sci-fi book came into my hands at a felicitous moment – a little earlier or a little later, and it would not have had the same effect. It was a serendipitous convergence of events. Another such convergence was my cooperation with the invention inspection of the Caspian Military Flotilla.

It did not take me long to become a patent specialist, and then I had a revelation: why not dig inside books on the methods of problem-solving. Well, first of all, I did not find many such books, which did not just surprise me – I was virtually dazed and confused. In 1946 I read a book by Orlov called *Inventor's Secrets.* We found books written by

[153] 1939 (M.O.).

Engelmeyer, read one, then another, then one more. Gradually I came to the conclusion that there was no inventing theory, that it was an amazing gap, and that it was either good luck or bad luck that I could not find such book. Naturally, I realized that I had to do this big thing. Create, select methods, design a theory, write it up, implement it.

...Produce an unequivocal definition of "invention" as the final result of a process purporting to remove a technical contradiction. What is to "solve a problem"? To solve a problem is to find and remove the underlying contradiction. All our knowledge, all that we learnt from books on inventing, depicted a struggle to eliminate contradictions. Our personal experiences, books, history – everything pointed in that direction.

There was a breaking point: the period from 1946 to 1948. It took forever for those two years to come and go. My life objectives were changing. Previously I was out to make an invention, a MAJOR invention, probably related to submarine vessels; in 1948 I realized that the rest of my life would rotate around the theory of inventive problem solving.

12. I continued to invent, but my priorities changed, as did my life plans. And it all happened over the course of two years. At first Shapiro followed my footsteps: he counted on a fast win, he hoped that the future TRIZ would rapidly gain acceptance and recognition. But in the course of our work we saw that rapid acceptance and recognition were not in the cards, that first we would have to deal with deeply-rooted prejudices. Prejudices that were extremely powerful and ancient, prejudices that could be traced back to primeval creative activities.

From the book by Genrikh Altshuller and Igor Vertkin *How to Become a Genius*:

13. Powerful inventors find powerful solutions to individual complex problems. Extra-powerful, super-powerful inventors formulate universal principles that underlie all solutions. A creative person must be capable of solving the most complex problems. Until recently there was no place where one could learn to do it. TRIZ principles can be applied in all walks of life. Today anybody can learn to solve creative problems. Any normal person can do that.

14. TRIZ is a young but evolving science. TRIZ is evolving not only deepwise (deepening), but also broadwise (expanding), creating bridgeheads for exact sciences. Today TRIZ is beginning to penetrate both science systems and art systems. There was a time when attempts to formulate the principles governing creative technical problem-solving appeared to be equally timid and hesitant.

And it took great courage to undertake, at the age of 20 (that is how old G. S. Altshuller was in 1946), a "heretical" task of formulating the "algorithm" of creation. Today TRIZ is already there, it is used to solve problems, it is being disseminated and researched.

15. For millennia, inventors have been waging a relentless war on problems, and each time they enter the battlefield armed with insignificant personal experience accumulated in the course of solving several dozen random problems (at best), and some knowledge. But experience and knowledge are of little use, as they lead to customary actions, while to solve a complex problem one needs to do something illogical, something extraordinary.

And inventions are made later than they could have been made. Because the existing inventing method is bad. It is worse than bad – it is wrong. Because it is the method of trial and error.

Naturally, the inventor does not mechanically go through limitless possibilities. First he tries things to which he is accustomed, things which seem to be logical and justified. When they fail, when after many months or years the inventor remains empty-handed, he resorts to trying abnormal, "wild", random things.

16. Similarly to Schweitzer's *Ehrfurcht vor dem Leben* (*Reverence for Life*), we need to speak of "reverence for time". Time that goes and never returns, time losing which is a crime at any age.

17. The first efforts to identify personal qualities typical for creative individuals were conducted with the participation of G.S.Altshuller and TRIZ trainers and developers V.M.Gerasimov, S.S.Litvin, I.M.Vertkin, B.L.Zlotin and A.V.Zusman, in 1984 and yielded the following 6 interrelated qualities:

Genrikh Altshuller, Raphael Shapiro and Valentina Nikolayevna Zhuravlyova (1933-2004, wife of Genrikh Saulovich, sci-fi writer).
Baku, 1959
www.altshuller.ru/photo/photo02.asp

Worthy objective	The individual must have a novel, previously unattained, significant, socially useful, and worthy objective (of system of objectives).
Plans	The individual must have a plan (or set of plans) designed to assure attainment of his/her objective(s) and to monitor implementation progress.
Capacity to work	The individual must be willing to perform, and actually perform, a huge volume of work to implement his/her plans.
Problem-solving method	The individual must have a method that can be used to solve problems that he/she may encounter on the way to his/her final destination.
Resilience	The individual must be capable of defending his/her ideas, dealing with public censure and incomprehension, "sticking to his/her guns," and remaining true to his/her ideals.
Effectiveness	The results attained (or their scope) must be commensurate with the original objective(s).

18. We must not lead an unconscious life. It is not worthy of us. Human beings are not leaves blown around by a hurricane. They must abide by human standards and rules, and at all times remain what they are. Human beings must compare themselves not to any random persons, such as neighbors living next door, but to people who have changed the course of history. And they must seek to become such people. Anyone can do that.

Anyone can ascend,
and to spend all one's life marking time
is a crime before oneself…

Steven Paul Jobs (1955-2011)
www.ferra.ru/techlife/news/2011/10/06/Steve-Jobs

11.6. *Steve Jobs*

Steve Jobs (magazine *Wired*, 1996: *The Next Insanely Great Thing* by Gary Wolf):

Creativity is just connecting things. When you ask creative people how they did something, they feel a little guilty because they didn't really do it, they just saw something. It seemed obvious to them after a while.

That's because they were able to connect experiences they've had and synthesize new things. And the reason they were able to do that was that they've had more experiences or they have thought more about their experiences than other people.

From the book[154] by Gene A. Landrum (creativity coach and well-known writer and biographer of creative people), where a separate chapter is dedicated to Steve:

1. At school he preferred to spend time with the older children. One of them, Steve Gary Wozniak, was four years his senior. Jobs was interested in electronics, and became ecstatic when he and Wozniak made a "blue box" which prevented the telephone company from recording long-distance calls. Wozniak manufactured such devices while he was studying at Berkley, while Jobs sold them still being a senior schoolboy. That role distribution presaged their future cooperation – three years later Jobs (20) and Wozniak (24) played out the same scenario at Apple.

2. He rejected generally accepted views, which guaranteed his future innovative success. He was a loner – too eccentric at times – and avoided other children. He was different from them, and he wanted to go his own way, which eventually led him to starting up Apple Computer, Inc.

3. Apple was already in operation in 1975, 1976 and 1977. The firm had no human or financial resources available to IBM, HP, Intel, DEC and many other firms which had all the needed to take the leading positions in the PC market. All the two Steves had was a dream. Still, they never gave up, they believed in success and, because of that, they always got what they wanted. They came very close to defeat in 1976, when Atari and HP almost completed a takeover of Apple Computer. However, the three firms believed that a US$100,000 price tag plus US$ 36,000 payoffs to the three title holders were unjustified. How lucky for the founders of Apple! After all, just four years later Jobs' dividends alone amounted to US$256 million.

The house of Paul and Clara Jobs, Steve's adoptive parents, at Crist Drive, Los Altos, California. In 1976 its garage became Steve Jobs and Steve Wozniak's workshop and the birthplace of the first personal computers and of the world's most expensive company (2011) – **Apple**.

4. Steve Jobs did not create any of the components making up microprocessor-based personal computers. He did not contribute to system design or software development.

[154] Gene N. Landrum *Profiles of Genius: Thirteen Creative Men Who Changed the World*. – NY, Prometheus Books, 1993

In fact, Jobs was not in any way involved in the technical part of Apple-1 and Apple-2 projects. All technologies were authored by Steve Wozniak. Jobs acted as a catalyst prompting the birth of a new market, he organized that precarious business – and he succeeded. Wozniak wanted to sell his inventions; instead, Jobs convinced him to sell the calculator to HP, and enter the risk-laden path of innovation.

5. Jobs became Apple's director in 1981, but his managerial career was (in the opinion of many who knew him) too authoritarian, erratic and unsustainable. The advent in 1983 of John Sculley marked the beginning of decline and ultimate fall of the temperamental and presumptuous Jobs, who was eventually fired in 1985.

6. Jobs did not take his dismissal close to heart. He immediately proceeded to develop a new concept envisaging implementation of state-of-the-art microprocessor technologies in education and production of work stations at his new company NeXT (Jobs stood at the helm of Apple one more time in 1997 – M.O.).

Steve Jobs[155]:

7. (Playboy, 1985) The most compelling reason for most people to buy a computer for the home will be to link it to a nationwide communications network. We're just in the beginning stages of what will be a truly remarkable breakthrough for most people — as remarkable as the telephone.

8. (Wired, 1996) When you're young, you look at television and think, There's a conspiracy. The networks have conspired to dumb us down. But when you get a little older, you realize that's not true. The networks are in business to give people exactly what they want. That's a far more depressing thought... We can have a revolution! But the networks are really in business to give people what they want. It's the truth.

9. (BusinessWeek, 1998) It's really hard to design products by focus groups. A lot of times, people don't know what they want until you show it to them.

10. (This and next paragraphs: Stanford University commencement address [June 12, 2005]) It wasn't all romantic. I didn't have a dorm room, so I slept on the floor in friends' rooms, I returned Coke bottles for the 5 cent deposits to buy food with, and I would walk the 7 miles across town every Sunday night to get one good meal a week at the Hare Krishna temple. I loved it. And much of what I stumbled into by following my curiosity and intuition turned out to be priceless later on.

[155] all pictures are from Wikipedia

11. Your work is going to fill a large part of your life, and the only way to be truly satisfied is to do what you believe is great work. And the only way to do great work is to love what you do. If you haven't found it yet, keep looking. Don't settle.

12. I didn't see it then, but it turned out that getting fired from Apple (1985 – M.O.) was the best thing that could have ever happened to me. The heaviness of being successful was replaced by the lightness of being a beginner again, less sure about everything. It freed me to enter one of the most creative periods of my life.

13. Your time is limited, so don't waste it living someone else's life. Don't be trapped by dogma – which is living with the results of other people's thinking. Don't let the noise of other's opinions drown out your own inner voice. *And most important, have the courage to follow your heart and intuition.* They somehow already know what you truly want to become. Everything else is secondary.

16. When I was young, there was an amazing publication called *The Whole Earth Catalog*, which was one of the bibles of my generation. It was created by a fellow named Stewart Brand not far from here in Menlo Park, and he brought it to life with his poetic touch. This was in the late 1960s, before personal computers and desktop publishing, so it was all made with typewriters, scissors and Polaroid cameras. It was sort of like Google in paperback form, 35 years before Google came along: It was idealistic, and overflowing with neat tools and great notions.

Stewart and his team put out several issues of The Whole Earth Catalog, and then when it had run its course, they put out a final issue. It was the mid-1970s, and I was your age. On the back cover of their final issue was a photograph of an early morning country road, the kind you might find yourself hitchhiking on if you were so adventurous. Beneath it were the words: "Stay Hungry. Stay Foolish." It was their farewell message as they signed off.

Stay Hungry. Stay Foolish.

And I have always wished that for myself. And now, as you graduate to begin anew, I wish that for you.

14. (FORTUNE magazine, November 9, 1998: The Three Faces Of Steve... by Brent Schlender & Steve Jobs) I'll tell you something else that makes you ***look at things differently***.

11.7. *Richard Branson*

Everyone needs
something to aim for.
You can call it a challenge,
or you can call it a goal.
It is what makes us human.
It was challenges
that took us
from being cavemen
to reaching for the stars.
Be innovative.
Nothing is impossible.
Think creatively.
The system is not sacred.
Find another way. [156]

Richard Branson

Sir Richard Charles Nicholas Branson (b. 1950) – the famous British entrepreneur, billionaire, founder of the airline company *Virgin Atlantic Airways* (1984) and others, including *Virgin Galactic* (2004), winner of world and other records in ballooning and crossing the Atlantic (in a balloon with hot air in 1986 and a speed boat *Virgin Atlantic Challenger II* in 1987).

Picture: http://zowienews.files.wordpress.com/2011/05/virgin-galactic-richard-branson.jpg

1. I seemed to be constantly breaking some unwritten rule about which I had no idea and, as a reward, was caned.

That was bad enough, but having to thank the headmaster politely for the privilege of having so much pain inflicted on my posterior was incredible.

A bigger problem was that I was dyslexic. Words were just a meaningless jumble to me, and however hard I struggled to read and spell, I couldn't for a long time, until I trained myself to concentrate over several years.

2. My first lesson in self-sufficiency was when I was about four years old. We'd been out somewhere and on the way back Mum stopped the car a few miles from our house and told me to find my own way home across the fields. She made it a game, one I was happy to play. It was an early challenge I've never forgotten.

As I grew older, these lessons grew harder.

3. Early one winter morning, when I was about twelve and home for half term from boarding school, Mum shook me awake and told me to get dressed. It was dark and cold, but I crawled out of bed. After breakfast in the kitchen – probably hot and nourishing porridge, to sustain me – I was given a packed lunch and an apple. 'I'm sure you will find some water along the way,' Mum said, as she waved me off on a fifty-mile bicycle ride to the south coast.

It was still dark when I set off on my own with a map in case I got lost.

---

[156] All paragraphs compiled here with very small rearrangement by M. O. from the book Branson, R. *Screw It, Let's Do It: Lessons in Life and Business (Expanded Edition)*. – Virgin Books, 2009.

I spent the night with a relative and returned home the next day. When I walked into the warm kitchen … I felt very proud, sure I would be greeted by cheers. Instead, Mum said, *'Well done, Ricky. Was that fun? Now run along, the vicar wants you to chop some logs for him.'*

4. To some people this might sound harsh. But the members of my family love and care for each other very much. Those early lessons, which increased as we grew, were because my parents wanted us to be strong and to rely on ourselves, to be free, independent spirits…Mum knew that losing wasn't fair, but it is life. It's not a good idea to teach children that they can win all the time. In the real world, people struggle and there are winners and losers and sometimes injustices that we have to rise above.

5. It was against the law, but I just thought I was bending the rules a bit and taking advantage… (due to sales of music records imported with manipulation in customs tax. – M.O.) It was a huge shock when I was arrested, driven to Dover and thrown into prison.

I couldn't believe it. I thought that only criminals were hanged up. But. Alone in that bleak cell lit by the unrelenting glare of a single bright light bulb, it slowly dawned on me that I wasn't a hippie pirate. This wasn't a game. And I *was* a criminal. My headmaster's words came back to me. When I left school, aged sixteen, he had said, 'Branson, I predict that you will either go to prison or become a millionaire.'

I wasn't a millionaire – but I was in prison. My parents had always drummed into me that all we had in life was our good name.

…In the morning, Mum came to the court to support me. The judge … set [a bail. – M.O.] at a whopping £30,000. I didn't have that kind of money… so Mum put up her home as security. Her trust in me was almost more that I could bear. She looked at me across the court and we both started to cry.

Customs agreed to settle the case by fining me a sum equal to three times my illegal profit. It came to a massive £45,000… It seemed a scary prospect to have to find but I wasn't angry. I had shown the law no respect and deserved to pay. ***No doing anything illegal*** has been my watchword since then.

…I will always remember her words on the train back to London. 'I know you've learned a lesson, Ricky. Don't cry over spilt milk. We've got to get on and deal with this head on.'

Your reputation is everything… *'Be fair in all your dealing. Don't cheat – but aim to win.'*

6. One of the things I try to do at Virgin is make people think about themselves and see themselves more positively. Some people call this 'reinventing yourself' – but I think that we all have inbuilt strengths that need to be discovered and brought out, not reinvented… I tell them, *'Believe in yourself. You can do it.'*

And I also say, *'Be bold but don't play gamble.'*

7. I believe in goals. It's never a bad thing to have a dream, but always I'm practical about it. I don't fantasize and sit daydreaming about the impossible. I set goals and then work out how to achieve them. Anything I want to do in life I want to do well and not half-heartedly.

I like to work holistically, within a team. I will never say, 'I can't do this because I don't know how to.' I'll ask people, look into it, find a way. Looking, listening, learning – these are things we should do all our lives, not just at school.

8. I was bold, yes – but not foolish.

9. It wasn't as if I were stealing all this; legally it was available under the rules of fair competition. But it was only by being persistent and innovative that we won the day.

11. The world is changing. Ideas and opportunities are expanding fast. Sometimes, your ideas don't take off. Even after careful research, not all ideas are good; sometimes your competitors have better ideas or they're there faster than you. The modern entrepreneur takes failure in his or her stride and moves on. You can learn from ideas that haven't worked and use it to know when to play the hand or when to fold.

Richard Branson and Burt Rutan (*Elbert Leander "Burt" Rutan*, b. 1943) – a prominent American innovator in the design of high-performance light aircraft, including aircraft *Model 76 Voyager* (the world record in 1986: 9-day non-stop flight around the globe); *Virgin Atlantic Global Flyer* (the world record in 2005: 3-day round the world non-stop flight, etc.); aircraft-carriers *WhiteKnightOne* and *WhiteKnightTwo*; rocket planes *SpaceShipOne* and *SpaceShipTwo*, and others.

Picture:http://s3-blogs.mentor.com/jvandomelen/files/2010/11/virgingalactic1.jpg

11. I believe we should all assess our lives from time to time.

Have we reached our goals? Are there things we can weed out that we don't need?

I'm not talking about throwing away old shoes or broken chairs. *I mean we need to lose our bad habits or lazy ways that hold us back and clutter our minds.*

12. The famous explorer Captain Robert Scott[157] was my grandfather's cousin. A man of great courage, he made two trips to the Antarctic as a step in his goal of being the first man to the South Pole... He reached the South Pole in 1912, but ... he was second. Roald Amundsen ... was the first. It was a terrible blow for Scott. Worn out and ill, he and his men died on the return journey.

... [people] just say, poor old Scott, he was brave, but lost the race.

But people don't remember that ***he had made the first balloon flight over Antarctica*** – an amazing and highly dangerous feat.

Having a go is the important thing, and even if you come second, third or fours, you'll know ***that you did your best.***

13. I was brought up to think we could all change the world. I believed that it was our duty to help others and to do good when we could; but it was never onerous.

14. We are now into the new millennium and the old ideas that industry is king and the devil take the hindmost are changing... ***The future is exciting. We could be on the threshold of a Renaissance not only of the way we all live but also of enterprise and invention.***

15. If you recognize something is a good idea, or if there's something in your personal life that you want to do, but aren't immediately sure how to achieve your goal, I don't believe that that little word *'can't'* should stop you. If you don't have the right experience to reach your goal, go in another direction, look for a different way in. There's always a solution to the most complex problem... Keep your eyes open. Look and learn.

16. Everyone needs to keep learning. Everyone needs goals. Each and every one of my lessons can be applied to all of us. Whatever we want to be, whatever we want to do, we can do it, because we can.

17. I learned many lessons: not only that if you want to so something you should just do it, but also to prepare well, have faith in yourself, help each other and most of all, never give up... keep trying and never to give up.

18. The best lesson I learned was to ***just do it***.

Whatever it is, however hard and daunting it might seem, as that ancient Greek, Plato, said, 'The beginning is the most important part of any work,' while the Chinese say, ' A journey of a thousand miles begins with a single step.'

The best of luck to you,

and have fun along your way.

Go ahead,

take that first step –

just do it.

[157] Look please an author's preface *'To Seek and Not to Yield'* at introductory page IV of this book.

11.8. *Anatoly Yunitskiy*

Anatoly Eduardovich Yunitsky (b. 1949) – inventor of *String Transport Systems* for the Earth and space (M.O. left, at *Rail Sky Way* Minsk's headquarters, Belarus, May 2015).

From conversations with A. E. Yunitsky and his personal archives:

1. Anatoly was six... Once his little sister Tamara saw other kids gorging themselves on ice-cream, their cheeks daubed all white, and asked him: Tolya, please pretty please buy me some ice-cream... The big brother's heart melted, and he did as he was asked. And he also topped the ice-cream with lemonade and brioche. Spending the three rubles that he had lifted off his mother. Then they lingered at the side of a busy road, afraid to cross it. After some wrangling they made a dash for it... stumbled, fell down... The brakes of a truck squealed, and a huge wheel stopped inches away from Tamara's head, with Anatoly looking on in dismay... That was a lesson for life: do not steal, or you will be punished. Or someone close to you will...

2. Tamara was five, Anatoly had just turned eight. At three in the morning he touched his sister on the shoulder to wake her up... The basket was full before they even noticed. To make all mushrooms fit in, Anatoly tucked in a row of twigs around the rim, and the basket suddenly became two times more capacious.

Then he put those mushrooms that did not fit in into a gray shopping bag – light enough to be carried by Tamara.

3. Anatoly learnt to milk cows when he was a small boy. Grownup women wondered how he managed to do it so quickly, and without a single drip missing the bucket. He also did it with two hands, like a dyed-in-the-wool dairy farmer. The funny thing is, nobody taught him to do that.

4. He liked to gather bilberries. He also learnt to do it with both hands. As a result, he was faster than grownup women and those who used "combs", ladle-like devices fitted with thing iron rods. Anatoly never resorted to the "comb". He saw how it ruined bilberry plants, tearing off leaves and stripping the twigs bare. The bilberries in his basket were clean, without a single leaf.

5. Then came fodder time. Cut grass was falling into the truck's body, and his task was to "dance" tramping the grass down and stuffing it into the corners. After a long summer day of such "dancing", his feet often cramped at night... And when it was time for corn harvesting, Anatoly went to work as an assistant combine harvester operator.

6. The first rockets designed and built by Anatoly were ... steam rockets, and they worked on water. Finding an empty shell somewhere in the forest, in a dugout or in a trench, he poured in water until the shell was two-thirds full. Then the mouth of the shell had to be sealed with a piece of wood. At first he could not manage it without losing his liquid "fuel", as to stop the shell he had to turn it over, emptying it of the water in the process. Later he found a better way to do it by skewering the water-filled shell onto a tree branch in an upright position, and then using a stone to ram the branch in. Then he cut the branch off with a knife and went on to build a "cosmodrome": the branch was stuck into the ground with the shell side up, and a "launching pad" of dry wood was made around it. A small fire was started, the water in the shell gradually warmed up, then boiled, and the pressure inside the shell quickly increased to a critical point. The shell then slipped off the branch and flew several hundred meters away, leaving behind a long steamy contrail.

7. The friends rummaged through ravines, found cartridges, broke off bullets and shook gunpowder out of the shells. From time to time they also came across old mines and artillery shells. Gunpowder was put into a special jar. That was for Anatoly. He had made a rocket, and it had to be launched. And it would not fly without gunpowder...

Our stands at the World industrial exhibition "Industriemesse'95" (20 years ago!) stood side by side. Hannover, Germany, 1995

8. To make a little more money, his mother, Yulia Stepanovna, part-timed as a charwoman at school. She had to scrub the floors in several classrooms. She alone could not make it, and her son helped her out.

The mother did not want him to do that dirty job, but he would not listen to her objections, and tagged along almost every night. When the head mistress came in one morning, she could not bring herself to step over the threshold – the floor was shining as if lit from within. She asked the teachers who had done it, and they answered: *It is the charwoman's kids, they toiled through the night together with her scrubbing the dirt off unpainted floorboards with knives...* With people having moved into new houses from dugouts only a short while ago, nobody even dreamed of painted floors...

9. The mother was not paid for the work she had done at the collective farm. She had to raise two children, but how could she pull it off if, at the end of the working year, they told her it was she who owned the farm, and not the other way round...

The family simply had no money to live on, and life in Kryuki became intolerable. In 1962 they went back to Kazakhstan, to stay with her other sister who lived not far from

Dzhezkazgan... One summer night the kids were sitting in a street cinema watching some film. It has already gotten dark. Suddenly they saw a burning candle rushing up into the sky to the left of the huge screen. It climbed higher and higher, and then disappeared in the dark sky. Everybody knew it was another rocket launch at Baikonur. And the entire world would learn about it the next day...

I want to say that Mother worked hard all her life. She dedicated herself to us, her children. She became a "shock worker of Communist labor", and her portrait was put up on the Board of Honor in the City of Nikolsky (now Satpayev) and also in the City of Dzhezkazgan...

10. Instead of fuel they used flammable combs, but they did not burn hot enough. They began to use motion picture film... Aircraft pilots "helped" them to design rockets by dropping silvery tape. Later he learnt that aircraft used that tape during military maneuvers to hide from radars. The tape provided excellent material for rocket bodies. All they had to do was to roll several layers of it around a "form" – a metal pipe – then glue it together so that the finished body could be pulled off the pipe, and finally make the rocket's combustible stuffing that would boost the rocket into "space."

11. When I was a student of Tyumen Engineering and Construction Institute, I liked sci-fi stories and fairy tales, especially *Adventures of Baron Münchhausen*. The thought that rocket fuel and rocket launches were environmentally hazardous had already visited me several times. And quite unexpectedly the baron gave me two ideas closely related to space exploration. The first was when he climbed off the Moon by a vine. That was Tsiolkovsky's idea! And the second was when he pulled himself out of the bog by his own hair braided into a pigtail! Can the cargo go into space "by itself"?! Anti-gravitation?! No, not yet. And the rest went against the laws of physics...

12. Friends asked me many times to manufacture fish-stunning explosive devices, and I told them that I would never do it, and they would have to go on using their fishing rods. They also invited me to their hunting trips, and I always turned those invitations down – I simply cannot bring myself to kill a living being...

13. I had been thinking of visiting my home village of Kryuki for ten years, and finally I made it there together with a TV crew... When the entire area had been covered with poisonous ashes from Chernobyl... Our old house had been robbed and burned down by a band of marauding looters.

14. Over the next fifteen years, the state, gangsters or "strategic partners" (who in reality proved to be no different from gangsters) did the same thing seven more times: they swindled me out of almost everything I had – money, land, property, companies... And they raked in a lot, too. Just intellectual property rights were valued by independent international appraisers at US$ 14 billion. Luckily, they could not take away the most important things – my freedom and my life.

15. Smoking? I quit. I said "no" one time – and that was it.

16. Anatoly mastered the Theory of Inventive Problem Solving (TRIZ) and made a rich treasure trove of practical patenting experience, just like Einstein who spent several years working in a Swiss patent office and later described that period with these words: *I learnt to separate the wheat from the chaff.* In all probability, without that background the great physicist would not have developed his relativity theory, nor would Yunitsky have designed his string transport system.

17. **Ideal Functional Model:** *solution of the problem lies within the Operating Zone – if we cannot move the center of mass, let it stay where it is.* And that center of mass must coincide with the Earth's center of mass, or there will be no equilibrium – either static or dynamic. There is only one technical solution here – the means of transportation must be in the form of a ring circling the planet. To go up into space, this ring must increase its diameter relying on internal forces. And the center of mass of the ring must always remain in the center of the planet.

18. String transport possesses colossal system-building capacity. For example, new settlements in Russia can be built in the most beautiful and environmentally sound localities. Because it will be possible to rapidly and reliably reach any metropolitan or recreational area all year round. *And nobody – from Chukotka and Vladivostok to Saint-Petersburg and Sochi – will be filling isolated, separated from Russia's common physical and psychological space.*

19. ...The first customer was the Ministry of Economy of Russia – it commissioned a working model of Yunitsky String Transport (YST, scale 1:10) for the Russian exposition at the World Fair EXPO 2005 in Nagoya, Japan. The working model was manufactured, tested and paid for by the customer. However, at the last meeting of the Russian steering committee one "government" scientific advisor made this peremptory statement: *String roads? Nobody in the world has them, and we want to go to Japan and make fools of ourselves!? Not on my life!*

Sergey Sibiryakov (left), economist, systems analyst and entrepreneur, discussing the prospects of promoting Yunitsky string technologies with Maksim Kalashnikov, futurist writer. Compilation of pictures from video interview (recommended to all my readers! – M.O.) under the heading: "Technopark: Innovations and Innovators," Issue No. 14, July 14, 2011, at new-core.ru/video/neyromir/strunnyitransportyunickogo-2

20. S. A. Sibiryakov (summary):

They would not let Eiffel build his tower. One problem cropped up after another! He said, I will build it with my own money, and let it stand for 30 years. If it is not used the way it should, I will go bust, but if it becomes profitable, it will mean that my technology is right. Just give me a 30-year concession to use it. The answer was: well, it is your money, go ahead and waste it.

Today the Eiffel Tower is, first, the most popular tourist attraction in Paris, second, the symbol of Paris, and third, Eiffel became one of the richest men in France.

This is what we now want to do in Moscow. We have investors! All we need is for the city administration to say – have a go at it!

21. In 1982 the *Technical Knowledge for Youth* magazine (Issue No. 6, pp. 34-36) published translation of the novel *The Fountains of Paradise* by the futurist writer Arthur

Clarke [158], and Anatoly Yunitsky, an engineer from Gomel, came out with the idea of going *Into Space … on a Wheel.* Then Anatoly wrote a letter to the famous master…

Arthur Clarke's response to Anatoly Yunitsky:

> *Dear Mr. Yunitsky! It was with great joy and satisfaction that I read your letter. It touched and excited me. It is wonderful that there are people living on our planet who are troubled by the technocratic development of mankind. They feel concerned and alarmed for the future of the planet – and it does not matter where they live, whether in the Soviet Union or Japan, America or France… I am very busy now, working on a new book, so my letter will be brief.*
>
> *In the nearest future, when I finish my work, I will write you a detailed letter. I have put your contact details and information about your idea into my computer. I believe that idea to be interesting and promising.*
>
> *Act! Victory favors those who are insistent and single-minded. I believe you are one of such people. I hug you.*
>
> *Respectfully yours, Arthur Clarke.*
>
> *P.S. Dear Anatoly! The answers to your other questions you will find in my recent interview. I think you will learn a little more about me. Write to me about yourself and about your new ideas – I will be happy to write back…*

22. In his book *Profiles of the Future* (1962), Arthur Clarke formulated the so-called *Clarke Laws* informing the evolution of modern science.

Law One: *When a distinguished but elderly scientist states that something is possible, he is almost certainly right. When he states that something is impossible, he is very probably wrong.*

Law Two: *The only way of discovering the limits of the possible is to venture a little way past them into the impossible.*

Law Three: *Any sufficiently advanced technology is indistinguishable from magic.*

And here I am saying "farewell" to my readers. And I am also saying to you: May there be more meetings! Including your meetings with Modern TRIZ!

And, naturally, I wish you good luck!

---

[158] Sir Arthur Charles Clarke (December 16, 1917, Minehead, Somerset, UK – March 19, 2008, Colombo, Sri Lanka) – English writer, scientist, futurologist and inventor, best known for his cooperation with Stanley Kubrick on the iconic science fiction movie *2001: A Space Odyssey* (1968); author of the idea for the creation of a group of geostationary satellites to provide global communication coverage (1945) and organize global weather forecasting (1954), enthusiastic supporter of the space elevator idea; in 1984 the *Technical Knowledge for Youth* magazine (M.O.: журнал "Техника – молодежи") commenced publication of Clarke's novel *2010: Odyssey Two.* The author dedicated the novel to Cosmonaut Aleksey Leonov and Academician A. D. Sakharov. The dedication to Sakharov was removed, but neither the translators nor the editors noticed that all Russian characters in the novel bore the names of Soviet dissidents who were well-known in the West. After the second installment was released, publication of the novel was discontinued by the censors; members of the magazine's editorial board were severely reprimanded; and Vasily Zakharchenko, editor-in-chief, was fired. Nevertheless, the novel was reprinted in Soviet times, complete with dedications to Leonov and Sakharov, in the same magazine by the same editor-in-chief that replaced Zakharchenko (S. V. Chumakov) in Issues Nos. 11-12 for 1989 and Nos. 1-5 for 1990. Incidentally, in Issue No. 5 for 1990 it was announced that the novel had been published in full; http://en.wikipedia.org/wiki/Arthur_C._Clarke.

11.9 *It's Your Choice!*

Dear Young Readers!

I hope that everything you read in this book, and especially in this chapter, unequivocally leads you to the conclusion that all good things are the product of your work, your personal perseverance, courage, and patience multiplied by your beliefs and dreams. Nothing to add, nothing to take away. *Sapienti sat.*[159]

Except that there is one more thing left unsaid. Health – you need to exercise *a lot* while you are still young and have enough time. It will help you to store power for the future.

Everybody knows that smoking is dangerous – but many people go on smoking. Quit it – forever!

Everybody knows that drugs are lethal, that they ruin your personality and make you a slave to this addition – and still there are people who, in an unthinking moment, decide that they will "just try." This is what death dealers are waiting for because such "tasters" always join their clientele and support the very existence of their criminal supply and distribution network. Say NO to yourself and to them!

Everybody knows that alcohol is hazardous, and still many people find no measure in consuming it, and many start doing it in their tender years.

There are many similar examples in what people do. Here is one more, and it is too important to keep silent about it. Everybody knows that excessively high speeds and inconsiderate driving are absolutely dangerous and cannot be permitted under any circumstances (!), and still many are not deterred by the risk of killing a chance pedestrian, other people in a similar vehicle, or themselves.

Why does it happen? The answer is simple: lack of elementary respect and self-respect, lack of the feeling of responsibility before others and before oneself, lack of the feeling of measure and of the understanding that if a certain line is crossed it may lead to ***irreversible consequences***, irreversible twists of fate, irreversible changes in other people's lives as well as one's own life.

There is also a lack of cordiality, compassion, and willingness to help others.

There is an abundance of ignorance, criminal foolishness, and caveman-like egoism.

Meanwhile, what we do need is to learn to make choices, to be responsible for those choices before other people and before ourselves, and to acquire an ability to manage our desires and abilities. We must build our own lives and not let ourselves to be manipulated by anybody whatsoever.

Fight for yourselves. Fight for people and things that you hold dear. And never give up following in the footsteps of the people about whom I told you in this book.

You can be yourselves. You can overcome all obstacles. You can reach your objectives. And join forces with those who think the same way you do. When you are together, you grow stronger.

Forward!

---

[159] *Sapienti sat* (Lat.) means: *A word to the wise is enough (for understanding).*

And now some words about this course and the fact that you will need to pay for it.

Let's be honest: All work must be paid for. All products and services have their value as determined by demand and quality.

We are offering you a worthy product. Invest into your future.

Find a way to earn as much as you really need. Never waste what you earn on trifles. You must not fail to help those who are weaker than you.

And before saying *'Good bye!'* I want to show you one "serious joke," which was created by students in 2011 and exhibited behind the glass in the main building of the Mariupol' State Technical University, Ukraine (s. picture). I am with this "robot" to demonstrate its height.

Inscription on the body-poster:

Student Senate.
Samizdat.[160]
Look and think!

His body is made of 874 cigarette packs which cost

UAH 7,943.

Could you find a better use for that money?

You can ask me: did YOU smoke? Yes, I did.

But fortunately I gave it up a very long time ago. I am only sorry that I smoked at all, and that I did not give it up earlier.

So, as currency course of December 2011, ***EUR 760*** or ***US$ 992*** turned into ***nothing*** and inflicted guaranteed ***harm*** on your physical and psychological health.

Just do not buy cigarettes or other unnecessary things for 6 months or for one year—and this course is yours!

Find a way to save or earn money to pay for this course!

Set your sights on worthy goals – and reach them. It is possible.

It's Your Choice!

You and only you must be the master of everything that happens in your lives.

[160] This Russian abbreviation-word "Samizdat" means "self-publishing."

Six important personal qualities according to TDCP:

Worthy objective	The individual must have a novel, previously unattained, significant, socially useful, and worthy objective (of system of objectives).
Plans	The individual must have a plan (or set of plans) designed to assure attainment of his/her objective(s) and to monitor implementation progress.
Capacity to work	The individual must be willing to perform, and actually perform, a huge volume of work to implement his/her plans.
Problem-solving method	The individual must have a method that can be used to solve problems that he/she may encounter on the way to his/her final destination.
Resilience	The individual must be capable of defending his/her ideas, dealing with public censure and incomprehension, "sticking to his/her guns," and remaining true to his/her ideals.
Effectiveness	The results attained (or their scope) must be commensurate with the original objective(s).

Genrikh Altshuller[161]

[161] Translated by author from G. S. Altshuller, I. M. Vertkin, *How to Become a Genius. Life Strategy of a Creative Personality*. – Minsk: Belarus, 1994. – (Альтшуллер Г.С., Верткин И.М. *Как стать гением. Жизненная стратегия творческой личности*. – Минск: Беларусь, 1994)

Book III
Primary Instruments
(Summary)

Main books (2015) by author to continue the ABC-TRIZ

S1. TRIZ

> 1. Identification and elimination of the systemic contradiction is the key to resolving the problem!
>
> 2. There is an endless multitude of inventive problems, while the number of systemic contradiction types is relatively small. **There are typical systemic contradictions – and typical techniques used for their elimination**. Problem resolution methods (techniques) can be identified by analyzing great inventions.
>
> 3. The strategy and tactics of directed problem resolution must rely on laws governing evolution of technical systems.
>
> G.S. Altshuller *How to Learn to Invent.* – Tambov, Tambov Book Publishers, 1961 (in Russian)

Structure of TRIZ

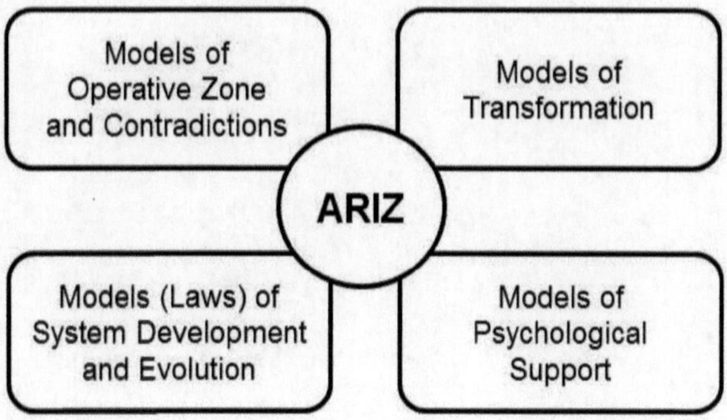

"...naturally[162], each problem is individual in its own way. In each problem there is something inimitable.

Analysis helps us to break through to the most important thing – *the systemic contradiction and its causes*. And the tables are turned immediately. It is now possible to conduct creative searches within the framework of a certain *rational scheme*.

There is no magic formula, but there are *techniques* which are sufficient in most cases."

G. S. Altshuller

Directed search does not exclude intuition. On the contrary, normalization of thinking processes creates a special "attitude" which is conducive to manifestation of intuition.

G. S. Altshuller *Invention Algorithm.* – Moscow, Moscow Worker Publishers, 1973 (in Russian)

[162] Compiled by M. O. from various sources with slight variations.

M.A. Orloff, *ABC-TRIZ*, DOI 10.1007/978-3-319-29436-0_12

S2. INVENTION IS A LOOK IN-DEPTH

Learn the "secret" of this poster: turn the sheet 90° clockwise and try to look through it rather than at it!

When you perceive 3-D depth of the image hidden in the picture, you are sure to experience the *miracle of understanding*! The same happens as you study TRIZ. But after you have independently completed at least several dozen assignments, after you have learnt to bravely tackle any challenging problem, *you will become initiated to the mystery and craft of TRIZ.*

S3. MENTALITY LEVELS AT IDEA GENERATING

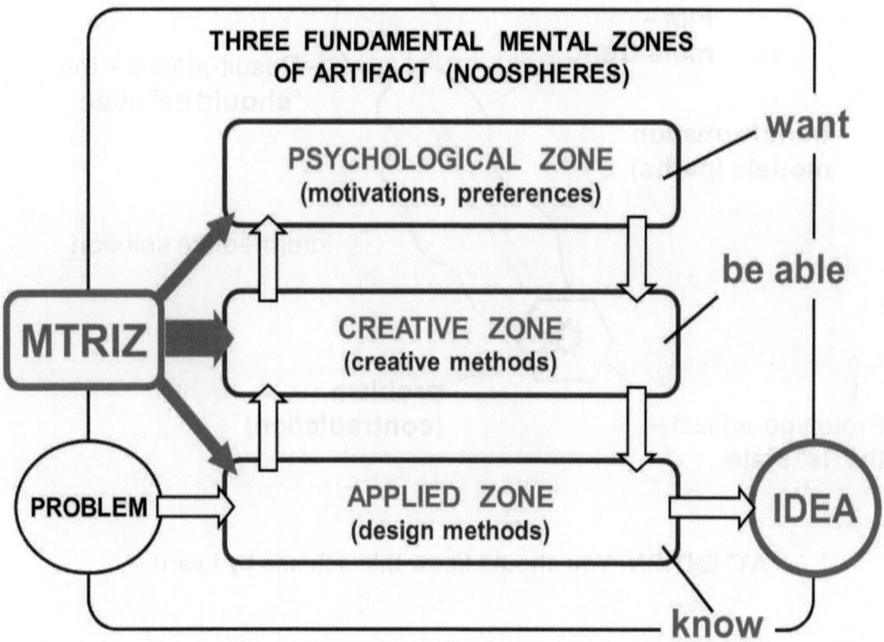

There are three "mental" zones (fig.) in design thinking:

1) the applied (design) zone;

2) the creative zone; and

3) the psychological zone.

Creating the efficient ideas uses all three "specialized" zones: professional (applied) and creative knowledge and skills as well as commitment and tenacity in the attainment of formulated objectives (emotions, psychology):

Know + Be Able + Want!

S4. DIRECTED SOLVING OF A PROBLEM WITH TRIZ

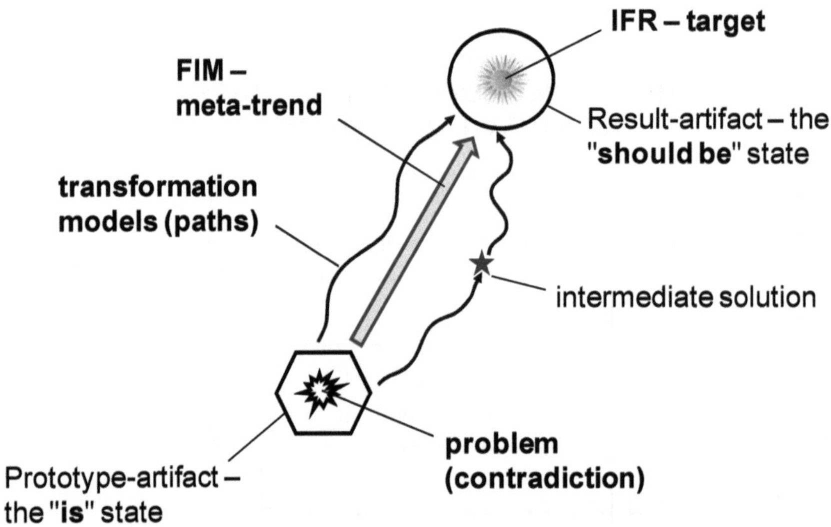

ATTENTION: You should know this scheme by heart!

The solving process starts with analysis of the initial problem situation and formulation of a contradiction modeling the problem. The contradiction resides in the prototype, which needs to be improved.

Then we need to establish the objective, the ideal final result (IFR [what properties must be obtained]) and formulate the tentative operating principle, the functional ideal model (FIM [how it must operate]). The FIM points to the meta-trend, the general direction in which we must move to arrive at the IFR.

The best possible IFR is this: The system BY ITSELF—based on self-organization—resolves the problem with available resources, which in some cases may undergo insignificant modifications.

Then we select transformation[163] models, i.e., specific paths that will take us to our destination.

Last but not least, we walk those paths, i.e., we invent specific solutions and change resources of the prototype artifact so as to finally arrive to the resultant, or "heir" artifact, which is consistent with the sought IFR.

Oftentimes once you have correctly formulated the IFR and the FIM, they will take you to the idea for the solution by themselves.

[163] TRIZ instruments constantly evolve; as this happens, updated materials are published on Modern TRIZ Academy Web sites.

S5. SYSTEM TENDENCY TO IDEALITY

$$E_{ideal} = \frac{Fnc = \text{Functionality}}{Phs = \text{Physicality}}$$

Functionality = {appointment, efficiency, operation, etc.} → 1 (normalized to 1)

Physicality = {size, weight, energy consumption, waste, pollution, etc.} → 0

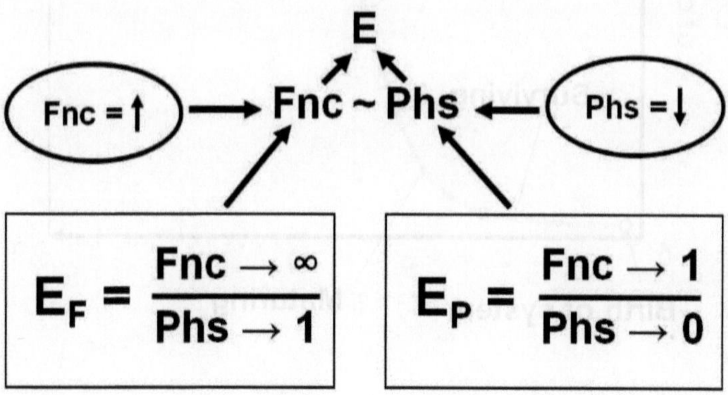

"Idealization" according to TRIZ means that, throughout their life cycles, all systems seek to improve the ratio of their functional abilities F to the costs of their realization P.

There are two possible meta-trends that may also form various combinations: According to the "Functional Expansion Law", it is possible:

1) to increase functionality $F \to \infty$ with limited investments $P \to 1$ (any parameter can almost always be rebased to 1) or,

2) according to the "physical compression law," it is possible to decrease costs $P \to 0$ with limited functionality $F \to 1$.

S6. LIMITATION OF MAIN SYSTEM CHARACTERISTIC

The increase of each characteristic is constrained by the physical properties of the system or by the limited range of implementation of the ideas embedded in the system. As a result, fully represented evolutional relationships have the form of a **logistical curve, or S-curve**.

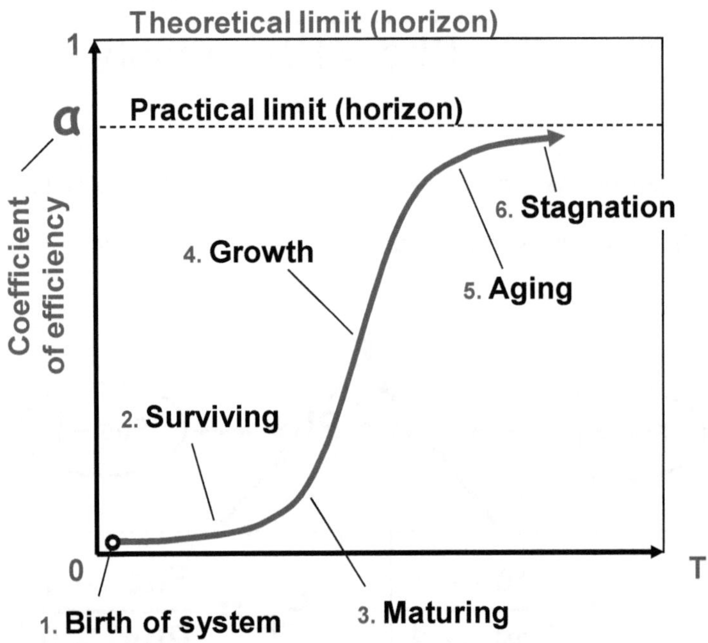

The development of any artifact is driven by its adaptation to the needs of different users. That is why there are so many models of cars, aircraft, mobile phones, television sets, computers, etc. Such transitions do not modify the type of the system. Accordingly, such transitions could be defined as **developmental**.

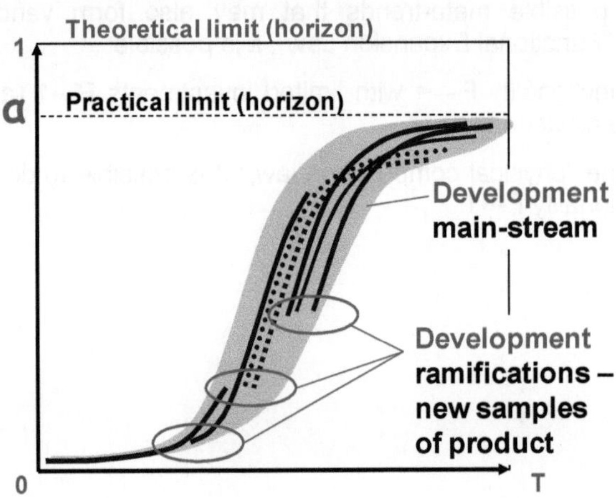

S7. BREAKTHROUGH OF MAIN SYSTEM CHARACTERISTIC

Each "point" of *ramification* and *bifurcation* is a conflict between the two main driving forces: growing needs (demand ["pull-forces"]) and growing abilities (supply ["push-forces"]). If abilities cannot meet growing needs, contradictions arise.

Those contradictions intensify when operative zone (OZ) resources prove to be deficient or depleted to the point of exhaustion and when transformation potential proves to be limited. **In such situations, TRIZ comes in very handy.**

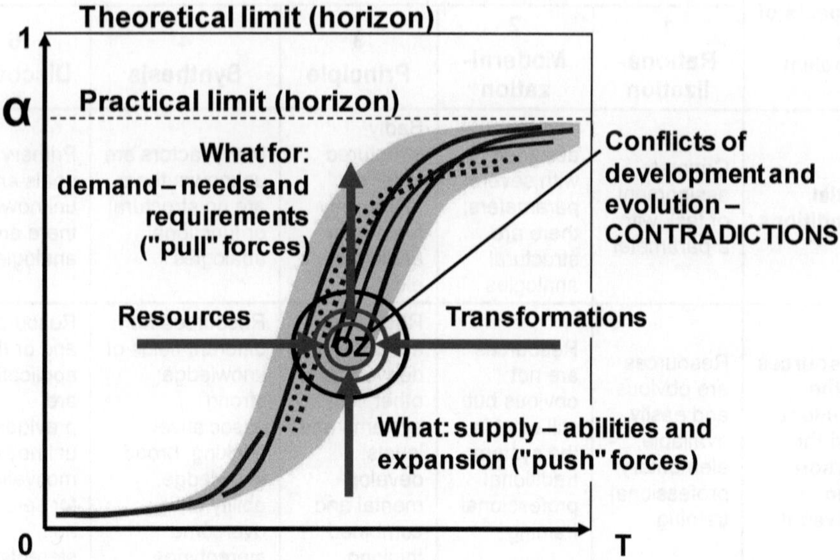

As a rule, after the developmental resources of a system of a given type have been exhausted, **there emerges a new type,** and systems of that type are used for the same purposes but have superior technical characteristics and efficiency. Such transition could be defined as **evolutional**.

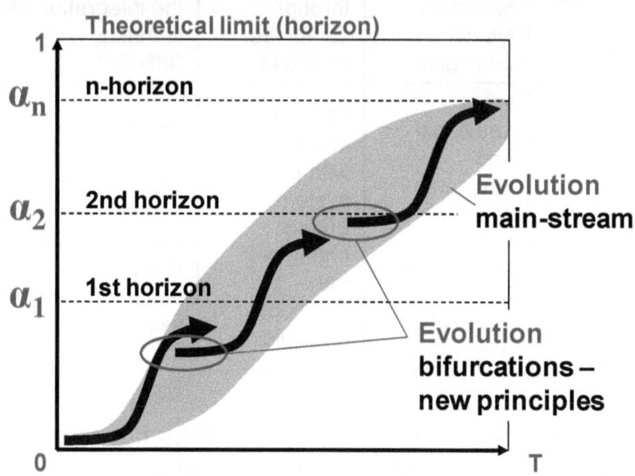

S8. LEVELS OF INVENTION

The steps of the developmental history of civilization—millions of inventions—are different from each other in their value. In the following table, you will find a classification of inventions according to their level and with reference to various characteristic indications.

Aspects of the problem	Levels of Invention				
	1 Rationa-lization	2 Moderni-zation	3 Principle	4 Synthesis	5 Discovery
Initial conditions	Concrete assignment of task with a parameter	Tasks assigned with several parameters; there are structural analogies	Badly structured "pile" of tasks; only functional analogies exist	Many factors are unknown; there are no structural or functional analogies	Primary goals are unknown; there are no analogies
Resources of the problem and the person who solves it	Resources are obvious and easily available; elementary professional training	Resources are not obvious but still exist in the system; traditional professional training	Resources are often derived from other systems and levels; develop-mental and combined thinking	Resources from different fields of knowledge; strong associative thinking, broad knowledge, ability to overcome stereotypes	Resources and or their application are previously unknown; motivation for selection, no stereotypes
Degree of difficulty	Tasks without conflicts	Standard problems	Nonstandard problems	Extreme problems	Unique problems
Trans-formation rules	Technical optimization as a solution	Technical solution based on typical analogies	Inventive solution through combined methods	Inventive solution through the integration of technical "effects"	Scientific and technical inventions
Level of innovation	Slight changes in the parameters of elements	Initial functional and structural solutions without changes in the functional principle	Meaningful inventions with positive system effects; changes in the functional principle	Intense inventions with system super-effects that lead to an essential change in neighbouring systems	Brilliant inventions with system super-effects that lead to an essential change in civilization

A creative idea is not obvious. It is an object that does not clearly exist in extant knowledge. It is an object that is created by human thinking.

S9. INVENTION COMPLEXITY

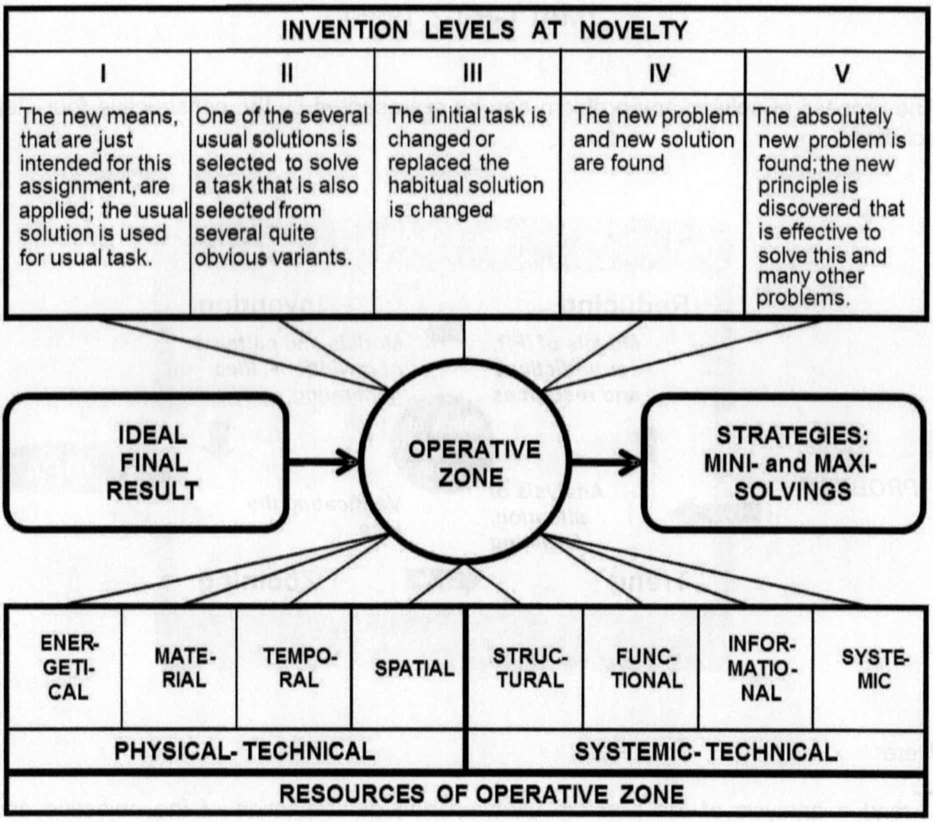

Meta-Strategy		Complexity of a problem
Minimum task	**Maximum task**	
The decisive resource is known (TRIZ instruments)	**The decisive principle is known** (TRIZ instruments)	**Low**
We can reveal new resource (TRIZ instruments)	**We can reveal new principle** (TRIZ instruments and methods for integrating alternative systems)	**Middle**
No resources can be revealed (Reformulation or substitution of the task)	**No principles can be revealed** (Transition to a synthesis of a system)	**High**

S10. META-ALGORITHM OF INVENTION "T-R-I-Z" (MAI T-R-I-Z 1995)

The process of solving any problem can be represented by the constructive four-step scheme.

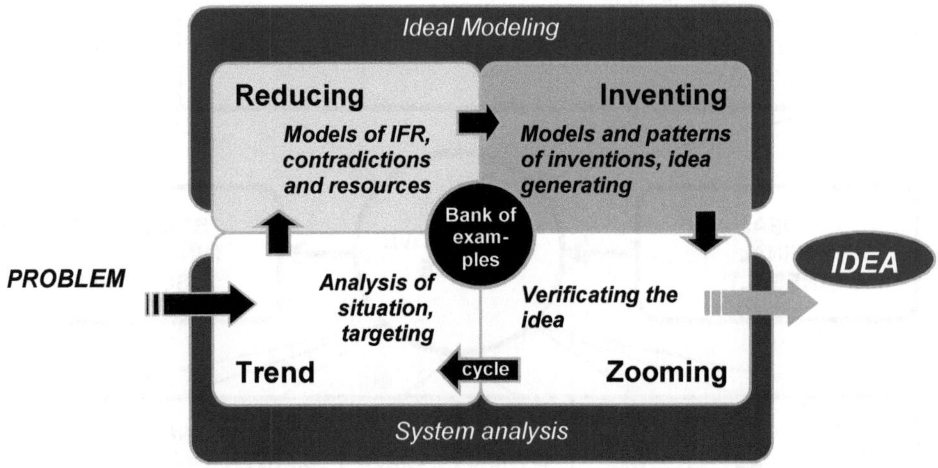

Here:

Trend = analysis of the problem situation and determination of the objective and direction of evolution;

Reducing = modeling of the problem in the form of a contradiction;

Inventing = generation of an idea with the help of transformation models;

Zooming = examination of ideas on different scales (from different levels [zooming]) and from different focal points (from multiple viewpoints and different aspects) to assess their efficiency.

The dash line points to cyclical repetition of the algorithm in situations where no satisfactory solution is obtained on completion of a cycle.

MAI T-R-I-Z 1995 conceptually is very close to the first algorithm of inventive problem solving ARIZ-1956.

S11. MODERN TRIZ: STANDARDIZATION OF TRAINING, PRACTICE, AND PROBLEM SOLVING ON THE BASIS OF MAI T-R-I-Z

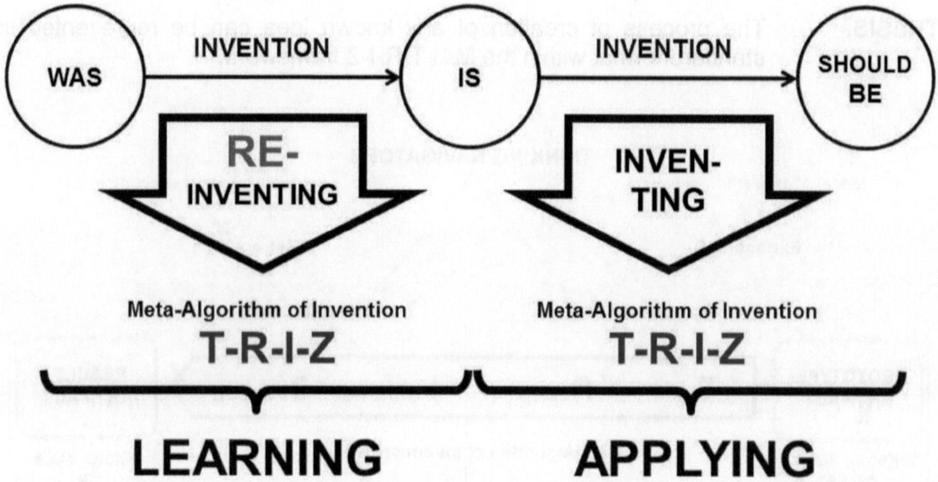

The same (standard) MAI T-R-I-Z scheme:

- **accelerates** basic TRIZ user training;

- **creates and reinforces** from the very beginning **correct skills** that can be used to solve TRIZ problems based on examples presented in the same standard format; and

- **accelerates efficient solution** of real-life problems due to the "automatism" developed and reinforced through the use of standard methods.

MAI T-R-I-Z:

Think about the solution rather than about the solution-finding method.

S12. REINVENTING ON THE BASE OF MAI T-R-I-Z

THESIS. The process of creation of any known idea can be represented in standard format within the MAI T-R-I-Z framework.

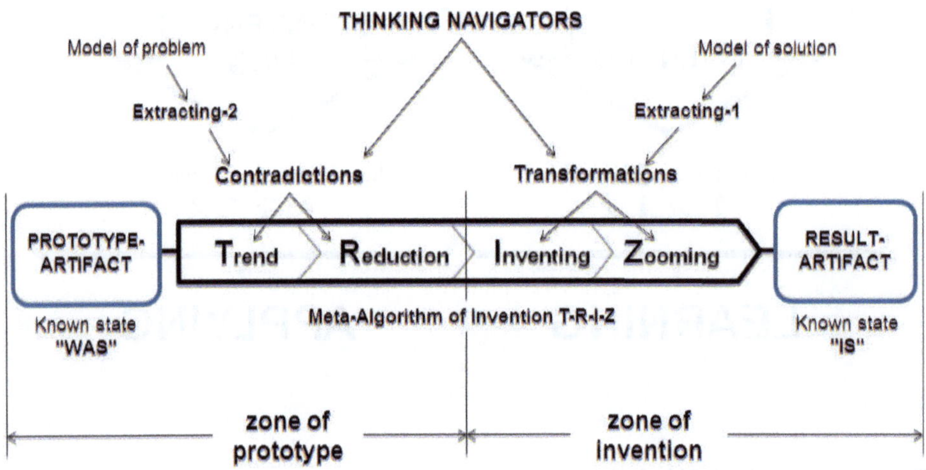

Extracting-1 = drawing out the transformation models from the result-artifacts.

Extracting-2 = drawing out the contradictions from the prototype-artifacts.

Reinventing = modeling of the process where the "was" prototype-artifact is transformed into an "is" invention-artifact.

The "was" state of the prototype-artifact is characterized by the existence of contradictions, which must be eliminated.

The "is" state of the result-artifact (invention-artifact) is free of initial contradictions and consistent with the new requirements posed to the artifact.

S13. INVENTING ON THE BASE OF MAI T-R-I-Z

THESIS. Invention of efficient ideas in any area of human activities can be implemented on the basis of MAI T-R-I-Z.

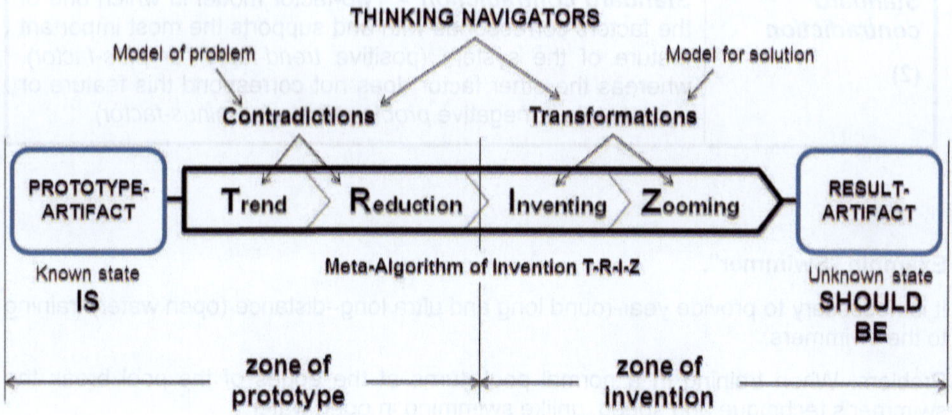

Inventing = the creation of efficient ideas to transform the "is" prototype-artifact, which contains the initial contradictions, into the "needs-to-be" target result-artifact (invention), which does not contain the initial contradictions and has the new required properties.

Understanding of Time = the problem solver is physically in the "is" state; he/she also possesses knowledge of the prototype-artifact.

The notion of the "needs-to-be" state, as well as all events and stages of the solution creation process, relate to the future result-artifact as a target of invention.

S14. STANDARD CONTRADICTION

Standard contradiction **(1)**	***Standard contradiction*** (in classical TRIZ: *technical contradiction*) = the binary (two-factor) model that reflects incompatible requirements between *two different functional features* of an object (or several conflicting objects).
Standard contradiction **(2)**	***Standard contradiction*** = Two-factor model in which one of the factors corresponds with and supports the most important feature of the system (positive *trend-factor* or *plus-factor*), whereas the other factor does not correspond this feature or counteracts it (negative *problem-factor* or *minus-factor*).

Example "Swimmer".

It is necessary to provide year-round long and ultra long–distance (open water) training to the swimmers.

Problem: When training in a normal pool, turns at the edges of the pool break the swimmer's technique and speed, unlike swimming in open water.

The original idea (optional): the special construction of pool (round, ellipsoid, or even figure-8 shape).

Standard contradiction (variant):

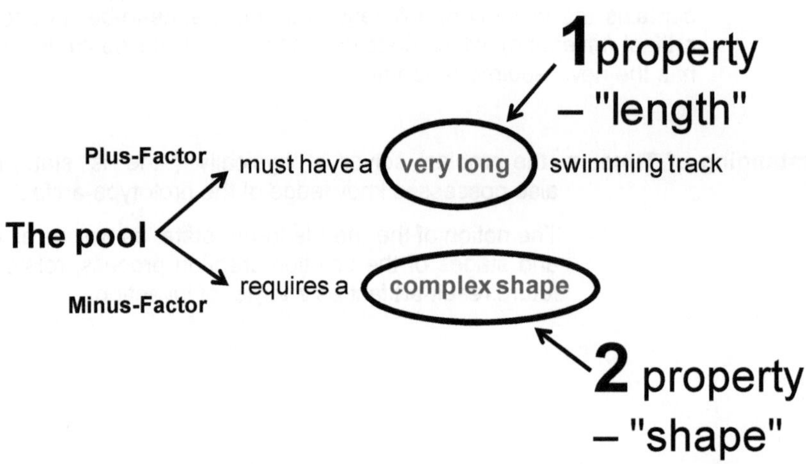

Text (formula) format: Pool ►endless swimming track **VS** complex shape

S15. RADICAL CONTRADICTION

Radical contradiction (1)	***Radical contradiction*** (in classical TRIZ: *physical contradiction*) = binary contradiction model in which contrary, i.e., excluding one another, requirements are being requested from one and the same feature of one and the same construct (component, resource, function, effect, condition, etc.).
Radical contradiction (2)	**Radical contradiction** = binary two-factor model where the first factor reflects one demand as a "plus-factor," and the second factor reflects the same demand as a "minus-factor" so that both factors represent the same property of the same construct (*component, resource, function, action, state, etc.*) but are incompatible.
Addition	In a radical contradiction: a) both factors may be necessary for the main useful function of the artifact; b) the "minus-factor" is the undesirable state of property which should be transformed into a plus-factor only; and c) the "minus-factor" opposes to develop the main property.

Example "Diver".

It is necessary to ensure the safe training of divers from a tower in the pool.

The problem: There is a significant risk of injury when enacting new pirouettes due to incorrect entry of the diver in the water and a dangerous hit on the water.

Radical contradiction (variant):

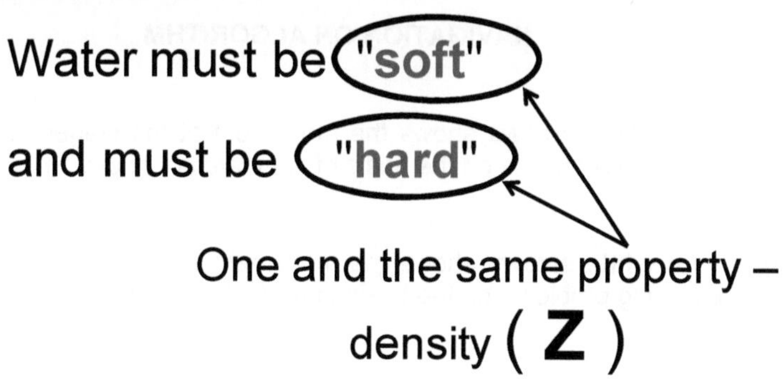

Formula format: Water must be ►soft (to spare the diver) **VS** hard (physical property)

S16. START: Simplest TRIZ-Algorithm of Resourceful Thinking

S16.1. START: integrated scheme

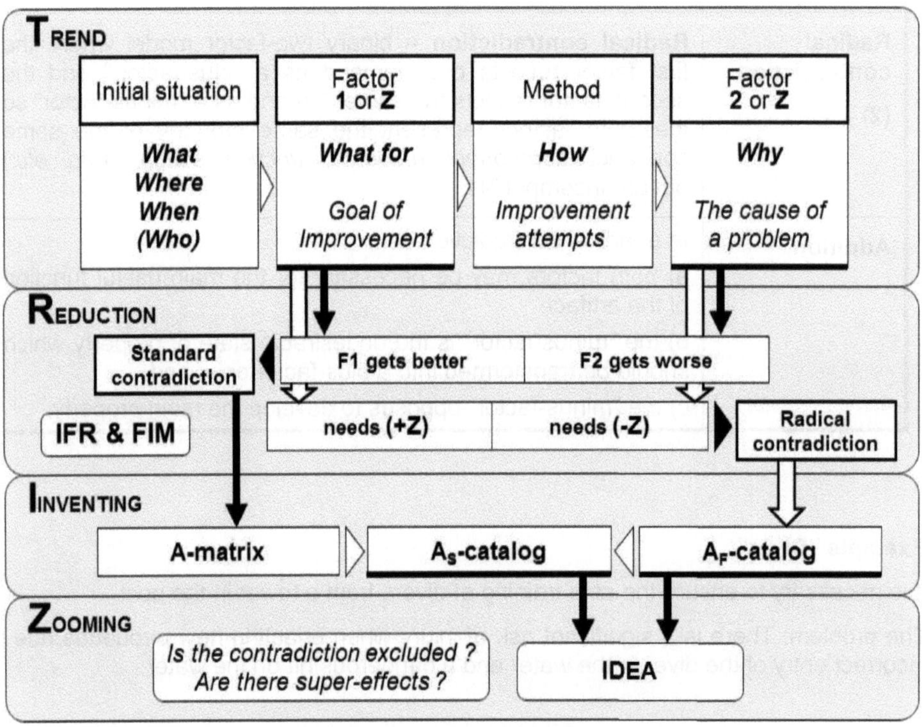

NAVIGATION ON ALGORITHM

The route with black arrows shows the movement at the stages of Reduction and Invention for solving problems on the basis of *standard contradiction*.

The route with white arrows shows the movement at the stages of Reduction and Invention for solving problems on the basis of *radical contradiction*.

Both routes start at the stage Trend and finish at the stage Zooming.

S16.2. START-pass through standard contradiction

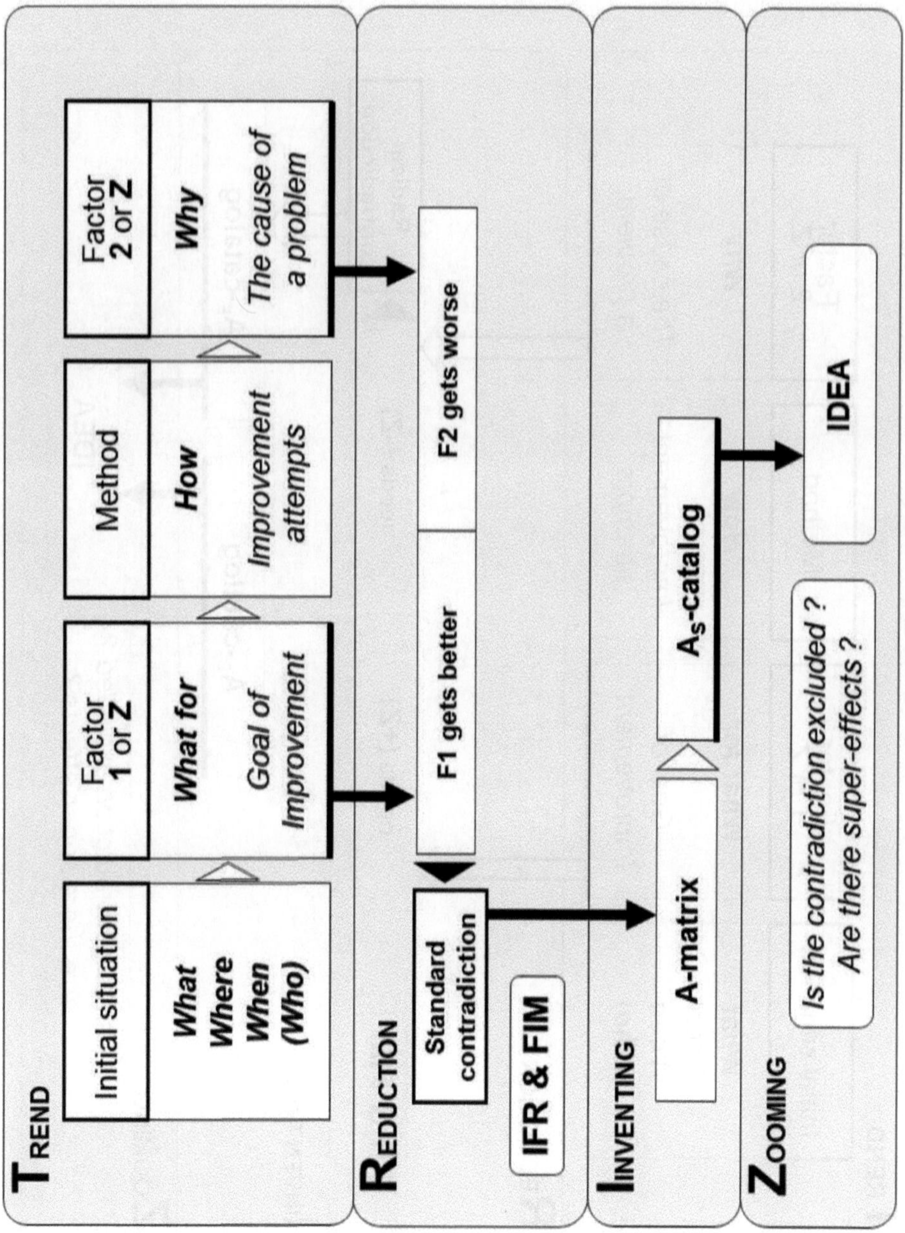

S16.3. START-pass through radical contradiction

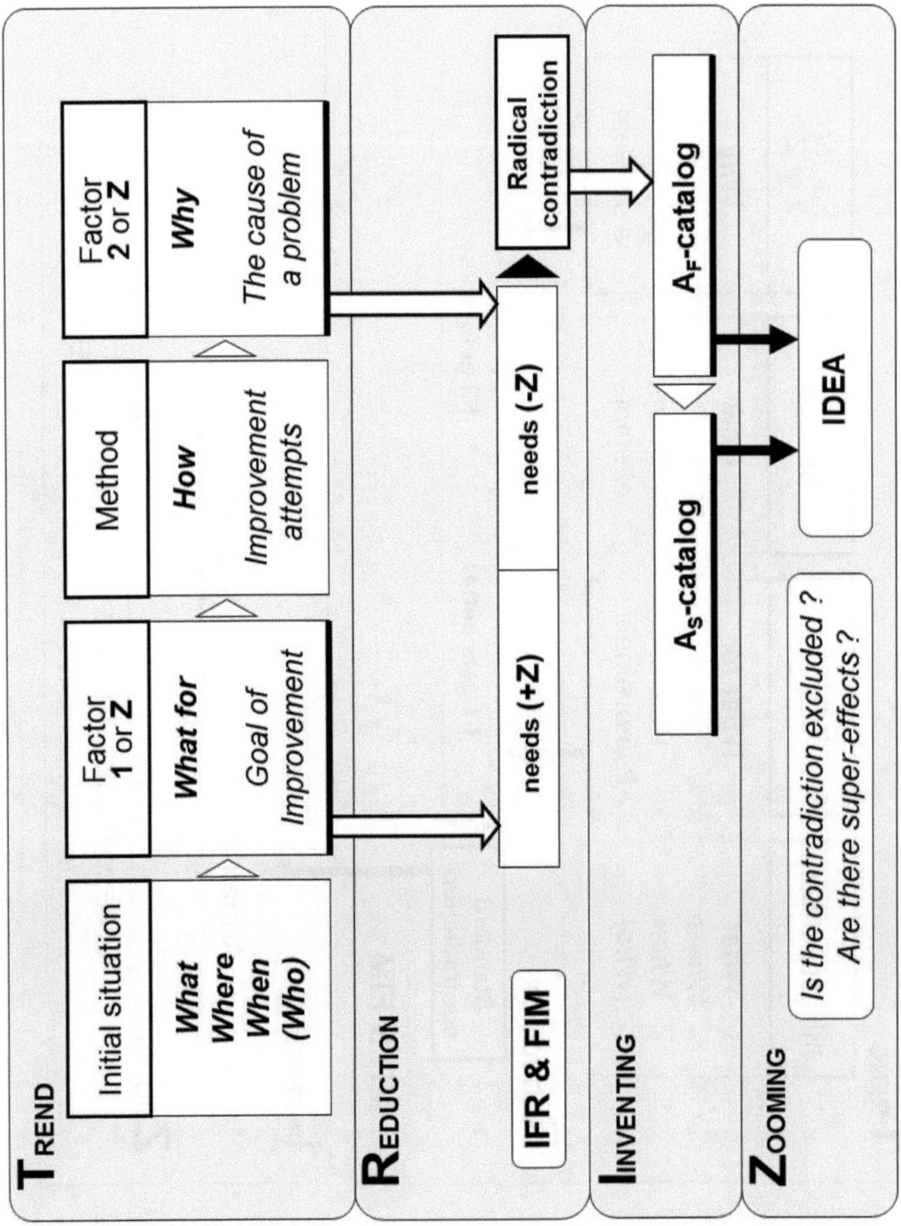

S17. OPERATIVE ZONE

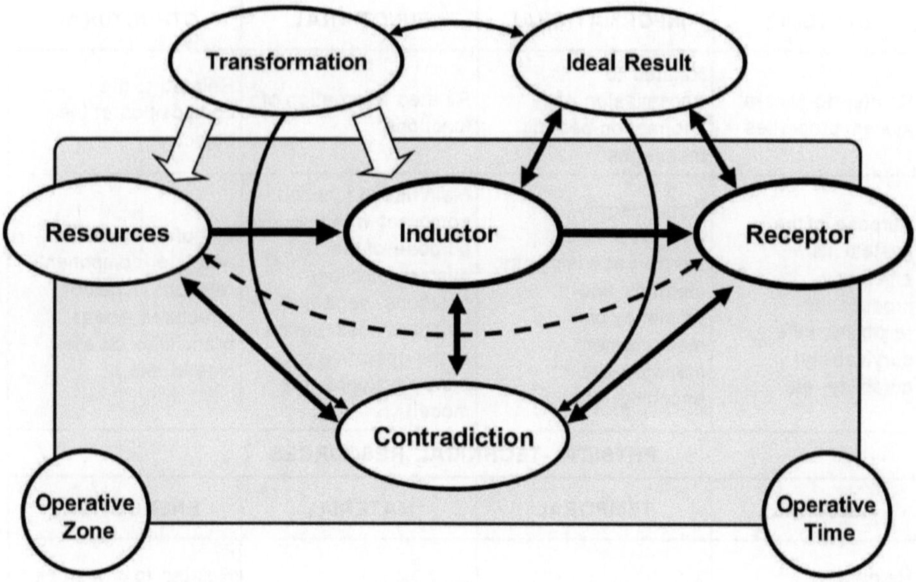

The OZ transformation process has the following stages:

1) selection of the inductor and receptor within the OZ;

2) definition of the contradictions that prevent acquisition of required properties by the receptor or that reflect other problems existing in the OZ;

3) determination of resources, both those already existing in the OZ and those that are required additionally;

4) formulation of the IFR and FIMs for the OZ; and

5) selection of transformation models subject to the following:

- Transformations are usually applied to the inductor, its own resources, and other resources available in, or introduced into, the OZ;

- realization of transformations is directed toward the replacement of contradictions with the ideal final result or the complete removal of contradictions.

S18. RESOURCES

SYSTEMIC-TECHNICAL RESOURCES			
SYSTEMIC	**INFORMATIONAL**	**FUNCTIONAL**	**STRUCTURAL**
Related to general system properties	Related to transmission of information-bearing messages	Related to creation of functions	Related to the composition of the object
Purpose of the system, its efficiency, productivity, reliability, safety, survivability, durability, etc.	Data integrity, accuracy, validity, interference immunity, methods and efficiency of measurement, management, encoding, etc.	Main useful function compliant with the purpose of the system, auxiliary functions, negative functions, description of the operative principle (functional model)	List of components and inter-component relations, types of structures (linear, branching, parallel, closed, etc.)
PHYSICAL-TECHNICAL RESOURCES			
SPATIAL	**TEMPORAL**	**MATERIAL**	**ENERGETICAL**
Related to geometric properties	Related to temporal evaluations	Related to properties of materials	Related to properties of energy and its manifestations
Shape of the object, dimensions of the object – length, width, height, diameter, etc., shape features – presence of cavities, projections, etc.	Frequency of events, duration of time intervals, duration of time lags/leads. Operative Time (OT): interval of existence of the problem situation	Chemical composition, physical properties, special engineering properties	Types of applied and measured energy, including mechanical, gravitational, thermal, electromagnetic and other forces; methods of energy utilization, etc.

Properties of the resource	**Value:** free → inexpensive → expensive
	Quality: harmful → neutral → useful
	Quantity: unlimited → sufficient → insufficient
	Readiness for application: ready → change in progress → creation in progress

S19. RESOLVING OF THE STANDARD CONTRADICTION WITH THE METHOD BICO (Binary In Cluster Out)

Matching and choice the adequate plus- and minus-factors for modeling the standard contradiction (e.g.):

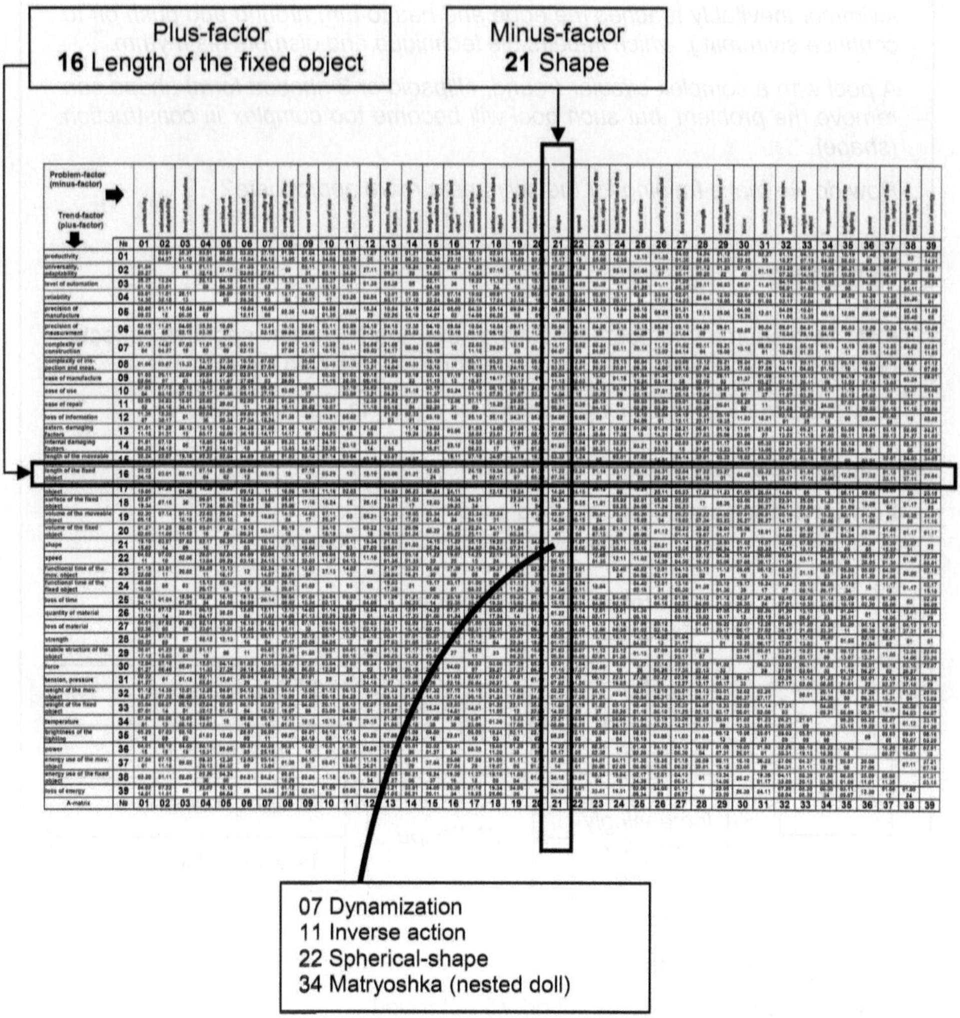

Formula of contradiction and navigators recommended by the A-matrix for solution:

16 **VS** 21 = 07, 11, 22, 34

S20. EXAMPLE "SWIMMER" (REINVENTING)

TREND

Open water bodies are used to train long-distance swimmers. In bad weather, training can become impossible. In a usual 50-meter pool, the swimmer inevitably reaches the edge and has to turn around and push off to continue swimming, which impairs his technique and disrupts his rhythm.

A pool with a complex circular (round, ellipsoid or 8-shaped form) shape can remove the problem, but such pool will become too complex in construction (shape).

How do we make training for the swimmers more appropriate?

REDUCING

FIM: X-resource, without producing the inadmissible negative effects, provides together with other existing resources obtaining the **IFR**:

[*adequate training of the long-distance swimmer in a pool*].

Standard Contradiction

Format of formula: Pool ► Long swimming track
VS Increasingly complex shape

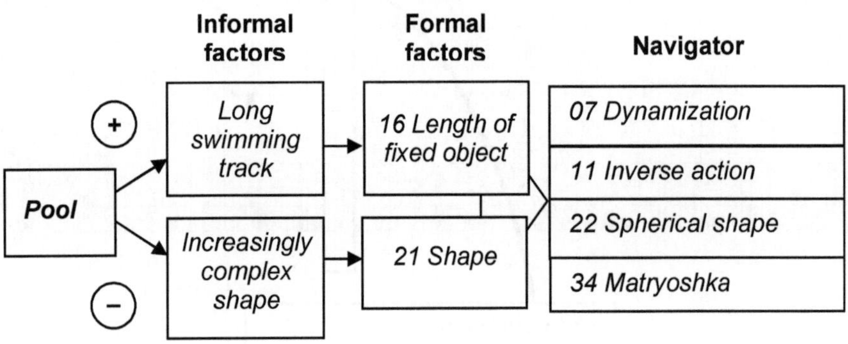

Radical Contradiction

Format of formula: Pool ► round **VS** not round

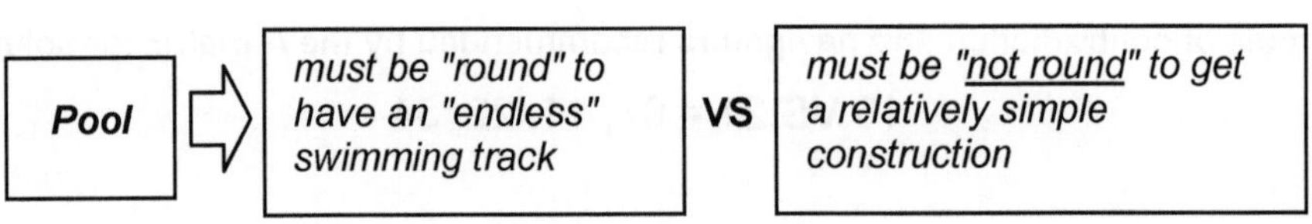

INVENTING

Key model: 11 Inverse action – *b) make a moveable part of an object or the environment fixed or a fixed part moveable.*

Key idea: *make the water moveable. Also:* ***07 Dynamization****.*

Take in account the ***22 Spherical shape*** *–a round pool proposed just as a first idea at brainstorming!*

ZOOMING

Have the contradictions been removed? – Yes. – ~~No~~.

Super-effects: *It becomes possible to manage training parameters (movement speed and phases) and adjust the swimmer's swimming technique "on the go" because the coach is located in the immediate vicinity of the swimmer.*

Negative effects: -

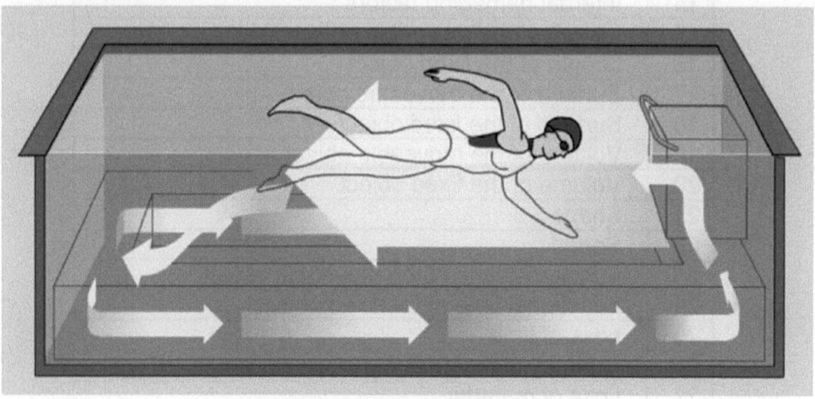

BRIEF DESCRIPTION

To provide long-distance swimmers with adequate training opportunities and simplify pool design, the water in the pool is made moveable in accordance with the navigators ***11 Inverse action*** *and* ***07 Dynamization.***

S21. A-MATRIX

S21.1. List of 39 Plus- and Minus-factors

#	Factor
01	Productivity
02	Universality, adaptability
03	Level of automation
04	Reliability
05	Precision of manufacture
06	Precision of measurement
07	Complexity of construction
08	Complexity of inspection and measurement
09	Ease of manufacture
10	Ease of use
11	Ease of repair
12	Loss of information
13	External damaging factors
14	Internal damaging factors
15	Length of the moveable object
16	Length of the fixed object
17	Surface of the moveable object
18	Surface of the fixed object
19	Volume of the moveable object
20	Volume of the fixed object
21	Shape
22	Speed
23	Functional time of the moveable object
24	Functional time of the fixed object
25	Loss of time
26	Quantity of material
27	Loss of material
28	Strength
29	Stabile structure of the object
30	Force
31	Tension, pressure
32	Weight of the moveable object
33	Weight of the fixed object
34	Temperature
35	Brightness of the lighting
36	Power
37	Energy use of the moveable object
38	Energy use of the fixed object
39	Loss of energy

S21.2. Table of A-matrix

Problem-factor (minus-factor) → / Trend-factor (plus-factor) ↓		productivity	universality, adaptability	level of automation	reliability	precision of manufacture	precision of measurement	complexity of construction	complexity of inspection and meas.	ease of manufacture	ease of use	ease of repair	loss of information	external damaging factors
	#	01	02	03	04	05	06	07	08	09	10	11	12	13
Productivity	01		03.01 04.27	35.37 01.10	03.01 02.30	09.03 06.02	03.02 15.04	37.19 04.18	01.06 13.05	01.04 05.18	03.04 34.08	03.09 02.29	11.07 36	21.01 11.18
Universality, adaptability	02	01.04 20.27		13.15 01	01.11 32.18	27.12	01.35 03.02	07.14 27.04	03	03.11 31	07.15 03.16	03.16 34.24	27.11	01.28 09.31
Level of automation	03	35.37 01.10	13.24 03.01		28.13 09	04.10 06.36	04.10 02.15	07.18 02	15.13 29	03.10 11	03.37 15.12	03.01 11	01.38	05.38
Reliability	04	03.01 14.30	11.01 32.18	28.11 13		28.09 03	09.12 28.36	11.01 03	13.17 04	04.02 24.17	13.19 17	03.28	02.04	13.01 05.17
Precision of manufacture	05	02.06 09.23	01.11 18	10.04 06.36	28.09 03		10.04 18.11	10.05 06	02.36	18.03	03.09 01.36	29.02	18.34 29	10.04 02.26
Precision of measurement	06	02.15 04.09	11.01 05	04.05 02.15	35.28 03.36	10.09 27		13.01 02.15	10.18 09.04	20.01 29.06	03.11 19.15	03.09 11.28	34.10 01.21	04.18 21.10
Complexity of construction	07	37.19 04	14.07 04.27	07.03 18	11.01 03	10.18 09	05.10 02.15		07.02 27.04	13.10 03.11	13.39 10.18	03.11	34.08 09.03	21.08 14.17
Complexity of inspection and measurement	08	01.06	03.07	15.33	13.17 04.32	27.34 24.09	10.18 09.04	07.02 27.04		35.04 28.14	05.35	37.10	01.38 13.21	21.08 14.04
Ease of manufacture	09	01.03 02.04	05.11 07	32.04 03	39.04 13.38	37.20 11.07	03.01 37.06	13.10 03	20.04 28.03		05.35 11.16	01.03 28.39	09.18 06.16	18.05
Ease of use	10	07.03 04	07.15 03.16	03.15 37.12	19.13 32.17	03.09 01.36	29.11 05.15	09.29 37.19	03.02	05.35 37		37.10 03.09	24.02 13.21	05.29 04.23
Ease of repair	11	03.09 02	34.03 24.16	15.01 34.11	28.02 03.16	29.02	02.05 11	01.03 11.28	01.24 19.11	03.01 28.02	03.37 10.07		12.39 11.10	01.02 05.16
Loss of information	12	11.36 07	18.35 39.17	01	02.04 36	27.24 09.34	09.03 27.34	20.11 18.24	01.38	09	13.21	05.02		21.02 03
External damaging factors	13	21.01 11.18	01.28 21.31	38.12 15	13.18 05.17	10.04 02.06	04.38 36.10	21.08 14.17	21.08 14.17	18.01 05	05.29 04.23	01.02 05	21.02 05	
Internal damaging factors	14	21.01 06.23	07.19 24.18	05	18.05 17.23	24.19 15.10	12.38 10	08.03 31	05.33 13.03	24.17 18.34	24.18 22.20	03.18	02.33 14	01.18 34
Length of the moveable object	15	22.24 04.14	22.07 03.16	19.18 10.16	02.22 14.17	02.04 14.27	04.09 24	03.08 10.18	01.03 10.18	03.14 19	07.14 01.24	03.04 02	03.18	03.07 19.18
Length of the fixed object	16	25.22 34.10	03.01	02.11	07.14 04	05.09 02	09.04 12	03.10	10	07.19 13	05.29	12	18.10	03.06
Surface of the moveable object	17	02.10 15.05	07.25	22.25 04.36	14.39	05.09	10.04 09.12	22.03 11	05.26 10.06	11.03 10.18	07.19 11.16	07.11 02.03	25.10	21.38 04.03
Surface of the fixed object	18	02.07 19.34	07.16	36	09.01 17.24	05.14 06.26	10.04 09.12	03.06 26	05.01 25.06	17.16	16.24	16	25.16	13.05 23.01
Volume of the moveable object	19	02.20 05.15	07.14	01.15 16.18	22.03 17.28	29.04 05.16	29.10 04	10.03	14.10 24	14.03 17	07.11 25.37	02	05.21	21.33 13.01
Volume of the fixed object	20	01.27 02.05	31.20 11.09	02.03 11.18	05.01 16	01.02 29	19.10 09.31	03.31	05.19 10	01	34.10 18.19	03	09.22 18	15.23 08.13
Shape	21	19.10 15.02	03.07 14	07.03 09	02.17 16	09.25 17	04.09 03	16.14 03.04	07.11 23	03.09 19.04	09.07 10	05.11 03	19.34 07.09	21.03 05.01
Speed	22	04.02 11	07.02 10	02.06	28.01 13.04	02.04 09.29	04.09 03.18	02.04 24.15	12.15 13.16	01.11 32.03	09.04 11.37	15.05 04.13	11.10	03.04 01.36
Functional time of the moveable object	23	01.19 22.08	03.01 11	20.02	28.05 11	12.13 16.17	12	02.24 14.07	08.14 23.01	13.03 24	37.13	14.02 13	02	21.07 38.04
Functional time of the fixed object	24	40.02 16.30	05	03	15.13 20.17	02.10 18	02.10 18	35.02 24	29.15 20.01	01.02	03	03	02	19.03 17.38
Loss of time	25	02.18 24.11	01.04	18.04 01.25	02.25 24	18.10 04.06	18.15 04.09	20.14	06.04 09.02	01.04 15.24	24.04 02.15	09.03 02	18.10 04.09	01.06 15
Quantity of material	26	11.14 12.13	07.12 14	32.01	06.12 04.17	38.25	12.05 04	12.11 13.02	12.13 14.06	14.03 01.13	01.14 02.29	05.09 02.29	18.04 01	01.38 14.31
Loss of material	27	04.01 02.36	07.02 05	01.02 06	02.14 23.01	01.02 18.31	16.15 31.04	01.02 04.18	01.06 02.11	07.15 38	09.04 05.18	05.01 15.13	09.03 02.12	38.21 25.17
Strength	28	14.01 02.22	07.12 09	07	28.12	12.13	12.13 16	05.11 04	13.12 07.17	28.12 02.09	09.17 04.05	13.28 12	04.10 02	06.01 27.03
Stable structure of the object	29	36.01 17.12	01.25 15.05	03.32 01	01.17 18	06	11	05.01 21.10	01.21 23.36	01.08	09.01 25	05.01 02.16	18.02 09	01.18 06.25
Force	30	12.04 01.27	07.19 06.40	05.01	12.01 11.33	04.14 27.26	01.02 36.18	10.01 02.06	26.27 02.08	07.27 06.03	03.04 12.29	07.03 28	09.34 02	03.01 17.06
Tension, pressure	31	02.22 01.27	01	01.18	02.11 08.01	12.01	20.04 29	08.03 01	05.26 27	03.01 16	28	05	04.05 34.18	21.05 27
Weight of the moveable object	32	01.12 18.27	14.35 07.32	10.01 06.08	12.28 03.13	04.01 10.06	04.13 01.10	10.25 26.15	04.14 10.09	13.04 03.26	01.12 05.18	05.13 04.28	02.18 01	21.33 06.13
Weight of the fixed object	33	03.04 07.01	08.07 14	05.10 01	02.04 32.12	02.03 01.19	06.10 04	03.02 10.23	29.04 19.07	04.03 09	20.11 03.09	05.13 04.28	02.07 01	05.08 21.27
Temperature	34	07.04 01	05.06 13	10.05 08.16	08.01 12.02	18	09.08 18	05.19 16	12.13 01.31	10.13	10.13	24.02 16	39.18	21.38 01.05
Brightness of the lighting	35	05.29 16	07.03 08	05.10 02	01.03	12.09	28.07 09	20.09 11	09.07	08.01 04.10	04.10 08	07.19 11.16	03.20	07.08
Power	36	04.01 15	08.19 15	04.05 19	08.18 10.31	09.05	09.07 05	40.08 25.15	08.01 16	10.02 15	10.01 02	01.05 02.15	02.08	08.21 31.05
Energy use of the moveable object	37	37.04 01	07.19 11.16	09.05	08.33 28.13	12.35 04.32	12.03 09	05.14 13.04	01.30	04.10 25	08.01	03.07 19.04	11.18 34.21	03.01 20.13
Energy use of the fixed object	38	03.20	01.11	02.05	02.26 36	04.24 06	04.01	04.24	08.01 16.29	03.24	11.18	01.19	05.02 34	02.05 21.27
Loss of energy	39	04.02 14.01	07.22 11.31	05	28.02 01	18.10 09.04	09	34.36	01.12 07.36	02.01	01.09 03	05.08	08.02	33.21 01.05
	#	01	02	03	04	05	06	07	08	09	10	11	12	13

Problem-factor (minus-factor) → / Trend-factor (plus-factor) ↓		internal damaging factors	length of the moveable object	length of the fixed object	surface of the moveable object	surface of the fixed object	volume of the moveable object	volume of the fixed object	shape	speed	functional time of the moveable object	functional time of the fixed object	loss of time	quantity of material
	#	**14**	**15**	**16**	**17**	**18**	**19**	**20**	**21**	**22**	**23**	**24**	**25**	**26**
Productivity	01	01.21 06.23	06.24 04.30	25.34 22.10	02.10 15.31	02.01 19.34	05.20 15.02	01.27 02.05	22.02 15.17	01.12 18	01.02 05.06	40.02 16.30	12.15	01.30
Universality, adaptability	02	18.24 28	01.03 14.05	03.01 16	01.25 14.34	07.16	07.01 14	07.31 16.12	07.27 03.32	01.02 22	11.03 01	05.16	01.04	12.01 07
Level of automation	03	05	22.11 04.19	36	19.22 11	11.10 24	01.11 16	10.11 18.31	07.09 03.11	04.02	20.39	02.16 11	18.04 01.25	01.11
Reliability	04	01.05 17.10	07.39 22.24	07.14 04.28	19.02 22.16	09.01 17.24	12.02 22.18	05.01 18	01.03 16.28	33.01 28.04	05.01 12.29	15.13 20.17	02.25 24	33.04 17.12
Precision of manufacture	05	24.19 15.10	02.04 14.27	05.09 02	04.38 14.09	05.14 06.26	09.04 05	29.02 01	09.25 17	02.04 09	12.13 17	18.04 15	09.10 04.06	09.25
Precision of measurement	06	12.38 23.02	04.10 35.16	09.04 12.16	10.04 09.12	10.04 09.12	09.11 20	10.18 11.31	20.04 09	04.11 09.18	04.20 09	02.10 18	18.15 04.09	05.20 09
Complexity of construction	07	08.03	03.08 10.18	10	22.03 11.16	20.26	15.10 20	03.16	14.11 04.07	15.02 04	02.24 04.07	02.11	20.14	11.12 13.02
Complexity of inspection and measurement	08	05.33	16.19 10.18	10	05.11 06.19	05.23 25.16	14.03 24.16	05.06 10.31	13.11 03.23	12.24 16.01	08.14 29.23	29.15 20.01	06.04 09.39	12.13 14.06
Ease of manufacture	09	33.18 23	03.14 11.19	07.19 13	11.03 10.37	16.17	11.14 03.17	01	03.04 11.13	01.11 32.03	13.03 24	01.16	01.04 15.24	01.36 03.18
Ease of use	10	31.18	03.19 11.37	03.24	03.19 11.16	06.16 07.23	03.16 01.07	24.06 23.31	07.15 14.04	06.11 15	14.12 32.29	03.16 29	24.04 02.15	37.01
Ease of repair	11	07.37 04	03.04 02.29	12.06 31	07.11 09	16.29	29.05 01.28	03	03.11 05.24	15.39	28.14 04.13	03	09.03 02.29	05.04 02.29
Loss of information	12	02.33 21	03.10	10	25.10	25.16	34.31	05.21	24.09	10.09	02	02	18.10 04.09	18.04 01
External damaging factors	13	11.18 19.24	19.03 23.24	03.06	21.03 38.04	13.05 23.01	21.36 27.01	15.23 08.13	21.03 12.01	33.21 01.04	21.07 38.04	19.03 17.38	01.06 15	01.38 14.31
Internal damaging factors	14		19.07 16.21	22.18	19.05 06.23	21.03 17	19.05 17	25.06 01.24	01.03	01.04 12.36	07.21 38.31	33.23 16.21	03.21	12.18 23.03
Length of the moveable object	15	19.07		19.11 25	07.19 24	34.07 03.24	34.19 04.01	19.31 08.24	03.32 02.14	11.24 32	08	02.01 03.08	07.05 14	14.01
Length of the fixed object	16	01.31	12.03 24		24.19 01	19.34 02.17	25.34 07	01.32 05.22	11.22 07.34	22.24 31	01.14 31	03.17 01	25.14 22	24.31 29.22
Surface of the moveable object	17	19.05 06.23	22.07 06.24	22.19 24.11		03.24 12.18	34.22 19.24	22.11	35.15 14.24	14.25 24.15	20.12	03.12 08	10.24	14.25 20.11
Surface of the fixed object	18	21.03 17	12.03	10.34 39.23	24.31 34		22.34 11	10.11 24	24.34 10	10.35	11.01	05.02 08.25	02.01 24.06	05.06 17.24
Volume of the moveable object	19	19.05 17.03	03.34 01.04	34.07 24	03.34 24.19	22.34 31		01.22 18	03.07 14.24	14.24 30.15	20.01 24	25.31 03	05.20 15.02	14.25 34
Volume of the fixed object	20	25.06 01.24	08.22	01.32 05.22	22.24 25.11	34.11 07	11.04 05.34		34.05 01	17.05 04	08.03 07.15	01.15 30	01.16 09.06	01.12
Shape	21	01.03	14.15 35.24	11.22 02.34	35.15 24.02	19.22 09.24	22.24 07.21	34.05 01		01.07 15.06	22.10 39.29	25.11 35.21	22.02 15.19	26.21
Speed	22	05.18 01.33	11.22 32	19.07 25	14.25 15	22.19 24	34.14 15	04.34	01.07 06.15		12.08 01.35	12.11	11.02	02.08 14.30
Functional time of the moveable object	23	33.23 16.21	05.08 39	37.39 08	12.19 08	39.37 08	02.05 08.25	02.25 37	22.10 04.29	12.01 35		02.40 24	40.02 04.06	12.01 02.17
Functional time of the fixed object	24	21	19.17 08	03.17 01	01.06 08.22	01.19 12.34	01.12 11.24	01.15 30	19.38 11.34	14.24 22.11	18.04		04.40 02.16	12.01 31
Loss of time	25	01.21 06.23	07.05 14	25.18 22.35	10.24 35.16	02.01 19.24	05.35 15.02	01.16 09.06	24.02 15.19	04.10 02.24	40.02 04.06	04.40 02.16		01.30 06.16
Quantity of material	26	12.01 17.23	14.22 01.06	07.31 01.24	07.22 14	05.06 17.24	07.40 14	01.30 31.03	01.22	01.14 15.04	12.01 02.17	12.01 31	01.30 06.16	
Loss of material	27	02.03 15.14	22.14 02.23	02.04 18	01.05 02.31	02.06 23.31	03.14 25.26	12.23 06.31	14.01 12.35	02.11 04.30	04.13 12.06	13.16 06.30	07.06 01.02	20.12 02.18
Strength	28	07.01 21.05	03.07 32.01	07.22 04.10	12.15 17.14	39.17 04	02.07 22.34	03.22 19.07	02.25 01.17	32.11 10.22	13.12 10	18.10 24	14.12 04.02	14.02 13
Stable structure of the object	29	01.17 13.23	11.07 03.04	27	05.28 11	23	04.02 08.23	15.04 01.17	21.03 06.24	38.07 04.06	11.13 02.01	23.12 01.36	01.13	07.09 01
Force	30	11.12 26.18	19.08 39.26	04.02	08.02 07	03.06 26.27	07.39 37.27	05.26 06.27	02.01 17.15	11.04 07.37	08.05	05.02 10	02.27 26	22.14 06.26
Tension, pressure	31	05.38 13.06	01.02 26	01.03 22.16	02.07 26.04	02.07 26.27	20.01 02	01.18	01.24 07.02	20 01.26	08.12 13	22.01 19.05	27.26 24	02.22 26
Weight of the moveable object	32	21.01 31.23	07.32 14.15	07.19 37.14	14.19 30.15	04.03 31.24	14.05 17.04	17.05 24.34	02.22 01.17	05.32 07.30	35.15 31.01	02.04	02.01 40.04	12.10 06.31
Weight of the fixed object	33	01.21 03.23	19.24	02.03 14.01	34.01	01.25 11.05	22.11 12	35.01 22.05	11.02 14.22	01.19 25	01.02 37	05.13 08.20	02.40 01.10	08.20 06.10
Temperature	34	21.01 05.18	07.08 39	07.08 39	12.01 23.06	01.30	15.23 17.06	01.20 24	22.21 08.09	05.04 26.25	08.11 23	08.06 26.17	01.04 33.06	12.19 25.23
Brightness of the lighting	35	01.08 09.23	08.09 16	22.01	08.09 10	19.24 03	05.11 02	22.11 02	09.25	02.11 08	05.08 20	20.02 04	08.03 10.19	03.08
Power	36	05.01 06	03.02 01.27	03.01 24	08.30	19.09 11.30	01.20 30	25.20 29	14.22 05.17	07.01 05	08.01 02.30	16	01.40 02.20	24.15 08
Energy use of the moveable object	37	05.01 20	37.04	05.08 07.04	07.08 29	01.05 19.07	01.11 06	12.15 01.34	37.05 14	32.07 01	04.01 20.06	04.11 02.08	01.30 08.06	15.36 16.06
Energy use of the fixed object	38	08.21 06	19.24 37	19.39 16	11.37 18	19.16 08	11.08 04	01.06	34.18	03.04	08.24 04	19.04 10	02.17 34.39	12.01 31
Loss of energy	39	33.01 05.21	34.05 20.11	20.30 34	07.10 19.25	19.34 25.06	34.06 36	34	24.18	16.01 30	33.01	19.31	02.06 09.34	34.06 29
	#	**14**	**15**	**16**	**17**	**18**	**19**	**20**	**21**	**22**	**23**	**24**	**25**	**26**

Problem-factor (minus-factor) ➡ Trend-factor (plus-factor) ⬇		loss of material	strength	stabile structure of the object	force	tension, pressure	weight of the moveable object	weight of the fixed object	temperature	brightness of the lighting	power	energy use of the moveable object	energy use of the fixed object	loss of energy
	#	27	28	29	30	31	32	33	34	35	36	37	38	39
Productivity	01	04.02 01.36	14.04 02.06	01.12 21.23	04.07 02.26	02.27 22	01.10 18.27	04.13 07.12	01.33 04.02	10.19 08.03	01.40 02	01.02 30.08	03	04.02 14.01
Universality, adaptability	02	07.02 05.11	01.12 09.20	01.25 22	07.19 40	01.16	03.20 07.32	08.07 14.16	13.05 12.01	20.21 10.03	08.03 14	08.01 14.11	16.03 37	06.07 03
Level of automation	03	01.02 06.35	29.11	06.03	05.01	11.01	04.10 06.01	04.10 01.02	10.05 08	32.09 08	04.05 13	05.09 11	01.03 05.11	36.04
Reliability	04	02.01 14.23	28.04	12.03 18.05	32.04 02.12	02.18 01.08	12.32 02.17	12.02 32.04	12.01 02	28.09 11	33.28 10.31	33.28 13.08	26.36	02.28 01
Precision of manufacture	05	01.31 02.18	12.13	25.06	04.08 15.26	12.01	04.09 11.06	10.01 13.39	08.10	12.09	09.05	09.05	02.18 35.40	11.09 05
Precision of measurement	06	02.16 31.04	04.20 09	09.01 11	09.05	20.04 09	09.01 10.04	04.01 29.10	20.08 04.18	20.03 09	12.20 09	12.20 09	18.16 24	10.09 13
Complexity of construction	07	01.02 04.14	05.11 04	05.21 19.08	10.16	08.03 01	10.25 15.26	05.10 01.23	05.19 11	18.19 11	40.08 25.15	13.05 14.04	04.02 11	02.01 11.05
Complexity of inspection and measurement	08	03.06 02.18	13.12 07.04	28.21 23.25	26.04 17.08	01.26 27.09	13.10 04.11	20.11 04.03	12.13 01.16	05.18 10	08.03 16.02	01.30	08.01 16	01.12 07.08
Ease of manufacture	09	07.15 38	03.12 02.09	28.11 03	01.37	01.08 03.27	04.14 07.16	03.13 26.11	13.10 06	04.18 13.03	13.03 37.18	04.10 13.03	03.24	08.01
Ease of use	10	04.09 05.18	09.17 12.04	09.01 25	04.11 01	05.09 37	29.05 11.07	20.11 03.29	10.13 11	11.19 03.18	01.15 05.02	03.11 18	18.37 11	05.08 11
Ease of repair	11	05.01 15.13	03.28 05.39	05.01	03.28 02	11	05.13 01.28	05.13 01.28	24.02	07.03 11	07.02 09.05	07.03 04.16	03.07 11.16	07.03 09.08
Loss of information	12	34.19 12.11	18.31 22.17	25.10 18.35	11.03	18.21	02.18 01	02.01 35	21.03 18	08	02.08	03.40 08	18	08.02
External damaging factors	13	38.21 08.17	06.01 27.03	01.18 25.06	11.01 23.06	21.05 27	21.33 13.23	05.21 11.18	21.38 01.05	03.08 09.11	08.21 31.05	03.18 20.13	02.05 21.27	33.21 01.05
Internal damaging factors	14	02.03 15	07.01 21.05	01.17 13.23	01.04 03.17	05.38 13.06	08.21 07.23	01.21 03.23	21.01 05.18	08.18 23.09	05.01 06	05.01 20	08.21 06	33.01 05.21
Length of the moveable object	15	24.14 36.02	32.01 14.15	03.32 07.15	19.02 24	03.32 01	32.07 14.15	03.19 07.24	02.07 08	09	03.01	32.01 18	24.03 18.07	34.05 01.23
Length of the fixed object	16	02.04 18.01	07.22 04.10	23.27 01	04.02	03.22 01	25.31 32.17	01.04 17.14	12.01 30.06	12.29	37.32	01.18 33.11	25.31 37.11	20.04
Surface of the moveable object	17	02.01 05.23	12.07 17.22	28.05 11.23	08.25 01.05	02.07 26.04	05.19 14.04	12.31 05	05.07 16	07.09 08.11	08.02 09.06	08.09	19.08 35	07.19 25.10
Surface of the fixed object	18	02.22 06.23	17	05.30	03.06 01.26	02.07 26.27	22.31 24.11	25.05 22.06	01.23 30	03.18 01.09	19.09	08.11 04	01.17	19.34 25
Volume of the moveable object	19	26.23 15.02	39.22 07.34	04.02 03.23	07.01 26.27	20.01 26.27	05.10 14.17	31.17 10	15.23 02.06	02.11 05	01.20 11.06	01	30.38 08	34.07 11.16
Volume of the fixed object	20	02.23 01.15	39.22 19.07	18.04 01.17	05.06 27	18.01	31.25 24.08	01.02 08.22	01.20 24	01.18 04.34	25.20	01.08 31.11	01.17	01.17
Shape	21	01.14 12.35	25.22 02.17	38.03 06.24	01.02 27.17	15.07 02.22	32.02 14.17	07.02 10.12	21.22 08.09	11.07 09	24.20 05	05.20 15.22	22.07 31	22
Speed	22	02.11 04.30	32.12 10.22	04.38 03.06	11.04 07.08	20.06 30.17	05.04 11.30	03.11	04.25 26.05	02.11 08	08.01 30.05	32.07 01.30	01.08	22.40 08.01
Functional time of the moveable object	23	04.13 12.06	13.12 02	11.12 01	08.05 16	08.12 13	08.35 15.31	31.15	08.01 23	05.08 24.01	08.02 01.30	04.20 01.06	20.06	02.18 01
Functional time of the fixed object	24	13.16 06.30	01.39	23.12 01.36	19.17 39	19.24 17	31.24 07	20.13 08.16	08.06 26.17	17.18 .34	16	11.17 18	01.17	02.17 15.18
Loss of time	25	01.06 02.23	14.12 04.06	01.12 21.35	02.27 26.35	27.26 24	02.40 27.01	02.40 10.35	01.14 33.06	03.08 10.19	01.40 02.20	01.30 08.06	03	02.35 06.09
Quantity of material	26	20.12 02.18	22.01 15.02	07.05 19.17	01.22 12	02.26 22.12	01.20 06.31	13.10 06.01	12.19 23	01.04 25.31	01	15.14 16.06	12.01 31	34.06 29
Loss of material	27		01.04 31.17	05.22 25.17	22.07 06.17	12.26 27.02	01.20 36.17	01.20 21.09	33.26 23.31	03.20 11	04.13 06.30	01.06 18.35	04.13 37.31	01.13 05.31
Strength	28	01.04 31.17		11.19 01	02.06 12.22	02.12 06.17	03.32 17.07	17.10 13.03	25.02 17	01.08	02.10 01.04	08.01 02	01	01
Stabile structure of the object	29	05.22 25.17	19.39 07		02.01 33.16	05.01 17	33.01 05.23	10.23 03.17	01.03 09	09.12 13.07	09.01 13.31	11.08	13.24 14.06	22.05 23.20
Force	30	32.01 17.35	01.02 22.13	01.02 33		06.33 28	32.03 27.06	06.11 03.04	01.02 33	11.08 01.18	08.01 06.27	08.19 02	03.16 26.27	22.07
Tension, pressure	31	02.26 12.27	39.06 12.17	01.38 05.17	26.01 33		02.26 27.17	11.14 02.06	01.23 08.05	18.37 21.01	02.01 22	22.18 02.27	19.24 37.18	05.26 29
Weight of the moveable object	32	35.01 12.31	04.13 06.17	03.01 08.23	32.02 06.27	02.26 27.17		08.01	20.14 24.30	08.03 09	37.26 06.31	01.37 15.31	01.03 04	20.05 15.08
Weight of the fixed object	33	35.32 11.25	04.05 02.13	10.23 03.17	32.02 08.01	11.14 02.06	17.31 03		04.08 09.21	01 08.09	07.08 06.21	12.19	06.08 04.03	06.08 04.07
Temperature	34	33.26 14.31	02.25 21.17	03.01 09	01.02 12.33	01.23 08.05	26.21 20.30	21.01 09		09.25 33.16	05.22 19.29	08.07 12.19	01.12	33.19 01.30
Brightness of the lighting	35	11.03	01.08	09.12 13	10.08 20	25.01 37	08.03 09	05.01 09	09.01 08		09	09.03 08	09.01 03.07	08.16 03.20
Power	36	04.13 06.30	10.02 04	01.09 07.31	10.05 26.01	21.02 01	32.26 30.31	08.10 19.13	05.22 19.29	16.20 08		16.20 08.27	08.07 16.03	02.01 30
Energy use of the moveable object	37	01.18 06.35	35.08 39.01	08.11 19.18	16.10 33.05	36.22 29	37.06 04.31	04.37 31.11	08.18 12.22	05.07 08	20.08 27.06		07.11	37.21 07.18
Energy use of the fixed object	38	04.13 06.31	01	13.24 14.06	26.27	19.39 01.17	04.11 32.08	08.39 20.13	01.08 33.26	08.05 01.09	35.08 11.01	05.08 11.35		01.31 03.18
Loss of energy	39	1.13 05.27	10	22.05 23.20	26.30	24.11	07.20 08.04	08.20 06.39	08.30 34	03.11 09.07	12.30	01.08 12	01.08 24	
	#	27	28	29	30	31	32	33	34	35	36	37	38	39

S22. As-CATALOGUE

S22.1. List of 40 navigators (specialized transformation models)

#	Navigator
01	Change in the aggregate state of an object
02	Preliminary action
03	Segmentation
04	Replacement of mechanical matter
05	Separation
06	Use of mechanical oscillations
07	Dynamization
08	Periodic action
09	Change in color
10	Copying
11	Inverse action
12	Local property
13	Inexpensive short-life object as a replacement for expensive long-life one
14	Use of pneumatic or hydraulic constructions
15	Discard and renewal of parts
16	Partial or excess effect
17	Use of composite materials
18	Mediator
19	Transition into another dimension
20	Universality
21	Transform damage into use
22	Spherical shape
23	Use of inert media
24	Asymmetry
25	Use of flexible covers and thin films
26	Phase transitions
27	Use of thermal expansion
28	Previously installed cushions
29	Self-servicing
30	Use of strong oxidants
31	Use of porous materials
32	Counterweight
33	Quick jump
34	Matryoshka (nested doll)
35	Unite
36	Feedback
37	Equipotentiality
38	Homogeneity
39	Preliminary counteraction
40	Uninterrupted useful function

S22.2. Table of As-catalogue

01. Change in the aggregate state of an object	a) This includes transitions into "pseudo-states" ("pseudo-liquid") and transitional states such as the use of the elastic properties of solid objects as well as simple transitions such as from a solid to a liquid state; b) changes in concentration or consistency, degree of flexibility, temperature, etc.
02. Preliminary action	a) Previous necessary (partial or complete) change of an object; b) prepare objects in advance so that they can be put to work from the best position and are available without loss of time.
03. Segmentation	a) Disassemble an object into individual parts; b) make it possible to disassemble an object; c) raise the degree of disassembly (reduction into parts) of an object.
04. Replacement of mechanical matter	a) Replace mechanical schemes with optical, acoustic, or olfactory schemes; b) use electrical, magnetic, or electromagnetic fields for the interaction of objects; c) replace static fields with dynamic ones, i.e., from temporally fixed to flexible fields, from unstructured fields to fields with a specific structure; d) use fields in connection with ferric-magnetic particles.
05. Separation	Separate the "incompatible part" ("incompatible property") from the object or—turned completely around—include the only really necessary part (necessary property) into the object.
06. Use of mechanical oscillations	a. Cause an object to vibrate; b) raise the frequency of the vibrations up to and including ultra-high frequencies if the object is already in motion; c) use resonating frequency, application of quartz vibrators; d) use ultrasound vibrations in connection with electromagnetic fields.
07. Dynamization	a) The characteristics of an object or an environment are changed to optimize every work procedure; b) disassemble an object into parts that are moveable among each other; c) make an object moveable that is otherwise fixed.
08. Periodic action	a) Transition from a continuous function to a periodic one (impulse); b) change periods if the function already runs that way; c) use breaks between impulses for other functions.

09. Change in color	a) Change the color of an object or its environment; b) change the level of the transparency of an object or its environment; c) use color supplements to observe objects or processes that are difficult to see; use color supplements to observe objects or processes that are difficult to see; d) add lighting if this kind of supplement is already in use.
10. Copying	a) Use a simplified and inexpensive copy instead of an inaccessible, complicated, expensive, inappropriate, or fragile object; b) replace an object or a system of objects with optical copies; use here a change in measure (blow-up or reduce the copy); c) if visible copies are used, they can be replaced with infrared or ultra-violet copies.
11. Inverse action	a) Instead of an action prescribed by the conditions of an assignment, complete a reverse action (heat an object instead of cooling it); b) make a moveable part of an object or the environment fixed or make a fixed part moveable; c) turn an object upside-down or around.
12. Local property	a) Change the structure of the object (the external environment, external influences) from the same to a different one; b) different parts of an object have different functions; c) every object should exist under conditions that correspond best to its functions.
13. Inexpensive short-life object as a replacement for an expensive long-life one	Replace an expensive object with a group of inexpensive objects without certain properties, for example, long life.
14. Use of pneumatic or hydraulic constructions	Use gaseous or fluid parts instead of fixed parts in an object, e.g., parts that can be blown up or filled with hydraulic fluid, air cushions, or hydrostatic or hydro-reactive parts.
15. Discard and renewal of parts	a) Parts that have fulfilled their task and are no longer part of an object should be disposed of (dissolved, evaporated, etc.); b) used parts of an object should be immediately replaced during work.
16. Partial or excess effect	When it is difficult to achieve the desired effect completely, we should try to achieve a bit less or a bit more; this can make the task much easier.

17. Use of composite materials	Move from homogeneous materials to combinations.
18. Mediator	a) Use another object to transfer or transmit an action; b) temporarily connect an object with another (easily separable) object.
19. Transition into another dimension	a) An object is shaped so that it can move or is placed not only in a linear fashion but also in two dimensions, i.e., on a surface; it is also possible to improve the transition from a surface to a three-dimensional space; b) do construction on several floors; tip or turn the object on its side; use the back of the space in question; c) use optical rays that strike a neighboring space or the back of the present space.
20. Universality	An object has several simultaneous functions so that other objects are not needed.
21. Transform damage into use	a) Use damaging factors, especially those from the environment, to achieve a useful effect; b) eliminate a negative factor by combining it with other negative factors; c) support the damaging factor until it no longer causes damage.
22. Spherical-shape	a) Change from linear parts of the objects to curved ones, from flat surfaces to spherical ones, from parts shaped like cubes or parallelepipeds to round structures; b) use rollers, balls, and springs; c) change to turning movements by using centrifugal force.
23. Use of inert media	a) Replace a normal medium with an inert one; b) complete a process in a vacuum.
24. Asymmetry	a) Move from a symmetrical shape of an object to an asymmetrical one; b) increase the degree if the object is already asymmetrical.
25. Use of flexible covers and thin films	a) Flexible covers and thin layers are used in place of the usual constructions; b) isolate objects from the external world with flexible covers or thin layers.
26. Phase transitions	Make full use of phenomena that occur during phase transitions such as a change in volume, radiation, absorption of warmth, etc.
27. Full use of thermal expansion	a) Make full use of the expansion (or reduction) of materials when heating them; b) Make use of materials with different coefficients of heat expansion.

28. Previously installed cushion	Increase the relatively low security of an object with safety measures in advance.
29. Self-servicing	a) The object services itself with auxiliary and repair functions; b) reuse waste (energy, material).
30. Use of strong oxidants	a) Replace normal air with an enhanced stream; b) replace an enhanced stream with oxygen; c) influence air or oxygen with ionizing rays; d) use of oxygen with ozone; e) replace ionized or ozone-oxygen with ozone.
31. Use of porous materials	a) Make an object porous or use supplementary porous elements (inserts, coverings, etc.); b) if the object already consists of a porous material, the pores can be filled with some kind of material in advance.
32. Counter-weight	a) Compensate for the weight of an object with its connection to another object with lifting power; b) compensate for the weight of an object using interaction with the external environment (for example, with aerodynamic or hydrodynamic forces).
33. Quick jump	Complete a process or some of its (damaging or dangerous) stages at high speed.
34. Matryoshka (nested doll)	a) An object is inside another object that is also inside another, etc.; b) an object runs through a hollow space in another object.
35. Unite	a) Unite similar objects or objects for neighboring operations; b) temporarily unite similar objects or objects for neighboring operations.
36. Feedback	a) Create a retroactive influence; b) change a retroactive influence that already exists.
37. Equipotentiality	Change work conditions so that it is not necessary to lift or lower an object.
38. Homogeneity	Objects that interact with the object in question must be made from the same material (or from one with similar properties).
39. Preliminary counter-action	If the conditions of a task require an action, then a opposite action should be taken in advance.
40. Uninterrupted useful function	a) Complete a job without interruptions where all parts work continuously at full capacity; b) eliminate idle running and interruptions.

S22.3. As-catalogue with brief examples

01. Change in the aggregate state of an object	a) This includes transitions into "pseudo-states" ("pseudo-liquid") and into transitional states such as the use of the elastic properties of solid objects as well as simple transitions such as from a solid to a liquid state; b) changes in concentration or consistency, in the degree of flexibility, in temperature, etc.
	(01-02) The German firm ZELTEC has developed a transparent cone to extract clean water from the ground **(navigator 01).** The cone is placed on the ground, and solar rays heat the surface. Water in the ground evaporates, precipitates on the walls of the cone, and flows into a collector placed around the cone at its bottom.

02. Preliminary action	c) Previous necessary (partial or complete) change of an object; and/or d) prepare objects in advance so that they can be put to work from the best position and are available without loss of time.
	(02-02) Nets are stretched around sheep in advance to collect wool that falls from the animals upon application of a special chemical preparation **(navigator 02).**

03. Segmentation	d) Disassemble an object into individual parts; e) make it possible to disassemble an object; and/or f) raise the degree of disassembly (reduction into parts) of an object.
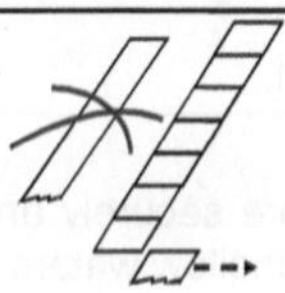	(03-02) To regenerate the cutting function of a knife, its blade is formed out of several segments that can be removed when they wear out. Small indentations are cut into the blade for this purpose **(navigator 03).**

04. Replacement of mechanical matter	e) Replace mechanical schemes with optical, acoustic, or olfactory schemes; f) use electrical, magnetic, or electromagnetic fields for the interaction of objects; g) replace static fields with dynamic ones, e.g., from temporally fixed to flexible fields, from unstructured fields to fields with a specific structure; and/or h) use fields in connection with ferric-magnetic particles.
	(04-02) The German firm Microtec has made a micro-miniature "submarine" for inspection and treatment of blood vessels. The external rotating magnetic field actuates the "propeller" **(navigator 04).**

05. Separation	Separate the "incompatible part" ("incompatible property") from the object or—turned completely around—include the only really necessary part (necessary property) into the object.
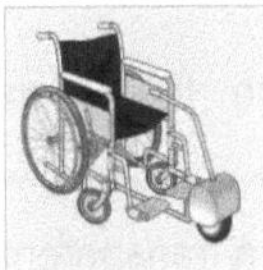	(05-01) A device is integrated into a wheelchair that is equipped with an electrical motor, an accumulator, and a guidance system in order to ensure a motorized forward movement of the chair (**navigator 05**).

06. Use of mechanical oscillations	e) Cause an object to vibrate; f) raise the frequency of the vibrations up to and including ultra-high frequencies if the object is already in motion; g) use resonating frequency, application of quartz vibrators; h) use ultrasound vibrations in connection with electromagnetic fields.
	(06-01) An American firm has produced special ultrasonic whistles that are mounted on the front bumper. At speeds over 50 kilometers per hour, under influence of the air stream the whistles emit ultrasonic signals which are frightening for animals, but inaudible for people **(navigator 06).** The higher the speed, the "louder" the signal.

07. Dynamization	d) The characteristics of an object or an environment are changed to optimize every work procedure; e) disassemble an object into parts that are moveable among each other; and/or f) make an object moveable that is otherwise fixed.
	(07-01) A new dynamic solution was found to more securely anchor yachts and improve their movement especially in shallow waters: The keel can be folded and turned using a hinge in the hull with flexible components **(navigator 07).**

08. Periodic action	d) Transition from a continuous function to a periodic one (impulse); d) change periods if the function already runs that way; and/or f) use breaks between impulses for other functions.
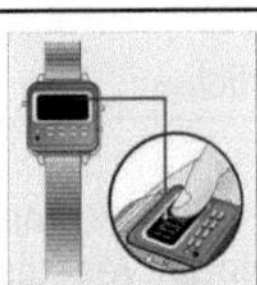	(08-02) The Japanese firm Seiko has produced an electronic watch that "falls asleep" (ceases to show time) if the watch remains motionless for >3 days. Internal memory continues to keep time, and the watch "wakes up" at a touch and starts to show the correct time right away **(navigator 08).**

09. Change in color	e) Change the color of an object or its environment; f) change the level of the transparency of an object or its environment; g) use color supplements to observe objects or processes that are difficult to see; and/or h) add lighting if this kind of supplements is already in use.
	(09-02) The French firm DAITEM proposed a new system to protect apartments and cars from thieves. The system uses a device that fills the apartment or car with dense, thick milky-white smoke upon detecting unauthorized entry **(navigator 09).** The device is activated when the burglar alarm system is tripped. The smoke does not damage either the apartment or the car.

10. Copying	d) Use a simplified and inexpensive copy instead of an inaccessible, complicated, expensive, inappropriate, or fragile object; e) replace an object or a system of objects with optical copies; use here a change in measurement (blow-up or reduce the copy); and/or f) if visible copies are used, they can be replaced with infra-red or ultra-violet copies.
	(10-01) A system has been developed to copy aquariums with many different species of fish. The system consists of a display, e.g., a computer monitor installed behind an aquarium with thin glass walls that imitates the movement of air bubbles and the natural movements of plants. The fish and their entire world are projected with video recordings on the computer screen (**navigator 10**). The combined planes create an illusion of existence of the fish in a big aquarium.

11. Inverse action	d) Instead of an action prescribed by the conditions of an assignment, complete a reverse action (heat an object instead of cooling it); e) make a moveable part of an object or the environment fixed or a fixed part moveable; and/or f) turn an object upside-down or around.
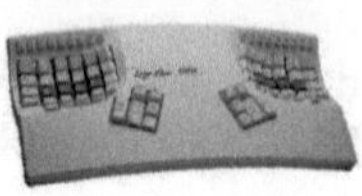	(11-02) The American firm Kinesis has developed a keyboard with left- and right-hand keys sunk in two separated recesses **(navigator 11).**

12. Local property	d) Change the structure of the object (the external environment, external influences) from the same to a different one; e) different parts of an object have different functions; and/or f) every object should exist under conditions that correspond best to its functions.
	(12-02) It has been suggested that pipes made of readily fusible material filled with fire-extinguishing liquid be installed above electrical cables to automatically "locate" and suppress electrical fires **(navigator 12).**

13. Inexpensive short-life object as a replacement for expensive long-life one	Replace an expensive object with a group of inexpensive objects without certain properties, for example, long life.
	(13-01) It has been suggested to create and apply small and inexpensive unmanned planes to record meteorological data above oceans **(navigator 13).**

14. Use of pneumatic or hydraulic constructions	Use gaseous or fluid parts instead of fixed parts in an object, e.g., parts that can be blown up or filled with hydraulic fluid, air cushions, or hydrostatic or hydro-reactive parts.
	(14-02) The Japanese firm Mugen Denko produced an inflatable life vest for motorcyclists **(navigator 14).**

15. Discard and renewal of parts	c) Parts that have fulfilled their task and are no longer part of an object should be disposed of (dissolved, evaporated, etc.); and/or d) used parts of an object should be immediately replaced during work.
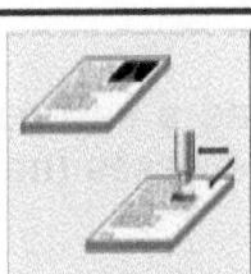	(15-02) The Japanese firm "Toshiba" developed computer printer ink that becomes almost completely colorless when heated **(navigator 15).**

16. Partial or excess effect	When it is difficult to achieve the desired effect completely, we should try to achieve a bit less or a bit more. This can make the task much easier.
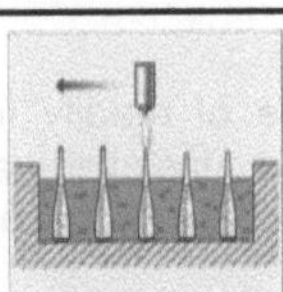	(16-02) Medicinal ampoules can be sealed off quickly by dipping them into a cooling liquid that protects them from a strong thermal field except for the tip, which subjected to intense heating (**navigator 16**).

17. Use of composite materials	Move from homogeneous materials to combinations.
	(17-02) In England, a multilayered curtain has been proposed. It contains pores of different sizes providing mechanical filtration of sound waves so that the aggregate pass band of a set of filters approximately corresponds to the spectrum of sea surf **(navigator 17).**

18. Mediator	c) Use another object to transfer or transmit an action; and/or d) temporarily connect an object with another (easily separable) object.
	(18-02) Single-cell microorganisms that glow in moving water were introduced into salt water pools to investigate the movements of dolphins (mediators according to **navigator 18**). The water flowing around the dolphins' bodies can then be recorded with high-precision video cameras.

19. Transition into another dimension	d) An object is shaped so that it can move or is placed not only in a linear fashion but also in two dimensions, e.g., on a surface. It is also possible to improve the transition from a surface to a three-dimensional space; and/or e) do construction on several floors; tip or turn the object on its side; use the back of the space in question; f) use optical rays that strike a neighboring space or the back of the present space.
	(19-02) Measuring with a normal measuring tape often means that the tape is used in different directions, either horizontally, vertically, or at various angles. This means the measurements are made in different dimensions (**navigator 19**). This makes it possible to construct closely connected three-dimensional representations of a room.

20. Universality	An object has several simultaneous functions so that other objects are not needed.
	(20-02) A robot with universal learning abilities has been developed at the TU Berlin to dismantle complex household appliances and industrial facilities that are no longer in use (solution according **navigator 20**).

21. Transform damage into use	d) Use damaging factors, especially those from the environment, to achieve a useful effect; e) eliminate a negative factor by combining it with other negative factors; and/or f) support the damaging factor until it is no longer causes damage.
	(21-01) Wood by-products (bark, shavings, and leftovers) could be used in large power plants. The production of extra electrical energy for saw mills or neighbouring facilities could compensate for the damage caused by large amounts of wood waste that collects in them **(navigator 21**).

22. Spherical-shape	d) Change from linear parts of the objects to curved ones, from flat surfaces to spherical ones; from parts shaped like cubes or parallelepipeds to round structures; a) use rollers, balls, and springs; and/or b) change to turning movements by using centrifugal force.
	(22-01) Scientists from the University of Potsdam (Germany) have developed a special stand on the basis of a rotating disk **(navigator 22).** The disk is fitted with protrusions that can be used as hand grips and foot rests. The axis of rotation of the disk can also be changed.

23. Use of inert media	c) Replace a normal medium with an inert one; and/or d) complete a process in a vacuum.
	(23-01) Powdered dry ice has long been used to disperse clouds; now German meteorologists in Germany have successfully used it to disperse fog at an airfield **(navigator 23).**

24. Asymmetry	c) Move from a symmetrical shape of an object to an asymmetrical one; and/or d) increase the degree of asymmetry if the object is already asymmetrical.
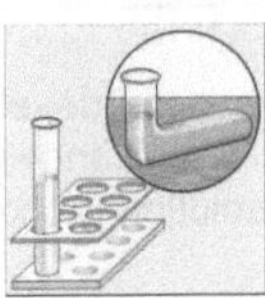	(24-01) A reagent tube needs to be developed with curved form and flat foundation so that it can stand on a table without a special frame; this is expected to simplify the use of such devices (**navigator 24**).

25. Use of flexible covers and thin films	c) Flexible covers and thin layers are used in place of the usual constructions; and/or d) isolate objects from the external world with flexible covers or thin layers.
	(25-01) German entrepreneur Peter Aschauer suggested a new rescue device – an avalanche airbag made of bright orange nylon **(navigator 25).** The airbag is carried in a small backpack and inflated by compressed nitrogen from a small cylinder activated by the user when he finds himself in danger of being buried under snow.

26. Phase transitions	Make full use of phenomena that occur during phase transitions such as a change in volume, radiation, absorption of warmth, etc.
	(26-01) To enable rapid cooling of injured areas, the French firm Crionic Medical offers gun-shaped portable devices; when the trigger is pulled, the gun emits a cloud of dry ice **(navigator 26).**

27. Use of thermal expansion	c) Make full use of the expansion (or reduction) of materials when heating them; and/or d) use materials with different coefficients of heat expansion.
	(27-01) The firm Sealed Air Corporation (USA) has developed highly elastic polyethylene bags that come in all sizes. Mechanical or thermal impact triggers the production of a polymer foam, which is then evenly distributed throughout the interior of the bag **(navigator 27).**

28. Previously installed cushions	Increase the relatively low security of an object with safety measures in advance.
	(28-02) Parachutes in the passenger section could help save passengers in plane crashes (**navigator 28**). The "superfluous" weight of cargo sections, wings, and engines could be separated and discarded so that the burden on the parachutes is reduced (as is their own weight). This could reduce the operating expenses attributable to such system.

29. Self-servicing	c) The object services itself with auxiliary and repair functions; and/or d) reuse waste (energy, material).
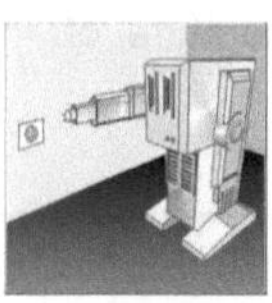	(29-01) Robots have a program that instructs them to search for a specified or any nearby electrical socket in order to charge their accumulators (**navigator 29**).

30. Use of strong oxidants	f) Replace normal air with an enhanced stream; g) replace an enhanced stream with oxygen; h) influence air or oxygen with ionizing rays; i) use of oxygen with ozone; and/or j) replace ionized or ozone-oxygen with ozone.
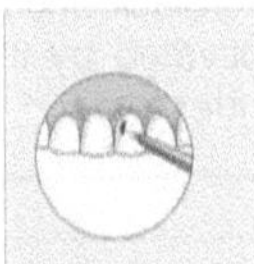	(30-01) The Swedish firm Mediteam has developed a method of treatment of caries without drilling holes by using a special ,mixture that dissolves the diseased tissue of the tooth in less than half a minute **(navigator 30).**

31. Use of porous materials	c) Make an object porous or use supplementary porous elements (inserts, coverings, etc.); and/or d) if the object already consists of a porous material, the pores can be filled with some kind of material in advance.
	(31-01) Experts of FAO (Food and Agriculture Organization of the United Nations) have developed and patented a way to sterilize coconut water without heating. The drink is filtered with micro-porous filters, which remove microbes and their spores **(navigator 31).**

32. Counterweight	c) Compensate for the weight of an object with its connection to another object with lifting power; and/or d) compensate for the weight of an object using interaction with the external environment (e.g., with aerodynamic or hydrodynamic forces).
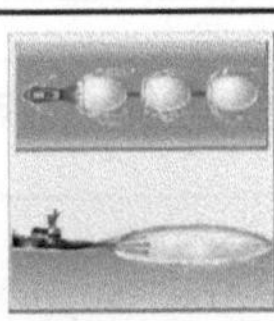	(32-02) Turkish engineers have offered a way to deliver lots of fresh water by sea transportation of long "chains" and "networks" of tanks made of polymeric film; each such tank contains 10,000 cubic meters of water. Because fresh water is lighter then sea water, the tanks stay afloat **(navigator 32).**

33. Quick jump	Complete a process or some of its (damaging or dangerous) stages at high speed.
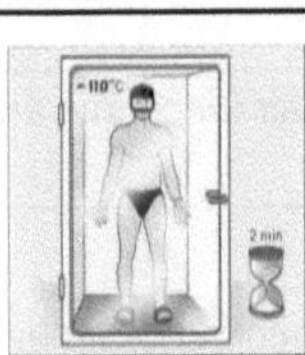	(33-02) An unusual way of treating arthritis and rheumatism is used by rheumatologists in the German city of Sendenhorst: The patient is placed in a refrigerating chamber with a temperature of –110 to –120°C for 1 to 2 minutes **(navigator 33).**

34. Matryoshka (nested doll)	c) An object is inside another object, which is also inside another object, etc.; and/or d) an object runs through a hollow space in another object.
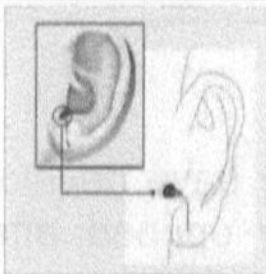	(34-02) A receiver and synthesizer are inserted into the ear to conveniently and discreetly hear radio programs, messages from mobile telephones, and other signals (**navigator 34**).

35. Unite	c) Unite similar objects or objects for neighboring operations; and/or d) temporarily unite similar objects or objects for neighboring operations.
	(35-02) Some 10,000 thermal microelements are installed on the underside of a watch to produce a sufficiently large electrical potential to power the watch; the microelements transform the difference between the temperature of the wrist and the temperature of the ambient environment into electricity (**navigator 35**).

36. Feedback	c) Create a retroactive influence; and/or d) change a retroactive influence that already exists.
	(36-02) Siemens has developed a computer program that determines the amount of cash to be placed inside each ATM based on previous usage data (**navigator 36**).

37. Equipotentiality	Change work conditions so that it is not necessary to lift or lower an object.
	(37-02) To make it easier for elderly and sick people to get into and out of the bathtub, is the tub is fitted with a door that can be opened when the tub is empty and is tightly sealed when the tub is filled with water **(navigator 37).**

38. Homogeneity	Objects that interact with the object in question must be made from the same material (or from one with similar properties).
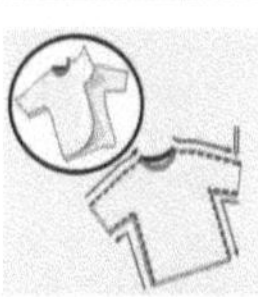	(38-01) In the Welding Institute in Cambridge (Great Britain), a shirt made from pieces of a synthetic fabric has been welded by laser **(navigator 38).**

39. Preliminary counteraction	If the conditions of a task require an action, then an opposite action should be taken in advance.
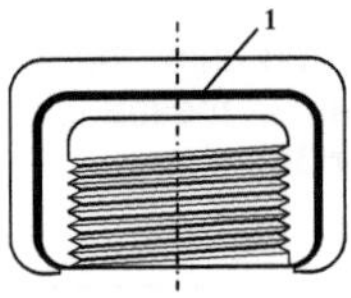	(39-02) English engineers form Salford University have developed a special cap for medicine tubes that cannot be opened after the expiration date (**navigator 39**).

40. Uninterrupted useful function	c) Complete a job without interruptions where all parts work continuously at full capacity; and/or d) eliminate idle running and interruptions.
	(40-02) The American firm Procter & Gamble has patented a way of treatment by inhalation during ironing by filling a steam generator of the iron with water containing several drops of medicine **(navigator 40).**

S23. RESOLVING THE RADICAL CONTRADICTION USING THE METHOD RICO (Radical In Cluster Out)

If you have one and the same property to be increased and decreased, the A-matrix cannot help because of one and the same plus- and minus-factors:

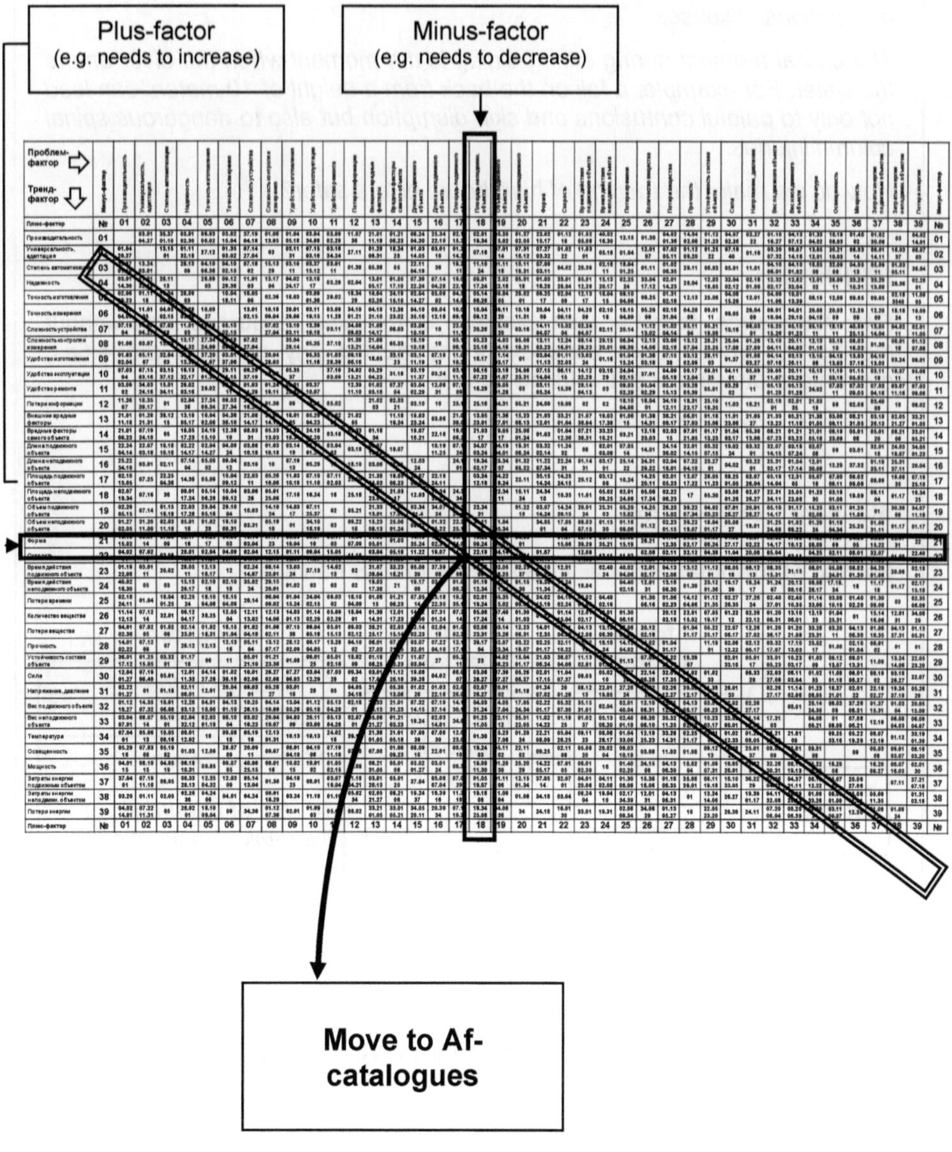

S24. EXAMPLE "DIVER" (REINVENTING)

S24.1. START-pass through Standard Contradiction

TREND

The training of high-board divers is associated with high risk of injuries and occupational diseases.

The critical moment during such training is the moment when the diver enters the water. For example, a fall on the back from a height of 10 meters can lead not only to painful contusions and skin disruption but also to dangerous spinal column injuries.

How do we make the training of high-board divers safer?

REDUCING

FIM: X-resource, together with available or modified resources, and without making the object more complex or introducing any negative properties, guarantees attainment of the following **IFR**:

[*safe entry into water from any height*]

Standard Contradiction

Format of formula: Jump ► Increased height **VS** traumas

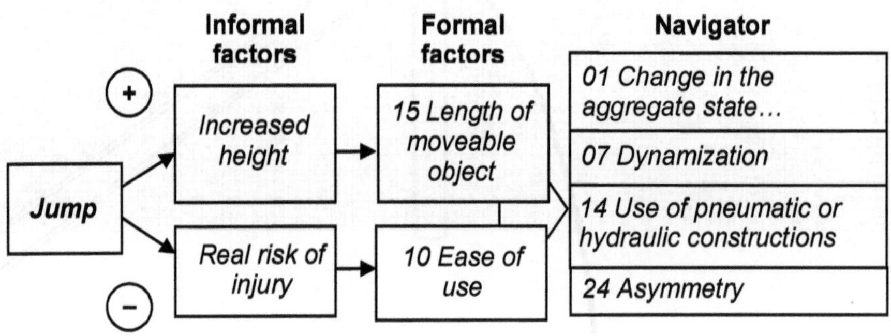

Radical Contradiction

Format of formula: Jump ► Increased height **VS** traumas

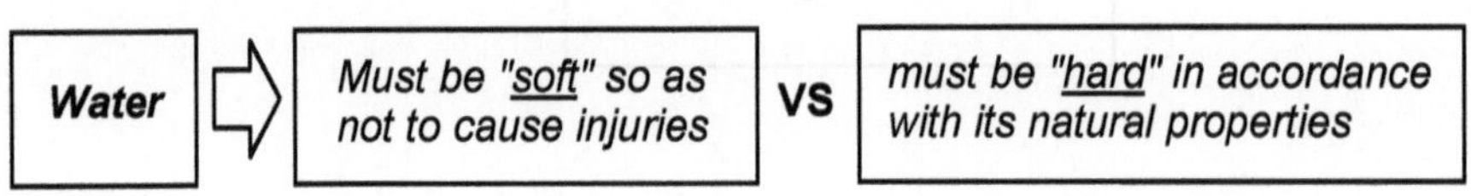

INVENTING

Key models: 01 Change in the aggregate state *– a) transitions into "pseudo-states"; b) changes in concentration;*

07 Dynamization *– c) make an object moveable that is otherwise fixed;*

14 Use of pneumatic or hydraulic constructions *– use gaseous or fluid parts instead of fixed parts in an object: inflatable…, air cushion…*

Key idea: *inject compressed air at the time of the jump, make an "air cushion" on the surface of the water.*

ZOOMING

Have the contradictions been removed? – Yes. – ~~No~~.

Super-effects: *safe training of children*

Negative effects: *increased complexity of construction*

http://www.myrthapoolsusa.com/eng/accessories-air-safety-cushion.htm

BRIEF DESCRIPTION

To make high-board diver training safe, a powerful jet of compressed air is injected into the water during the jump, creating an "air cushion" on the surface of the water.

Specialized navigators used: ***01 Change in the aggregate state****,* ***07 Dynamization,*** *and* ***14 Use of pneumatic or hydraulic constructions****.*

S24.2. START-pass through Radical Contradiction

TREND

The training of high-board divers is associated with high risk of injuries and occupational diseases.

The critical moment during such training is the moment when the diver enters the water. For example, a fall on the back from a height of 10 meters can lead not only to painful contusions and skin disruption but also to dangerous spinal column injuries.

How do we make training of high-board divers safer?

REDUCING

FIM: X-resource, together with available or modified resources, and without making the object more complex or introducing any negative properties guarantees attainment of the following **IFR**:

[*safe entry into water from any height*]

Radical Contradiction (two forms):

Textual:

Water has to be ►**soft** (to spare diver) **VS hard** (physical property)

Graphical:

Water	⇨	*Must be "soft" so as not to cause injuries*	**VS**	*Must be "hard" in accordance with its natural properties*

INVENTING

***Key specialized models from Afs-catalogue of four fundamental transformations: 01 Change in the aggregate state** – a) transitions into "pseudo-states"; b) changes in concentration;* ***14 Use of pneumatic or hydraulic constructions** – use gaseous or fluid parts instead of fixed parts in an object: inflatable…, air cushion…*

Key idea: *Inject compressed air at the time of the jump; make a* ***"preliminary installed air cushion"*** *on the surface of the water.*

ZOOMING

Have the contradictions been removed? – Yes. ~~No.~~

Super-effects: 1) safe training of children; 2) inviting the visitors for active recreation in diving pool by jumping from the big height!

Negative effects: increased complexity of construction.

http://www.natare.com/natare-pools/competition-and-training-pools

BRIEF DESCRIPTION

To make high-board diver training safe, a powerful jet of compressed air is injected into the water during the jump, thus creating an "air cushion" on the surface of the water.

Navigators used: ***01 Change in the aggregate state*** *and* ***14 Use of pneumatic or hydraulic constructions*** *to realize fundamental transformation in* ***structure*** *and* ***material/energy****.*

S25. Afs-CATALOGUE

Fundamental transformation	Relationship to As-navigators
Separation of conflicting properties in space One part of system space has property **A**, while another part of system space has property **not-A**.	**05 Separation:** Remove the disruptive part, emphasize the part needed. **10 Copy:** Use simplified and inexpensive copies. **19 Transition to another state:** Increase the freedom of an object, use construction in several layers, use lateral and other surfaces. **22 Spherical shape:** Transition to curved surfaces and shapes; use of wheels, balls, or springs. **24 Asymmetry:** Transition to asymmetrical shapes, increase asymmetry. **25 Use of flexible covers and thin layers:** Use flexible covers and thin layers instead of normal constructions. **34 Matryoshka:** Store an object inside another one in stages; place an object in the hollow space of another one.
Separation of conflicting properties in time: During one time interval, the system has property **A**; during another time interval, it has property **not-A.**	**02 Preliminary action:** Run the necessary effect partially or completely; arrange the objects so that they can go to work faster. **07 Dynamization:** Make an object (or its parts) moveable; optimize the characteristics of the process (of an object) in every stage. **08 Periodic action:** a) Transition from a continuous functioning to a periodic one (impulse); b) change the periods if the functioning already runs that way; c) use the breaks between impulses for other functions. **18b Mediator:** Connect an object with another (easily removable) one for a specific time. **28 Previously installed cushion:** Consider possible disruptions in advance. **33 Quick jump:** Accelerate a process strongly so that damages do not appear. **35b Unite:** Temporally connect similar or neighboring operations with each other. **39 Preliminary counteraction:** An opposite effect must be completed in advance to run the primary effect. **40 Uninterrupted useful function:** Eliminate idle time and interruptions so that all parts of an object function at full capacity.
Separation of conflicting properties in structure: some elements of the system have property	**03 Segmentation:** Disassemble an object into its parts; increase the degree of "disassembly." **04c Replacement of mechanical matter**: Replaceme static fields with dynamic ones, from temporally fixed to flexible fields, from unstructured fields to fields with a specific structure. **11 Inverse action:** Take the opposite action from the one that is apparent given by the conditions at hand.

A, while other elements or the system as a whole have (has) property **not-A**.	**12 Local property:** Transition from a similar to a different structure so that every part can complete its function under the best condition. **15 Disposal and regeneration of parts:** A used-up part can be disposed of or regenerated during its function. **18 Mediator:** a) use another object to transfer or transmit an action; b) temporarily connect an object with another (easily separable) object. **32a Counterweight**: Compensate for the weight of an object with its connection to another object with lifting power. **35a Unite:** Connect similar objects or objects for neighboring operations with each other. **36 Feedback**: a) create a retroactive influence; b) change a retro-active influence that already exists.
Separation of conflicting properties in material/energy: for one purpose, the material has property **A**; for another purpose, it has property **not-A.**	**01 Change of the aggregate state of an object:** Change the concentration or consistency, full use of properties like the elasticity of materials, etc. **14 Use of pneumatic or hydraulic constructions:** Use gaseous or fluid parts instead of fixed parts in an object: parts that can be blown up or filled with hydraulic fluid, air-cushions, hydrostatic or hydro-reactive parts. **17 Use of materials that consist of several components:** Transition from similar materials to those consisting of several components. **15 Discard and renewal of parts**: a) Parts that have fulfilled their task and are no longer part of an object should be disposed of (e.g., dissolved, evaporated, etc.); b) used parts of an object should be immediately replaced during work. **16 Partial or excess effect:** When it is difficult to achieve the desired effect completely, we should try to achieve a bit less or a bit more. **23 Use of inert media:** Replace a medium with something inert. Let processes run in a vacuum. **26 Use of phase transitions:** Full use of phenomena that occur during phase transitions: changes in volume, in the radiation or absorption of heat. **27 Use of heat expansion:** full use of the heat expansion of materials, the use of materials with different heat expansion. **29b Self-servicing: Make** use of the waste of material and energy. **30 Use of strong oxidants:** Replace air with oxygen, influence air with ion beams, use of ozone. **31 Use of porous materials:** Shape an object in a porous way, fill porous parts with some kind of material. **38 Homogeneity:** Manufacture objects that influence each other from one and the same material.

S26. Af-CATALOGUE FOR FOUR FUNDAMENTAL MODELS

1. Separation in space	One part of system space has property **A**, while another part of system space has property **not-A**.

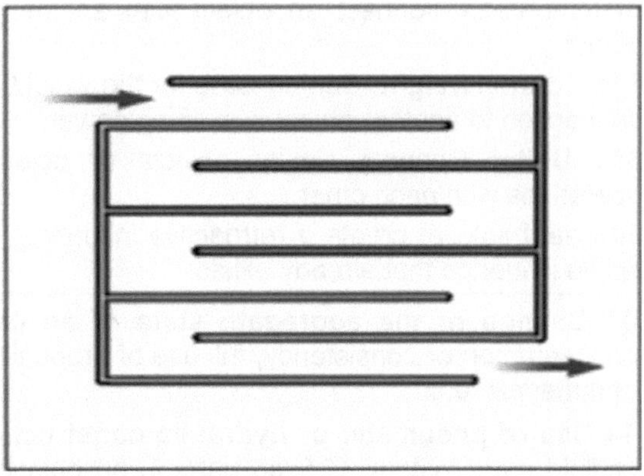

(**QUEUE ORDERING**) To physically eliminate the possibility of a chaotic crowd emerging in front of a building (museum, theater, exhibition hall), the area is reorganized by erecting dividing constructions that form a narrow corridor, which people can negotiate only by walking one after another (**transformation in space**).

2. Separation in time	During one time interval, the system has property **A**; during another time interval, it has property **not-A**.

(**SAME-LEVEL ROAD CROSSING**) To physically eliminate the possibility of vehicles colliding as they traverse traffic routes at the same level, conflicting traffic flows are divided in time by using special signal systems (such as traffic lights or traffic wardens) that regulate the order of movement (**transformation in time**).

3. Separation in structure	Some elements of the system have property **A**, while other elements or the system as a whole have (has) property **not-A**.

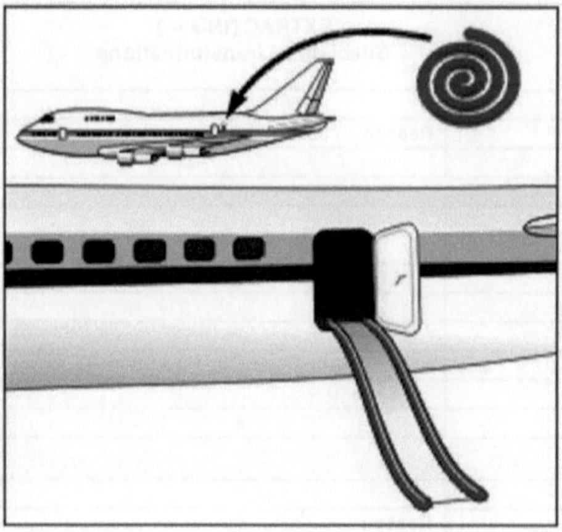

(**EMERGENCY ESCAPE CHUTE**) To create an emergency escape chute that is durable and "inflexible" when in use and rolled-up and "flexible" when not in use, it is made in the form of several joined elastic pipes that are inflated by compressed air (**transformation in structure**).

4. Separation in material	For one purpose, the material has property **A**; for another purpose, it has property **not-A**.

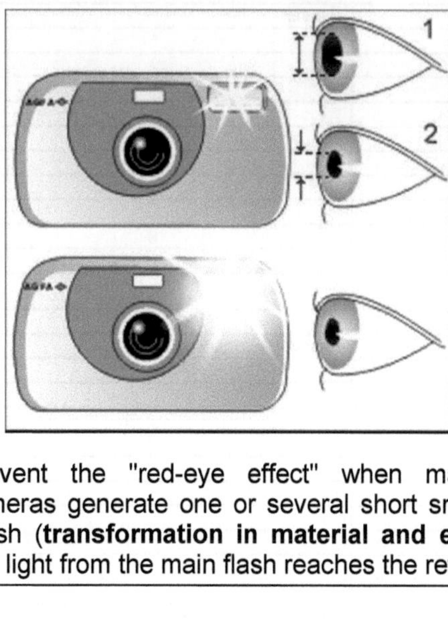

(**PRE-FLASH**) To prevent the "red-eye effect" when making flash-assisted photographs, many cameras generate one or several short smaller flashes before setting off the main flash (**transformation in material and energy**); as a result, pupils contract, and less light from the main flash reaches the retina.

S27. Simple form for Extracting-1

Attention: All forms could be reproduced from this and the following pages or downloaded from the academy's Web site

EXTRACTING – 1
Specialized transformations

Artifact 1:	
Navigator	**Reason**

Artifact 2:	
Navigator	**Reason**

Artifact 3:	
Navigator	**Reason**

Artifact 4:	
Navigator	**Reason**

S28. Examples at form of Extracting-1

Artifact 1: The Scrunchie "Operation" (the fastening of a hair tail with an elastic hair tie – "scrunchie")	
Navigator	**Reason**
02 Preliminary action	**Key idea:** to get an elastic hair tie – "scrunchie" – on the hand
18 Mediator	hand is a mediator to place scrunchie temporaraly
34 Matryoshka (nested doll)	a scrunchie isput on the hand, and is inside of scrunchie

Artifact 2: The Corner of a Personal Planner's Page (line perforating for accurately tearing off the corner of page)	
Navigator	**Reason**
02 Preliminary action	**Key idea:** preliminary perforating the right line at the corner of page
03 Segmentation	devide the page into two parts, one for writing and one to be teared off
05 Separation	separate (assignment, definition) small corner to tear off
12 Local property	allocation a right line of small perforating holes
16 Partial or excess effect	perforating holes to prepare easy tearing off

Artifact 3: Inflatable Crowd (to fill large spaces with people, movie-makers have invented use of inflatable dummies)	
Navigator	**Reason**
03 Segmentation	devide a crowd into two parts with real people and dummies
10 Copying	dummies copy and substitute the people
13 Inexpensive short-life object…	use of dummies is cheaper
14 Use of pneumatic… constructions	**Key idea:** to use inflatable dummies
15 Discard and renewal of parts	easy replace of any kind of dummy
20 Universality	dummy can reproduce many functions and postures
32 Counter-weight	inflatable dummy is light to transportation and installing

Artifact 4: Ice for a Drink (use hermetically encapsulated ice to cool drinks)	
Navigator	**Reason**
05 Separation	**Key idea:** to separate an ice from drink liquid
08 Periodic action	ice could be periodicalle melted and frozen
10 Copying	to protect ice with casing (as cooling the bottle in cold water = prototype)
11 Inverse action	opposed to prototype to use ice in bottle (or in some capsule) for water cooling
18 Mediator	**Key idea:** to transmit cooling through the wall of capsule
25 Use of flexible covers…	to use flexible (as well as inflexible in practice!) casing to incapsulate ice
34 Matryoshka (nested doll)	**Key idea:** to put ice inside capsule

S29. "START-form" for reinventing

NUMBER ______ **TITLE** ____

TREND

REDUCING

FIM: X-resource, without producing the inadmissible negative effects, provides together with other existing resources obtaining **IFR** [____].

Standard Contradiction

Informal factors **Formal factors** **Navigators from A-matrix / Extracting**

+

−

(Extra) Navigators from Extracting

Radical Contradiction

must be **VS**

INVENTING

ZOOMING

Are contradictions removed? – Yes. – No.

Super-effects: ______________________________

Negative effects: ______________________________

BRIEF DESCRIPTION

Picture

Trainee ______________________________________ Date ________________

S30. Example "Ice for a Drink" at "START-form"

TREND

To keep the drink (juice, cocktail, etc.) cool for a longer period of time, you can add pieces of ice. However, with the ice melting the taste of the drink is changing, as the relative content of water in the dwindling amount of the drink is increasing. How do you prepare an iced drink so that its taste remains unchanged in the course of consumption?

REDUCING

FIM: X-resource, without producing the inadmissible negative effects, provides together with other existing resources obtaining **IFR**:

[*taste of the drink, which has been cooled (with ice), must remain unchanged*].

Standard Contradiction

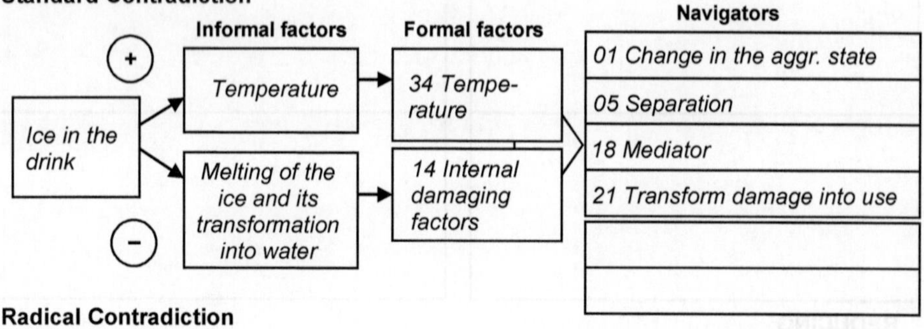

Radical Contradiction

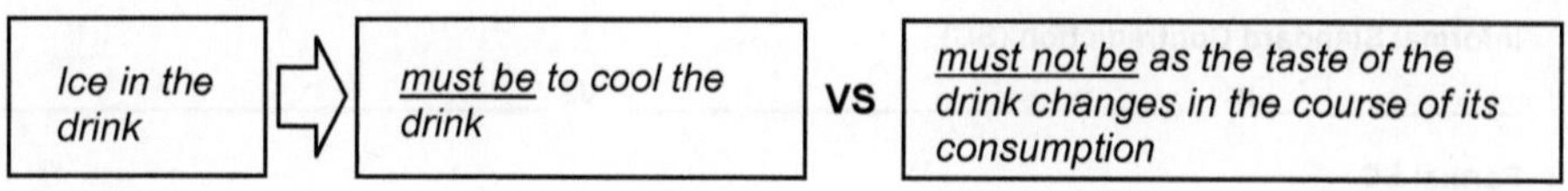

INVENTING

Key model: 18 Mediator – *a) Use another object to transfer or transmit an action.*

Key idea: *encapsulate the ice – the drink becomes cooler, and no water gets into the drink.*

Additional solution option based on the Radical Contradiction – on the (material) segmentation navigator and specialized navigator 38 Homogeneity: interacting objects should be made from the same materials – ice figurines should be made from the same drink (which is possible not for all drinks).

ZOOMING

Have the contradictions been removed? – Yes. – ~~No~~.

Super-effects: *it becomes possible to decorate the drink with nicely shaped figurines.*

Negative effects: -

BRIEF DESCRIPTION

Use hermetically encapsulated ice to cool drinks according to navigator 18 Mediator. Ice figurines can be made from the same drink in line with model 38 Homogeneity.

S31. Brief Junior-form for Extracting and Reinventing

ARTIFACT

WAS: Prototype-artifact	IS: Result-artifact
illustration	illustration
WAS-description	BECAME-description

REDUCING

Informal Standard Contradiction (SC)

________ ▶ ________________ VS ________________

Formal SC

________ ▶ ________________ VS ________________ = ________

Radical Contradiction (RC)

________ ▶ ________________ VS ________________

INVENTING

LC	Navigator	Function

Transformations description

Student ________________ Date ________

S32. Example "Fischer' Dowel" on brief Junior-form

ARTIFACT | **ARTUR AND KLAUS FISCHER DOWELS**

WAS: Prototype-artifact	**IS: Result-artifact**
	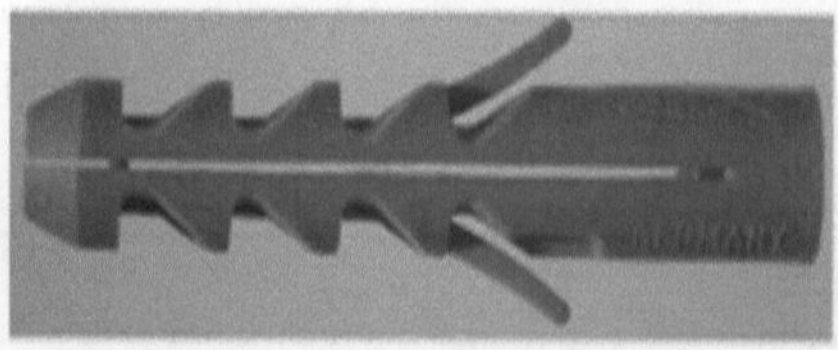
In the past, if we wanted to drive a screw into a brick or concrete wall, we made small, wooden wall plug (screw anchor, dowel) and hammered it into pre-drilled hole. Unfortunately, sometimes the plugs crumbled and fell out of the wall. What can we do?	The plastic dowel invented in 1958 by Professor Artur Fischer, founder of the Fischer Group, has become a truly global product. Mass production of this product was organized under the guidance of Professor Klaus Fischer (A. Fischer's son), also a well-known and rather prolific inventor. The dowel made of elastic nylon – then a very new material.

REDUCING

Informal Standard Contradiction (SC)

Wall plug ► increasing the depth or diameter of the hole **VS** decreasing in reliability

Formal SC

Wall plug ► **16** Length of the fixed object **VS 04** Reliability **= 04, 07, 14**

Radical Contradiction (RC)

Wall plug ► "hard" (to tightly fill the hole) **VS** "soft" (to let the screw in)

INVENTING

LC	Navigator	Function
++	01 Change in the aggr. state of an object.	a) ... use of elastic properties of solid objects: ***the body of the wall plug is made of elastic plastic***
++	07 Dynamization	a) characteristics of an object or an environment are changed to optimize every work procedure – ***mobile wall plug's body!***
+	Fundamental change in Structure	Additionally to change in material: ***slots, crests, wings, elastic parts – to assure initial movement and relable fixing the screw***

Dominant models (marked with two pluses) used to resolve the SC: **07 Dynamization** and **04 Replacement of mechanical matter** (according to A-matrix), and **01 Change in the aggregate state of an object** (from Extracting). Invention of the plastic dowel totally agrees with these navigators: the body of the dowel is made of elastic plastic (model 01) so that it expands, and its diameter increases, as the screw goes through it (models 04 and 07). **Fundamental models: Spatial** – the parts are mobile and immobile, while the system as a whole is elastic; **Structural** – cuts, crests, wings; **Material** – the new plastic material (nylon) which is elastic, resilient and tensile.

Student	M.O.	Date	03.10.2015

S33. "Junior-form" for reinventing (two pages)

ARTIFACT

Result-artifact description in comparing with prototype

illustrations: artifact-prototype –
artifact-result –

EXTRACTING:

illustration

illustration

WAS: artifact-prototype

Problem → Trend → Reduction → Inventing → Zooming → **Idea**

IS: artifact-result

Extracting of transformations

LC	Navigator	Function

Extracting of contradictions

Contradiction type	Description
Standard	
Radical	

Student

Date

REINVENTING:

TREND

REDUCTION

Ideal Final Result:

STANDARD CONTRADICTION (SC) – Extracting-2S

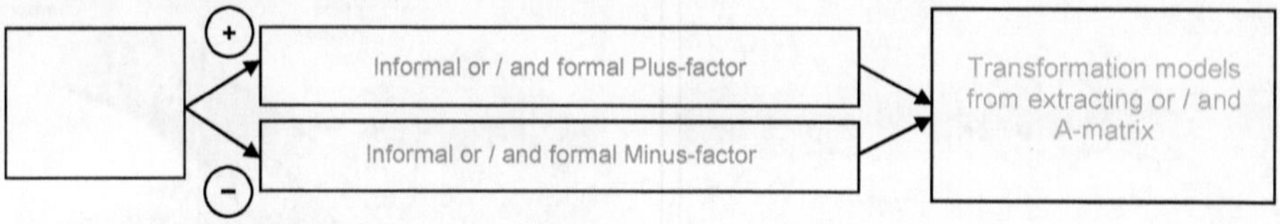

RADICAL CONTRADICTION (RC) – Extracting-2R

VS

INVENTING

ZOOMING

Are the contradictions eliminated? – Yes (or NO?)

Super-effects:

Negative effects:

Development trends:

Change of surrounding systems:

Advanced application:

Student

Date

S34. Example "Leonardo da Vinci's Bridge" on a "Junior-form"

ARTIFACT Leonardo da Vinci's Bridge

Description Leonardo da Vinci invented a rotating bridge for military use. Own troops could use the bridge joining the banks of a river to safely cross it. When necessary, the bridge could be "retracted" (turned away) to prevent enemy troops from quickly crossing over. Besides, the bridge could be used to cross rivers in peacetime, and enabled the passage of river-going vessels.

The bridge was rotated by cables around an axial pylon. It was also supported by cables and could use boats or empty barrels as additional supports.

Illustrations: prototype artifact – Russian floating bridge (http://les.novosibdom.ru/node/468); resultant artifact – *Leonardo da Vinci Machines* (www.labirint.ru/books/121061).

EXTRACTING

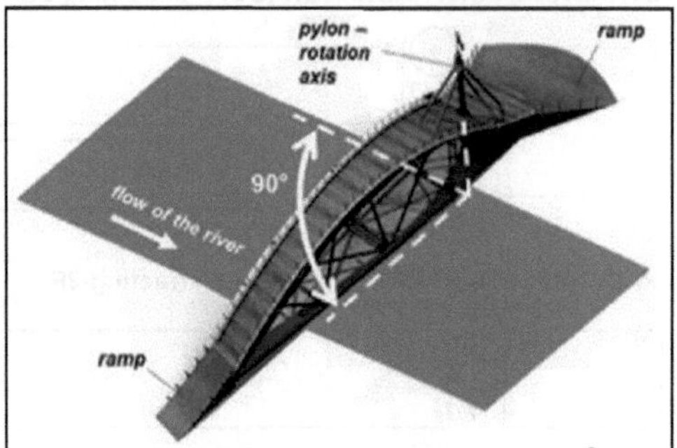

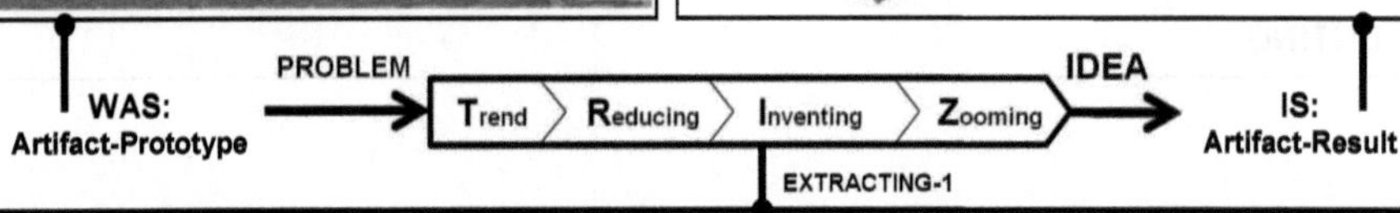

LC	No.	Navigator	Function
++	07	Dynamization	c) make an object moveable that is otherwise fixed
	12	Local property	c) every object should exist under conditions that correspond best to its functions
++	19	Transition into another dimension	a) an object is shaped so that it can move or is placed not only in a linear fashion, but also in two dimensions, meaning on a surface
	32	Counter-weight	a) compensate for the weight of an object with its connection to another object with lifting power

REINVENTING

TREND Known bridges (floating and prefabricated bridges) cannot be used in situations where it may be necessary to promptly provide a crossing and then quickly "retract" the bridge to allow the passage of vessels. Construction of a draw bridge across a relatively wide river presents numerous technical challenges. In such cases the bridge is usually built with multiple spans, whereupon each span can be turned into a small draw bridge. However, such bridges are stationary, and cannot be easily dismantled.

A different type of bridge is needed where a crossing has to be made at an arbitrary spot, provided, of course, that the width of the river is not too large, say, under 100 meters. What can you suggest?

REDUCING **Macro-FIM:** The X-resource, without making the system excessively complex and without causing inadmissible negative effects, ensures, together with other available resources, the obtaining of the

IFR: [a crossing that can be quickly provided and quickly removed]

STANDARD CONTRADICTION (SC) – Extracting-2S

Bridge

(+) *A crossing that can be quickly provided and quickly removed*

(–) *Immobility, complexity, large weight of the structure*

From Extracting-1

07 Dynamization
12 Local property
19 Transition into another dimension
32 Counter-weight

RADICAL CONTRADICTION (RC) – Extracting-2R

Bridge ⇒ *must be to provide crossing* **&** *must not be to allow passage to awaiting vessels*

INVENTING

To attain Objective 1 (acceleration of provision and removal of the crossing), the bridge is made moveable (capable of horizontal rotation) – dominant models *07 Dynamization* and *19 Transition into Another Dimension.*

To attain Objective 2 (realization of two bridge positions – across the river and alongside the bank), the bridge is rotated by pull cables around an axial pylon – model *12 Local Property.*

To attain Objective 3 (bridge weight balancing), a counterweight is fixed under a moveable ramp in the shorter section – model 32 *Counter-Weight.*

Fundamental models: *Temporal* – the bridge performs its two main functions during different time intervals; *Structural* – certain elements are introduced to enable the bridge to rotate.

Scientific Effects: Application: rotary pulleys (for example, by analogy with windmills); cable-pulling hoists; balance scales; internal enforcement ribs (beams, girders) to make the structure sturdier.

ZOOMING

Contradictions Removed: Yes.

Super-Effects: Can be used for military purposes (deny crossing to enemy troops).

Negative Effects: Relative complexity of construction.

Development Trend: Automation of the crossing construction/removal operation.

Changes in Ambient Systems: Bridge abutments and ramps must be aligned with the banks.

Beauty of Solution: 100 — Highest grade. Substantiation:
1) such structure was not known before, there are no direct counterparts; 2) functionality is changed very fast.

S35. As-catalogue with examples of extracting and reinventing

01. Change in the aggregate state of an object	(a) This includes transitions into "pseudo-states" ("pseudo-liquid") and into transitional states such as the use of the elastic properties of solid objects as well as simple transitions such as from a solid to a liquid state; (b) changes in concentration or consistency, in the degree of flexibility, in temperature, etc.

EXAMPLE. Water from the ground.

Short description

The German firm ZELTEC has developed a transparent cone to extract clean water from the ground **(navigator 01).** The cone is placed on the ground, and solar rays heat the surface. Water in the ground evaporates, precipitates on the walls of the cone, and flows into a collector placed around the cone at its bottom.

EXTRACTING

R	#	Navigator	Reasoning
++	*01*	***Change in the aggregate state of an object***	*Dominating navigator*
+	09	Change in color	Transparent cone
+	12	Local property	The water condensing on the walls is gathered in a special collector
+	19	Transition into another dimension	Use of the inner side of the cone
	29	Self-servicing	The water gathers in the collector by itself

REINVENTING

Trend

Collecting sufficient water is a serious problem in hot regions. A certain amount of water is present in the ground even in desert sand. How can we access this water at minimal expense?

Reducing

Maxi-FIM: the OZ itself achieves the result

[*water appears by itself from the ground or the air*].

Micro-FIM:

An X-resource in the shape of *particles of material or energy* is present in the OZ and ensures the achievement of the **Maxi-FIM.**

Model of the SC:

Effect, State, Object						
Collecting water from the ground and/or air						
(+)-Factor			**(-)-Factor**			**Cluster**
Use in different regions	02	Universality	Minimization of the use of energy	37	Energy use	**01,** 08, 11, 14

Inventing

Idea: use ***navigator 01*** to force solar energy to evaporate water from the ground. To achieve this result, a transparent cone has been developed that can be transported, unrolled, and deployed at the necessary site. A collector is installed on the inner wall of the cone. Evaporated water that condenses on the inner side of the cone flows down into the collector.

Zooming

The SC is solved.

The OZ itself uses solar energy to automatically extract water from the ground and/or air.

02. Preliminary action	a) Previous necessary (partial or complete) change of an object; b) prepare objects in advance so that they can be put to work from the best position and are available without loss of time.

EXAMPLE. Net to collect wool.

Short description

Nets are stretched around sheep in advance to collect wool that falls from the animals upon application of a special chemical preparation **(navigator 02).**

EXTRACTING

R	#	Navigator	Reasoning
+	***02***	***Preliminary action***	*Dominating navigator*
+	05	Separation	The falling wool is collected in the net
	10	Copying	Copy of a mesh shopping bag
	29	Self-servicing	The wool gathers in the net by itself
+	34	Matryoshka (nested doll)	The sheep is "inserted" into the net

REINVENTING

Trend

In an attempt to reduce the work load associated with shearing sheep, Australian scientists have developed a protein, which, when applied to sheep wool, creates attenuated sections at the base of mature hairs, thus causing them to fall off at the same time. Then the wool continues to grow until the protein is applied again. However, when the wool falls onto the ground, it loses quality and is difficult to collect. How can we collect "self-shearing" wool from thousands of sheep?

Reducing

OT: from the moment the "shearing substance" is applied until the wool falls off the sheep.

Macro-FIM: The X-resource—without causing unacceptable negative effects—ensures, together with the other available resources, attainment of the following result:

[*collecting all wool from the body of the animal*].

Model of the SC

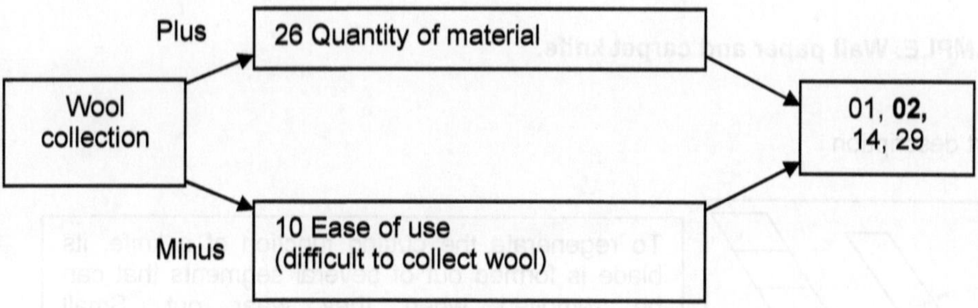

The model is focused on eliminating the contradiction between the increased "convenience" of the new shearing method and the low productivity of colleting the "sheared" wool.

Inventing

Idea according to ***navigator 02*:** Stretch a net around the sheep in advance to collect the wool. After the net is removed and emptied, it can be used again.

Zooming

The SC is solved completely!

Additional system effect 1: Stress for the animals when shearing with mechanical or electrical shears is reduced greatly.

Additional system effect 2: Difficult work when shearing sheep is replaced by a simpler and more productive navigator—automated application of the "shearing substance" and automated stretching and removal of the nets.

03. Segmentation	a) Disassemble an object into individual parts; b) make it possible to disassemble an object; c) raise the degree of disassembly (reduction into parts) of an object.

EXAMPLE. Wall paper and carpet knife.

Short description

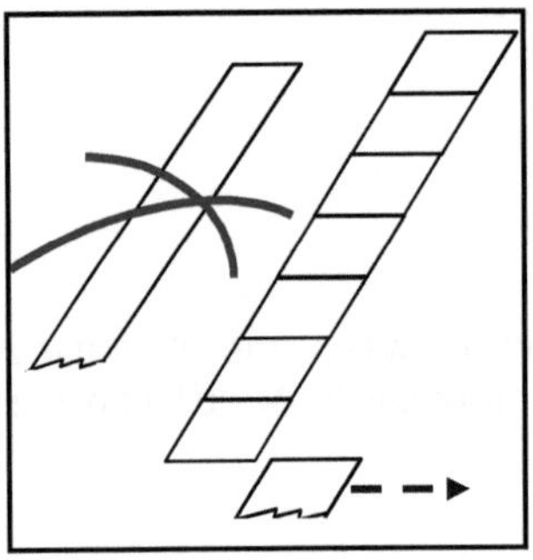

To regenerate the cutting function of a knife, its blade is formed out of several segments that can be removed when they wear out. Small indentations are cut into the blade for this purpose **(navigator 03).**

EXTRACTING

R	#	Navigator	Reasoning
+	02	Preliminary action	The indentations are made in advance.
++	***03***	***Segmentation***	*Dominating navigator*
	15	Discard and renewal of parts	The worn blade is thrown away, and the cutting edge is renewed.

REINVENTING

Trend

Previously the entire knife was tossed out after its blade got blunt. Under normal household conditions, it was very difficult to regenerate a sharp cutting edge. How can we increase the service life of a knife?

Reducing

OR: the cutting edge of a knife.

Macro-FIM: The X-resource—without causing unacceptable negative effects—ensures, together with the other available resources, attainment of the following result: [*regenerating the cutting edge of a knife*].

Model of the SC:

Effect, State, Object						
Retention of a knife's sharp cutting edge						
(+)-Factor			**(-)-Factor**			**Cluster**
This means the reconstruction of its form	21	Form	Knife becomes blunt	13	External damaging factors	01, **03,** 05, 21

Inventing

In accordance with ***navigator 03***, the knife is separated into several parts; each part can be separated from the remaining sections after it is used. A sharp edge forms at the line of separation that is part of the next section of the knife. This section can also be broken off from the remaining parts after it is worn out.

Zooming

The SC is solved!

04. Replacement of mechanical matter	a) Replace mechanical schemes with optical, acoustic, or olfactory schemes; b) use of electrical, magnetic, or electromagnetic fields for the interaction of objects; c) replacement of static fields with dynamic ones, from temporally fixed to flexible fields, from unstructured fields to fields with a specific structure; d) use of fields in connection with ferric-magnetic particles.

EXAMPLE. "Submarine" for blood vessels.

Short description

The German firm Microtech has made a micro-miniature "submarine" for the inspection and treatment of blood vessels. The external rotating magnetic field actuates the "propeller" **(navigator 04).**

EXTRACTING

R	#	Navigator	Reasoning
+	***04***	***Replacement of mechanical matter***	*Dominating navigator*
+	05	Separation	The catheter itself is inside the blood vessel, but it is controlled from the outside.
+	10	Copying	Copy of a submarine
	12	Local property	The catheter delivers the medicine precisely to the targeted spot.
+	19	Transition into another dimension	Possibility of movement along a complex trajectory

REINVENTING

Trend

Inspection and treatment of blood vessels with the aid of a catheter is a painful and traumatic procedure because it is difficult to push the catheter through the vessel. Is it possible to automate this operation?

Reducing

OZ and the operating spatial resource: internal space of the blood vessel. The problem is that the catheter gets snagged on the walls of the vessel with two negative consequences: (1) it gets stuck; and (2) it can injure the vessel. Usually the head of catheter is the part used to perform the surgery or deliver the medicine (operating unit).

Macro-FIM: The X-resource—without causing unacceptable negative effects—ensures, together with the other available resources, attainment of the following result:

[*light and precise advance of the operating unit into the zone of surgery (treatment) or inspection*].

Standard contradiction:

Effect, State, Object						
Delivery of the tool or medicine into the blood vessel						
(+)-Factor			**(+)-Factor**			**Cluster**
-	10	Ease of use	Injury	27	Loss of material	**04,** 05, 09, 18

Inventing

Idea according to ***navigator 04***: Instead of the catheter advanced by "rough" intrusion, it is proposed to use a self-moving operating unit. For this purpose the operating unit is made in the form of a "submarine." This is a capsule with the "propeller" set in action by an external magnetic field.

Navigator 05 is also present: The operating unit is isolated from the catheter and introduced into the operating space!

Zooming

Standard contradiction is removed.

The advancement of the "operating unit" has become less traumatic.

Very strong system super-effect: The operating unit has become universal because it is possible to ensure the arrangement on it of different tools.

05. Separation	Separate the "incompatible part" ("incompatible property") from the object or– turned completely around– include the only really necessary part (necessary property) into the object.

EXAMPLE. Motorized wheelchair.

Short description

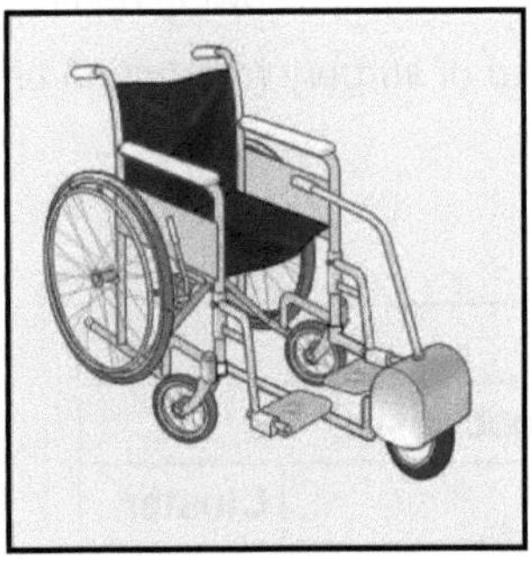

A device is integrated into a wheelchair that is equipped with an electrical motor, an accumulator, and a guidance system to ensure the motorized forward movement of the chair (**navigator 05**).

EXTRACTING

R	#	Navigator	Reasoning
+	02	Preliminary action	The "tractor" is prepared in advance.
+	***05***	***Separation***	*Dominating navigator*
+	10	Copying	Copy of a "tractor" with "trailer"
	12	Local property	The large motorized wheelchair is divided into parts.
	15	Discard and renewal of parts	The "tractor" can be detached and attached.

REINVENTING

Trend

Chairs for the independent forward movement of disabled persons require greater energy when traveling over relatively large distances, e.g., when someone wants to drive through a park. The installation of motors in wheelchairs makes the system much more complicated and increases both the weight (thus making the chair less maneuverable indoors) and the costs.

Reducing

OT: time spent on excursions outdoors (in parks and woods).

Macro-FIM: The X-resource—without causing unacceptable negative effects—ensures, together with the other available resources, attainment of the following result: [*motorized movement of wheelchairs outdoors*].

Model of the SC:

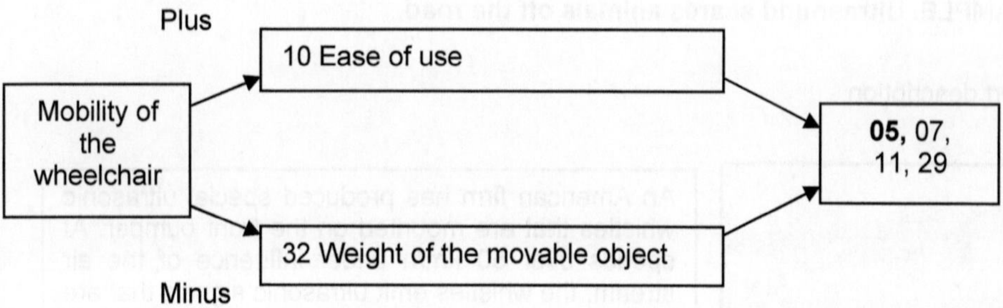

Inventing

Idea according to ***navigator 05***: A unit is mounted on the wheelchair just for excursions that consists of an electrical motor with an accumulator and a guidance system for its forward movement. In this way the necessary property of motorized forward movement exists only during excursions.

Zooming

The SC is solved!

Super-effect: The attached motor unit can also be used indoors.

06. Use of mechanical oscillations	a) Cause an object to vibrate; b) raise the frequency of the vibrations up to and including ultra-high frequencies if the object is already in motion; c) use of the resonating frequency, application of quartz vibrators; d) use of ultra-sound vibrations in connection with electromagnetic fields.

EXAMPLE. Ultrasound scares animals off the road.

Short description

An American firm has produced special ultrasonic whistles that are mounted on the front bumper. At speeds over 50 km/h under influence of the air stream, the whistles emit ultrasonic signals that are frightening for animals but inaudible for people **(navigator 06).** The higher the speed, the louder the signal.

EXTRACTING

R	#	Navigator	Reasoning
+	02	Preliminary action	The "whistles" are installed in advance.
+	05	Separation	The signal moves ahead of the vehicle.
+	***06***	***Use of mechanical oscillations***	*Dominating navigator*
	07	Dynamization	The higher the speed, the "louder" the signal.
+	10	Copying	A noise is used to scare off animals,

REINVENTING

Trend

Animals become accustomed to traffic along large highways and often appear on the road. They do not fear sound and light signals. Is it possible to prevent the appearance of animals on the road or remove them from the road as soon as possible when a vehicle approaches?

Reducing

OZ: space in front of the moving vehicle. The more rapidly the vehicle moves, the stronger the level of the "frightening" signal must be to be effective at larger distances. The signal must operate reliably and without human intervention, e.g., automatically.

Macro-FIM: The X-resource—without causing unacceptable negative effects—ensures, together with the other available resources, attainment of the following result:

[*scaring animals off the road without inflicting harm on people*].

Standard contradiction:

Effect, State, Object						
Scaring animals off the road						
(+)-Factor			**(+)-Factor**			**Cluster**
Speed of movement	22	Speed	Not automated	03	Level of automation	02, **06**

Inventing

It is known that animals react to the ultrasound. Consequently, it is possible to equip vehicles with an ultrasound generator. However, this must be done so as to ensure that the generator operates automatically and generates a stronger signal at higher speeds.

Idea according to ***navigator 06***: An American firm produced special ultrasonic whistles that are mounted on the front bumper Under the action of air flow at speeds >50 km/h, he whistles emit an ultrasonic signal that scares animals off the road but is inaudible for people. The higher the speed, the louder the signal. Of course this whistle works automatically.

Zooming

Standard contradiction is removed. Simple and effective solution is created.

07. Dynamization	a) The characteristics of an object or an environment are changed to optimize every work procedure; b) disassemble an object into parts that are moveable among each other; c) make an object moveable that is otherwise fixed.

EXAMPLE. Moveable keel of a yacht.

Short description

A new dynamic solution was found to more securely anchor yachts and improve their movement especially in shallow waters: The keel can be folded and turned using a hinge in the hull with flexible components **(navigator 07).**

EXTRACTING

R	#	Navigator	Reasoning
+	03	Segmentation	The keel and the hull are separated.
+	***07***	***Dynamization***	*Dominating navigator*
+	10	Copying	Copy of an oar
+	12	Local property	The keel is raised in shallow waters.
	14	Matryoshka (nested doll)	The keel is "inserted" into the hull.

REINVENTING

Trend

The "high," long, and thin keel of yacht becomes an obstacle when mooring the yacht in shallow waters. However, such keel, with its large area, ensures course stability when the yacht is navigating deep waters.

Reducing

OZ: keel of the yacht

OT: mooring.

Maxi-FIM: The operating zone by itself ensures attainment of the following result:

[*"large" and "absent" keel of yacht*].

Model of the SC:

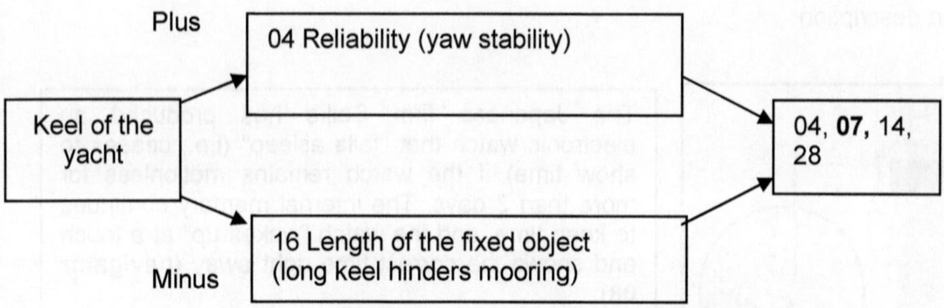

Inventing

Construction where the keel was drawn (***07 dynamization***) in through a central slot in the hull of the yacht, and construction with two moveable keels at the sides of the yacht already existed. An American ship builder discovered two new solutions: 1) a keel deflecting on a hinge joint inside the hull of the yacht; and 2) a double keel with separate hinged components.

Zooming

The SC is solved by means of a periodic change in the keel's state, i.e., lifted for mooring and dropped when sailing.

Both of these new solutions also provide a considerable systemic super-effect: There is no slot in the most important part of the hull, so stability of the yacht is not compromised.

08. Periodic action	a) Transition from a continuous functioning to a periodic one (impulse); b) change the periods if the functioning already runs that way; c) use the breaks between impulses for other functions.

EXAMPLE. Sleeping watch.

Short description

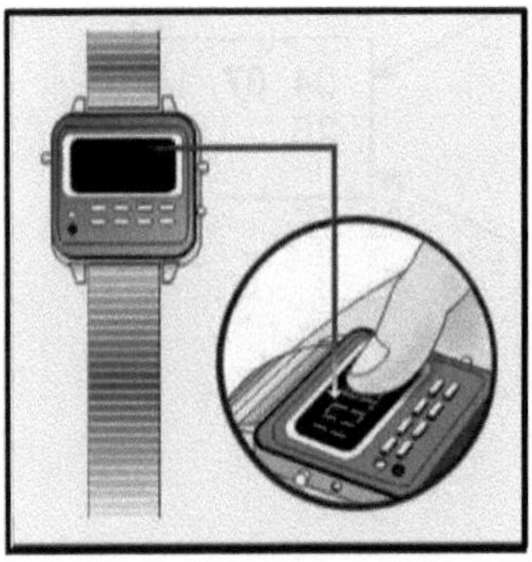

The Japanese firm Seiko has produced an electronic watch that "falls asleep" (i.e., ceases to show time) if the watch remains motionless for more than 3 days. The internal memory continues to keep time, and the watch "wakes up" at a touch and shows the correct time right away **(navigator 08).**

EXTRACTING

R	#	Navigator	Reasoning
	01	Change in the aggregate state of an object	The "sleeping" state is introduced.
+	***08***	***Periodic action***	*Dominating navigator*
+	10	Copying	Human sleep
+	16	Partial or excess effect	Removal of excessive action—time indication when the watch is not being used
	19	Self-servicing	The watch saves energy by itself.

REINVENTING

Trend

All movable electronic devices have a common problem: relatively short battery life. How can we prolong operation of such devices despite the imperfection of power elements, e.g., those used in watches?

Reducing

OT: the time interval when we read the time or, to put it simply, when we look at the watch to know the precise time.

Micro-FIM: The X-resource in the form of particles of substance or energy is located inside the operating zone and ensures, together with the other available resources, attainment of the following result:

[*power-saving battery operation*].

Standard contradiction:

Effect, State, Object						
Length of electronic wrist watch operation						
(+)-Factor			**(+)-Factor**			**Cluster**
Infrequent replacemen t of batteries and normal operation of the watch	10	Ease of use	–	39	Loss of energy	05, **08**, 11

Inventing

Idea according to ***navigator 08***: Because the operating time, i.e., the time when the watch is used for its designated purpose, is limited to that short interval of time when we look at the dial and read its data, the watch can "sleep" and not display any information at any other time. In fact, more energy is used to power the display than to power the time-keeping electronic circuit.

Navigator ***11 Inverse action*** is present as well: In normal watches (both mechanical and electronic), the time is displayed continuously, and much energy is expended. In contrast, the new watch does the opposite thing and operates periodically, i.e., only when used.

Zooming

Standard contradiction is removed.

Important systemic super-effect: Reduction of the quantity of replaceable batteries decreases environmental pollution.

09. Change in color	e) Change the color of an object or its environment; f) change the level of the transparency of an object or its environment; g) use color supplements to observe objects or processes that are difficult to see; h) add lighting if this kind of supplements is already in use.

EXAMPLE. Fog in an apartment or an automobile.

Short description

The French firm DAITEM proposed a new system to protect apartments and cars from thieves. The system uses a device that fills the apartment or car with dense thick milky-white smoke upon detecting unauthorized entry **(navigator 09).** The device is activated when the burglar alarm system is tripped off. The smoke does not damage the apartment or the car.

EXTRACTING

R	#	Navigator	Reasoning
+	05	Separation	Required property is introduced into the operating space at the required time.
++	*09*	***Change in color***	*Dominating navigator*
++	10	Copying	Smoke, fog
+	28	Previously installed cushion	Protective device installed in advance
	29	Self-servicing	The operating space protects itself.

REINVENTING

Trend

Almost all known burglar alarm systems have one common deficiency: Even after the alarm is tripped, the burglar still can steal things from the apartment or steal the car (if the alarm system does not block the engine). It is possible to reduce the damage inflicted by a burglar by somehow impeding his actions?

Reducing

Alarm is convenient in the operation, but the criminal can still steal something important in the short time after it goes off,.

OZ: protected space (space inside the apartment or car). The important parameter is the OT: Protective action must be quick to start, and then it must last until the arrival of the owner or the police.

Macro-FIM: The X-resource—without causing unacceptable negative effects—ensures, together with the other available resources, attainment of the following result:

[*rapid and sustained production of insurmountable obstacles impeding the criminal*].

Standard contradiction:

Effect, State, Object						
Difficulty of acting in closed space						
(+)-Factor			**(+)-Factor**			**Cluster**
–	10	Ease of use		27	Loss of material	04, 05, **09,** 18

Inventing

Idea according to ***navigator 09***: The French firm DAITEM proposed a new system to protect apartments and cars from thieves. The system uses a device that fills the apartment or car with dense thick milky-white smoke upon detecting unauthorized entry (navigator 09). The device is activated when the burglar alarm system is tripped. The smoke does not damage the apartment or the car.

Zooming

Standard contradiction is removed.

10. Copying	d) Use a simplified and inexpensive copy instead of an inaccessible, complicated, expensive, inappropriate, or fragile object; e) replace an object or a system of objects with optical copies; use here a change in measurement (blow-up or reduce the copy); f) if visible copies are used, they can be replaced with infra-red or ultra-violet copies.

EXAMPLE. Aquarium with a million fish.

Short description

A system has been developed to copy aquariums with many different species of fish. The system consists of a display, for example, a computer monitor installed behind an aquarium with thin glass walls that imitates the movement of air bubbles and the natural movements of plants. The fish and their entire world are projected with video recordings on the computer screen (**navigator 10**). The combined planes create an illusion of existence of the fish in a big aquarium.

EXTRACTING

R	#	Navigator	Reasoning
+	03	Segmentation	The prototype aquarium is divided into parts.
+	05	Separation	The fish are "separated" from the water and transferred "into the monitor."
++	*10*	*Copying*	*Dominating navigator*
+	12	Local property	Two different parts together imitate an aquarium.
	14	Use of pneumatic or hydraulic constructions	Air bubbles in the water

REINVENTING

Trend

How can we quickly change the contents of an aquarium to increase environmental diversity and boost the number of fish?

Reducing

OZ: volume of the aquarium.

We could suggest that a more complicated construction would increase the ease of changing the contents presented. This is how we can formulate the appropriate version of the problem: a reduction of the construction's complexity while retaining the ease of use of the aquarium.

Macro-FIM: The X-resource—without causing unacceptable negative effects—ensures, together with the other available resources, attainment of the following result:

[*convenient replacement of the contents of the aquarium*].

Model of the SC:

Effect, State, Object						
Aquarium						
(+)-Factor			**(+)-Factor**			**Cluster**
–	07	Complexity of the construction	–	10	Ease of use	**10,** 13, 18, 39

Inventing

Idea: A system was suggested that copies an aquarium with the species of fish desired. The system consists of a display, for example, a computer monitor behind a thin-walled aquarium with a small amount of water that imitates the natural movement of air bubbles and plants. The fish and their natural environment can be presented with a video recording (***navigator 10***). The combined levels create an illusion of existence of the fish in a large aquarium.

Zooming

The SC is solved.

Drawback: The aquarium can only be observed from one direction.

Extremely powerful systemic super-effect 1: Lack of true fish that would otherwise be confined against their will in the aquarium.

Powerful systemic super-effect 2: It would be simple to create aquariums of any size, for example, for hotel lobbies, kindergartens, and schools.

Powerful systemic super-effect 3: There is a possibility of demonstrating different inhabitants of water environments.

11. Inverse action	d) Instead of an action prescribed by the conditions of an assignment, complete a reverse action (heat an object instead of cooling it); e) make a moveable part of an object or the environment fixed or a fixed part moveable; f) turn an object upside-down or around.

EXAMPLE. Keyboard with recessed keys.

Short description

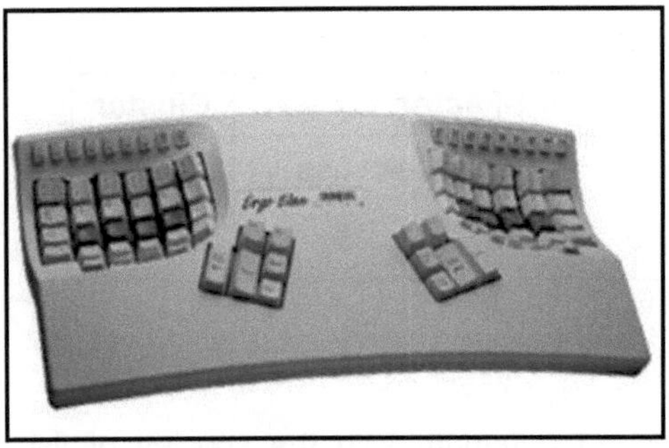

The American firm Kinesis has developed a keyboard with left- and right-hand keys sunk into two separated recesses **(navigator 11).**

EXTRACTING

R	#	Navigator	Reasoning
++	03	Segmentation	The keyboard is divided into functional areas
++	***11***	***Inverse action***	*Dominating navigator*
+	12	Local property	Each key is located in the most convenient position\, and is recessed
	34	Matryoshka (nested doll)	The keys are "inserted" into the keyboard
+	35	Unite	Key groups are joined to be used by the left and right hands

REINVENTING

Trend

A normal keyboard has an even surface on which the index finger and the little finger have to move the farthest distance to reach their keys. Can we improve this design so that each finger functions under the same conditions?

Reducing

OZ: set of keys on the keyboard.

Main problem – Different distances must be traveled by different fingers to reach their assigned keys on an ordinary keyboard.

Macro-FIM: The X-resource—without causing unacceptable negative effects—ensures, together with the other available resources, attainment of the following result:

[*optimal distance to be traveled by each finger to reach "its" keys*].

Model of the SC:

Effect, State, Object						
Keys on the keyboard						
(+)-Factor			**(+)-Factor**			**Cluster**
–	10	Ease of use	–	15	Length of the moveable object	03, **11**, 19, 37

Inventing

According to ***navigator 11***, the alphanumeric keys are placed into two recessed spaces: one for the left hand and one for the right hand. This shortens the distance traveled by each finger. The functional keys are operated by the thumbs. The entire hand rests on a pad constructed diagonally upward so that the burden on each wrist is substantially reduced.

The solution features an efficient interpretation of ***navigators 03*** and ***19***.

Zooming

The SC is solved.

Drawback: It takes a while to adapt from a standard keyboard to the new configuration.

Systemic super-effect 1: The width of the keyboard is reduced. This makes it possible to move the mouse pad closer to the hand and increase available space on the desk.

12. Local property	d) Change the structure of the object (the external environment, external influences) from the same to a different one; e) different parts of an object have different functions; f) every object should exist under conditions that correspond best to its functions.

EXAMPLE. Extinguishing electrical fires.

Short description

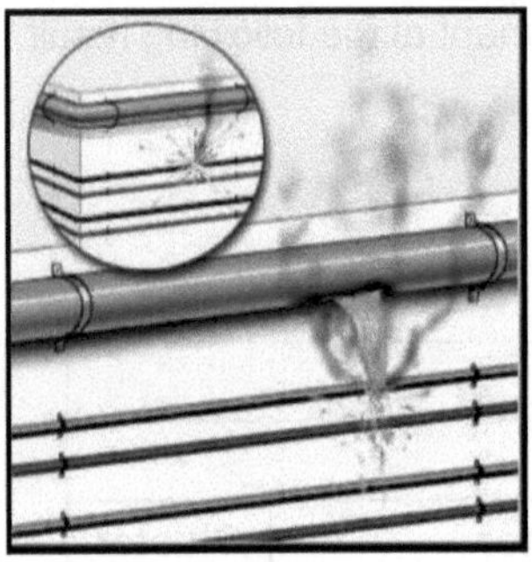

It has been suggested that pipes made of readily fusible material filled with fire extinguishing liquid be installed above electrical cables to automatically "locate" and suppress electrical fires **(navigator 12).**

EXTRACTING

R	#	Navigator	Reasoning
++	01	Change in the aggregate state of an object	The pipe filled with the "extinguisher" melts
+	18	Mediator	The pipe delivers the "extinguisher" to the place where it is needed most.
	19	Transition into another dimension	The pipe is located above the cable.
++	***12***	***Local property***	*Dominating navigator*
++	28	Previously installed cushion	The "extinguisher" is placed above threatened cables in advance.

REINVENTING

Trend

A difficult problem with industrial energy systems is locating and quickly extinguishing fires in high-voltage cables. These cables can be several kilometers long.

Reducing

OZ: the entire length of an electrical cable.

Major problem - localization of fire and delivery to the point or burning or overheating of a large quantity of cooling and extinguishing liquid.

A variation of an ideal construction for an extinguishing system can be represented with several FIMs.

Macro-FIM: The X-resource—without causing unacceptable negative effects—ensures, together with the other available resources, attainment of the following result: [*exact localization of the hot spot and suppression of the fire*].

Maxi-FIM: The operating zone by itself ensures attainment of the following result: [*the extinguishing material gets to the burning zone itself*].

Micro-FIM: The X-resource in the form of particles of substance or energy is located inside the operating zone and ensures the attainment of the **Maxi-FIM.**

Model of the SC:

Effect, State, Object						
Localization and suppression of electrical fires						
(+)-Factor			**(+)-Factor**			**Cluster**
–	06	Precision of measurement	–	16	Length of the fixed object	04, 09, **12,** 16

Inventing

Navigator 12, the key navigator, dictates the need to precisely apply the extinguishing liquid (special liquid solution) in the conflagration zone. A method was developed that places a tube (pipe), which contains the extinguishing fluid and is made of a readily fusible material, above the cable along its entire length. Increasing temperature causes the tube to automatically melt directly over the overheated area, and the extinguishing liquid flows directly into the danger zone.

Zooming

The SC is solved.

Systemic super-effect 1: minimal time to locate the hot spot

Systemic super-effect 2: the use of extinguishing material sets off the fire alarm

13. Inexpensive short-life object as a replacement for expensive long-life one	Replace an expensive object with a group of inexpensive objects without certain properties, for example, long life.

EXAMPLE. A small flying laboratory.

Short description

It has been suggested to create and apply small and inexpensive unmanned planes to record meteorological data above oceans **(navigator 13).**

EXTRACTING

R	#	Navigator	Reasoning
++	10	Copying	Copy of a "large" airplane
+	11	Inverse action	Make miniature sensors and install them on a small airplane instead of filling a "large" airplane with "large" instruments.
	12	Local property	A small airplane is enough.
++	***13***	***Inexpensive short-life object as a replacement for expensive long-life one***	*Dominating navigator*
+	16	Partial or excess effect	The use of an excessively "large" airplane is prevented.

REINVENTING

Trend

It is very difficult to measure temperature and air pressure and record wind direction and speed from satellites, and it is too expensive to constantly keep "weather ships" in the sea. What can be done taking into account the large territory to be covered by observation?

Reducing

OZ: enormously large territory above the ocean. The important parameter is OT: regular measurement of parameters at several points above the ocean on a 24-hour basis.

Macro-FIM: The X-resource—without causing unacceptable negative effects—ensures, together with the other available resources, attainment of the following result:

registration of meteorological data at several points above the ocean continuously on a 24-hour basis].

Standard contradiction:

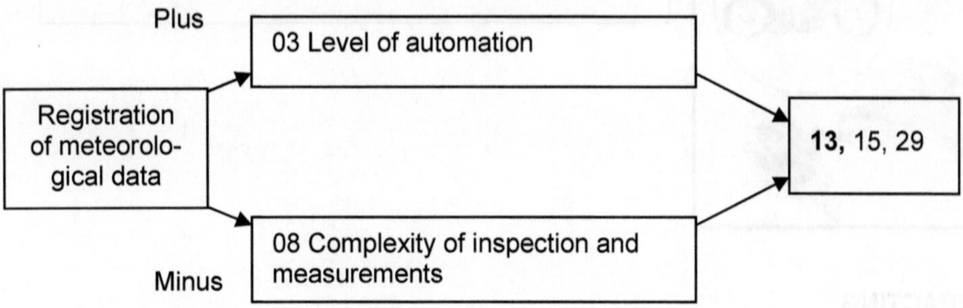

Inventing

Idea according to ***navigator 13***: To register meteorological data above the oceans, it has been suggested to use small and inexpensive unmanned planes ("drones") that could form a dynamic network of "laboratories" above the ocean and be started simultaneously from several points. Dozens of such aircraft could be in flight simultaneously at different heights along their assigned routes.

Zooming

Standard contradiction is removed.

Example. On August 21, 1998, the American automatic meteorological balloon "Laima" (named after the Lithuanian-Latvian goddess of happiness) independently flew from Newfoundland to Scotland. This laboratory aircraft, with a weight of 13 kilograms and a wing span of 3 meters, consumed approximately 8 liters of fuel over the duration of the flight. "Laima" covered the distance of 3200 km in 26 hours.

14. Use of pneumatic or hydraulic constructions	Use gaseous or fluid parts instead of fixed parts in an object, i.e., parts that can be blown up or filled with hydraulic fluid, air cushions, or hydrostatic or hydro-reactive parts.

EXAMPLE. Inflatable life vest for motorcyclists.

Short description

The Japanese firm Mugen Denko produces an inflatable life vest for motorcyclists **(navigator 14).**

EXTRACTING

R	#	Navigator	Reasoning
+	01	Change in the aggregate state of an object	The vest becomes "bloated" with air.
+	12	Local property	Protection in most exposed places
++	***14***	***Use of pneumatic or hydraulic constructions***	*Dominating navigator*
	18	Mediator	Protection from major injuries
++	28	Previously installed cushion	In the direct sense of the word

REINVENTING

Trend

The problem of saving lives and minimizing injuries sustained by victims of motorcycle accidents has no reliable solution. Is it possible to improve available methods or suggest new ideas?

Reducing

OZ: protected body parts of the motorcyclist.

The important parameter is OT: The safety device must operate as quickly as possible after the beginning of the driver's fall.

Macro-FIM: The X-resource—without causing unacceptable negative effects—ensures, together with the other available resources, attainment of the following result:

[*rapid and reliable protection of the falling motorcyclist*].

Standard contradiction:

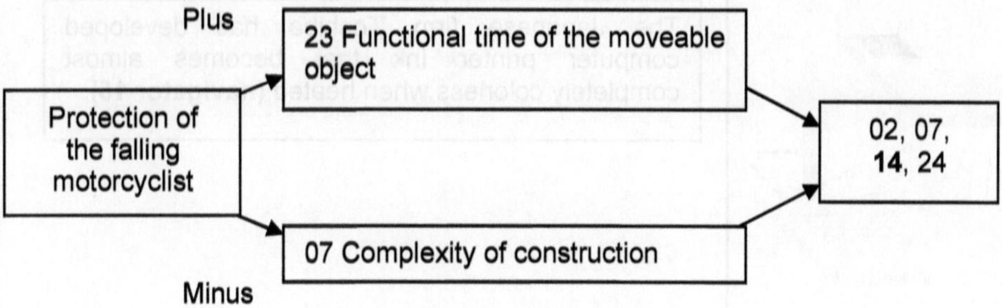

Inventing

Idea according to ***navigator 14***: Japanese firm "Mugen Denko" produces life vests for motorcyclists. The operating principle is the same as in inflatable life vests used in marine rescue. The inflatable life vest is fastened to the motorcycle by a cable, which is connected to the valve of a small compressed carbon dioxide tank. If the motorcyclist begins to fall from his vehicle, the cable stretches and opens the valve, and the life vest is immediately inflated with gas thus protecting the motorcyclist from injuries.

Zooming

Standard contradiction is removed.

Note: The cable must not become activated, and must automatically be detached, when the motorcyclist gets off the motorcycle after finishing a normal trip.

15. Discard and renewal of parts	c) Parts that have fulfilled their task and are no longer part of an object should be disposed of (e.g., dissolved, evaporated, etc.); d) used parts of an object should be immediately replaced during work.

EXAMPLE. Disappearing ink.

Short description

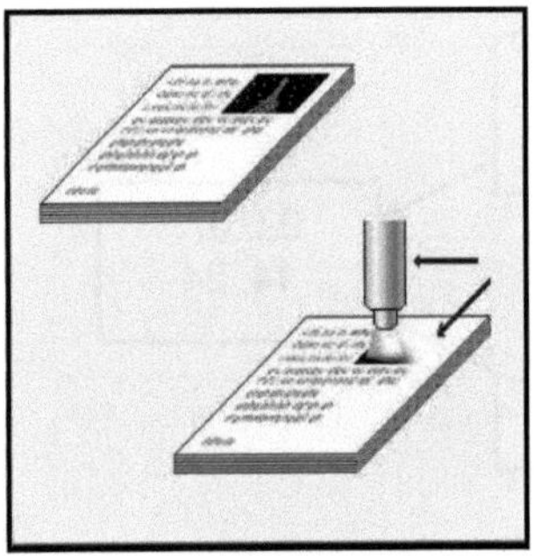

The Japanese firm Toshiba has developed computer printer ink that becomes almost completely colorless when heated **(navigator 15).**

EXTRACTING

R	#	Navigator	Reasoning
+	01	Change in the aggregate state of an object	Change in the state of the ink
++	09	Change in color	Discoloration of the ink
+	12	Local property	Only the ink disappears.
++	***15***	***Discard and renewal of parts***	*Dominating navigator*
+	26	Phase transitions	A special physical effect is used.

REINVENTING

Trend

How can we reduce the quantity of scrap paper?

Reducing

It is necessary to ensure the repeated use of paper. However, it is not simple to remove the coloring substance that has penetrated the fibers in the paper. Known attempts have rendered the paper useless.

OZ: surface of the paper after printing.

Micro-FIM: The X-resource in the form of particles of substance or energy is located inside the operating zone and ensures, together with the other available resources, attainment of the following result:

[*clean sheet of paper*].

Standard contradiction:

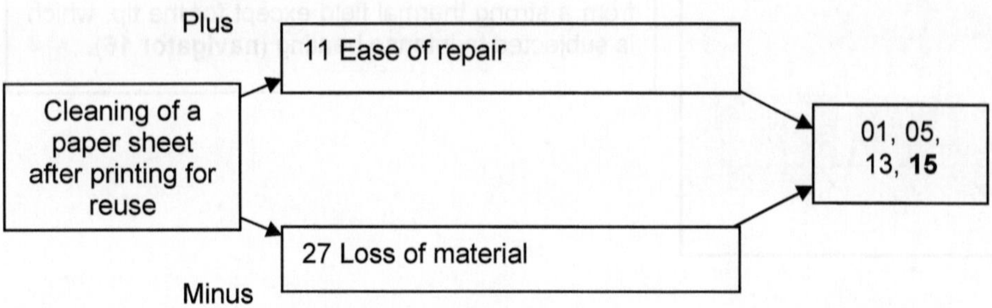

Inventing

Idea according to ***navigator 15***: The Japanese firm Toshiba has developed computer printer ink that almost completely decolorizes when heated to 180 °C. The paper becomes suitable at least for rough copies.

Zooming

Standard contradiction is removed.

Super-effect 1: The company "Toshiba" plans to produce special heat chambers using "worthless" heat generated by office equipment, for example, by air conditioners.

Super-effect 2: On an industrial scale, this technology for preliminary paper cleaning will considerably facilitate processing scrap paper that comes from the offices.

16. Partial or excess effect	When it is difficult to achieve the desired effect completely, we should try to achieve a bit less or a bit more. This can make the task much easier.

EXAMPLE. Melting ampoules.

Short description

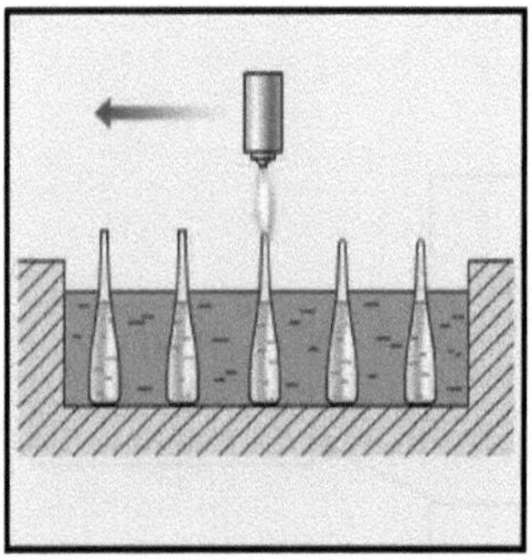

Medicinal ampoules can be sealed off quickly by dipping them into a cooling liquid that protects them from a strong thermal field except for the tip, which is subjected to intense heating (**navigator 16**).

EXTRACTING

R	#	Navigator	Reasoning
	10	Copying	Copy of a system where protection from overheating is provided by a cooling fluid
++	12	Local property	Only the tips of the ampoules are heated
++	***16***	***Partial or excess effect***	*Dominating navigator*
+	18	Mediator	The cooling fluid (water) draws excessive heat away from the ampoule
+	28	Previously installed cushion	The ampoules are preinstalled in the cooling fluid

REINVENTING

Trend

Sealing ampoules has been a difficult problem for a long time because heating their upper sections led to the heat being distributed over their entire surface, which compromised the substance they contained. How can we improve the process of ampoule sealing?

Reducing

OZ: material at the tip of the ampoule (glass), body of the ampoule (glass), the inner space of the ampoule with medicinal substance, the thermal field for the heating (flame of a burner, laser beam), air.

Main problem – prevention of heat distribution onto the entire ampoule.

OT: time of heating plus time of heat transfer; in any case, it is reasonable to shorten the OT.

Maxi-FIM: The operating zone by itself ensures attainment of the following result:

[*rapid sealing of the ampoule without damaging the medicine within*].

Micro-FIM: The X-resource in the form of particles of substance or energy is located inside the operating zone and ensures attainment of the **Maxi-FIM.**

Model of the SC:

Effect, State, Object						
Heating the tip of the ampoule to the melting point						
(+)-Factor			**(+)-Factor**			**Cluster**
–	34	Temperature	–	07	Complexity of construction	05, **16**, 19

Inventing

The key method is ***method 16***, where excessive thermal field is fed into the heating zone while the body of the ampoule is immersed in the cooling fluid.

This is an efficient interpretation of ***method 05***, where the effect of the increasingly intensive thermal field is changed from deleterious (when the ampoule is cooled by ambient air) to useful (when the ampoule is cooled by special liquid).

Zooming

The SC is solved.

System super-effect 1: The operating time is shortened.

System super-effect 2: The medicine within the ampoule can no longer be damaged.

17. Use of composite materials	Move from homogeneous materials to combinations.

EXAMPLE. Drapes against noise.

Short description

In England, a multi-layered curtain has been proposed. It contains pores of different sizes providing mechanical filtration of sound waves so that the aggregate pass band of a set of filters approximately corresponds to the spectrum of sea surf **(navigator 17).**

EXTRACTING

R	#	Navigator	Reasoning
+	03	Segmentation	An ordinary curtain is divided into multiple layers.
++	12	Local property	Pores with different diameters neutralize sound waves of certain lengths
++	***17***	***Use of composite materials***	*Dominating navigator*
++	31	Use of porous materials	The layers have pores with different diameters.
+	25	Use of flexible covers and thin films	The composition contains many thin layers

REINVENTING

Trend

Street noises are especially bad because of sharp, unexpected sounds. Ordinary drapes provide insufficient noise protection. How can we decrease the harmful effect of noise?

Reducing

OZ: window area.

Additional heavy drapes are installed for noise protection. However, they are inconvenient, and they decrease illumination.

Micro-FIM: The X-resource in the form of particles of substance or energy is located inside the operating zone and ensures, together with the other available resources, attainment of the following result:

[*the "softening" of street noise which penetrates the apartment through the window*].

Standard contradiction:

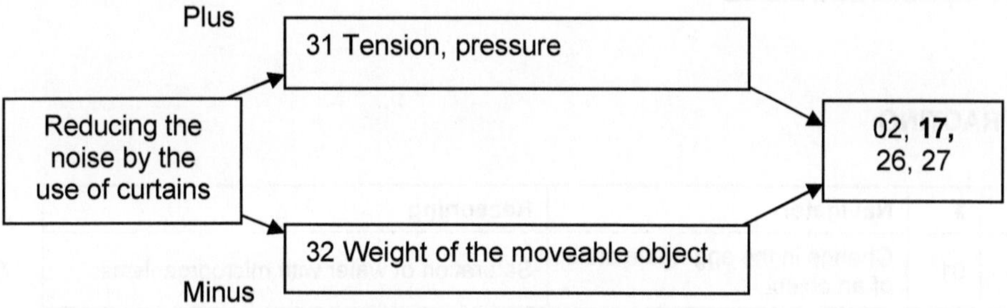

Inventing

Idea according to ***navigator 17***: In England, a curtain has been proposed consisting of multiple thin transparent layers. It contains pores of different sizes providing mechanical filtration of sound waves so that the aggregate pass band of a set of filters approximately corresponds to the spectrum of sea surf

Zooming

Standard contradiction is removed.

18. Mediator	c) Use another object to transfer or transmit an action; d) temporarily connect an object with another (easily separable) object.

EXAMPLE. Dolphins in a glowing stream.

Short description

Single-cell micro-organisms that glow in moving water were introduced into salt water pools to investigate the movements of dolphins (mediators according to **navigator 18**). The water flowing around the dolphins' bodies can then be recorded with high-precision video cameras.

EXTRACTING

R	#	Navigator	Reasoning
+	01	Change in the aggregate state of an object	Saturation of water with microorganisms
+	09	Change in color	Apparently another dominating navigator
++	12	Local property	Color is changed where needed.
++	***18***	***Mediator***	*Dominating navigator*
+	26	Phase transitions	A special physical effect is used.

REINVENTING

Trend

How can we make water flowing around the body of a dolphin visible and record it using photo or video equipment? The application of special dyes to the bodies of these animals is work intensive and often dangerous for the dolphins.

Reducing

OZ: water in the space close to swimming dolphins.

Maxi-FIM: The operating zone by itself ensures attainment of the following result:

[*underwater flows that result from the movement of dolphins are made visible*].

Micro-FIM: The X-resource in the form of particles of substance or energy is located inside the operating zone and ensures attainment of the **Maxi-FIM.**

Model of the SC:

Effect, State, Object						
Water in the immediate vicinity of the dolphin's body						
(+)-Factor			**(+)-Factor**			**Cluster**
Surface of the dolphin's body	02	Universality	Long-term and not dangerous	04	Reliability	01, 11, **18**

Inventing

Glowing single-cell micro-organisms were introduced in advance into a salt-water pool (mediators according to ***navigator 18***). These organisms glow in moving water. Recordings made with high-precision cameras make it possible to determine how water flows around the bodies of the dolphins.

It is interesting to note that ***navigator 02*** (change in the properties of water itself) and ***navigator 11*** (opposite effect: dye the water, not the dolphin) also offer good interpretations.

Zooming

The SC is solved.

Powerful systemic super-effect 1: Nothing happens to the dolphins.

Powerful systemic super-effect 2: The method is also universal for other objects that move in water.

19. Transition into another dimension	d) An object is shaped so that it can move or is placed not only in a linear fashion but also in two dimensions, i.e., on a surface. It is also possible to improve the transition from a surface to a three-dimensional space; e) do construction on several floors; tip or turn the object on its side; use the back of the space in question; f) use optical rays that strike a neighboring space or the back of the present space.

EXAMPLE. 3D-measuring tape.

Short description

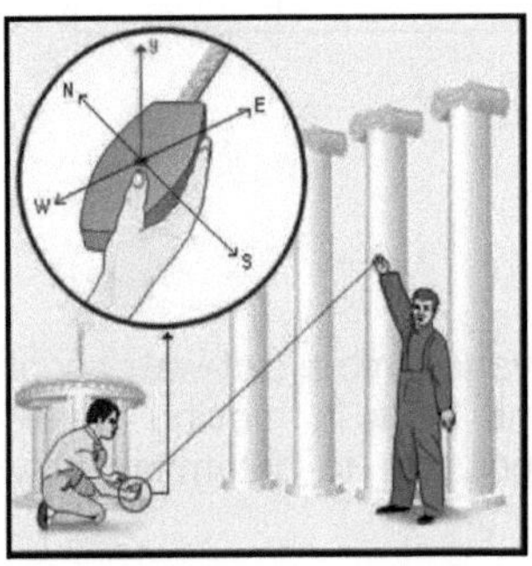

Measuring with a normal measuring tape often means that the tape is used in different directions, either horizontally, vertically, or at various angles. This means the measurements are made in different dimensions (**navigator 19**). This makes it possible to construct closely connected three dimensional representations of a room.

EXTRACTING

R	#	Navigator	Reasoning
++	04	Replacement of mechanical matter	No "mechanical" measurements are taken; instead, sensors measure angular rotation and tilt of the body of the tape.
+	12	Local property	Measuring change in the position of the body of the tape
++	***19***	***Transition into another dimension***	*Dominating navigator*
	29	Self-servicing	The tape measures change in the position of its body by itself.

REINVENTING

Trend

A wealth of measurements of linear length and collective angles are made to construct three-dimensional images of complicated objects (tunnels, caves, industrial facilities) in which certain primary lines are closely connected to each other. These operations are completed with different instruments; they are also work intensive and relatively imprecise.

Reducing

OZ: body (the entire construction) of the entire roll of measuring tape in a working state.

Maxi-FIM: The operating zone by itself ensures attainment of the following result:

[*measuring the orientation of the measuring tape with respect to the given direction*].

Micro-FIM: The X-resource in the form of particles of substance or energy is located inside the operating zone and ensures attainment of the Maxi-FIM.

Model of the SC:

Effect, State, Object						
Determining the orientation of the tape						
(+)-Factor			**(+)-Factor**			**Cluster**
Imprecise	06	Precision of measurement	Labor-intensive	10	Ease of use	03, 11, 15, **19**

Inventing

A measuring tape was designed at the Massachusetts Institute of Technology that automatically measures the length of an unrolled part of the tape, registers with a compass the direction (***navigator 19***) in which the body of the tape was turned, and uses a sensor to determine the angle to the horizontal at which the measurement was taken. The parameters are then stored in a portable computer that can construct on its screen a three-dimensional representation of the objects measured.

Zooming

The SC is solved.

20. Universality	An object has several simultaneous functions so that other objects are not needed.

EXAMPLE. A robot dismantles old washing machines.

Short description

A robot with universal learning abilities has been developed at the TU Berlin to dismantle complex household appliances and industrial facilities that are no longer in use (solution according **navigator 20**).

EXTRACTING

R	#	Navigator	Reasoning
++	10	Copying	Copying specialist actions
+	12	Local property	Operations are adjusted automatically.
++	*20*	***Universality***	*Dominating navigator*
+	36	Feedback	The robot can adjust its actions based on results of examination.
	40	Uninterrupted useful function	The robot can work without breaks.

REINVENTING

Trend

How can we simplify the dismantling of household appliances, such as washing machines of various kinds, and industrial devices that are no longer in use?

Reducing

OZ: Dismantling of obsolescent devices of all kinds.

Maxi-FIM: The operating zone by itself ensures attainment of the following result:

[*optimal technology to dismantle old machines*].

Micro-FIM: The X-resource in the form of particles of substance or energy is located inside the operating zone and ensures attainment of the Maxi-FIM.

Model of the SC:

Effect, State, Object						
Technology to dismantle old machines						
(+)-Factor			**(+)-Factor**			**Cluster**
–	02	Universality	–	01	Productivity	01, 04, **20,** 27

Inventing

It has been suggested that universal robots be used with programming that makes it possible to adapt them to new operational navigators according to ***navigator 20*** and to teach them new jobs.

Zooming

The SC is solved.

Powerful systemic super-effect 1: universality of robots and the possibility of using them in constructions with almost any imaginable complexity.

Indirect powerful systemic super-effect 2: Improving the process of dismantling older devices and facilities makes it possible to reuse large elements of these devices or to transform them into a different form instead of melting them down. This also helps to reduce environmental pollution.

21. Transform damage into use	d) Use damaging factors, especially damaging influences from the environment, to achieve a useful effect; e) eliminate a negative factor by combining it with other negative factors; f) support the damaging factor until it is no longer causes damage.

EXAMPLE. Power plant with wood byproducts.

Short description

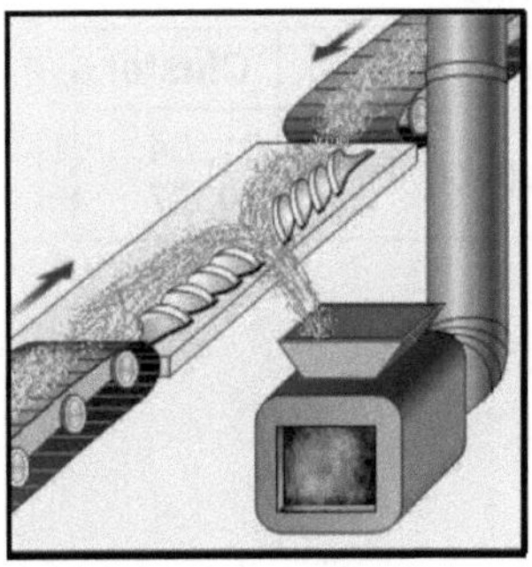

Wood by-products (bark, shavings, and leftovers) could be used in large power plants. The production of extra electrical energy for saw mills or neighbouring facilities could compensate for the damage caused by the large amounts of wood waste that collects in them **(navigator 21**).

EXTRACTING

R	#	Navigator	Reasoning
+	01	Change in the aggregate state of an object	Use incineration of chips and sawdust.
++	10	Copying	Burning wood
	12	Local property	Chips and sawdust are turned into fuel.
++	***21***	***Transform damage into use***	*Dominating navigator*

REINVENTING

Trend

A large amount of wood waste collects near saw mills and wood processing plants that can then decompose. How can we reduce this damage to the environment?

Reducing

OZ: natural environment in which waste from wood processing is introduced as a negative influence, i.e., the wood waste itself.

Maxi-FIM: The operating zone by itself ensures attainment of the following result:

[*reduction of damaging influence on the environment caused by waste from wood processing*].

Micro-FIM: The X-resource in the form of particles of substance or energy is located inside the operating zone and ensures attainment of the Maxi-FIM.

Model of the SC:

Effect, State, Object						
Waste from wood processing						
(+)-Factor			**(+)-Factor**			**Cluster**
The natural environment is damaged by wood processing waste	14	Internal damaging factors	Additional processing impairs productivity	01	Productivity	01, 06, **21,** 23

Inventing

We suggest that waste from wood processing be burned in local power plants according to ***navigator 21***.

Zooming

The SC is solved.

The production of extra energy compensates saw mills and wood-processing plants for the damage caused by large amounts of wood waste that collects near them.

22. Spherical shape	d) Change from linear parts of objects to curved ones, from flat surfaces to spherical ones; from parts shaped like cubes or parallelepipeds to round structures; e) use rollers, balls, and springs; f) change to turning movements by using centrifugal force.

EXAMPLE. Rock-climbing in a sports hall.

Short description

Scientists from the University of Potsdam (Germany) have developed a special stand on the basis of a rotating disk **(navigator 22).** The disk is fitted with protrusions that can be used as hand grips and foot rests. The axis of rotation of the disk can also change position.

EXTRACTING

R	#	Navigator	Reasoning
+	06	Use of mechanical oscillations	Rocking and rolling (stoops and bends)
++	07	Dynamization	The disk rotates and tilts.
++	12	Local property	Change of the disk profile
++	***22***	***Spherical shape***	*Dominating navigator*
	24	Asymmetry	Protrusions are asymmetric.

REINVENTING

Trend

How can we improve the training of rock-climbing athletes if stationary simulators offer few different training routes?

Reducing

OZ: surface of the training simulator.

OT: time of training; Athletes get accustomed to the stationary "wall" very quickly. To eliminate this disadvantage, protrusions are manually rearranged. This is a labor-intensive and ineffective process.

Macro-FIM: The X-resource—without causing unacceptable negative effects—ensures, together with the other available resources, attainment of the following result:

[*ability to use any route type and unlimited training time*].

Standard contradiction:

Effect, State, Object						
Simulator for rock-climber training						
(+)-Factor			**(+)-Factor**			**Cluster**
Greater (unlimited) length of training route is necessary	15	Length of the moveable object	Short routes have poor variability	02	Universality, adaptability	03, 07, 16, **22**

Inventing

Idea according to ***navigator 22***: Scientists from the University of Potsdam (Germany) have developed a special stand on the basis of a rotating disk (***navigator 22***). The disk is fitted with protrusions that can be used as hand grips and foot rests. The axis of rotation of the disk can also change position.

Navigator 7 is also present.

Zooming

Standard contradiction is removed.

Development of idea: The stand can also be a good attraction in a theme park.

23. Use of inert media	e) Replace a normal medium with an inert one; f) complete a process in a vacuum.

EXAMPLE. Fog dispersal.

Short description

Powdered dry ice has long been used to disperse clouds. Now German meteorologists have successfully used it to disperse fog at an airfield **(navigator 23).**

EXTRACTING

R	#	Navigator	Reasoning
++	01	Change in the aggregate state of an object	Freezing of small droplets in the air
+	05	Separation	The drops are "separated" from the air by precipitation.
+	21	Transform damage into use	Cool the air even more
++	***23***	***Use of inert media***	*Dominating navigator*
+	26	Phase transitions	The drops draw heat from the air.

REINVENTING

Trend

It is known that aircraft landing is complex and dangerous in foggy weather. Is it possible to disperse the fog?

Reducing

OZ: space above the takeoff and landing strip at the airport.
OT: time of aircraft landing.

Micro-FIM: The X-resource, in the form of particles of substance or energy, is located inside the operating zone and ensures, together with the other available resources, attainment of the following result:

[*fog dispersal*].

Standard contradiction:

Effect, State, Object						
Dispersal of fog						
(+)-Factor			**(+)-Factor**			**Cluster**
Space above the landing strip at the airport	20	volume of the fixed object	The fog persists because of high humidity and low air temperature	13	External damaging factors	08, 13, 15, **23**

Inventing

Idea according to ***navigator 23***: Powdered dry ice has long been used to disperse clouds. Now German meteorologists have successfully used it to disperse fog at an airfield. A minute later, small icy crystals appear in air and soon fall to the ground.

Other navigators present: *08 periodic action, 13 inexpensive short-life object as a replacement for expensive long-life one, 15 discard and renewal of parts.*

Zooming

Standard contradiction is removed.

24. Asymmetry	e) Move from a symmetrical shape of an object to an asymmetrical one; f) increase the degree of asymmetry if the object is already asymmetrical.

EXAMPLE. Horizontal reagent tube.

Short description

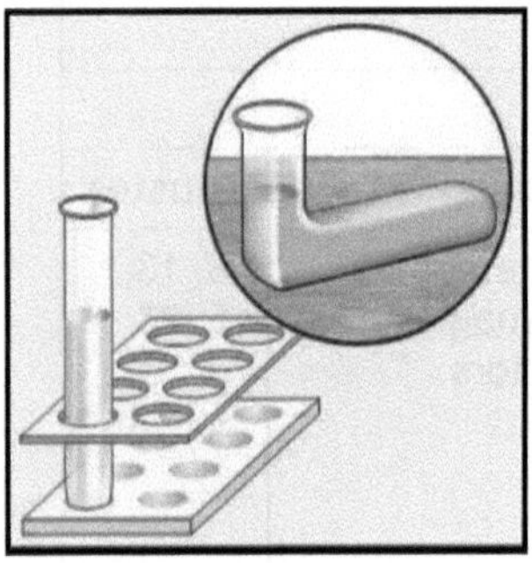

A reagent tube needs to be developed with both a curved form and flat foundation so that it can stand on a table without a special frame. This is expected to simplify the use of such devices (**navigator 24**).

EXTRACTING

R	#	Navigator	Reasoning
+	05	Separation	The object is stabilized.
++	10	Copying	Copy of a vase or cup
+	19	Transition into another dimension	The reagent tube is bent within a "plane" rather than drawn along a "line."
++	***24***	***Asymmetry***	*Dominating navigator*
+	29	Self-servicing	The reagent tube stands by itself.

REINVENTING

Trend

Typical reagent tubes are cylinder-shaped and therefore cannot stand on the table when open because they then spill their liquid contents. However, it is often quite useful to have several reagent tubes close at hand while working. The holding frames for such tubes take up a lot of space. How can we solve this problem?

Reducing

OZ: reagent tubes, frames to hold reagent tubes in a vertical position.

Maxi-FIM: The operating zone by itself ensures attainment of the following result: [*discarding the frame while retaining the ability to use open reagent tubes*].

Micro-FIM: The X-resource, in the form of particles of substance or energy, is located inside the operating zone and ensures attainment of the Maxi-FIM.

Model of the SC:

Effect, State, Object						
Reagent tube						
(+)-Factor			**(+)-Factor**			**Cluster**
It has insufficient bearing area	18	Surface of the fixed object	It cannot stand without support	10	Ease of use	16, **24**

Inventing

We suggest that an "asymmetrical" reagent tube be developed according to navigator 24. Such tube must have both a curved form and flat foundation so that it can stand on the table without a special holding frame.

Zooming

The SC is solved.

Drawback: This kind of tube can suddenly fall over, which is not possible when using a holding frame.

25. Use of flexible covers and thin films	e) Flexible covers and thin layers are used in place of the usual constructions; f) isolate objects from the external world with flexible covers or thin layers.

EXAMPLE. Rescue in an avalanche.

Short description

German entrepreneur Peter Aschauer suggested a new rescue device—an avalanche airbag made of bright orange nylon **(navigator 25).** The airbag is carried in a small backpack and inflated by compressed nitrogen from a small cylinder activated by the user when he finds himself in danger of being buried under snow.

EXTRACTING

R	#	Navigator	Reasoning
+	10	Copying	Copy of a life buoy or life vest
++	14	Use of pneumatic or hydraulic constructions	The cushion is filled with compressed air.
++	***25***	***Use of flexible covers and thin films***	*Dominating navigator*
+	28	Previously installed cushion	Can be recognized as the second dominating navigator
+	33	Quick jump	The cushion is inflated very quickly.

REINVENTING

Trend

Each year dozens of mountaineers and skiers die in the mountains because of snow avalanches. When an avalanche is set off unexpectedly, there is very little time to perform maneuvers aimed at securing one's safety. Can we increase the chances of survival?

Reducing

OZ: space inside the avalanche.

OT: time of motion with the avalanche. The known complex and relatively effective constructions are too heavy.

Micro-FIM: The X-resource, in the form of particles of substance or energy, is located inside the operating zone and ensures, together with the other available resources, attainment of the following result:

[*rescue in an avalanche*].

Standard contradiction:

Effect, State, Object						
Rescue in an avalanche						
(+)-Factor			**(+)-Factor**			**Cluster**
–	07	Complexity of construction	–	32	Weight of the movable object	10, 15, **25**, 26

Inventing

Idea according to ***navigator 25***: German entrepreneur Peter Aschauer suggested a new rescue device—an avalanche airbag made of bright orange nylon (***navigator 25***). The airbag is carried in a small backpack and inflated by compressed nitrogen from a small cylinder activated by the user when he finds himself in danger of being buried under snow.

Navigator ***10 copying*** is present too: The principle of water rescuing with the aid of inflatable life-vests is used.

Zooming

Standard contradiction is removed.

26. Phase transitions	Make full use of phenomena that occur during phase transitions such as a change in volume, radiation, or absorption of warmth, etc.

EXAMPLE. Cooler gun.

Short description

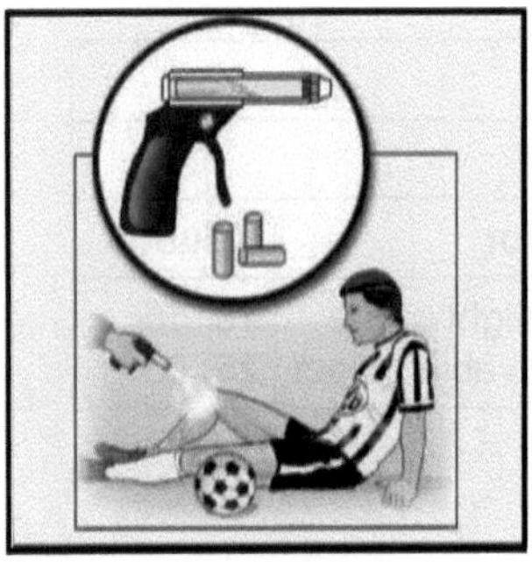

To enable rapid cooling of injured areas, the French firm Cryonic Medical offers gun-shaped portable cooling devices. When the trigger is pulled, the gun emits a cloud of dry ice **(navigator 26)**.

EXTRACTING

R	#	Navigator	Reasoning
+	10	Copying	The "gun" shape is very convenient.
++	14	Use of pneumatic or hydraulic constructions	Compressed air is used.
++	***26***	***Phase transitions***	*Dominating navigator*
+	28	Previously installed cushion	The "gun" is loaded in advance.

REINVENTING

Trend

It is known that many injuries i.e., bruises, sprains, dislocations, require cooling to obtain analgesic and anti-edematous effects. However, the storage of cooling agents and their precise delivery to the injured area call for the use of devices that are quite inconvenient. How can we increase the efficiency of cooling-agent application?

Reducing

OZ: place of injury.

OT: immediately after occurrence of an injury event.

Macro-FIM: The X-resource—without causing unacceptable negative effects—ensures, together with the other available resources, attainment of the following result:

[*convenient device for precise delivery of cooling agent to the injured spot*].

Standard contradiction:

Effect, State, Object						
Delivery of cooling agent to the injured spot						
(+)-Factor			**(+)-Factor**			**Cluster**
–	05	Precision of manufacture	–	18	Surface of the fixed object	05, 06, 14, **26**

Inventing

Idea according to ***navigator 27***: To enable rapid cooling of injured spots, the French firm Cryonic Medical offers gun-shaped portable cooling devices. When the trigger is pulled, the gun emits a cloud of dry ice.

Navigator *14 Use of pneumatic or hydraulic constructions* is also present here.

Zooming

Standard contradiction is removed.

Effectiveness: Within half a minute, this "shot" of dry ice decreases the temperature of the injured spot from 33 °C to 2 °C ..

27. Use of thermal expansion	e) Use expansion (or reduction) of materials when heating them; f) use materials with different coefficients of heat expansion.

EXAMPLE. Magic packing.

Short description

The Firm Sealed Air Corporation (USA) has developed highly elastic polyethylene bags that come in all sizes. Mechanical or thermal impact triggers the production of polymer foam, which is then evenly distributed throughout the interior of the bag **(navigator 27).**

EXTRACTING

R	#	Navigator	Reasoning
++	01	Change in the aggregate state of an object	Increased volume of material
+	04	Replacement of mechanical matter	A "clever" material is used instead of various fillers.
+	05	Separation	Introduction of a new property: complete filling of the entire empty space
+	10	Copying	Copying of inflatable structures
++	**27**	***Use of thermal expansion***	*Dominating navigator*

REINVENTING

Trend

The problem of packing of fragile items (for example, those made of glass) and "inconvenient" items i.e., those that have many protruding parts, is well-known. How can we make the material used to pack such items dense but harmless to them?

Reducing

OZ: space inside the packing box.

The application of ordinary packing materials, such as granules, requires the additional fixation of articles, e.g., in the course of prolonged transportation, articles move to the walls of the box and are not protected by packing material.

Micro-FIM: The X-resource, in the form of particles of substance or energy, is located inside the operating zone and ensures, together with the other available resources, attainment of the following result:

[*reliable and "cautious" fixation of items in the box*].

Standard contradiction:

Effect, State, Object						
Dense packing of fragile or "inconvenient" items for transportation						
(+)-Factor			**(+)-Factor**			**Cluster**
–	19	Volume of the movable object	–	31	Tension, pressure	01, 20, 26, **27**

Inventing

Idea according to ***navigator 26***: The firm Sealed Air Corporation (USA) has developed highly elastic polyethylene bags that come in all sizes. Mechanical or thermal impact triggers the production of polymer foam, which is then evenly distributed throughout the interior of the bag.

Navigators 01 and 20 are also used. The foam in a "fluid" state is transformed into the "solid" state (navigator 01 change in the aggregate state of an object). The foam makes it possible to pack items of arbitrary form (navigator ***20 universality***).

Zooming

Standard contradiction is removed.

28. Previously installed cushions	Increase the relatively low security of an object with safety measures in advance.

EXAMPLE. Parachutes for aircraft.

Short description

Parachutes in the passenger section could help save passengers in plane crashes (**navigator 28**). The "superfluous" weight of cargo sections, wings, and engines could be separated and discarded so that the burden on the parachutes is reduced (as is their own weight). This could reduce the operating expenses attributable to such system.

EXTRACTING

R	#	Navigator	Reasoning
++	03	Segmentation	Division of the aircraft into parts
+	05	Separation	Precious cargo is "separated" from the falling aircraft.
++	10	Copying	Copy of "human" parachutes
++	***28***	***Previously installed cushion***	*Dominating navigator*
++	34	Matryoshka (nested doll)	The parachutes are hidden inside the body of the aircraft.

REINVENTING

Trend

The main goal in air accidents is to save the passengers. How can we reach this goal?

Reducing

OZ: passenger section of a crashing aircraft with passengers.

We note the following formulation of the ideal solution because we cannot count on any external standard factors.

Maxi-FIM: The operating zone by itself ensures attainment of the following result:

[*safe landing for the passengers of a crashing aircraft*].

Micro-FIM: The X-resource, in the form of particles of substance or energy, is located inside the operating zone and ensures attainment of the Maxi-FIM.

Model of the SC:

Effect, State, Object						
Passenger section of a crashing aircraft with passengers						
(+)-Factor			**(+)-Factor**			**Cluster**
Passengers are rescued	04	User friendliness	Large size of the aircraft	16	Length of the fixed object	04, 07, 14, **28**

Inventing

A parachute system is installed in the aircraft in advance according to ***navigator 28***.

This navigator is used together with the following additional navigators:

05 separation – The passengers (located in the passenger module) are separated from the distressed aircraft;

03 segmentation – The aircraft is divided into the salvaged part and the nonsalvaged part;

29 self-servicing – The module rescues itself.

In the future, this example may be supplemented with minus factor *32 Weight of the moveable object*, and relevant additions may be made to the A-matrix.

Zooming

The SC is solved.

29. Self-servicing	e) The object services itself with auxiliary and repair functions; f) reuse waste (energy, material).

EXAMPLE. Self-recharging robots.

Short description

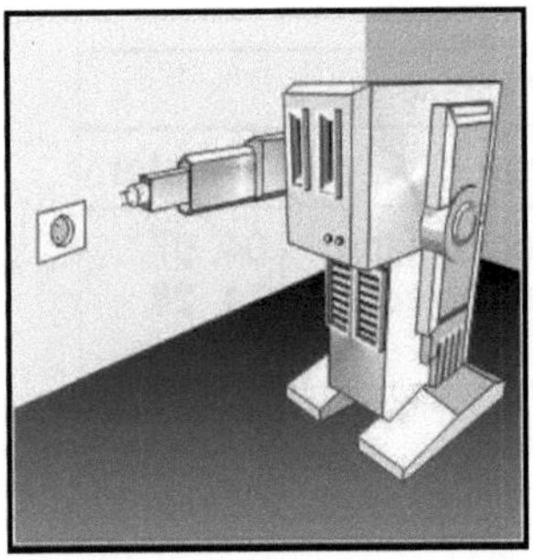

Robots have a program that instruct them to search for a specified or any nearby electrical socket to charge their accumulators (**navigator 29**).

EXTRACTING

R	#	Navigator	Reasoning
+	05	Separation	The required function is assigned to the robot.
++	10	Copying	Copying of human action
+	28	Previously installed cushion	The robot is "trained" to plug itself in advance.
++	***29***	***Self-servicing***	*Dominating navigator*

REINVENTING

Trend

The accumulators of industrial and household robots require periodic recharging. How can we simplify this task for the owner of the robot?

Reducing

OZ: robot with rechargeable accumulators.

Maxi-FIM: The operating zone by itself ensures attainment of the following result: [*recharging of accumulators*].

Micro-FIM: The X-resource, in the form of particles of substance or energy, is located inside the operating zone and ensures attainment of the Maxi-FIM.

Model of the SC:

Effect, State, Object						
Robot with rechargeable accumulators						
(+)-Factor			**(+)-Factor**			**Cluster**
–	10	Ease of use	–	07	Complexity of construction	19, **29**

Inventing

According to ***navigator 29***, robots can be programmed to look for a specified or any nearby electrical sockets by themselves.

Navigators 28, 10, 19, and 20 could be constructively interpreted.

Zooming

The SC is solved.

This principle is often used in real life.

30. Use of strong oxidants	g) Replace normal air with an enhanced stream; h) replace an enhanced stream with oxygen; i) influence air or oxygen with ionizing rays; j) use of oxygen with ozone; k) replace ionized or ozone-oxygen with ozone.

EXAMPLE. Dissolution instead of drilling a tooth.

Short description

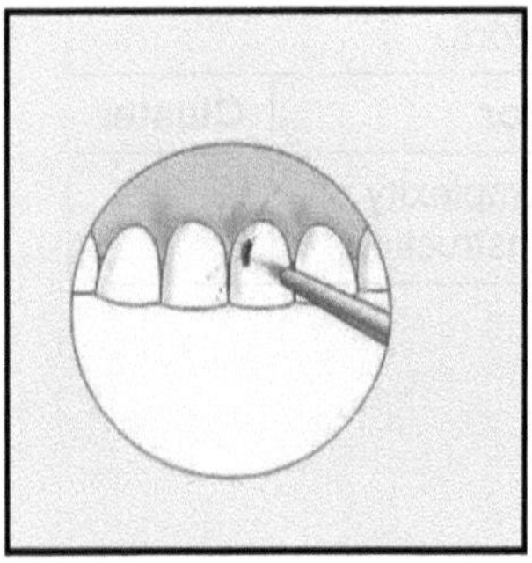

The Swedish firm"Mediteam has developed a method of treatment of caries without drilling hole,by using a special mixture that dissolves the diseased tissue of the tooth in less than half a minute **(navigator 30).**

EXTRACTING

R	#	Navigator	Reasoning
++	01	Change in the aggregate state of an object	Tooth material is dissolved
+	10	Copying	Copying of the effect that acid has on tissue
+	12	Local property	Local action is assured
+	21	Transform damage into use	Useful application of "solvents"
++	***30***	***Use of strong oxidants***	*Dominating navigator*

REINVENTING

Trend

Usually it is necessary to bore the tooth to remove the fragile diseased tissue with the microbes and to put in a filling, which is applied to the cleaned and stable foundation. Can we change this painful operation and reduce its duration?

Reducing

To reduce the treatment time, it is possible to increase the speed of drilling; however, this can lead to unwarranted removal of healthy tissue.

OZ: sore place on the tooth.

OT: removal of affected tissue.

Macro-FIM: The X-resource—without causing unacceptable negative effects—ensures, together with the other available resources, attainment of the following result:

[*painless and precise removal of affected tissue*].

Standard contradiction:

Effect, State, Object						
Removal of diseased tissue from a tooth						
(+)-Factor			**(+)-Factor**			**Cluster**
-	22	Speed	-	27	Loss of material	02, 04, 11, **30**

Inventing

Idea according to ***navigator 30***: The Swedish firm Mediteam has developed a method of treatment of caries without drilling holes by using a special mixture that dissolves the diseased tissue of the tooth in less than half a minute.

Navigator *04 Replacement of mechanical matter* is also present here.

Zooming

Standard contradiction is removed.

Very important super-effect: Only diseased tissue is removed rapidly and painlessly with simultaneous protection of healthy tissue.

31. Use of porous materials	e) Make an object porous or use supplementary porous elements (inserts, coverings, etc.); f) if the object already consists of a porous material, the pores can be filled with some kind of material in advance.

EXAMPLE. Sterilization of coconut water.

Short description

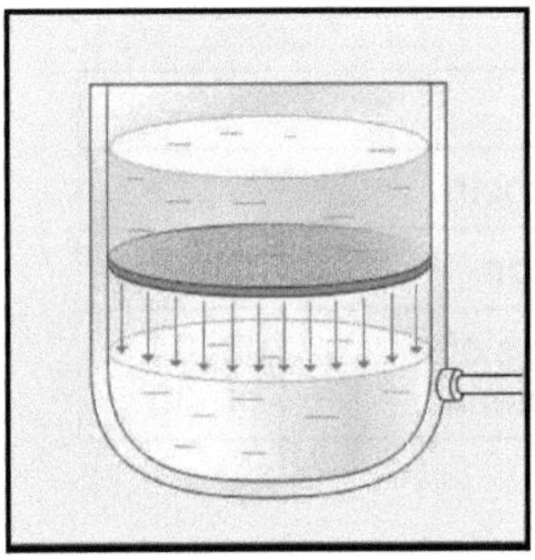

Experts from the FAO (Food and Agriculture Organization of the United Nations) have developed and patented a way to sterilize coconut water without heating. The drink is filtered with microporous filters that remove microbes and their spores **(navigator 31).**

EXTRACTING

R	#	Navigator	Reasoning
	04	Replacement of mechanical matter	Modification of the membrane structure
+	05	Separation	Hazardous substances are "separated" from coconut water.
++	10	Copying	Copy of a filter
	12	Local property	Possibility to select the pore diameter
++	***31***	***Use of porous materials***	*Dominating navigator*

REINVENTING

Trend

Each unripe coconut contains approximately 1.5 liters of transparent, cool, sour-sweet liquid that quenches thirst and contains a number of vitamins, potassium and sugar. This is the so-called “coco water.” As the coconut is ripening, the water is saturated with fat droplets and turns into coco milk.

Coco water does not yield itself to pasteurization. It sours when heated; therefore it cannot be stored in jars and bottles for subsequent sale.

Reducing

OZ: glass for treating coco water.

Macro-FIM: The X-resource—without causing unacceptable negative effects—ensures, together with the other available resources, attainment of the following result:

[*sterilization of coco water*].

Standard contradiction:

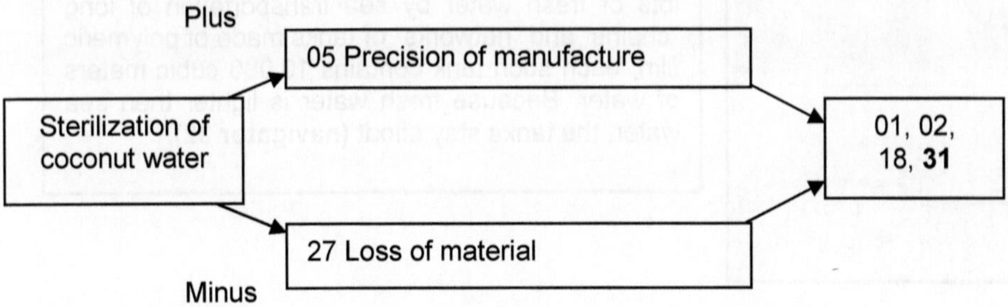

Inventing

Idea according to ***navigator 31***: Experts from the FAO (Food and Agriculture Organization of the United Nations) have developed and patented a way to sterilize coconut water without heating. The drink is filtered with microporous filters that remove microbes and their spores.

Navigators *01 Change in the aggregate state of an object* and 18 *Mediator* are present in the solution.

Zooming

Standard contradiction is removed.

32. Counterweight	e) Compensate for the weight of an object with its connection to another object with lifting power; f) compensate for the weight of an object using interaction with the external environment (for example, with aerodynamic or hydrodynamic forces).

EXAMPLE. Transportation of water by sea.

Short description

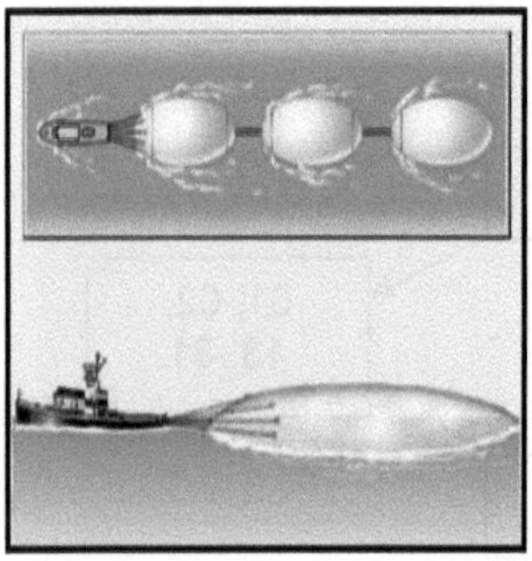

Turkish engineers have offered a way to deliver lots of fresh water by sea transportation of long "chains" and "networks" of tanks made of polymeric film; each such tank contains 10,000 cubic meters of water. Because fresh water is lighter then sea water, the tanks stay afloat **(navigator 32).**

EXTRACTING

R	#	Navigator	Reasoning
++	10	Copying	Copy of floating ice
++	11	Inverse action	Instead of filling ships with water, put "water into water."
++	14	Use of pneumatic or hydraulic constructions	Use of the difference between fresh and sea water density
++	***32***	***Counterweight***	*Dominating navigator*
	35	Unite	String the "tanks" so that they form a garland.

REINVENTING

Trend

Problems with drinking water in many arid regions of the world are well known. It is necessary to organize transportation by sea of large quantities of fresh water.

However, this delivery requires large tankers. How can we ensure delivery of large volumes of fresh water?

Reducing

OZ: container for storage and transportation of a large quantity of water.

Micro-FIM: The X-resource, in the form of particles of substance or energy, is located inside the operating zone and ensures, together with the other available resources, attainment of the following result:

[*delivery of a large quantity of fresh water*].

Standard contradiction:

Effect, State, Object						
Delivery of a large quantity of fresh water						
(+)-Factor			**(+)-Factor**			**Cluster**
–	32	Weight of the movable object	For moving lots of water	30	Force	02, 06, 27, **32**

Inventing

Idea according to ***navigator 32***: Turkish engineers have offered a way to deliver lots of fresh water by sea transportation of long "chains" and "networks" of tanks made of polymeric film; each such tank contains 10,000 cubic meters of water. Because fresh water is lighter then sea water, the stay keep afloat.

Zooming

Standard contradiction is removed.

33. Quick jump	Complete a process or some of its (damaging or dangerous) stages at high speed.

EXAMPLE. Anti-sauna: treatment at –110 °C.

Short description

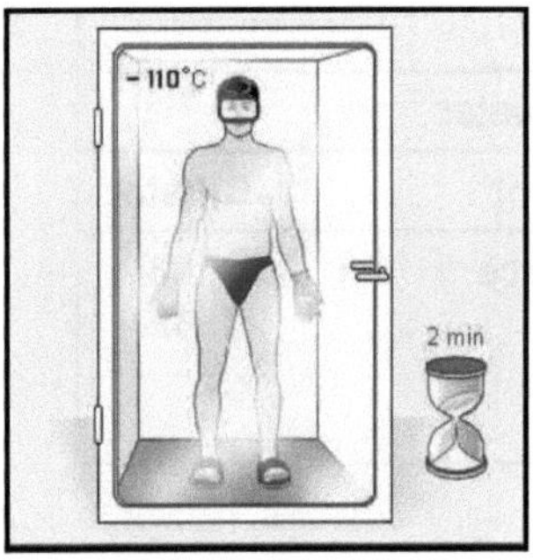

An unusual way of treating arthritis and rheumatism is used by rheumatologists in the German city of Sendenhorst: The patient is placed in a refrigerating chamber with a temperature of –110 °C to –20 °C for 1 to 2 minutes **(navigator 33).**

EXTRACTING

R	#	Navigator	Reasoning
+	11	Inverse action	Use of refrigeration instead of heating.
++	16	Partial or excess effect	Refrigeration at a temperature of –110 °C!
++	21	Transform damage into use	Strong but short-lived refrigeration proves useful.
+	26	Phase transitions	Cooling of skin surface
++	***33***	***Quick jump***	*Dominating navigator*

REINVENTING

Trend

The therapeutic effect of low temperatures is well known; for example, people bathe in cold water, dip into ice holes, and roll in the snow after sauna. It is also known that exposure to low temperatures may result in frost-bite and have other negative effects. How can we avoid those negative effects and maximize the therapeutic effect?

Reducing

OZ: human skin, air with very low temperature (in a special room).

OT: time of exposure to low temperature (cold air).

Macro-FIM: The X-resource—without causing unacceptable negative effects—ensures, together with the other available resources, attainment of the following result:

[*therapeutic action of cold air*].

Standard contradiction:

Effect, State, Object						
"Treatment" in low temperature						
(+)-Factor			**(+)-Factor**			**Cluster**
Very low	34	Temperature	Danger of frostbite	27	Loss of material	14, 26, 31, **33**

Inventing

Idea according to ***navigator 33***: An unusual way of treating arthritis and rheumatism is used by rheumatologists in the German city of Sendenhorst: The patient is placed in a refrigerating chamber with a temperature of minus 110 °C to 120 °C for 1 to 2 minutes. The skin is cooled to +2 °C. The session rids the patient of pain for a relatively long period.

Zooming

Standard contradiction is removed.

Super-effect: The use of conventional medicines is reduced between treatment sessions.

Deficiency: The danger of freezing remains; therefore before entering the chamber, the patient puts on a respirator and protects parts of his or her body.

Development of the idea: Sport physicians have also used this method of treatment. It has been established that brief exposure to ultra-low temperatures is equivalent to a good muscle massage; it also improves blood supply and increases endurance of the athlete by 20 %.

34. Matryoshka (nested doll)	e) An object is inside another object that is also inside another object, etc.; f) an object runs through a hollow space in another object.

EXAMPLE. Radio receiver in the ear.

Short description

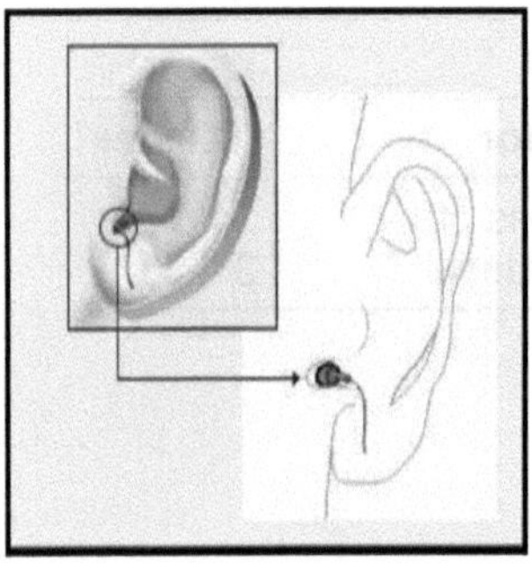

A receiver and synthesizer are inserted into the ear in order to conveniently and discreetly hear radio programs, messages from mobile telephones, and other signals (**navigator 34**).

EXTRACTING

R	#	Navigator	Reasoning
+	05	Separation	Remove pesky wires.
++	12	Local property	Place the object as close to the "operating zone" as possible.
++	19	Transition into another dimension	Reverse use: Instead of expanding the object "into a line," collapse it "into a point."
++	***34***	***Matryoshka (nested doll)***	*Dominating navigator*

REINVENTING

Trend

How can we discreetly listen to radio programs or mobile telephones?

Reducing

OZ: human ear, headset, wire from the receiver or cell phone to audio oscillators in the headset.

The problem is that wires from a Walkman, radio, or mobile telephone are visible. This reveals the fact that an audio device is in use. Freedom of movement is also limited when wires are present because they can be damaged or destroyed.

Maxi-FIM: The operating zone by itself ensures attainment of the following result: [*perception of audio signals with a device to receive and play them*].

Micro-FIM: The X-resource, in the form of particles of substance or energy, is located inside the operating zone and ensures attainment of the Maxi-FIM.

Model of the SC:

Effect, State, Object						
Ear						
(+)-Factor			**(+)-Factor**			**Cluster**
To introduce a new function of discreet hearing	01	Productivity	Long wires decrease required functionality	16	Length of the fixed object	10, 22, 25, **34**

Inventing

The idea for a solution is based on ***navigator 34***: The receiver and synthesizer are hidden in the ear.

Zooming

The SC is solved.

Systemic super-effect: This principle can be used widely when creating universal receivers that are compatible with different transmitters installed in Walkmans, CD players, mobile telephones, etc.

35. Unite	e) Unite similar objects or objects for neighboring operations; f) temporarily unite similar objects or objects for neighboring operations.

EXAMPLE. 10,000 thermal elements in a watch.

Short description

Ten thousand thermal microelements are installed on the underside of a watch to produce an sufficiently large electrical potential to power the watch. The microelements transform the difference between the temperature of the wrist and the temperature of the ambient environment into electricity (**navigator 35**).

EXTRACTING

R	#	Navigator	Reasoning
+	05	Separation	The required property (generation of energy) is "introduced" into the system.
++	10	Copying	Ten thousand copies of thermal couples are used.
+	19	Transition into another dimension	Instead of a "point" storage battery, the sources of energy are located within a "plane."
++	26	Phase transitions	A special physical effect is used.
++	***35***	***Unite***	*Dominating navigator*

REINVENTING

Trend

Can we produce watches in which batteries do not have to be replaced so often? Is it possible to improve the concept of an auto battery that weighs 10 kg?

Reducing

OZ: watch case, internal energy source of the watch, human skin under the watch, environment (air).

Macro-FIM: The X-resource—without causing unacceptable negative effects—ensures, together with the other available resources, attainment of the following result: [*guaranteed electric power supply to the wrist watch without frequent replacement or recharging of the battery*].

Model of the SC:

Effect, State, Object						
Energy production for watches						
(+)-Factor			**(+)-Factor**			**Cluster**
–	23	Functional time of the movable object	–	32	Weight of the movable object	08, 15, 31, **35**

Inventing

Ten thousand thermal microelements are installed on the underside of a watch to produce an sufficiently large electrical potential to power the watch. The microelements transform the difference between the temperature of the wrist and the temperature of the ambient environment into electricity (***navigator 35***). The system powers the watch itself when it is carried around the wrist and charges the internal accumulator. The accumulator powers the watch when it is not carried.

Zooming

The SC is solved.

36. Feedback	g) Create a retroactive influence; h) change a retroactive influence that already exists.

EXAMPLE. Money for the ATM.

Short description

Siemens has developed a computer program that determines the amount of cash to be placed inside each ATM based on previous usage data (**navigator 36**).

EXTRACTING

R	#	Navigator	Reasoning
	04	Replacement of mechanical matter	Use of automation
+	10	Copying	Copying of human action
++	11	Inverse action	Receive required information from the automated device instead of going out to the device to check for such information.
+	12	Local property	Each automated device "knows" its state.
++	*36*	***Feedback***	*Dominating navigator*

REINVENTING

Trend

Too much cash lying idle inside an ATM is not a good thing. There are more than 50,000 ATMs in Germany, and the amount of cash in each such ATM can be as high as 400,000 Euros. This is usually too much because customers cannot use it all before the ATM is refilled. This money (20,000 to 40,000 Euros/ATM on average) is removed from circulation. Can we deposit an optimal amount of cash in each ATM?

Reducing

OZ: money in the ATM.

Could we increase the economic effectiveness of ATMs as well as their user-friendliness? Imprecise estimates of the amount of cash needed are the main problem in this case.

Maxi-FIM: The operating zone by itself ensures attainment of the following result: [*optimal amount of cash in ATMs*].

Micro-FIM: The X-resource, in the form of particles of substance or energy, is located inside the operating zone and ensures attainment of the Maxi-FIM.

Model of the SC:

Effect, State, Object						
Money deposited for use in an ATM						
(+)-Factor			**(+)-Factor**			**Cluster**
–	10	Ease of use	–	05	Precision of manufacture	01, 03, **36**

Inventing

Siemens has developed a computer program that determines the amount of cash to be placed inside each ATM based on previous usage data.

Zooming

The SC is solved.

Estimates indicate that this program allows to return to circulation more than 50 million Euros/year.

An obvious development of this solution based on navigator 36 *Feedback* would be the installation of sensors to estimate the remaining money and a signal for new deposits in the ATM.

37. Equipotentiality	Change work conditions so that it is not necessary to lift or lower an object.

EXAMPLE. Door in... a bath.

Short description

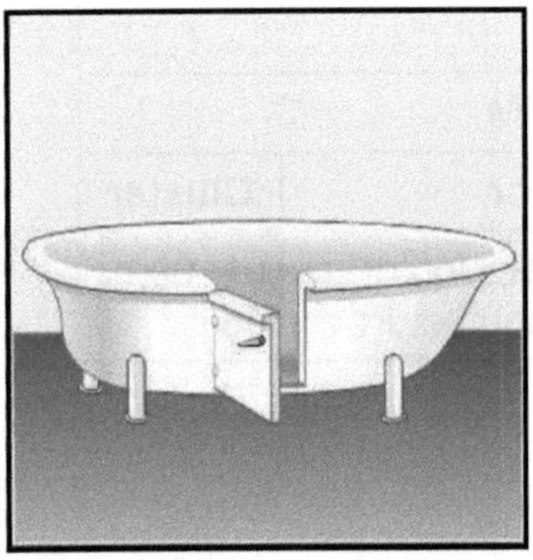

To make it easier for elderly and sick people to get in and out of the bathtub, the bathtub is fitted with a door that can be opened when the tub is empty and is tightly sealed when the tub is filled with water **(navigator 37).**

EXTRACTING

R	#	Navigator	Reasoning
	02	Replacement of mechanical matter	A "dynamic" element is introduced into the conservative structure of the bathtub.
++	19	Transition into another dimension	Reverse use of an existing technique: Move within a plane instead of rising above it (bathtub).
++	11	Inverse action	First get inside the bathtub, then fill it with water.
++	12	Local property	The door is located at the required height.
++	***37***	***Equipotentiality***	*Dominating navigator*

REINVENTING

Trend

It is difficult and dangerous for elderly and sick people to get into and out of the bathtub without outside aid. How can we facilitate this process?

Reducing

OZ: bathtub.

OT: time when the person gets into or out of the bathtub.

Macro-FIM: The X-resource—without causing unacceptable negative effects—ensures, together with the other available resources, attainment of the following result:

[*safe and comfortable way to get in and out of the bathtub*].

Standard contradiction:

Effect, State, Object						
Safe and comfortable way to get in and out of the bathtub						
(+)-Factor			**(+)-Factor**			**Cluster**
–	32	Weight of the movable object	–	37	Energy use of the moveable object	01, 15, 31, **37**

Inventing

Idea according to ***navigator 37***: To make it safer and easier for elderly and sick people to get in and out of the bathtub, the bathtub is fitted with a door that can be opened when the tub is empty and is tightly sealed when the tub is filled with water.

Zooming

Standard contradiction is removed.

38. Homogeneity	Objects that interact with the object in question must be made from the same material (or from one with similar properties).

EXAMPLE. The shirt can be glued or welded (electric welding!)

Short description

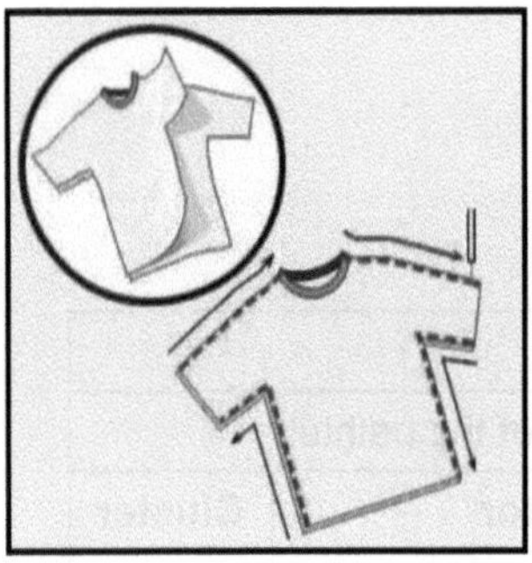

In the Welding Institute in Cambridge (Great Britain), a shirt made from pieces of a synthetic fabric has been welded by laser **(navigator 38).**

EXTRACTING

R	#	Navigator	Reasoning
++	01	Change in the aggregate state of an object	Melting of the material
+	10	Copying	Copying of metal welding
++	*38*	***Homogeneity***	*Dominating navigator*

REINVENTING

Trend

Is it possible to make a shirt... without using thread?

Reducing

Cloth usually tears in places where its pieces are sewed together with thread. It happens under the influence of excessive tearing force, especially in work clothing made of artificial fabrics.

OZ: clothing piece joints.

OT: time when the pieces are joined.

Micro-FIM: The X-resource, in the form of particles of substance or energy, is located inside the operating zone and ensures, together with the other available resources, attainment of the following result:

[*reliable joining of pieces in an article of clothing*].

Standard contradiction:

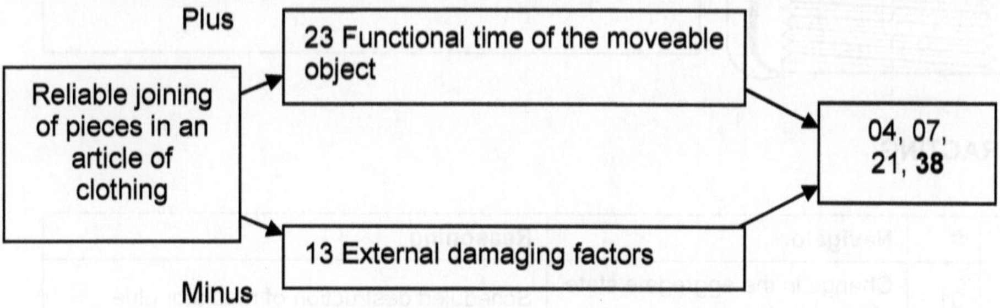

Inventing

Idea according *to* ***navigator 38***: At the Welding Institute in Cambridge (Great Britain), a shirt made from pieces of a synthetic fabric has been welded together by laser.

Navigator *04 Replacements of the mechanical medium* is also present: The ability of pieces of material to be joined by welding or gluing, instead of sewing with thread, has been accomplished.

Zooming

Standard contradiction is removed.

39. Preliminary counteraction	If the conditions of a task require an action, then a opposite action should be taken in advance.

EXAMPLE. Cap for medicine.

Short description

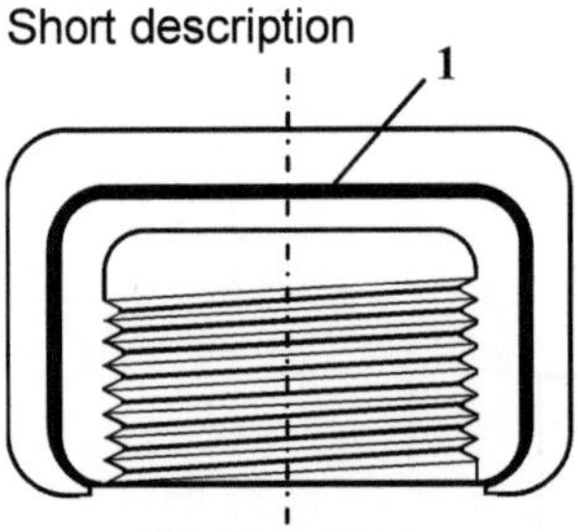

English engineers form Salford University have developed a special cap for medicine tubes that cannot be opened any more after the expiration date (**navigator 39**).

EXTRACTING

R	#	Navigator	Reasoning
++	01	Change in the aggregate state of an object	Scheduled destruction of mediator glue
+	11	Inverse action	Instead of being glued to the bottle, the inner cap is "unglued" from the outer cap, which then starts to freely rotate.
+	28	Previously installed cushion	Impossibility to unscrew the cap prevents the use of expired medicine.
++	***39***	***Preliminary counteraction***	*Dominating navigator*

REINVENTING

Trend

How can we ensure that medicine is not taken by accident after its expiration date?

Reducing

OZ: container with medicine, specifically a glass tube, cap screwed onto the tube, medicine in the tube, environment (air).

We can interpret the problem in the following way: It would be useful to not make the construction of the tube more complicated. Some kind of unknown X-resource that prevents the use of the medicine after its expiration date, but does not compromise user-friendliness of the tube, is needed.

Written warnings on the tube are uncomfortable because the user has to always check whether the medicine has expired or not.

Maxi-FIM: The operating zone by itself ensures attainment of the following result: [*impossibility of taking medicine accidentally after its expiration date*].

Micro-FIM: The X-resource, in the form of particles of substance or energy, is located inside the operating zone and ensures attainment of the Maxi-FIM.

Model of the SC:

Effect, State, Object						
Sealed medicine tube						
(+)-Factor			**(+)-Factor**			**Cluster**
We need a simple way to prevent access to medicine	07	Complexity of construction	Using the tube must not become less convenient	10	Ease of use	10, 18, **39**

Inventing

English engineers from Solford University have developed a special cap for medicine tubes that cannot be opened any more after the expiration date. Screw windings are cut into the tube and into an inner cap that, with time ,slowly melt through the effect of a layer of polymer glue 1. The condition of the glue changes with time, and its effect on the inner cap can therefore be regulated to match time to expiry. When the glue has dissolved completely, the outer cap turns freely and therefore *"intelligently" resists all efforts* to open the tube (***39 preliminary counteraction***). This is the signal that the medicine should no longer be used.

Navigators 10 and 18 are also at work here.

Zooming

The SC is solved.

40. Uninterrupted useful function	l) Complete a job without interruptions where all parts work continuously at full capacity; m) eliminate idle running and interruptions.

EXAMPLE. Therapy by means of an iron.

Short description

The American firm Procter & Gamble has patented a way of treatment by inhalation during ironing by filling a steam generator of the iron with water containing several drops of medicine **(navigator 40).**

EXTRACTING

R	#	Navigator	Reasoning
++	10	Copying	Copying of inhalation
+	11	Inverse action	The medicine comes to the user, instead of the user going to the medicine
+	12	Local property	The medicine is delivered precisely where it is needed
++	29	Self-servicing	The medicine comes to the user by itself (see 11)
++	*40*	***Uninterrupted useful function***	*Dominating navigator*

REINVENTING

Trend

When treating catarrhal diseases, inhalation of therapeutic substances with hot humid air is frequently recommended. Sometimes inhalation requires considerable time. Could it be combined not only with watching the television but also with some suitable domestic work?

Reducing

During any work, the medicine must reach respiratory organs (mouth, nose) with warm humid air (vapor, fog).

OZ: airspace around the face.

OT: time of inhalation, combined with suitable domestic work.

Micro-FIM: The X-resource, in the form of particles of substance or energy, is located inside the operating zone and ensures, together with the other available resources, attainment of the following result:

[*delivery of medicine to respiratory organs with hot humid air*]

Standard contradiction:

Effect, State, Object						
Delivery of an inhaled medicine						
(+)-Factor			**(+)-Factor**			**Cluster**
To spray a medicine in great volume	36	power	-	07	Complexity of construction	08, 15, 25, **40**

Inventing

Idea according to ***navigator 40***: The American firm Procter & Gamble has patented a way of treatment by inhalation during ironing by filling the steam generator of the iron with water containing several drops of medicine.

Zooming

Standard contradiction is removed.

Super-effect: linen acquires pleasant smell.

S36. Af-catalogue (with reinventing)

1. Separation in space	One part of system space has property **A**, whereas another part of system space has property **not-A**.

EXAMPLE. QUEUE ORDERING.

Brief Description

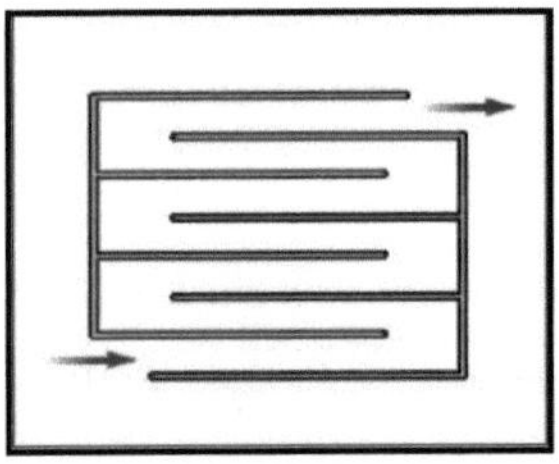

To physically eliminate the possibility of a chaotic crowd emerging in front of a building (museum, theater, exhibition hall), the area is reorganized by erecting dividing constructions that form a narrow corridor through which people can navigate only by walking one after another (***transformation in space***).

REINVENTING

Trend

Conflicts and certain threats arise when people try to get through some narrow passage (for example, a door or a gate) out of turn. In such situations, people are not always capable of generating or supporting orderly movement. How can we encourage them to move in turn?

Reducing

OZ (Operative Zone): large congregation of people in front of an entrance (for example, entrance to a building).

OT (Operative Time): interval of time from the moment when people begin to congregate to the moment when they pass through the entrance (i.e., enter the building).

RC Model:

Property: movement (of objects) to the entrance must be orderly (they must move in turn)

→ The property "orderliness" ***must be*** *to* avoid fighting and jostling in front of the entrance to the building.

→ The property "orderliness" ***is absent because*** certain egotistical undisciplined people provoke emergence of a disorderly crowd, which causes fighting and jostling.

Macro-FIM: *The X-resource*—without causing unacceptable negative effects—ensures, together with the other available resources, attainment of the following result: [*orderly queue movement*].

Maxi-FIM: The operating zone ***by itself*** ensures attainment of the following result:

[*orderly queue movement*].

Micro-FIM: *The X-resource*, in the form of *particles of substance or energy*, is located inside the operating zone and ensures the obtaining of the **Maxi-FIM.**

Inventing

The requirement of ***impossibility to move outside of the queue*** can be realized only if the freedom of movement of mobile objects in space is restricted. Accordingly, pursuant to the fundamental technique of ***transformation in space***, it is recommenced to spatially divide conflicting objects so that they cannot interact in a conflicting fashion. With respect to generating or supporting orderly movement in a queue, this may mean the erection of dividing constructions, such as barriers, walls, shrubs, etc. to create a narrow "corridor" leading to the entrance. If we want the space in front of the entrance to hold a large number of people, such "corridor" can be shaped in the form of a "serpentine" covering the entire area.

The X-resource in this example is the dividing construction.

Zooming

The RC has been eliminated.

One part of the system—that located outside of the conflict zone—possesses one property (free nonorganized space), whereas the other part of the system—that located inside the conflict zone—possesses the opposite property (space organized by dividing and guiding constructions).

This principle is widely used in real life.

It should be noted that the key role in this solution is played by structural transformation.

2. Separation in time	During one time interval the system has property **A**, whereas during another time interval it has property **not-A**.

EXAMPLE. SAME-LEVEL ROAD CROSSING.

Brief Description

To physically eliminate the possibility of vehicles colliding as they traverse traffic routes at the same level, conflicting traffic flows are divided in time by using special signal systems (such as traffic lights or traffic wardens) which regulate the order of movement (***transformation in time***).

REINVENTING

Trend

There is a risk of vehicles colliding as they traverse traffic routes, for example, at road intersections. How can we prevent such collisions?

Reducing

OZ: intersection of traffic routes, conflicting vehicles.

OT: interval of time when vehicles moving at traversing courses arrive at an intersection of traffic routes at the same time.

In this example, it is possible to formulate several radical contradictions. However, whatever the wording, we must define the actors and the conditions in which the contradiction evolves.

RC Model:

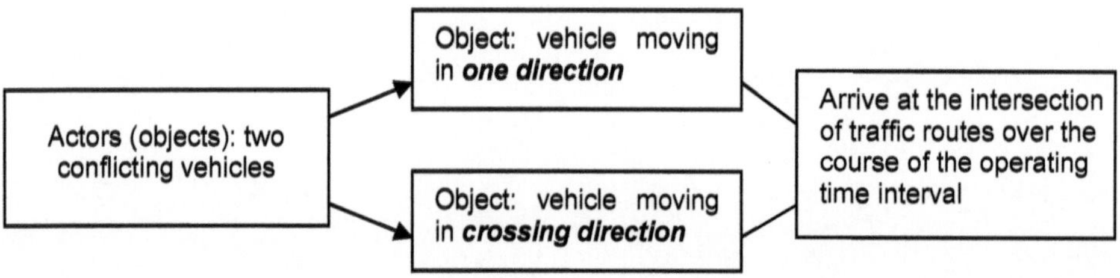

Macro-FIM: *The X-resource*—without causing unacceptable negative effects—ensures, together with the other available resources, attainment of the following result: [*impossibility of vehicles colliding with each other*].

Inventing

The requirement of ***impossibility of physical interaction*** can be realized only in the absence of spatial contact between objects. Accordingly, pursuant to the fundamental technique of ***transformation in time***, it is recommenced to temporally divide conflicting objects so that they cannot spatially interact in a conflicting fashion.

With respect to traffic routes, this may mean that conflicting objects must take turns in traversing the intersection. To make that possible, intersections are equipped with traffic lights.

The X-resource in this example is the traffic light and… the rules dictating the need to traverse the intersection in accordance with certain traffic light signals (informational resource).

Zooming

The RC has been eliminated.

The system possesses one property (permits the passage of vehicles through the intersection in one direction) during one time interval, and the opposite property (permits the passage of vehicles through the intersection in another direction) during another time interval.

This principle is widely used in real life. For example, aircraft, trains, and sea vessels are assigned certain time intervals during which they can take off, land, or arrive at (or depart from) train stations and sea ports, etc.

It should be noted that the key role in this solution is played by structural transformation.

3. Separation in structure	Some elements of the system have property **A**, whereas other elements or the system as a whole have (has) property **not-A**.

EXAMPLE. EMERGENCY ESCAPE CHUTE.

Brief Description

To create an emergency escape chute that is durable and "inflexible" when in use and rolled-up and "flexible" when not in use, the chute is made in the form of several joined elastic pipes which are inflated by compressed air (***transformation in structure***).

REINVENTING

Trend

The escape chute must be sufficiently long and durable to enable the passengers to leave the plane when it makes an emergency landing quickly and without injuries. For example, it can be made in the form of a spare ladder. But when not in use, such chute (even if the ladder is collapsible) takes up too much space inside the plane. In addition, it may be rather heavy. How can we design a more compact and (to the extent possible) lightweight escape chute?

Reducing

OZ: emergency escape chute. OT: interval of time when the chute is used by the passengers to leave the plane.

"Preoperating" and "postoperating" time: interval of time when the chute is not in use and is stored onboard the plane.

RC Model:

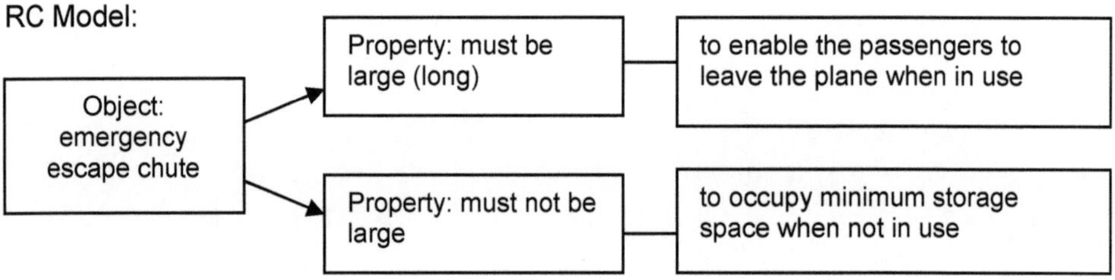

Macro-FIM: *The X-resource*—without causing unacceptable negative effects—ensures, together with the other available resources, attainment of the following result: [*emergency escape chute that is compact and light when not in use and long and durable when in use*].

Maxi-FIM: The operating zone **by itself** ensures attainment of the following result: [*convertible emergency escape chute with the properties defined in Macro-FIM*].

Inventing

The requirement whereby the ***object must concurrently possess incompatible properties*** can be realized only by simultaneous transformation of multiple resources, one of which plays the dominating role. With respect to the emergency escape chute, this means that it must be made of several joined elastic pipes ("flexible" construction elements). When not in use, the chute has a comparatively small weight, and is stored as a relatively compact roll. When in use, the chute is inflated with compressed air, becomes long *and* durable ("inflexible"), and enables passengers to escape the plane.

The X-resource in this example is the new elastic and dynamized structure of the chute, its new material and shape, and compressed air filling the elastic pipes.

Zooming

The RC has been eliminated.

The system as a whole (the chute as an aggregation of joined elastic pipes) possesses one property ("flexible," i.e., capable of being stored in a roll) for one purpose as well as the opposite property ("inflexible," i.e., can be unrolled for use, pipes are filled with compressed air) for another purpose.

Super-effect: Such chute has a much smaller weight compared with previously used constructions.

It should be noted that the key role in this solution is played by transformation of spatial, substantial, and energetical resources.

4. Separation in material	For one purpose, the material has property **A**; for another purpose, it has property **not-A**.

EXAMPLE. PRE-FLASH.

Brief Description

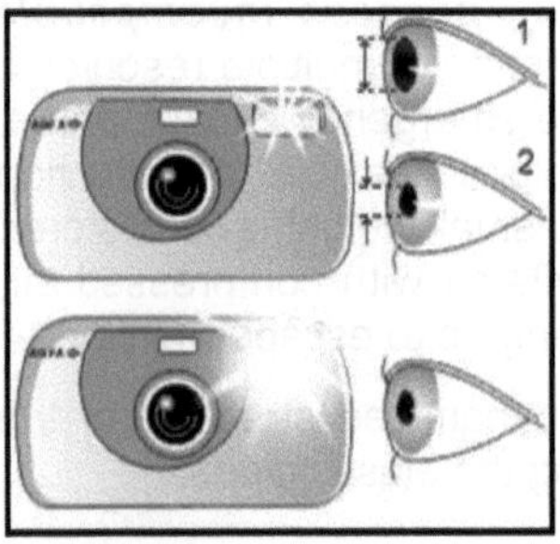

To prevent the "red-eye effect" when taking flash-assisted photographs, many cameras generate one or several short smaller flashes before setting off the main flash (***transformation in substance and energy***). As a result, pupils contract, and less light from the main flash reaches the retina.

REINVENTING

Trend

For a long time, flash-assisted color photography has been plagued by the "red-eye effect" where pupils come out filled with red color. How can we get rid of this problem?

Reducing

OZ: pupil and light flow from the flash.

OT: operating time of the main flash.

Macro-FIM: *The X-resource*—without causing unacceptable negative effects—ensures, together with the other available resources, attainment of the following result: [*normal (black) color of the pupil*].

RC Model:

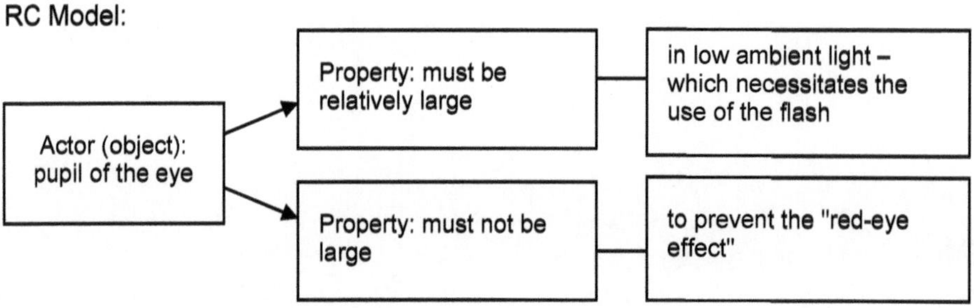

Inventing

The requirement whereby the ***object must concurrently possess incompatible properties*** can be realized only by simultaneous transformation of multiple resources, one of which plays the dominating role. Pursuant to the fundamental ***transformation in substance (energy)*** technique, when taking flash-assisted photographs, many cameras generate one or several short smaller flashes (pre-flashes) before setting off the main flash. These pre-flashes make the pupils contract, thereby reducing the amount of light reaching the retina and reflected by it.

The X-resource in this example is the light flow from pre-flashes that bring the pupils into the desired (contracted) state before the beginning of the operating time interval.

Zooming

The RC has been eliminated.

The relevant element, or the entire system (pupil), possesses one property ("large" or dilated in normal state) for one purpose and/or during one time interval, and the opposite property ("small" or contracted by the time the main flash goes off) for another purpose and/or during another time interval.

It should be noted that efficient solution has been obtained by performing a transformation in time where part of the action is transferred to "preoperating" time.

The solution also uses transformation of the spatial resource.

S37. Terms and abbreviations

Actor (inductor, receptor), *147*
A-matrix, *25, 390, 392*
ARIZ – Algorithm of Inventive Problem Solving, *21, 33*
Artifact, *23, 24, 88*
A-Studio, 22
As-catalog, *26, 394, 395,399, 428*

BICO – Binary Input Cluster Output, *74, 387*
Brainstorming, 31

Contradiction, *16, 19, 20, 41, 51*
general, 159
radical, 20, 47, 381
standard, 20, 45, 380

Development, *51, 217*

Evolution, *218*
Extracting, *59, 61*

Factors, *26, 76, 390*
minus-, 26, 76, 390
plus-, 26, 76, 390
FIM – Functional Ideal Model (Modeling), *160*
Fundamental Transformations, *83, 414, 415, 508*

Idealization, *222*
IFR – Ideal Final Result, *24, 160*
Inductor, *147*
Invention, *19*
Inventive Problem, *19*

Levels of invention, *224, 374*

Macro-FIM, *161*
Main inductor, 147
Main receptor, *147*
MAI T-R-I-Z – Meta-Algorithm of Invention, based on the 4 stages T-R-I-Z, *35, 39, 377*
Trend (Diagnostics), 35, 39, 377
Reducing (Reformation), 35, 39, 377
Inventing (Transformation), 35, 39, 377
Zooming (Verification), 35, 39, 377

Maxi-FIM, *161*
Mental Zones, *139*
Micro-FIM, *161*
Maximum task, *375*
Minimum task, *375*

Noosphere, *139*

OR – Operative Resource, *152, 153*
OT – Operative Time, *153*
OZ – Operative Zone, *141, 143, 385*

Problem, *19*

Receptor, *147*
Reinventing, *87*
Resources, *152, 153, 386*
energetical, 153, 386
functional, 153, 386
informational, 153, 386
material, 153, 386
physical-technical, 153, 386
spatial, 153, 386
systemic, 153, 386
systemic-technical, 153, 386
structural, 153, 386
temporal, 153, 386
RICO – Radical Input Cluster Output, *81, 409*

S-curve, *217*
Specialized Transformations, *26, 394*
START T-R-I-Z, *71*
System, *217*
bifurcation, 218
development, 217, 218
ramification, 218
evolution, 218

TRIZ – Theory of Inventive Problem Solving, *20, 367*
TRIZ Laws, *217*
TRIZonal concept, *138*

9-screen model, *221*

S38. Main Web sites of AIMTRIZ

www.mtriz.com

www.modern-triz-academy.com

www.gramtriz.com

MIX
Papier aus verantwortungsvollen Quellen
Paper from responsible sources
FSC® C105338
www.fsc.org

If you have any concerns about our products,
you can contact us on
ProductSafety@springernature.com

In case Publisher is established outside the EU,
the EU authorized representative is:
Springer Nature Customer Service Center GmbH
Europaplatz 3, 69115 Heidelberg, Germany

Printed by Libri Plureos GmbH
in Hamburg, Germany